Space Weather Impact on GNSS Performance

Vladislav Demyanov · Yury Yasyukevich ·
Maria A. Sergeeva · Artem Vesnin

Space Weather Impact on GNSS Performance

Springer

Vladislav Demyanov
Institute of Solar-Terrestrial Physics
Siberian Branch of Russian Academy
of Sciences
Irkutsk, Russia

Irkutsk State Transport University
Federal Agency of Russian Railroads
Irkutsk, Russia

Maria A. Sergeeva
LANCE, SCiESMEX, Instituto de
Geofisica, Unidad Michoacan
Universidad Nacional Autonoma de Mexico
Morelia, Michoacán, Mexico

CONACYT, Instituto de Geofisica, Unidad
Michoacan
Universidad Nacional Autonoma de Mexico
Morelia, Michoacán, Mexico

Yury Yasyukevich
Institute of Solar-Terrestrial Physics
Siberian Branch of Russian Academy
of Sciences
Irkutsk, Russia

Artem Vesnin
Institute of Solar-Terrestrial Physics
Siberian Branch of Russian Academy
of Sciences
Irkutsk, Russia

ISBN 978-3-031-15873-5 ISBN 978-3-031-15874-2 (eBook)
https://doi.org/10.1007/978-3-031-15874-2

This Springer imprint is published by the registered company Springer Nature Switzerland AG
The registered company address is: Gewerbestrasse 11, 6330 Cham, Switzerland

Contents

Abbreviations

AATR	Along Arc TEC Rate
AltBOC	Alternative Binary Offset Carrier
ATFM	Air Traffic Flow Management
BPSK	Bi-Phase Shift Keying
C/A	Coarse Acquisition
CBOC	Composite Binary Offset Carrier
CDMA	Code Division Multiple Access
CL	Coarse Long (code)
CM	Coarse Modified (code)
CME	Coronal Mass Ejection
CNR	Carrier-to-Noise Ratio
CONUS	Continental United States
CORS	Continuously Operating Reference Stations
COSPAR	Committee on Space Research
DCB	Differential Code Biases
DGPS	Differential GPS
DIX	Disturbance Ionosphere Index
DLL	Delay Lock Loop
DMSP	Defense Meteorological Satellite Program
DOP	Dilution Of Precision
DROTI	Derivative of ROTI
EGNOS	European Geostationary Navigation Overlay System
EIA	Equatorial Ionization Anomaly
EIT	Extreme Ultra Violet Imaging Telescope
FDMA	Frequency Division Multiple Access
GBAS	Ground-Based Augmentation System
GDOP	Geometric Dilution Of Precision
GEC	Global Electron Content
GEO	Geostationary Earth Orbit
GIM	Global Ionospheric Map
GIVE	Grid Ionospheric Vertical Error

GLONASS	Global Navigation Satellite System
GLOT	GLONASS Timescale
GMT	Greenwich Mean Time
GNSS	Global Navigation Satellite System (as a complex of several global positioning systems)
GPS	Global Positioning System (NavStar)
GPST	GPS Timescale
GSO	Geo Synchronous Orbit
GST	Greenwich Standard Time
GUS	GEO Uplink Sub-system
HAL	Horizontal Alert Limit
HDOP	Horizontal Dilution Of Precision
HF	High Frequencies
HP	High Precision
HPL	Horizontal Protection Level
ICAO	International Civil Aviation Organization
ICD	Interface Control Document
IERS	International Earth Rotation and Reference Systems Service
IF	Intermediate Frequency
IGP	Ionospheric Grid Point
IGS	The International GNSS Service
ILWS	International Living With a Star (Program)
IMF	Interplanetary Magnetic Field
IOD	Issue Of Data
IP	Initial Phase (of magnetic storm)
IPP	Ionospheric Pierce Point
IS	Interface Specification
ISC	the International Science Council
ISES	International Space Environment Service
ISWI	International Space Weather Initiative
ITRF	International Terrestrial Reference Frame
LAAS	Local Area Augmentation Systems
LBAS	Local Based Augmentation Systems
LOF	Lowest Observed Frequency
LOL	Loss Of Lock
LP	Localizer Performance
LPV	Localizer Performance with Vertical Guidance
LPV200	Localizer Performance with Vertical Guidance till 200 foot decision height
LT	local Time
MEO	Medium Earth Orbit
MIT	Main Ionospheric Trough
MOF	Maximum Observed Frequency
MOPS	Minimum Operational Performance Standard
MP	Main Phase (of magnetic storm)

MS	Master Station
MUF	Maximum Usable Frequency
NANU	Non-Available Non-Used
NOAA	National Oceanic and Atmospheric Administration, the U.S.
OCS	Operation Control Segment
OIS	Oblique Ionospheric Sounding
OS	Open Service
PDOP	Positioning Dilution Of Precision
PLL	Phase Lock Loop
PPP	Precise Point Positioning
PRN	Pseudo Random Noise (code)
PSK	Phase Shift Keying
QZSS	Quasi-Zenith Satellite System
R&D	Research and Developing
RAIM	Receiver Autonomous Integrity Monitoring
RF	Radio Frequency
RINEX	Receiver INdependent EXchange (Format)
RMS	Root Mean Square
RNP	Required Navigation Parameters
RO	Reference Oscillator
ROT	Rate Of TEC change
ROTI	Rate of TEC change Index
RP	Recovery Phase (of magnetic storm)
RPS	Recording and Processing Software
RRC	Range Rate Corrections
RSDCM	Russian System of Differential Correction and Monitoring
RT	Real Time
RTK	Real Time Kinematic
RTS	Receiver Time Scale
RTSW	Real Time Solar Wind (System)
s.f.u	Solar flux unit
SA	Selective Availability
SBAS	Satellite-Based Augmentation Systems
SCOSTEP	Scientific Committee on Solar-Terrestrial Physics
SEM	Space Environmental Monitor
SI	Sudden Impulse
SIS	Signal In Space
SISRE	Signal In Space Ranging Error
SLR	Satellite Laser Ranging
SMR	Signal-to-Multipath Ratio
SOHO	Solar Heliospheric Observatory
SP	Standard Precision
SRTS	System Reference Time Standard
SS	Space Segment
SSC	Sudden Storm Commencement

SSN	Sun Spot Number
STS	Satellite Time Scale
SU	Soviet Union
SV	Satellite Vehicle
SW	Space Weather
TAI	Timescale Atomic International
TEC	Total Electron Content
TECU	Total Electron Content Unit
TID	Travelling Ionospheric Disturbances
TOW	Time Of Week
UDRE	User Differential Range Error
UHF	Ultra High Frequency
UIRE	User Ionosphere Range Error
UNCOPUOS	United Nations Office for the Peaceful Uses of Outer Space
USNO	United States Naval Observatory
UT	Universal Time
UTC	Universal Time Coordinated
VAL	Vertical Alert Limit
VDOP	Vertical Dilution Of Precision
VIS	Vertical Ionospheric Sounding
VLBI	Very Long Baseline Interferometry
VPL	Vertical Protection Level
WAAS	Wide Area Augmentation Systems
WMO	World Meteorological Organization
WMS	WAAS Master Station
WRS	WAAS Reference Station

Chapter 1
Introduction

The up-to-date strategy in developing international organizations and communities (RTCA, ICAO, etc.) is evolutional transition to global navigational satellite system (GNSS) as a basic means to solve numerous diverse problems of navigation, geodesy, surveying, etc. Similar intentions underlie the state radio navigation plans of the world leading countries and inter-government radio navigation programs (ESA-ICD 2005, RNP RF 2019, DOT-VNTSC-OST-R-15-01 2020). Thereby, updating the GNSS national segments (both ground- and space-based) has been permanently urgent. Solving this problem is directly related to maintaining a high level of technologic and economic development of the countries operating the GNSS (USA, China, Russia, European Union, India, Japan).

One should note that, by now, potentials of national GNSSs–GPS, GLONASS, Galileo, BeiDou, and others—where the 1970s–1990s technological ideas underlay, urge their further development with allowance for up-to-date requirements. One of the GNSS important trends is developing technological and organizational provisions aimed to offset negative effects from space weather. Here, *space weather* stands for geomagnetic and sporadic disturbances of near-Earth space (NES), as well as for bursts of powerful solar radio emissions, which may cause direct GNSS signal suppression (Demyanov and Yasyukevich 2021). Chap. 3 describes the space weather concept in greater detail.

Before 2000, there were no sufficient studies investigating space weather effects on the GNSS operation quality. Scarce papers by individual authors dealt, primarily, with traditionally problematic areas for decimeter radio wave propagation in the equatorial and polar ionosphere (Crane 1977; Aarons 1982; Conker et al. 2003). Direct suppression of navigation satellite signals by solar radio bursts, according to the estimates by the known American researcher Klobuchar, was generally addressed as a rare and exotic phenomenon (Klobuchar et al. 1999).

With the turn of the twenty-first century, when the demand for satellite navigation technologies dramatically increased, it became clear that space weather effects may appear a serious constraint when developing GNSS potentials. Numerous studies

V. Demyanov et al., *Space Weather Impact on GNSS Performance*,
https://doi.org/10.1007/978-3-031-15874-2_1

established multiple cases of hard failures in the GNSS operation and their applications (Conker et al. 2003; Afraimovich and Perevalova 2006; Astafyeva et al. 2014, Dabove et al. 2020, Demyanov and Yasuikevich 2021).

Thus, there is a necessity to provide the required navigation standards for the GNSSs and their applications under unfavorable space weather effects.

On the other hand, now, one faces unprecedentedly wide capabilities to use GNSS signals for investigating near-Earth space, atmosphere, ocean, and Earth surface. In particular, an important scientific trend is to remotely sound the Earth's ionosphere and atmosphere. The chain of couplings (Sun–interplanetary medium–magnetosphere–ionosphere–atmosphere–Earth surface) is an integral physical system (Asmar et al. 2019).

Therefore, the atmosphere and its upper part, the ionosphere, may be addressed as sensitive indicators for the status of those geospheres. This may be used to study atmospheric effects of such phenomena as powerful earthquakes, tsunami, volcano eruptions, etc. Building global maps for the total electron content (TEC) in the ionosphere and its variations enables climate investigations into the ionosphere and ionospheric inhomogeneities on the global scale (Hernandez-Pajares et al. 2009; Li et al. 2020). Today, several geophysical GNSS-based research trends have been developing: radio occultation GNSS measurements (GNSS-RO), GNSS-reflectometry of the Earth surface and ocean (GNSS-R), GNSS-based sounding of the ionosphere and atmosphere (Jin et al. 2014).

Based on the above, one may state that **GNSS technologies have outgrown their initial purpose (providing navigation itself) and become one of the driving factors in the international technological progress**. Thereby, providing the required quality of GNSS measurements to solve applied and research issues acquires a special urgency. Hence, there comes a fundamental scientific problem. One should reveal and produce quantitative–qualitative estimates for prevalent radio physical and radio technical mechanisms and typical regularities of GNSS characteristic degradation under the impact from space weather.

The goal of this monograph is to provide an extensive worldwide notion about the up-to-date status of GNSS scientific and technological development and investigations into the above fundamental and applied problems. We discuss possible trends to improve scientific-methodical and hardware-software GNSS means (as well as their applications) with allowance for space weather extreme factors.

References

Aarons J (1982) Global morphology ionospheric scintillations. Proc IEEE 70(4):360–378. https://doi.org/10.1109/PROC.1982.12314

Afraimovich EL, Perevalova NP (2006) GPS-monitoring of the earth's upper atmosphere. Irkutsk: GU NC RVH VSNC SO RAMN, p 480. ISBN 5-98277-033-7 (in Russian)

Asmar SW et al (2019) Future of planetary atmospheric, surface, and interior science using radio and laser links. Radio Sci 54(4):365–377. https://doi.org/10.1029/2018RS006663

Astafyeva E, Yasyukevich Y, Maksikov A, Zhivetiev I (2014) Geomagnetic storms, super-storms, and their impacts on GPS-based navigation systems. Space Weather 12(7):508–525. https://doi.org/10.1002/2014SW001072

Conker RS, El-Arini MB, Hegarty CJ, Hsiao T (2003) Modeling the effects of ionospheric scintillation on GPS/satellite-based augmentation system availability. Radio Sci 38(1):1001. https://doi.org/10.1029/2000RS002604

Crane RK (1977) Ionospheric scintillation. Proc IEEE 65(2):180–199. https://doi.org/10.1109/PROC.1977.10456

Dabove P, Linty N, Dovis F (2020) Analysis of multi-constellation GNSS PPP solutions under phase scintillations at high latitudes. Appl Geomat 12:45–52. https://doi.org/10.1007/s12518-019-00269-4

Demyanov V, Yasyukevich Y (2021) Space weather: risk factors for global navigation satellite systems. Solar-Terrestrial Physics 2:28–47. https://doi.org/10.12737/stp-72202104

ESA-ICD (2005) European space agency: Galileo—the European programme for global navigation services, 2nd edn. ESA Publications Division, Noordwijk, The Netherlands. http://ec.europa.eu/dgs/energy_transport/galileo/documents/brochure_en.htm

Federal Radionavigation Plan (2019) DOT-VNTSC-OST-R-15-01. Published Date: 2020 Jan 10. https://rosap.ntl.bts.gov/view/dot/43623

Hernandez-Pajares M, Juan JM, Sanz J et al (2009) The IGS VTEC maps: a reliable source of ionospheric information since 1998. J Geod 83(3):263–275. https://doi.org/10.1007/s00190-008-0266-1

Jin S, Cardellach E, Xie F (2014) GNSS remote sensing: theory, methods and applications, remote sensing and digital image processing. 19:276. https://doi.org/10.1007/978-94-007-7482-7

Klobuchar JA, Kunches JM, Van Dierendonck AJ (1999) Eye on the ionosphere: potential solar radio burst effects on GPS signal to noise. GPS Solutions 3(2):69–71

Li Z, Wang N, Hernández-Pajares M et al (2020) IGS real-time service for global ionospheric total electron content modeling. J Geod 94:32. https://doi.org/10.1007/s00190-020-01360-0

Radio-navigation Plan of Russian Federation. RNP RF. Moscow (2019) https://internavigation.ru/wp-content/uploads/2021/10/%D0%A0%D0%9D%D0%9F-%D0%A0%D0%A4-2019-2024_%D0%9F%D1%80.3296_04.09.19.pdf

Chapter 2
GNSS Overview

Vladislav Demyanov

2.1 GPS, GLONASS, BDS, and Galileo

As of early 2022, there are four Global Navigation Satellite Systems (GNSS) in the world: GPS (USA), GLONASS (Russia), Galileo (European Union), BeiDou/Compass (China). By their composition, structure, and the concept of positioning solutions, these systems are mainly identical.

The GPS, GLONASS, and Galileo satellites are located in medium-Earth orbits (MEOs). The orbital constellation of BeiDou spacecraft, unlike three other GNSS, involves not only MEO satellites, but also satellites with geostationary (GEO) and geosynchronous (GSO) orbits. Such a satellite configuration provides the best coverage of the Asian-Pacific region with the radio-navigation field.

Figure 2.1 presents the generalized GNSS block diagram. The system comprises three basic subsystems (segments): a satellite segment (satellite vehicles constellation), a ground operation control segment, and a user segment (navigation receivers, user's equipment).

To provide a higher positioning accuracy, system integrity check, and availability of the required navigation parameters (RNP), there are supplements in the system composition: local area and wide-area differential ground- and space-based systems. For better understanding the main issue of this book, let's us considering all these GNSS segments in more details.

2.1.1 GNSS Operation Control Segment

The GNSS Operation Control Segment (OCS) involves the system Master Station (MS), a ground network of Monitor and Control Stations, and a System Reference Time Standard (SRTS). The subsystem is to:

(1) generate and correct the system timescale, predict the deviation of satellite Reference Oscillators (system time–frequency provision);

V. Demyanov et al., *Space Weather Impact on GNSS Performance*,
https://doi.org/10.1007/978-3-031-15874-2_2

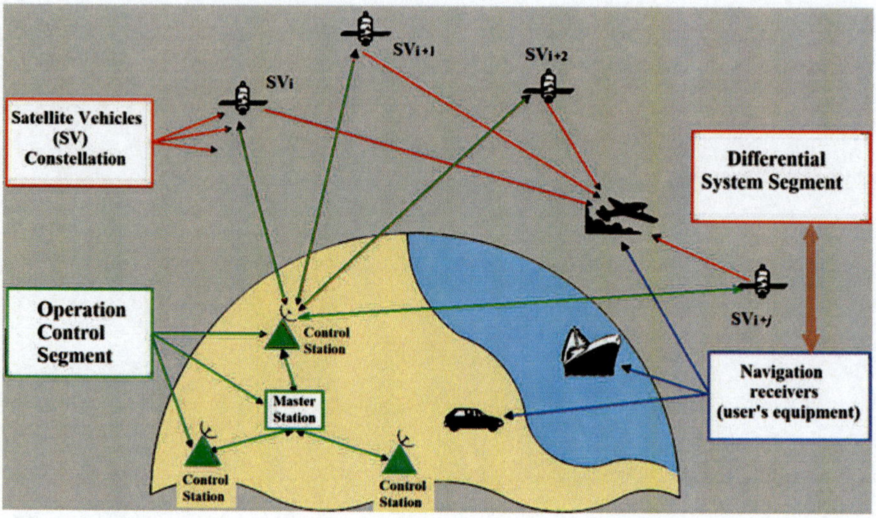

Fig. 2.1 Typical GNSS structure. Re-produced, modified and translated from Russian figure based on figure from the published book (Perov and Kharisov 2010)

(2) estimate, predict, and correct the ephemerides for navigation satellites (system ephemeris provision);
(3) check GNSS satellite operation quality and alert users to anomalous signals from individual satellites (check the GNSS space segment and radio-navigation field integrity);
(4) monitor the radio navigation field conditions meeting the requirements for accuracy and continuity of positioning solutions (system integrity check);
(5) update periodically the correcting and service information onboard satellites;
(6) control (via radio) the space-vehicle subsystem operation.

A GNSS user positioning accuracy is directly related to the control subsystem reference oscillator quality. The SRTS generates the GNSS common system timescale, relative to which all the GNSS radio measurements are taken. Correcting periodically onboard navigation satellite reference oscillator and GNSS user receiver reference oscillator is also performed relative to the system timescale.

The SRTS reference oscillator generates a high-stable reference signal of the following form:

$$u_{RO}(t) = U_0 \cdot \sin((2\pi \cdot f_0 \cdot t + \varphi_0) + \varphi(t) + \delta\varphi(t)), \tag{2.1}$$

where U_O, f_O is the nominal amplitude and frequency of the reference signal; φ_O is the initial phase of the reference signal; $\varphi(t)$ is systematic phase deviation from the reference signal; $\delta\varphi\ (t)$ are random variations and noise of the reference signal phase.

To eliminate systematic phase deviation $\varphi(t)$, one monitors continuously the reference frequency and periodically corrects the GNSS system timescale relative to the

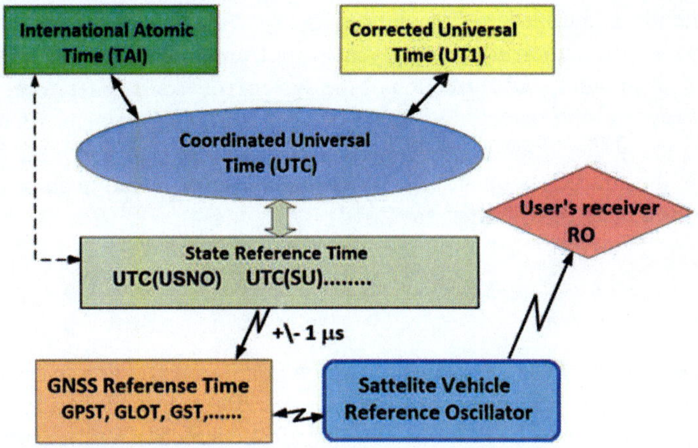

Fig. 2.2 Hierarchy and interaction among time-and-frequency standards

GNSS country-operator's state time (e.g., UTC for the USA, UTC (SU) for Russia, etc.). Figure 2.2 illustrates this process, showing a generalized scheme for interaction of time-and-frequency standards at different levels.

Three timescales compose the time-and-frequency standard system: the international atomic timescale (TAI); the astronomical timescale corrected with allowance for periodic motions of the Earth poles (UT1), and the combined timescale of the Coordinated Universal Time (UTC). The UTC is periodically synchronized with the two first basic scales and is the basis to form the state timescales (UTC, UTC(SU), etc.).

System timescales for different GNSS (GLOT, GPST, GST, etc.) are corrected contemporally with the planned UTC correction, every six months (at midnight on Jan 1 and Jul 1 UTC). At that, the systematic error of binding the GNSS system timescale to the state time should not exceed 1 μs (Perov and Kharisov 2010).

References on different GNSS system timescales have the known relation among themselves. In particular, the GLONASS and GPS system time scales are related as follows (Xu 2007)

$$\begin{aligned} UTC &= GLOT + \Delta\tau - 3h \\ GPST &= GLOT + m - 3h, \end{aligned} \tag{2.2}$$

where *GLOT* is the GLONASS system time; *GPST* is the GPS system time; $\Delta\tau$ is the *GLOT* deviation from UTC ($\leq$1 μs); *m* is the *leap seconds* parameter that shows the mistiming between *GLOT* and *GPST*, that has accumulated for the current epoch since the GPS start (1980 Jan 6).

By the GNSS timescale, there is a periodic sync of timescales for onboard time-and-frequency standards, as well as predictions for short-period drifts of those scales until the next sync (with a $\geq$ 4-h interval).

Like already mentioned, systematic deviation of the reference oscillator phase (frequency) may be sufficiently precisely measured, predicted, and eliminated during continuous observation and correction of the system timescale. Therefore, the main factor that restricts the potential accuracy of the radio measurements, is a random phase (frequency) deviation of the reference oscillator $\delta\varphi(t)$. We express a random deviation of the nominal frequency $\delta f_0(t)$ through the corresponding phase deviation $\delta\varphi(t)$ as follows (Perov and Kharisov 2010)

$$\delta f_0(t) = \frac{1}{2\pi \cdot f_0} \cdot \frac{d\delta\varphi(t)}{dt} \tag{2.3}$$

Implementing instantaneous values $\delta f_0(t)$ over an accumulating interval τ characterizes a short-time frequency instability of the reference oscillator. Because this process is random, it may be described by statistical characteristics, in particular, by the Alan variation as follows (Kaplan 1996):

$$\sigma_{AL}^2(\tau) = 2 \cdot \int_{f\,\min}^{f\,\max} S_{\delta f_0}(f) \cdot \frac{\sin^4(\pi \cdot f \cdot \tau)}{(\pi \cdot f \cdot \tau)^2} df, \tag{2.4}$$

where $S_{\delta f_0}(f)$ is the spectral density of the frequency $\delta f_0(t)$ variations; f is the reference oscillator frequency instantaneous value, Hz; $f_{\max}$, $f_{\min}$ is the range of medium frequencies of random processes that create the reference oscillator frequency instability (phase white noise, phase flicker-noise, frequency white noise, frequency flicker-noise, Wiener frequency noise) (Perov and Kharisov 2010).

Since the GNSS epoch onset, there has been continuous update of the reference oscillator, as a result of which their short-time instability has constantly been decreasing. Figure 2.3 illustrates this, presenting the week-averaged root-mean-square (RMS) estimates for the GPST variations relative to the UTC (black curve).

One can see that updating the SRTS reference oscillator enabled to reduce the GPST variations down to, on average, ~ 0.1–3 ns (equivalent ranging error being <

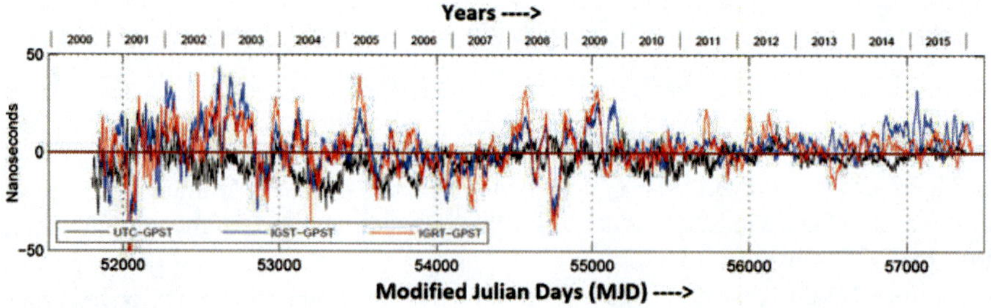

Fig. 2.3 GPST variations relative to the UTC, due to the GPS reference oscillator instability (black curve) (Villiger and Dach 2021)

1 m). Presently, using high-stable SRTS reference oscillators allows one to achieve the relative frequency instability $\sigma_{AL}(\tau) \sim 10^{-14} \div 10^{-15}$ (Exertier et al. 2018). At the same time, due to constantly growing needs of GNSS users and requirements for the positioning accuracy, permanent efforts have focused on developing optical frequency standards (optical clock) with the relative frequency instability of ~ 10^{-16} and higher (at the $\tau = 1 \div 100$ s accumulation time) (Riehle 2017).

An important function of the ground control system is to determine GNSS ephemerids. The quality of performing this determines the potentially available accuracy of positioning solutions within the GNSS operation region. The main method for positioning solutions in a GNSS is the ranging method. As reference points for ranging measurements, one uses navigation satellites, whose coordinates vary constantly during their orbital motions. By using ground-based metrological and computing tools, the ground control segment facilities periodically measure and define the satellites' ephemerids and predict them for the set extrapolation period (4–12 h).

One performs the main check and correction for GNSS satellites' motion orbital elements from the GNSS state-operator territory over a limited time. The duration of satellite interaction with a monitor stations between the acquisition range entry ($\tau 1$) and exit ($\tau 2$) may be estimated as follows (Perov and Kharisov 2010)

$$T_{VIEW} = (\tau 2 - \tau 1) = \frac{T_{SV}}{\pi} \cdot \left(\arccos\left[\left(\frac{R_E \cdot \cos\alpha}{(R_E + h_{SV})} \right) \right] - \alpha \right), \tag{2.5}$$

where T_{SV} is the satellite rotation period, hr; R_E is the Earth radius, km; α is the minimal admissible angle altitude of the satellite (determined from admissible refraction distortions and fading of the satellite signal; at typical calculations, it equals 0–10°); h_{SV} is the satellite orbital height over the Earth surface, km.

For MEO GNSS with the orbit height of ~ 20,000 km and at $\alpha = 0°$, the duration of the communication session between a monitor station and a satellite is $T_{VIEW} \approx$ 300 min, i.e. about 5 h. Because of the restricted T_{VIEW}, the process of ephemeris correction and prediction is split into two stages:

(1) Metric facilities of monitor stations (ranging lasers or measuring stations) measure ranges to visible satellites, compute their coordinates, and orbit elements (ephemeris).
(2) By using the known analytical laws of satellite orbital motion and corrections for disturbing effects, one computes prognostic values for ephemerids or coordinates of satellites at reference instants (ephemeris reference time).

The ephemeris predictions obtained thereby may be used in a GNSS user receiver until the next improved prediction arrives.

Between the ephemeris reference instants, the computed values for satellite coordinates will differ from the real due to insufficiently accurate estimate for parameters of disturbing effects. At the GNSS early stage, when only regular system monitor stations were used, errors in determining satellite orbital positions between ephemeris corrections reached units of meters. Figure 2.4 illustrates the evolution of errors in

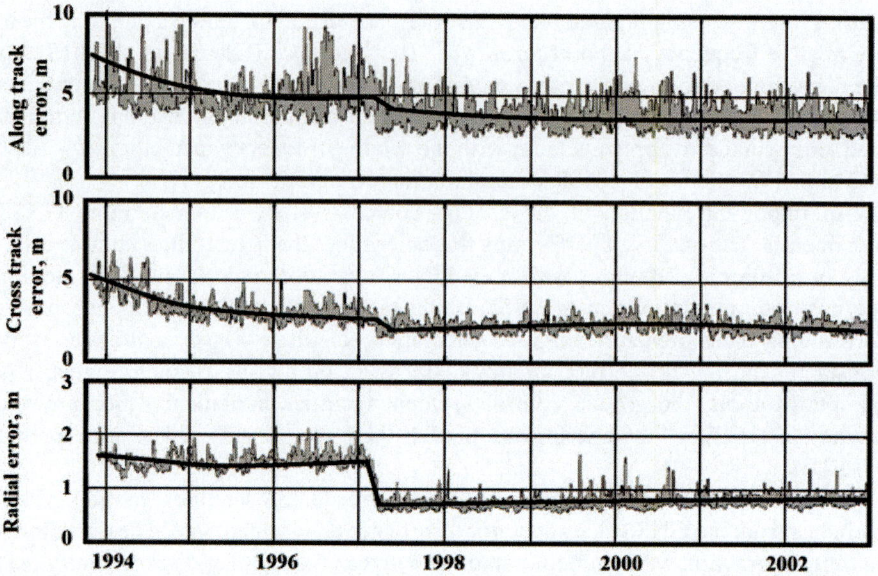

Fig. 2.4 Evolution of errors in GPS satellite coordinate solutions, by using ephemeris predictions. Re-produced and modified figure based on figure from the published paper (Warren and Raquet 2003)

determining the GPS satellite orbital coordinate components over 1994–2003. Here, presented are mean diurnal values for radial, along-track, and cross-track components of errors in positioning all the observable satellites (Warren and Raquet 2003).

Obviously, the satellite coordinate determination error by only the facilities from a limited territory is sufficiently high and contains a systematic error.

Presently, there is a number of ground-based reference networks for geophysics and geodynamic monitoring (IGS, UNAVCO, etc.) and international analytical centers (CODE, ESA, USNO, etc.) (Bruyninx et al. 2012, Herring et al. 2017, Villiger and Dach 2021). Thereby, there emerged new capabilities to broadcast the ephemerids improved by those additional facilities to all the GNSS users concerned. Due to this, the error in the GNSS satellite orbital positioning may be reduced to tens of mm (Villiger and Dach 2021).

2.1.2 Satellite Constellations and Signals

A space segment (SS) is a system of reference points to measure ranges. The GNSS satellite orbital configuration is such, that it ensures observation from no less than four navigation satellites in any point of the Earth. Table 2.1 presents the configuration and characteristics of satellite constellations, that differ for individual GNSS.

Table 2.1 Characteristics for GNSS orbit constellations (as of 2022 Jan 10)

System parameter	GLONASS	GPS	Galileo	BeiDou
Satellites number	25	32	26	GEO–8 GSO–12 MEO–29
Satellites in operation	22	29	22	44
Orbit planes	3	6	3	3
Nominal orbital altitude, km	19,100	20,181.6	23,222	21,528
Orbit inclination, deg	64,8	55	56	55 (MEO)
Separation of orbits by longitude, deg	120	60	120	120
Shift of satellites by latitude argument, deg	45	45	45	45
Mean revolution period (hrs: min)	11:15	11:57	14:04	12:53 (MEO)
Frequency bands	L1, L2, L3	L1, L2, L5	E1 (= L1), E5, E6	B1 (= L1), B2, B3

The space segment provides continuous broadcast of signals, whose quadrature components involve ranging codes of Coarse/Acquisition (C/A) and Precise (P) accuracy, service, control, and correcting information. The basis for forming the carries of all the signal components in each GNSS satellite is the common onboard frequency oscillator.

The up-to-date GNSS satellites continuously broadcast signals at three fundamental coherently-bound frequencies. Figure 2.5 presents the scheme for the distribution of frequency and sub-bands, and designates signal components emitted at those frequencies by different GNSS satellites. To distinguish the signals from individual satellites among themselves, the GPS, Galileo, and BeiDou use the code division multiple access (CDMA) with one common carrier for all signals in each sub-band.

The GLONASS satellite signals use the frequency division multiple access (FDMA) in each sub-band and have the same ranging code for every satellite. Using FDMA makes the GLONASS more anti-jamming. However, the FDMA technology does not provide a rational use of a frequency band, and may cause problems of electromagnetic compatibility with other radio-frequency facilities (radio astronomy, broadband satellite communication). Thereby, in 2021, launched were new-generation GLONASS satellites that broadcast (along with the former FDMA) CDMA signals for civil users (ICD GLONASS CDMA L3, 2016; ICD GLONASS CDMA L1, 2016). New signals are broadcast at the same central frequencies of the L1, L2, and L3 sub-bands. Figure 2.5 illustrates this, showing new GLONASS signal frequency bands with green rectangles.

Let us address the GNSS signal composition in greater detail. When describing analytically the signals, we take into account that the carrier frequency of these signals

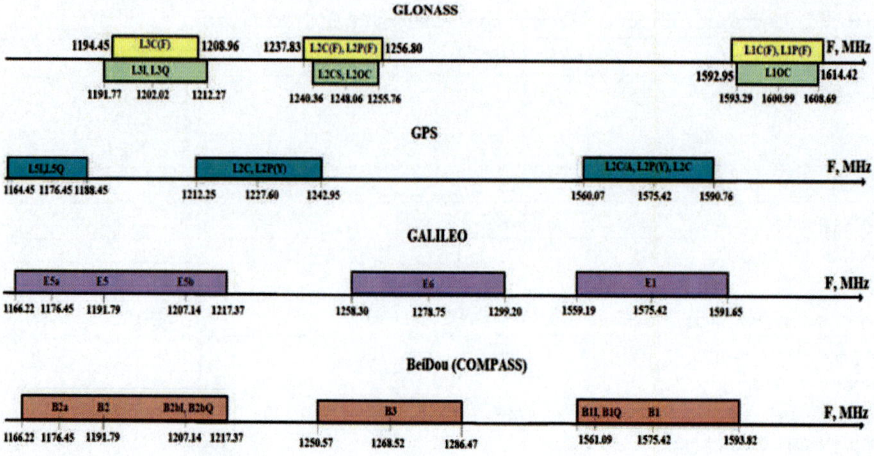

Fig. 2.5 GNSS frequencies and signal components

is subject to the Phase Shift Keying (PSK). Due to this, ranging codes and service information are broadcast to the user receiver. A general analytical presentation for a phase (angular) modulated signal has the following view (Gonorovsky 1986)

$$S_{FM}(t) = A \cdot \sin(\omega_0 t + m \cdot \theta(t)), \tag{2.6}$$

where ω_0 is the cyclic frequency of the carrier, m is the scaling coefficient, $\theta(t)$ is the broadcast message.

Taken into account (2.6), one may present a PSK signal as follows:

$$\begin{aligned} S_{FMN}(t) &= A \cdot \sin(\omega_0 t + \Gamma_k \cdot \varphi_0) \\ \Gamma_k &= \Gamma_{k-1} \oplus \theta_k, \end{aligned} \tag{2.7}$$

where φ_0 is the carrier initial phase, k is the bit number in the broadcast message, θ_k is the bit content in the broadcast message, Γ_k is a PSK discrete scaling parameter determined by the broadcast message.

GPS signals

Each GPS satellite signals contain unique ranging codes: the Coarse/Acquisition code (C/A-code) and the Precision code (P-code). Radio signals from the second-generation GPS satellites (Block II, IIA, IIR) occupy two frequency bands (20.46 MHz each), with the fixed 1575.42 MHz (L1) and 1227.6 MHz (L2) central frequencies (ICD-GPS-200 C 1993). The phase of carrier frequencies L1 and L2 is modulated by two binary code pseudorandom noise codes (PRN). Considering the general analytical presentations (2.6) and (2.7), GPS signals of the frequency sub-bands (S_{L1} and S_{L2}) may be presented as follows:

$$S_{L1}(t) = A_1 \cos(\omega_1 t + [C1P(t) \oplus D(t)] \cdot \varphi_0) + A_1 \sin(\omega_1 t + [C1C(t) \oplus D(t)] \cdot \varphi_0) \tag{2.8}$$

$$S_{L2}(t) = A_2 \cos(\omega_2 t + [C2P(t) \oplus D(t)] \cdot \varphi_0), \tag{2.9}$$

where $A_{1,2}\cos\big(\omega_{1,2}t\big)$ is the carrier signal at the frequency f_1 or f_2; φ_0 is the carrier signal initial phase, $C1P(t)$ is the P-code encrypted by the additional *W*-code, in order to eliminate its unapproved use; *C1C(t)* is the C/A-code; *D(t)* is the navigation and service information synchronized for a broadcast with the C/A-code bits.

Hereinafter, for convenience, we will term the signal component ranging codes (*C1P*, *C2P*, etc.) through the designations used in the RINEX 3.04 format to term the corresponding code pseudo-ranges measured by those codes (Gurtner and Estey 2009).

Note, also, that using the functions sin(x) and cos(x) to describe the in-phase (I) and quadrature (Q) components of the signal may be chosen arbitrarily. The quadrature of the signal components implies that the latter (i.e., C/A-code plus the service message, and P(Y)-code plus the service message) should have a phase shift by $\pi/2 \pm 100$ mrad among themselves, regardless of the carrier initial phase (ICD-GPS-200 C 1993).

The new-generation GPS satellites (IIR-M, IIF, Block III) broadcast signals at the same L1 and L2 central frequencies, but within the 30.69 MHz band. The band expansion is caused by that the L2 signals contain (along with the P(Y)-code) an improved ranging code of higher accuracy for civil users, L2C (L2 Civil). The L2C-code is generated based on the CM and CL-codes. The CM-code has the 511.5-kHz pulse repetition rate and the 20-ms duration (10,230 bits). The CL-code has the same repetition frequency, but its duration is 1.5 s (7,676,250 bits).

Besides, new GPS satellites broadcast a signal within the L5 frequency band with the 1176.45 MHz band center and with the 24 MHz frequency band. The L5 carrier is modulated by two codes: I5 and Q5. Both codes have the 10.23 MHz repetition frequency and the 1-ms duration (10,230 bits). In this case, the I5-code is additionally modulated by the navigation information.

The L2 and L5 signals are described as follows (IS-GPS-200 J 2018, IS-GPS-705 A 2010):

$$S_{L2}(t) = A_2 \cos(\omega_2 t + [C2P(t) \oplus D(t)] \cdot \varphi_0) + A_2 \sin(\omega_2 t + [C2L(t) \oplus Dc(t)] \cdot \varphi_0) \tag{2.10}$$

$$S_{L5}(t) = A_5 \cos(\omega_5 t + [C5I(t) \oplus D_5(t)] \cdot \varphi_0) + A_5 \sin(\omega_5 t + C5Q(t) \cdot \varphi_0), \tag{2.11}$$

where *C2L(t) is* the common-access improved ranging code (CM and CL broadcast alternatively with the bit temporal separation); *C5I(t)* and *C5Q(t)* are independent, mutually synchronized, in-phase and quadrature components of the P-code; *Dc(t),*

$D_5(t)$ are navigation and service information synchronized for broadcast with *C2L(t)*- and *C5I(t)*-code bits, respectively.

Note that the structure of the S_{L2}(t) *I* component may be broadcast in one of three versions (IS-GPS-200 J 2018):

(1) Combination of C/A-code and *D(t)*;
(2) C/A-code with no *D(t)*;
(3) Time-separated bit-by-bit broadcast of the CM-code with the *Dc(t)* service message and the CL-code (as described above).

The control subsystem may activate (or deactivate) the W-coding mode (anti-spoofing mode). Besides, to decrease the accuracy of coordinate solution by unapproved users, there is Selective Availability (SA) mode. When including this mode into the navigation message, one intentionally introduces a pseudorandom error to the system time and to the satellites ephemerides, which results in an essential decrease in the accuracy of navigation solutions (since 2000 May, this mode has been deactivated).

GLONASS signals

Signals from all the second-generation GLONASS satellites (Glonass-M) contain common C/A and P-codes. The Interface Control Document (ICD GLONASS 2002) defines individual nominal carrier frequencies (lettered frequencies) for GLONASS satellites in the L1 and L2 sub-bands as:

$$f_{k,1} = f_{0,1} + k\Delta f_1; \qquad f_{k,2} = f_{0,2} + k\Delta f_2 \tag{2.12}$$

where *k* is the numbers of lettered frequencies of the broadcast signals in L1 and L2 (one uses the letter numbers from k = −7 to k = + 6); f_{01} = 1602 MHz is the L1 band center; f_{02} = 1246 MHz is the L2 band center; Δf_1 = 562,5 kHz and Δf_2 = 437,5 kHz is the separation of lettered frequencies in the L1 and L2 bands.

For the GLONASS operation, there are 21 lettered frequencies in each L1 and L2 sub-band. For a more rational frequency use, applied are the same lettered frequencies for the satellites that are on the opposite sides of the Earth (antipodal satellites).

The Glonass-M lettered frequencies in the L1 and L2 sub-bands are modulated by two binary pseudorandom noise codes as follows (ICD GLONASS 2002):

$$\begin{aligned} S_{L1,k}(t) = A_1 \cos\left(\omega_{1,k}t + [\mathrm{C1}P(t) \oplus D(t)] \cdot \varphi_0\right) \\ + A_1 \sin\left(\omega_{1,k}t + [\mathrm{C1C}(t) \oplus D(t)] \cdot \varphi_0\right) \end{aligned} \tag{2.13}$$

$$\begin{aligned} S_{L2,k}(t) = A_2 \cos\left(\omega_{2,k}t + [C2P(t) \oplus D(t)] \cdot \varphi_0\right) \\ + A_1 \sin\left(\omega_{2,k}t + [C2C(t) \oplus D(t)] \cdot \varphi_0\right) \end{aligned} \tag{2.14}$$

where *k* is the number of lettered frequency; $A_{1,2}\cos\left(\omega_{1,2,k}t\right)$ is the carrier signal at the corresponding lettered frequency; *C1P(t)* and *C2P(t)* are the encrypted P-codes;

C1C(t) and *C2C(t)* is the common-access ranging code of standard accuracy (C/A-code); *D(t)* is the navigation and service information synchronized for broadcast with the C/A-code bits.

When using the FDMA, the C/A-code signal occupies a 1-MHz frequency band, while the P-code signal occupies 10 MHz around the corresponding lettered frequencies.

The third-generation GLONASS satellites (Glonass-K, Glonass-K2) broadcast CDMA signals that are described as follows (ICD GLONASS CDMA 2016, ICD GLONASS CDMA L1 2016, ICD GLONASS CDMA L3 2016):

$$\begin{aligned} S_{L1}(t) = {} & A_1 \cos(\omega_1 t + C4B(t) \cdot \varphi_0) \\ & + A_1 \sin(\omega_1 t + [C4A(t) \oplus D_C(t)] \cdot \varphi_0) \end{aligned} \tag{2.15}$$

$$\begin{aligned} S_{L2}(t) = {} & A_2 \cos(\omega_2 t + C6B(t) \cdot \varphi_0) \\ & + A_2 \sin(\omega_2 t + [C6A(t) \oplus D_C(t)] \cdot \varphi_0) \end{aligned} \tag{2.16}$$

$$\begin{aligned} S_{L3}(t) = {} & A_3 \cos(\omega_3 t + C3I(t) \cdot \varphi_0) \\ & + A_3 \sin(\omega_3 t + C3Q(t) \cdot \varphi_0), \end{aligned} \tag{2.17}$$

where $C4\mathrm{A}\ (t)$, $C6A(t)$, $C3Q(t)$ are open-access C/A-codes; $C4B(t)$ and $C6B(t)$ are encrypted P-codes; $D_C(t)$ is the navigation and service information synchronized for broadcast with the common-access code bits.

Galileo signals

Galileo satellites broadcast signals in four frequency sub-bands: E5a, E5b, E6, and E1. These signals use various types of phase-shift keying, have different structures, and, therefore, occupy frequency bands of different widths. Table 2.2 illustrates this (OS SIS ICD 2021).

The E5 signal is formed by temporal crowding of the E5a and E5b *I* and *Q* components at the output of the common multiplexer. The E5a and E5b structure has the following view (OS SIS ICD 2021):

$$S_{L5,a}(t) = A_5 \cos\bigl(\omega_5 t + \bigl[C5I(t) \oplus D_{C,a}(t)\bigr] \cdot \varphi_0\bigr)$$

Table 2.2 Frequency characteristics for Galileo signals

Signal	Carrier frequency, MHz	Receiver reference bandwidth, MHz
E1	1575,420	24,552
E6	1278,750	40,920
E5	1191,795	51,150
E5a	1176,450	20,460
E5b	1207,140	20,460

$$+ A_5 \sin(\omega_5 t + C5I(t) \cdot \varphi_0) \tag{2.18}$$

$$S_{L7,b}(t) = A_7 \cos\bigl(\omega_7 t + \bigl[C7I(t) \oplus D_{C,b}(t)\bigr] \cdot \varphi_0\bigr)$$
$$+ A_7 \sin(\omega_7 t + C7I(t) \cdot \varphi_0), \tag{2.19}$$

where $C5I(t)$ and $C7I(t)$ are open-access C/A-codes; $D_{C.a}(t)$ and $D_{C.b}(t)$ are the navigation and service information synchronized for broadcast with the $C5I(t)$ and $C7I(t)$ code bits, respectively.

The E1 and E6 signals are formed of two opposite-phase carrier components when using the composite binary offset carrier (CBOC) modulation and have the following description (OS SIS ICD 2021):

$$S_{L1}(t) = A_{1,1} \cos(\omega_1 t + [C1A(t) \oplus D_C(t)] \cdot \varphi_0)$$
$$- A_{1,2} \cos(\omega_1 t + C1B(t) \cdot (\varphi_0 \pm \pi)) \tag{2.20}$$

$$S_{L6}(t) = A_{6,1} \cos(\omega_6 t + [C6A(t) \oplus D_A(t)] \cdot \varphi_0)$$
$$- A_{6,2} \cos(\omega_6 t + C6B(t) \cdot (\varphi_0 \pm \pi)), \tag{2.21}$$

where $C6A(t)$, $C1A(t)$ and $C1B(t)$ are open-access ranging codes; $D_A(t)$ and $D_C(t)$ are the navigation and service information synchronized for broadcast with the $C6A(t)$ and $C1A(t)$ code bits; $C6B(t)$ is the pilot signal for broadcast of the ranging code on the E6B-component (broadcast of this code is planned in the future); $A_{x,1}$ and $A_{x,2}$ are the E1 and E6A- and Bcomponent amplitudes that contain (each) 50% of the carrier power (the symbol x corresponds to the L1 or L6 carrier).

BeiDou signals

The BeiDou signals are broadcast at three fundamental frequency sub-bands B1, B2, and B3. To increase the system users' possibilities, there are additional signal components B1C, B2a, and B2b are introduced into the signals of the new-generation BDS-3 satellites. Table 2.3 provides the frequency characteristics for the BeiDou signals (BDS-SIS-ICD-B1I-3.0 2019, BDS-SIS-ICD-B3I-1.0 2018, BDS-SIS-ICD-B2a-1.0 2017, BDS-SIS-ICD-B1C-1.0 2017, BDS-SIS-ICD-PPP-B2b-1.0 2020).

Table 2.3 BeiDou signal frequency characteristics

Signal	Carrier frequency, MHz	Reference bandwidth (at 1 dB), MHz
B1	1561,098	4,092
B2	1207,140	20,460
B2b	1207,140	20,460
B1C	1575,420	32,736
B2a	1176,450	20,460
B3	1268,520	20,460

The structure of the signals broadcast by the BDS-2 satellites within the B1 and B2 sub-bands is formed of the carrier I and Q components and has the following view:

$$\begin{aligned} S_{\mathrm{B1}}(t) &= A_{B1}\cos\left(\omega_{B1}t + \left[C2I(t) \oplus D_{B,1}(t)\right]\cdot\varphi_0\right) \\ &\quad + A_{B1}\sin\left(\omega_{B1}t + \left[C2Q(t) \oplus D_{B,2}(t)\right]\cdot\varphi_0\right) \end{aligned} \tag{2.22}$$

$$\begin{aligned} S_{\mathrm{B2}}(t) &= A_{B2}\cos\left(\omega_{B2}t + \left[C7I(t) \oplus D_{2,I}(t)\right]\cdot\varphi_0\right) \\ &\quad + A_{B2}\sin\left(\omega_{B2}t + \left[C7Q(t) \oplus D_{2,Q}(t)\right]\cdot\varphi_0\right), \end{aligned} \tag{2.23}$$

where $C2I(t)$, $C2Q(t)$ are open-access C/A-codes; $C7I(t)$ and $C7Q(t)$ are precise codes; $D_{B,1}(t)$ and $D_{B,2}(t)$ are the navigation and service information synchronized for broadcast with the $C2I(t)$ and $C2Q(t)$ code bits; $D_{2,I}(t)$ and $D_{2,Q}(t)$ are the navigation and service information synchronized for broadcast with the $C7I(t)$ and $C7Q(t)$ code bits.

The B3 sub-band signals contain the I component solely and have the following description (BDS-SIS-ICD-B3I-1.0 2018)

$$S_{\mathrm{B3}}(t) = A_{B3}\cos\left(\omega_{B3}t + \left[C6I(t) \oplus D_{B,3}(t)\right]\cdot\varphi_0\right) \tag{2.24}$$

where $C6I(t)$ is the open-access P-code; $D_{B,3}(t)$ is the navigation and service information synchronized for broadcast with the $C6I(t)$ code bits.

The B1C signal is generated of two I and one Q components. The first I component is modulated by the ranging code, navigation, and service information. The second and third Q components contain the pilot signal that, if necessary, may be replaced with the additional ranging code. The B1C signal structure is described as follows (BDS-SIS-ICD-B1C-1.0 2017)

$$\begin{aligned} S_{\mathrm{B1C}}(t) &= A_A\cos(\omega_{B1C}t + [C1A(t) \oplus D_{C1A}(t)]\cdot\varphi_0) \\ &\quad + A_P\sin(\omega_{B1C}t + A1P(t)\cdot\varphi_0) + A_D\cos(\omega_{B1C}t + B1P(t)\cdot\varphi_0) \end{aligned} \tag{2.25}$$

where $C1A(t)$ is the open-access P-code; $D_{C1A}(t)$ is the navigation and service information pieces synchronized for broadcast with the $C1A(t)$ code bits; $A1P(t)$ and $B1P(t)$ are pilot components; A_A, A_P and A_D are the amplitudes of components correlated as 1/4, 29/44, and 1/11, respectively.

The B2a signal is generated of two Q components, the first of which is subject to Bi-Phase Shift Keying (BPSK) by the ranging code, navigation, and service information. The second Q component contains only the pilot modulation that, if necessary, may be replaced with a modulation by the ranging code. The B2a signal structure has the following view (BDS-SIS-ICD-B2a-1.0 2017):

$$S_{\mathrm{B2}a}(t) = A_{2a}\cos(\omega_{B2a}t + [C5D(t) \oplus D_{C5}(t)]\cdot\varphi_0)$$

$$+ A_{2a} \sin(\omega_{B2a}t + C5P(t) \cdot \varphi_0), \tag{2.26}$$

where $C5D(t)$ is the open-access P-code; $D_{C5}(t)$ is the navigation and service information synchronized for broadcast with the $C5D(t)$ code bits; $C5P(t)$ is the pilot component.

And, finally, the B2b signal is to ensure the Precise Positioning Service (PPS). The signals of this type are broadcast by the geostationary (GEO) BDS-2 and BDS-3 satellites and are used to broadcast of high-precision corrections to errors of ephemeris and time–frequency services when using all the GNSS systems (including BeiDou) (BDS-SIS-ICD-PPP-B2b-1.0 2020). The corrections may be used by GNSS users, who are within the operation zone in the Asian-Pacific region (with the zone center in China). The B2b signal structure is similar to that of the B2a addressed above. The BDS-3 broadcast signals has the following description:

$$S_{B2b}(t) = A_{2b} \cos(\omega_{B2b}t + [C7I(t) \oplus D_{C7}(t)] \cdot \varphi_0)$$
$$+ A_{2b} \sin(\omega_{B2b}t + C7P(t) \cdot \varphi_0), \tag{2.27}$$

where $C7I(t)$ is the open-access P-code; $D_{C7}(t)$ is the navigation and service information (including high-precision corrections) synchronized for broadcast with the $C7I(t)$ code bits; $C7P(t)$ is the pilot component.

GNSS satellite observation geometry

Another important feature of the GNSS orbital constellations is the observed spatial configuration at different geographic latitudes. The number of satellites with different elevation angles depends essentially on the latitude. Therefore, even at an even distribution of satellites in orbits round the Earth, the local geometric configuration of the navigation constellation depends significantly on the user's location.

Figures 2.6 and 2.7 show the daily median number of visible satellites for different navigation systems at different latitudes. The authors computed that number for the prime meridian, based on the known information about the satellite ephemerids from GPS, GLONASS, Galileo, and BeiDou.

Figure 2.6 indicates that the number of observed satellites decreases with an increase in the elevation angle. This is related to that, at the satellites' orbit 55° inclination, the satellites with > 60° elevation angles become non-accessible at high latitudes. Partly, this observability decrease is offset due to observations by over-the-pole satellites at high latitudes. It is interesting to note the effect of increase in the number of satellites with the 30–50° elevation angles, observed at high latitudes.

From Fig. 2.7, one can see that the azimuthal distribution of satellites is sufficiently even. The principal difference for the west and east sectors is related to the BeiDou constellation unevenness. This implies, on average, 5–7 additionally visible satellites in the east direction.

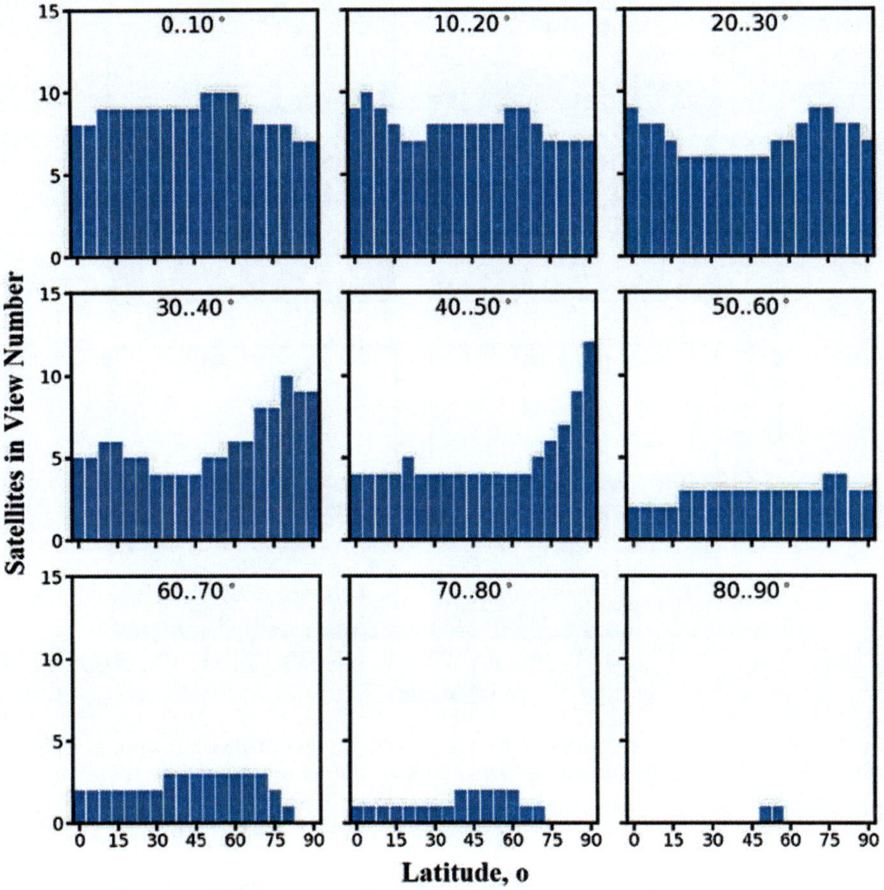

Fig. 2.6 Median number of GPS, GLONASS, Galileo, and BeiDou satellites at the set elevation angles depending on the latitude

The positioning error substantially depends on the navigation constellation geometry. A measure for satellite geometry is the PDOP (Positioning Dilution of Precision) factor. As long as ranging measurement errors are equally precise and non-correlated, the PDOP relates the positioning error to the ranging measurement error as follows (Kaplan 1996):

$$\sqrt{\sigma_X^2 + \sigma_Y^2 + \sigma_Z^2} = PDOP \cdot \sigma_{\Delta R}, \tag{2.28}$$

where $\sigma_{\Delta R}$ is the RMS of the total equivalent ranging error; σ_X, σ_Y, σ_Z are the RMS errors in determining rectangular geocentric coordinates of the GNSS user.

Here, observation of satellites at high elevation angles is particularly critical. Excluding the equation of observations from such satellites from the navigation

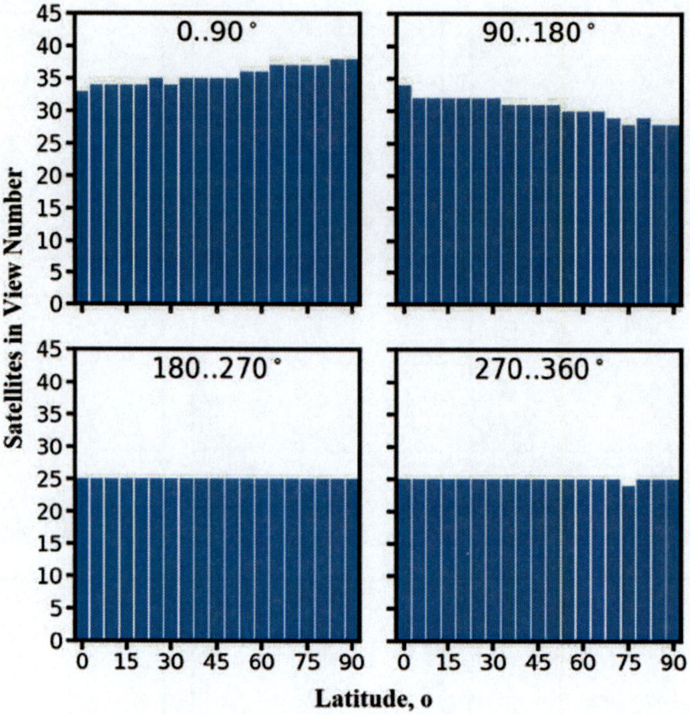

Fig. 2.7 Median number of observed GPS, GLONASS, Galileo, BeiDou satellites at different latitudes in various azimuthal directions (0–90°, 90–180°, 180–270°, 270–360°)

function composition leads to an essential deterioration of the geometrical factor. At the same time, excluding *near-the-horizon* satellites affects, as a rule, not so significantly. Above, we showed the dependence of the number of observed GNSS satellites with high elevation angles on the observation point latitude. Thereby, one may also expect the PDOP dependence on the latitude.

Figure 2.8 illustrates the daily-averaged PDOP dependence on the user's latitude for GPS and GLONASS (Demyanov et al. 2010). The calculations were also performed for the prime meridian based on the known ephemerid information of the GPS and GLONASS orbital constellations.

Apparently, the geometric factor for GPS and GLONASS deteriorates with the growth in the observer's latitude. However, in the GLONASS system, the geometric factor is better, at high latitudes (>60°), than that of the GPS. This is due to a larger orbital inclination, and, hence, more observed GLONASS satellites with high elevation angles at high latitudes.

Thus, despite the existing requirement to the equal positioning accuracy within the global GNSS coverage zone, there is, indeed, a regular positioning error caused by the PDOP dependence on the system user's latitude. By analogy, one may extend the same considerations to the Galileo and BeiDou.

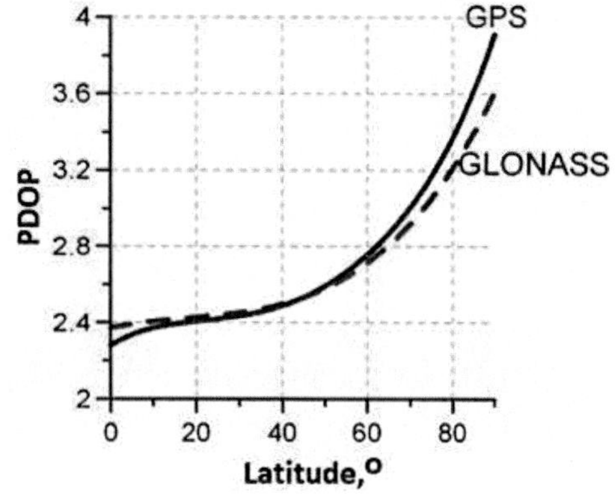

Fig. 2.8 Geometric factor dependence on the observation latitude for GPS and GLONASS

Synchronizing Reference Oscillators of GNSS satellites

The next relevant issue concerning GNSS orbital constellations is synchronizing satellite Reference Oscillators relative to the system timescale. Synchronizing Reference Oscillators may be active (GLONASS) or passive (GPS).

At the passive sync, the operation control subsystem computes the satellite onboard scale time from the data in the navigation service message. Afterward, the parameters for the timescale drift model for that satellite are computed and broadcast to the satellite via the radio channel. The satellite, then, broadcasts the parameters to users in the service message.

The active sync directly obtains the current timing from the satellite upon request from the Operation Control Segment and corrects (also directly) the Reference Oscillator. The active sync is more precise, but requires installing additional hardware onboard the satellite and within the OCS.

During the sync of Reference Oscillators, one should, first of all, estimate the computed time by the satellite Reference Oscillator. Because the correction is performed when interfacing the ground and satellite segments remote from each other, one should account for an additional signal lag in the propagation medium and due to biases in the onboard hardware as following:

$$t_{SV} = t_{SV}^{RAD} + \Delta t_{PROP} + \Delta t_{REL} + \Delta t_{ATM} + \Delta t_{EQ}, \tag{2.29}$$

where t_{SV}^{RAD} is the satellite signal radiation instant (by the Reference Oscillator); Δt_{PROP} is the signal propagation time to the receiver; Δt_{REL} is the Reference Oscillator offset as a consequence of relativistic and gravitation effects; Δt_{ATM} is a signal refraction lag when propagating through the atmosphere; Δt_{EQ} is a signal delay in the satellite and receiver hardware.

Upon estimating the satellite time t_{SV}, there comes its comparison with the corresponding timing of the GNSS system scale (T_{SYS}) at the Operation Control Segment of GNSS. As a result, there emerges an estimate for the mistiming between the onboard and the system timescales, $\Delta t_{SV} = t_{SV} - T_{SYS}$. The complexity of

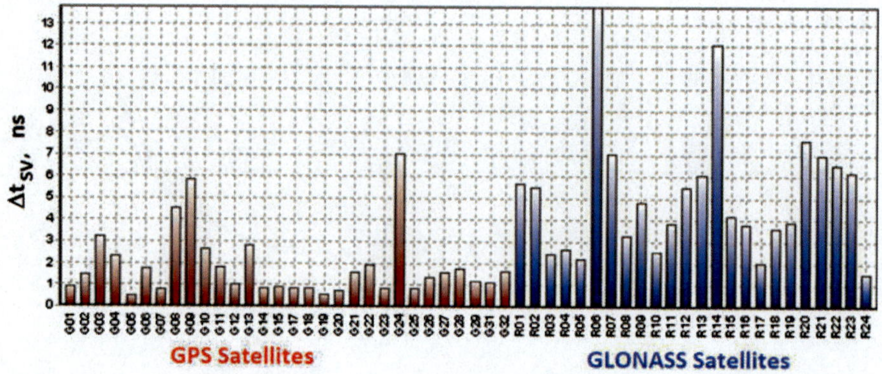

Fig. 2.9 RMS sync errors for GPS (red) and GLONASS (dark blue) reference oscillators by the OCS facilities. The data were obtained based on measurements from the IGS network of 60 stations within two days in 2013 December (Report C # 337/13, 2013. https://www.glonass-iac.ru/en/sostavOG/)

correcting the Reference Oscillator is caused by insufficient knowledge of values and spatial–temporal behavior of the errors in the t_{SV} estimate (2.29).

An essential improvement in estimating the sync errors (2.29) may be achieved by organizing long-term observations of mistiming between the system and Reference Oscillators. However, direct measurements with solely GNSS OCS metrological facilities are performed within a limited territory over a limited time (2.5). Therefore, the accuracy of such measurements is restricted.

Figure 2.9 demonstrates the Reference Oscillator sync errors for GPS and GLONASS satellites by the OCS facilities, relative to the updated estimates t_{SV} from the results of IGS data processing. The sync error for the Reference Oscillator is ~ 0.5–7 ns (equivalent ranging error being 0.15–2.1 m) for GPS and ~ 1.5–15 ns (0.45–4.5 m) for GLONASS.

Due to deploying global networks for geophysical and geodynamic monitoring and owing to increasing the number of international analytical data processing centers, there is a possibility to significantly improve the mistiming between the GNSS onboard and system timescales.

Figure 2.10 presents the updated estimates for sync errors of the GPS Reference Oscillators relative to the system time. These estimates were obtained from the results of processing the IGS measurements for 2020 Nov 19 at the WHU Analysis Center (Villiger and Dach 2021).

The results presented in Fig. 2.10 demonstrate considerable possibilities of the up-to-date global network metrological facilities to provide high-precision sync of GNSS Reference Oscillators.

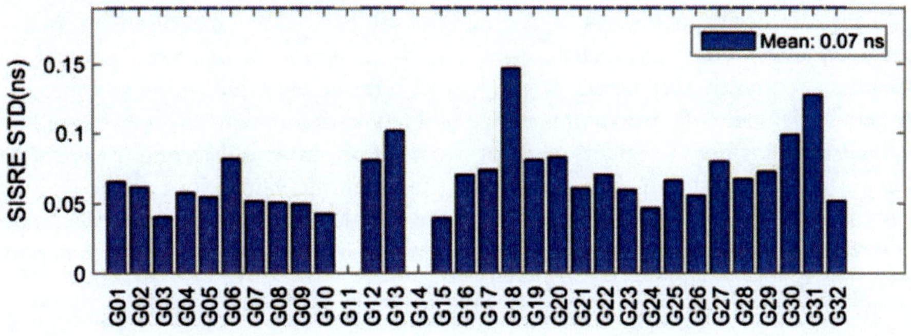

Fig. 2.10 Residual mean sync error for the GPS reference oscillators relative to the system time (Villiger and Dach 2021)

2.1.3 User Receiver

The GNSS user receiver performs passive ranging to visible satellites and determines its own radial velocity, as well as receives and processes navigation and service information that is broadcast from a satellite. As a result of processing the obtained measurements, navigation and service information, the user receiver solves positioning. The positioning task is to estimate the current components of the user (object) extended state vector:

$$\Pi = \left|X, Y, Z, t, t', X', Y', Z'\right|^T, \tag{2.30}$$

where *X, Y, Z* are the user's spatial coordinates; X', Y', Z' are the user velocity vector components; *t* is the GNSS current system time; $t\prime$ is the correction for the user timescale, relative to the GNSS system timescale.

The above elements of the vector Π (except for *t*) cannot be measured directly. But they can be determined indirectly through measuring the satellite signal parameters, such as code pseudo-delay and the signal pseudo-phase (τ, $\Delta\varphi$), as well as the Doppler pseudo-frequency shift (f_D). Here, the prefix *pseudo* denotes a parameter measured with an error. We will term the true values for the above parameters code delay, carrier phase, and the Doppler frequency drift, respectively.

Once the above parameters are measured, one may compute the geometry corresponding to them: code (R_C), and phase (L) pseudo-ranges and pseudo-velocity (V_p), as following

$$\begin{aligned} R_C &= c \cdot \tau \\ L &= c \cdot \Delta\varphi \big/ (2\pi \cdot f) \\ V_R &= f_D \cdot \lambda \end{aligned} \tag{2.31}$$

where λ and f are the wavelength and the satellite signal frequency.

The radio signal parameters (τ, Δφ, and f_D) subject to direct measurements in the user navigation receiver are referred to as *radio-navigation parameters*. Their corresponding geometric parameters (R_C, L, V_p) are referred to as *navigation parameters*. A geometrical site of space points with equal navigation parameters are referred to as *a position surface*. The intersection of two position surfaces determines *a position line*—a geometric site of the points having two definite values for two navigation parameters. The *location* is determined by the coordinates for the intersection point of three position surfaces (location spatial coordinates) or two position lines (location on a plane) (Perov and Kharisov 2010).

To find a position (i.e. positioning solution), one uses functional relations between navigation parameters (RC, L, Vp) and the **Π** vector components (2.30). Those functional relations are referred to as *navigation functions*. There are several known methods to determine spatial coordinates of GNSS user, based on different navigation functions (Perov and Kharisov 2010). Table 2.4 presents a short overview and characteristics for those methods.

The main method for positioning solution, that is implemented in GNSS user receivers, is the *pseudo-ranging* method. The latter is based on passive measurements of pseudo-range (code or phase) between the user and each observable satellite. The pseudo-range here serves as a navigation parameter, and each position surface is a sphere with the radius ρ_i and the center coinciding with the phase center of the satellite antenna. The equation for such a sphere has the following view:

Table 2.4 Methods to find a positioning solution

Positioning method	Advantages	Disadvantages
Ranging (pseudo-ranging)	Possibility to measure object state vector parameters in real time (RT) Possibility to measure the shift between the receiver timescale (RTS) and the satellite timescale (STS)	Necessity in redundancy of navigation information to determine deviations of the RTS Necessity in high-precision sync between the STS and the GNSS system timescale
Differential-ranging	No necessity in redundancy of navigation information No need in high-precision sync between the RTS and STS	Impossibility to measure the shift between the RTS and STS
Radial-velocity (pseudoradial-velocity)	High precision and simplicity in measuring the object velocity vector components	Impossibility to measure coordinates in RT High requirements for the satellite and receiver reference oscillator frequency stability
Differential-radial-velocity	Insensitivity to the satellite and receiver reference oscillator frequency instability	Impossibility to measure the shift between the satellite and receiver timescales

$$\rho_i = r_i + c \cdot \Delta t_u + \varepsilon_i = \sqrt{(x_i - X)^2 + (y_i - Y)^2 + (z_i - Z)^2} + c \cdot \Delta t_u + \varepsilon_i \quad (2.32)$$

where r_i is the geometric range to ith satellite; $c \cdot \Delta t_u$ is the ranging error caused by the receiver timescale drift relative to the GNSS system timescale; ε_i are random ranging errors caused by other factors for the ith satellite; x_i, y_i, z_i are the known coordinates for the ith satellite; X, Y, Z are the unknown GNSS user coordinates.

The functional relation (2.32) between the navigation parameters and the GNSS user state vector enables to determine a corresponding navigation function for the positioning solution:

$$\Delta \mathbf{x} = \mathbf{H}^{-1} \times \Delta \boldsymbol{\rho}, \quad (2.33)$$

where $\mathbf{H} = \begin{vmatrix} a_{x,1} & a_{y,1} & a_{z,1} & 1 \\ a_{x,2} & a_{y,2} & a_{z,2} & 1 \\ \dots & \dots & \dots & \dots \\ a_{x,n} & a_{y,n} & a_{z,n} & 1 \end{vmatrix}$ is the matrix of direction cosines of the SV navigation constellation; $\Delta \boldsymbol{\rho} = \begin{vmatrix} \Delta \rho_1 \\ \Delta \rho_2 \\ \dots \\ \Delta \rho_{1n} \end{vmatrix}$ is the error vector for pseudo-ranges; $\Delta \mathbf{x} = \begin{vmatrix} \Delta X \\ \Delta Y \\ \Delta Z \\ \dots \\ \Delta \mathbf{t}_u \end{vmatrix}$ is the error vector for the GNSS user state vector estimate.

Solving the set of Eqs. (2.33) should be obtained so that to minimize the discrepancy $\Delta \mathbf{x}$ between the true (most probable) value of the user coordinates and their estimate. The accuracy and robustness of the positioning solution depends on the solution method used in the user receiver. To solve this problem, there are sufficiently many methods developed. The main of them are: the group of iterative methods and group of Kalman filtering (Xu 2007). Below, we will address this problem in greater detail.

There emerges a natural question: how can one define the maximally available positioning accuracy, i.e., a *potential accuracy of the positioning solution*? Unless the geometry effect is addressed, the positioning accuracy is restricted by *an error in measuring a radio-navigation parameter*. Its value should be defined as a minimally achievable RMS measurement error in the following view

$$\sigma = \frac{1}{\beta \cdot CNR} \quad (2.34)$$

where $\beta = 1/\tau$ is the satellite signal spectrum effective width, Hz; τ is the ranging code chip length, s; *CNR* is the carrier to noise ratio at the input of the radio-navigation parameter meter (PLL or DLL), dBW.

Thus, the potential positioning accuracy depends, first of all, on the characteristics for the ranging code and on the carrier to noise ratio at the signal reception point. For example, when measuring the C/A code range, these parameters are $\tau_i = 1\ \mu\text{s}$, $\beta = 1/\tau_i = 1$ MHz, a typical $CNR = 40$ dBW. Then, the code bias lag measurement error is $\sigma = 25$ ns (the equivalent ranging error being 7,5 m). At the typical mean value for the geometric factor PDOP = 2, the corresponding RMS positioning error (2.30) is 15 m.

When measuring a range by using the *P(Y)*-code ($\tau_i = 0,\ 1\ \mu\text{s}$), and at the same *CNR*, the positioning error will be $\sigma = 2.5$ ns (the equivalent ranging error being 0.75 m). At the PDOP = 2, this corresponds to the positioning error of 1.5 m.

In real measurements of the radio-navigation parameters, the measurement accuracy estimate is done considering the basic characteristics of a meter. In this case, the RMS error in measuring the code pseudo-range (m) and the carrier phase (rad) may be estimated as follows (Kaplan 1996):

$$\sigma_{Rc} = L_C \cdot \sqrt{\frac{4 \cdot F1 \cdot d^2 \cdot \Delta F_{DLL}}{10^{0.1 \cdot CNR}} \cdot \left(2(1-d) + \frac{4 \cdot F2 \cdot d}{T \cdot 10^{0.1 \cdot CNR}}\right)} \tag{2.35}$$

$$\sigma_{\varphi} = \sqrt{\frac{\Delta F_{PLL}}{CNR} \cdot \left(1 + \frac{1}{2T \cdot CNR}\right) + \left(m \cdot \frac{\sigma_F(\tau) \cdot f}{\Delta F_{PLL}}\right)^2}, \tag{2.36}$$

where $Lc = c/F_{\text{PRN}}$ is the ranging code wavelength, m; c is the speed of light in vacuo; F_{PRN} is the ranging code pulse repetition rate, imp/s (e.g., F_{PRN} is $1.023 \cdot 10^6$ for the GPS C/A-code, $10.23 \cdot 10^6$ imp/s for the GPS P(Y)-code); *F1, F2* are the parameters depending on the type of the correlator and the code delay discriminator (typical values being *F1* = 0.5, *F2* = 1); *d* is the delay correlator spacing between early and late code replicas, in fractions of the ranging code chip length (typical values being $d = 0{,}1$ or 0,5); ΔF_{DLL}, ΔF_{PLL} are the noise bandwidth of the code delay and the carrier phase lock loop; *CNR* is the carrier to noise ratio at the meter input, dBW; T is the accumulation time in the correlator, ms; $\sigma_F(\tau)$ is the short-time instability of the receiver reference oscillator; f is the signal carrier frequency, Hz; m is the coefficient depending on the lock lop order ($m = 144$ for the second-order loop and $m = 160$ for the third-order loop).

From expressions (2.35–2.36), one can see that the accuracy of the positioning solution depends significantly on the features of the user receiver radio-frequency (RF) front-end (expressed through the *CNR*), on the receiver reference oscillator instability, on the type of the phase and code pseudo-delay discriminator, and their settings.

Another important specific feature for signal processing in a receiver is related to a necessity of estimating the radio navigation parameters under the conditions of encrypting the ranging code (i.e. under spoofing mode). We illustrate this process by the most known example of the GPS P(Y)-code signal processing. To prevent an unapproved access to the ranging P-code, the latter is subject to additional encrypting by the closed *W*-code (anti-spoofing operation). Figure 2.11 illustrates this.

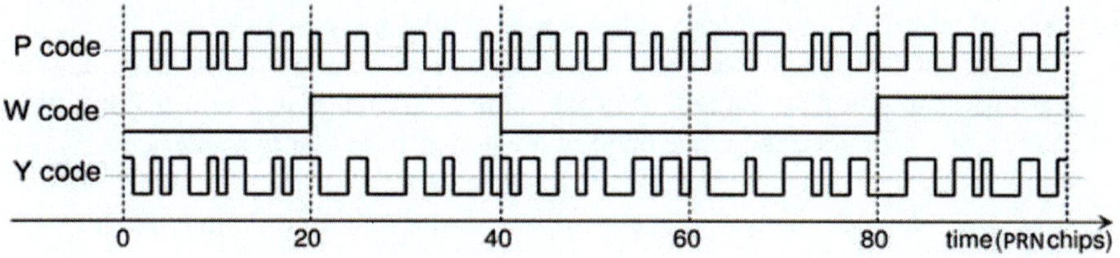

Fig. 2.11 Generating P(Y) encrypted code

Because general-use receivers have no access to the *W*-code, one applies a codeless or a semi-codeless technique to detect the information on the code bias and the carrier phase from the components of the signals containing the P(Y)-code.

The essence of the *codeless processing technique* is as follows. At the user receiver signal processor input, there come signal samples of the *i*th satellite, transported, in advance, onto the intermediate frequency (IF) (Linty 2010).

$$S_{L2}(n) = A_{2,n}\cos(\omega_{IF}nT_S + [P(Y)(n) \oplus D(n)] \cdot \varphi_0) + \varepsilon(n), \tag{2.37}$$

where *n* is the signal sample number; $A_{2,n}$ is the level (amplitude) of a signal sample; T_S is a sampling period of the incoming signal; ω_{IF} the cyclic IF containing the Doppler frequency drift; $\varepsilon(n)$ is a sample of the additive in-band interference.

Because the P(Y)-code is a sequence of discrete phase values switched between the opposite sign values (« − 1 » or « + 1»), squaring the signal (2.37) leads to a removal of such a phase code from it. Along with this, removed will be also the codes of the navigation and service information *D(n)*. Thus, upon squaring, a signal like the following is generated:

$$S_{L2}^2(n) = A_0 + A_{2,n}\cos(2\omega_{IF}nT_S + 2 \cdot \varphi_0) + \varepsilon^2[n]. \tag{2.38}$$

Processing such a signal allows one to obtain the corresponding estimates for the carrier phase and for the Doppler frequency drift $\left(\hat{f}_d, \hat{\varphi}\right)$. Herewith, a fine phase tuning of the receiver reference oscillator within an element of the ranging code is done coarsely by available measurements of the open C/A-code. This occurs, because, as a result of the above operation, the ranging code structure in the measurement channel will be lost.

A code division of signals from separate satellites, in this case, is also impossible, and it is performed by the Doppler frequency that is estimated in the C/A-code processing channel. Thus, the measurements of the carrier phase of the Doppler frequency in the encrypted signal processing channel appears to be bound to the measurements in the C/A-code processing channel. Therefore, the accuracy of the phase measurements in the codeless processing mode drops and will not be higher, than that in the C/A-code channel.

Except this problem, from (2.38) it follows that using the described procedure results in the following negative consequences (Linty, 2010):

1. In band additive interferences at the tracking inlet increase as $\varepsilon^2[n]$;
2. Because the tracking noise band should have a sufficient width to provide with a robust frequency lock, then, to process the $S_{L2}^2(n)$ signal, like it follows from (2.38), one will need the noise band twice wider, than that to process the $S_{L2}(n)$ signal.
3. Precise P(Y)-code pseudo-ranging appear impossible and defined with C/A – code tracking accuracy.

As per (2.35, 2.36), the dispersion for the estimate of a phase and an pseudo-delay is determined by the meter input *CNR*. The presence of the above negative features of codeless processing, as well as cross-correlation effects, results in an essential *CNR* deterioration at the input of the tracking meter. To account for this, one uses an equivalent *CNR*, when estimating the error dispersion of the radio-navigation parameter filtering:

$$CNR_E = CNR \cdot R^2(\varepsilon_\tau) \cdot sinc^2\big(0.5 \cdot \varepsilon_f \cdot T,\big) \tag{2.39}$$

where T is the accumulation time in the meter integrator; $R(\varepsilon_\tau)$ is the ranging code autocorrelation function; ε_f is the signal frequency estimate error, Hz; ε_τ is the pseudo-delay estimate error; *CNR* is a carrier to noise ratio at the meter input with allowance for the receiver thermal noise only.

Estimating the carrier phase with the reference oscillator frequency lock by the coarse C/A-code used as an a priori estimate in the P(Y)-code tracking channel will lead to that the value $R(\varepsilon_\tau)$ will be insignificant. Therefore, owing to the essential difference in the structures of the C/A- and P(Y)-codes, there will emerge an essential *CNR* decrease in the channel with using a codeless processing.

Figure 2.12 shows the dependence of RMS of estimating the carrier phase on the *CNR* at the meter input, when using a codeless meter and a standard meter applying the known P-code structure. The dotted line shows the same dependences obtained by approximating the experimental observations. The presented data are obtained through the statistical test method in (Linty 2010). Like one may make certain, to achieve the equal accuracy of estimating the phase at the input of a codeless meter, one should provide a *CNR* by ~ 13 dB higher, as compared with a standard P-code meter.

The presence of the above disadvantages in the algorithms of codeless processing the encrypted code signals led to developing more perfect methods of semi-codeless processing. *A semi-codeless processing (Z-tracking)* implies that, before estimating radio-navigation parameters, the known P-code and service information are removed from the incoming signal (2.37). Thus, in the input signal, there is only the unknown W-code left:

$$S_W(n) = a_{S[W],n} \cos(\omega_{IF} n T_S + [W(n)] \cdot \varphi_0) + \eta[n]. \tag{2.40}$$

The P-code proper is generated by the known law as a linear combination of four known code sequences X1A, X1B, X2A, and X2B. The P-code total sequence for

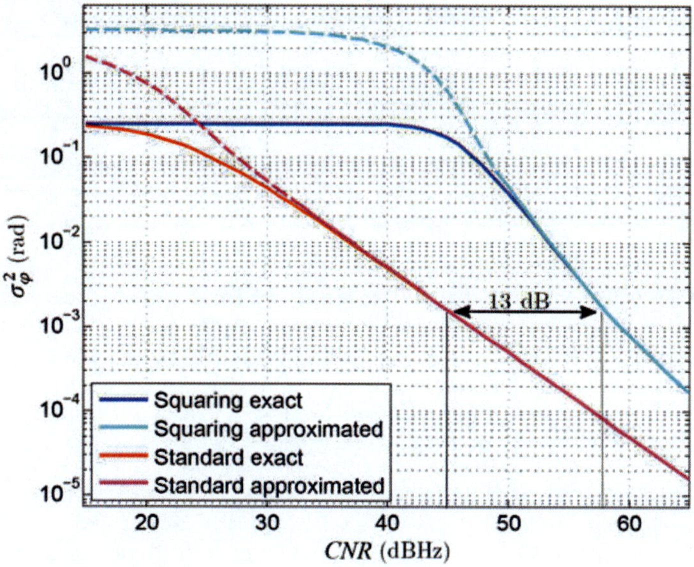

Fig. 2.12 Dependence of the carrier phase estimate RMS on the *CNR* at the input of a carrier phase tracking loop (Linty 2010)

each *i*th satellite looks like (Psiaki et al. 2003):

$$P_i = X1 \oplus X2_i, \tag{2.41}$$

where $X1 = X1A \oplus X1B$, $X2 = X2A \oplus X2B$, and the code pattern $X2_i$ is generated by removing the first bits corresponding to the number of the *i*th GPS satellite from the $X2$.

The exact W-code structure is unknown, but the code is known to be generated synchronously with the X1A-code, i.e., it is related to the P-code through a certain relation. The X1A pulse repetition rate is 10,23 MHz, and code sequence length is 4092 pulses. Psiaki et al. (2003) note that, first, P-code pulse formation falls to one W-code pulse. At such a relation, *M* pulses of the W-code are generated. Further, there follow *N* pulses of the *W*-code, to each of which P-code B pulses fall. After 4092 code pulses advance, the sequence recurs. Figure 2.13 illustrates this.

The values for the parameters *A*, *B*, *M*, and *N* are exactly unknown, but A≈B≈20 (Psiaki et al. 2003). Thereby, one can recover the view for the auto-correlated function of the signal $S_W(n)$ (2.40). The function may be expressed as follows (Psiaki et al. 2003).

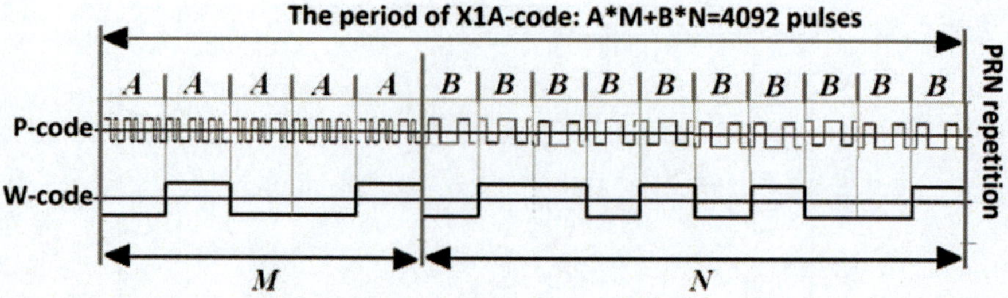

Fig. 2.13 Relation between the encrypted W-code and open P-code

$$R(\Delta T) = \begin{cases} a^2_{s[W]} + \sigma^2_{\eta,W} & |\Delta T = 0 \\ a^2_{s[W]} \cdot \frac{(M+N)\cdot|\Delta T|}{4092} & |1 \le \Delta T \le A \\ \frac{a^2_{s[W]}N}{4092}(B - |\Delta T|) & |(A+1) \le \Delta T \le B \\ 0 & |(B+1) \le |\Delta T|, \end{cases} \tag{2.42}$$

where the ΔT temporal shift is measured in the number of P-code elements.

Accumulating the $S_W(n)$ time series allows one to establish the values for A, B, M, and N at a satisfactory accuracy, as long as the observational interval has the duration not shorter, than the W-code cycle duration (>100 ms). Afterward, one may implement estimating the pseudo-delay in the P-code tracking channel. Due to this, there emerges an additional *CNR* gain in the encrypted signal tracking channel. The gain can be estimated as follows (Kaplan 1996):

$$CNR_C = 10 \cdot lg\left(\frac{F_{PRN}}{2 \cdot \Delta F_{DLL}}\right), \tag{2.43}$$

where F_{PRN} is the pulse repetition rate in the code ranging sequence (see (2.35)); ΔF_{DLL} is the noise bandwidth of the code pseudo-delay tracking loop.

When using the above methods to process the *CNR* in the encrypted signal tracking channel will drop as compared with the case, when the encrypted code structure is known. Chen et al. (2005) indicates that using codeless processing algorithms involves a decrease in the value for the equivalent CNR by 27–30 dB, and semi-codeless—by 14–17 dB. One should account for this feature, when studying the ionosphere and the Earth upper atmosphere by using GNSS signals containing the encrypted ranging code.

2.2 Main Error Sources in Ranging and Positioning

2.2.1 Code-Pseudo-Range and Carrier Phase Errors

To conveniently analyze the errors that are a part of a pseudo-range and of a pseudo-phase, one should make a definition for a number of fundamental values. To such values, we refer the time t, frequency $f(t)$, and phase $\varphi(t)$ in application to the carrier signal generated by the satellite reference oscillator. All the measurements of radio-navigation parameters in the GNSS hardware are taken relative to the system time. Therefore, it is important to determine the correlation between those measurements' errors and the timescales' errors that originate onboard a satellite and in a navigation receiver.

The frequency is related to the carrier phase as follows (Teunissen and Kleusberg 1998):

$$f(t) = d\varphi(t)/dt \tag{2.44}$$

$$\varphi(t) = \int_{t0}^{t1} f(t)d\tau + \varphi(t_0), \tag{2.45}$$

where $\varphi(t_0)$ is the carrier initial phase at the t_0 instant; $f(t)$ is an instantaneous value for the carrier frequency measured at t instants.

The signal phase may be converted into running time $t_j(t)$ sample on the timescale of the jth navigation receiver by subtracting the initial phase from it, with a subsequent division by the nominal carrier frequency f_0:

$$t_j(t) = (\varphi(t) - \varphi(t_0))/f_0 = \frac{\varphi(t)}{f_0} - t_j(t_0). \tag{2.46}$$

As long as the carrier instantaneous frequency equals the nominal value, then from (2.45) and (2.46) it follows:

$$t_j(t) = t - t_j(t_0). \tag{2.47}$$

During real measurements, one should account for the reference oscillator frequency instability. Here, we mean the total frequency instability of reference oscillators of a satellite and of a receiver. This frequency is reflected in total phase measurements of a jth receiver. Considering this, the time measured on the receiver scale will be equal to:

$$t_j(t) = \frac{1}{f_0}\int_{t0}^{t1} f(t)dt - t_j(t_0). \tag{2.48}$$

As long as one uses the concept of frequency deviation ($\delta f(t) = f_0 - f(t)$) to account for the reference oscillator frequency instability, then one may write:

$$t_j(t) = \frac{1}{f_0} \int_{t0}^{t1} (f_0 + \delta f(t))dt - t_j(t_0). \tag{2.49}$$

Let us introduce a *time deviation* $\delta t_j(t)$ that originates when estimating the present instant on the receiver scale owing to a short-time frequency instability of the satellite reference oscillator and of the user receiver:

$$\delta t_j(t) = \int_{t0}^{t1} \frac{\delta f(t)}{f_0} dt. \tag{2.50}$$

With allowance for this, one may write a final relation between the time obtained by measuring the signal carrier phase of the receiver reference oscillator and the real time t:

$$t_j(t) = t + \delta t_j(t) - t_j(t_0) = t + \Delta t_j(t). \tag{2.51}$$

Viz., the time measured in the receiver by the carrier phase is the sum of the real time, time deviation, and a summand that accounts for the non-zero initial phase of the receiver reference oscillator signal.

For convenience of further reasoning, we combined the last two terms in the right member of Eq. (2.51). Thereinafter, we will term this time measurement error component *a receiver timescale error* $\Delta t_j(t)$. Through similar reasoning, accounting for the satellite reference oscillator frequency instability, one may also introduce the concept *of a satellite timescale error*, $\Delta t_i(t)$. Herewith, we accept time samples at the GNSS reference oscillator output as *the real system time, t*. This timescale also involves the errors of deviation from the UT timescale (see Sect. 2.1.1), but, in the further reasoning, we will not take this into account.

Additionally, one should consider that, during the propagation of a signal from the ith satellite to the jth receiver $\left(\Delta t_j^i\right)$, due to the rotation of the Earth and the related rectangular geocentric coordinate system, the satellite real coordinates do not anymore correspond to those that were valid at the signal emission instant. Let us denote the satellite coordinates at the signal emission instant t_s as x_i^s, y_i^s, z_i^s, and the receiver coordinates as X_j^k, Y_j^k, Z_j^k. Then the geometric distance between the ith satellite and the jth receiver, that was recorded at the instant t_k by the receiver clock, equals (Povalyaev 2008).

$$\tilde{R}_j^i = \left[\left(x_i^s - X_j^k\right)^2 + \left(y_i^s - Y_j^k\right)^2 + \left(z_i^s - Z_j^k\right)^2\right]^{0.5}. \tag{2.52}$$

Let us admit that the satellite location points with coordinates x_i^s, y_i^s, z_i^s and x_i^k, y_i^k, z_i^k at the t_s and t_k instants are close. With allowance for this, we expand Function (2.52) into a Taylor series in the point x_i^k, y_i^k, z_i^k, and consider only the series terms not above the first degree, then

$$\tilde{R}_j^k \approx \tilde{R}_j^i - \frac{x_i^k - X_j^k}{\tilde{R}_j^i}(x_i^k - x_i^s) - \frac{y_i^k - Y_j^k}{\tilde{R}_j^i}(y_i^k - y_i^s) - \frac{z_i^k - Z_j^k}{\tilde{R}_j^i}(z_i^k - z_i^s) \quad (2.53)$$

In Series (2.53), one may not consider the last summand, because the Z-axis of the rectangular geocentric coordinate system transits through the North Pole. Therefore, the Earth rotation does not lead to a change in the corresponding satellite coordinate. With allowance for this, one can express (from 2.53) the correction for the ranging caused by the Earth rotational displacement for the Δt_j^i time interval (Povalyaev 2008).

$$\Delta R_i^s = \tilde{R}_i^k - \tilde{R}_i^s = \left(X \cdot y_i^k - Y \cdot x_i^k\right)\frac{\Omega_\oplus}{c}, \quad (2.54)$$

where $\Omega_\oplus = 7{,}292{,}115 \cdot 10^{-11}$ rad/s is the Earth rotation angular velocity; c is the speed of light in vacuo.

Thus, the range $\hat{R}_j^i$ measured at the t_k instant by the receiver clock should be corrected as $\hat{R}_j^k = \tilde{R}_j^i - \Delta R_i^s$ before using. Clearly, one will observe this error dependence on the observational latitude, with minimal being at the Pole, and maximal being at the equator.

Now, let us switch directly to analyzing the composition of the measured code pseudo-range. The pseudo-range $\rho_i^j(t)$ equals the difference in the time measured on the receiver timescale $t_j(t)$ at the signal incoming instant and on the ith satellite timescale at the signal emission instant $t_i(t)$, multiplied by the nominal speed of light in vacuo c:

$$\rho_i^j(t) = c\left(t_j(t) - t_i\left(t - \tau_i^j\right)\right) + \varepsilon_i^j, \quad (2.55)$$

where τ_i^j is a delay for the signal propagation from the ith satellite oscillator to jth receiver meter; ε_i^j is the pseudo-range errors.

The receiver scale time $t_j(t)$ and the satellite onboard-scale time $t_i\left(t - \tau_i^j\right)$ are equal to the GNSS system time plus the corresponding timescale errors (2.51), i.e.

$$t_j(t) = t + \Delta t_j(t) \quad (2.56)$$

$$t_i\left(t - \tau_i^j\right) = t - \tau_i^j + \Delta t_i\left(t - \tau_i^j\right). \quad (2.57)$$

Substituting these equations into (2.55) enables to obtain the following view of a pseudo-range:

$$\rho_i^j(t) = c\tau_i^j + c\left(\Delta t_j(t) - \Delta t_i\left(t - \tau_i^j\right)\right) + \varepsilon_i^j. \quad (2.58)$$

The signal propagation time τ_i^j may be presented as a total of three independent components: a signal delay during its generation and emission by the satellite antenna; a time delay for the signal propagation from the transmitting antenna to the receiving one; a signal delay at its receiving and processing in the receiver:

$$\tau_i^j = dt_i + \delta\tau_i^j + dt_j, \tag{2.59}$$

where dt_i is an additional signal delay on the paths from the reference oscillator output to the emitting satellite antenna phase center; dt_j is a signal delay on the path from the receiving antenna phase center to the input of the receiver tracking meter; $\delta\tau_i^j$ is the ranging error due to the effect of the satellite signal refraction in the ionosphere and in the Earth atmosphere.

The signals of the first and second frequency retard differently in the satellite hardware and in the receiver. The difference between the total signal delay in the satellite and in the receiver at the first and second frequency is referred to as *differential code biases* (*DCB*):

$$DCB = \left(dt_{i,f1} + dt_{j,f1}\right) - \left(dt_{i,f2} + dt_{j,f2}\right). \tag{2.60}$$

The error originating at the satellite signal propagation in space, $\delta\tau_i^j$, is related to that the signal propagation velocity in it differs from the velocity in vacuo. This difference is considered by the propagation medium refractive index (n), therefore, one may write

$$c \cdot \delta\tau_i^j = \int_s n \cdot ds. \tag{2.61}$$

This integral can be split into three separate components:

$$c \cdot \delta\tau_i^j = \int_s ds + \int_s (n-1) \cdot ds + \left(\int_s n \cdot ds - \int_s ds\right). \tag{2.62}$$

Here, the first summand is an ideal geometric distance measured between the satellite antenna at the broadcast instant and the receiver antenna at the signal reception instant. Because there may be several signal propagation paths and an interference among the latter, then, indeed, the first summand is the sum of a geometric distance ($\rho_{0,i}^j$) and a multipath-induced error (dm_i^j).

The multipath-induced error (*multipathing*) may be induced by a signal reflection from satellite conducting surfaces (satellite multipathing, normally insignificant) or reflecting surfaces near an antenna (receiver multipathing, the main contributor). With allowance for this, we have:

$$\int_s ds = \rho_{0,i}^j + dm_i^j. \tag{2.63}$$

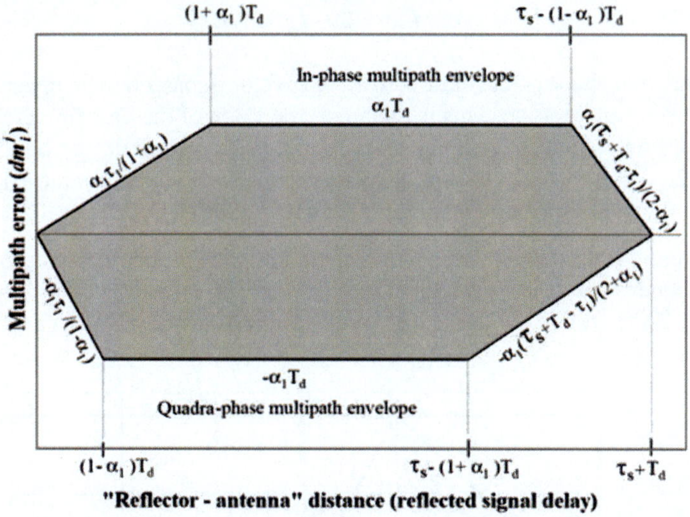

Fig. 2.14 Multipath error envelope characteristics (Ray 2000). Re-produced and modified figure based on the part of the figure published under open assess (https://prism.ucalgary.ca/handle/1880/42732)

To estimate the multipath contribution to the pseudo-range error, one uses a characteristic termed *a multipath error envelope.* To understand the fundamental regularities of this error generation, we address a simplest case, when there is a direct signal arriving from a satellite at an antenna, and one re-reflected signal that arrives at the same antenna with a τ_1 delay relative to the direct signal. Figure 2.14 presents the multipath noise envelope for this case (Ray 2000).

Figure 2.14 introduces the following designations: α_1 is the reflection coefficient (that depends on the conductance of the reflecting surface near the receiver antenna and on the characteristics of the reflected signal wave polarization); T_d is code delay correlator spacing; τ_s is the ranging-code chip length, sec; τ_1 is a delay time between the reflected and the direct signal at the reception point, sec.

Based on analyzing the presented characteristic, one may indicate the following features for the multipath error generation in the receiving antenna location:

- The multipath error forms in a local area around the receiving antenna. This area size directly depends on the ranging-code chip length (duration) τ_s. The reflected signals that arrived at the meter input with a delay $\tau_1 > T_d$ (i.e. beyond the discrimination characteristic operation domain) do not contribute significantly to the multipath error;
- The multipath error value directly depends on the reflection coefficient α_1, i.e. on the relation between the direct and the reflected signal amplitudes at the reception point. This relation is associated with the reflection coefficient through the *Signal-to-Multipath-Ratio* (*SMR*), like

$$SMR = 20 \cdot lg\left(\alpha_1^{-1}\right). \tag{2.64}$$

- The particular shape of the multipath error envelope characteristic depends on the delay discriminator type. This is related to the value of the discriminator spacing T_d, that should not be beyond the discrimination characteristic operation domain. Figure 2.15 presents the view for delay discriminators' typical characteristics, and their analytical description is in Table 2.5 (Ray 2000).

Generating the discrimination characteristic in a code pseudo-delay discriminator occurs by superposing the signal direct replica (Prompt) with the same-signal replica delayed by T_d (Late) and with the same-signal replica leading by T_d (Early). In the

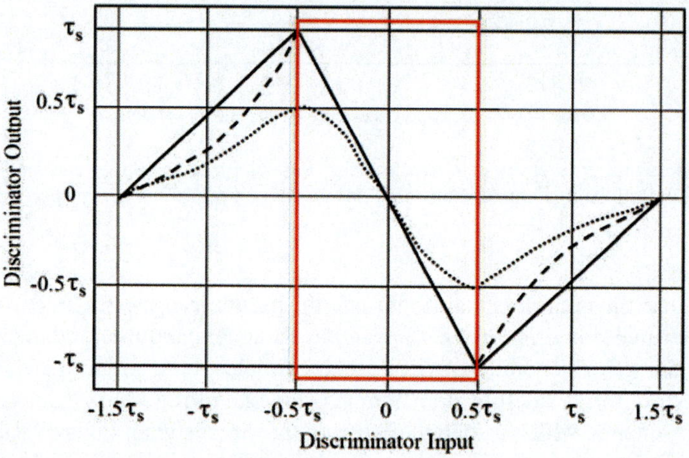

Fig. 2.15 Typical discrimination characteristics for code pseudo-delay discriminators. Red rectangle highlights the discrimination characteristic operation domain (Ray 2000). Re-produced and modified figure based on the part of the figure published under open assess (https://prism.ucalgary.ca/handle/1880/42732)

Table 2.5 Various Types and Characteristics of code Delay Lock Loop (DLL) Discriminators (Ray 2000)

Discriminator type		Algorithm
Coherent		$Dc = IE - IL$
Non-coherent	Early minus late power	$Dn = \left(IE^2 + QE^2\right) - \left(IL^2 + QL^2\right)$
	Early minus late envelope	$Dn = \sqrt{IE^2 + QE^2} - \sqrt{IL^2 + QL^2}$
	Dot-product	$Dn = IP(IE - IL) + QP(QE - QL)$

analytical description of the discrimination characteristics (Table 2.5), those replicas are denoted as: *IP*, *QP* (I and Q components of the direct replica); *IE, IL, QE, QL* (I and Q signal replicas with a early and delay, respectively).

Let us analyze the composition of the signal and its replicas at a reception point, accounting for the action of the direct and the re-reflected signals. The signal at the reception point (with allowance for multipath phenomena) represents a superposition of the direct and several re-reflected signals. For simplicity, we will regard that the carrier frequency of those signals (ω_0) is the same, the phase modulation of the service information carrier is not accounted for, then the signal at the reception point looks like (Ray 2000)

$$S_\Sigma(t) = A \cdot \sum_0^N \alpha_i \cdot cos(\omega_0 t + \theta(t - \tau_i)\varphi_i), \tag{2.65}$$

where A is the direct signal amplitude; α_i are the re-reflection coefficients for each ith signal; N is a total number of re-reflected signals; τ_i is the ith signal delay relative to the direct signal; φ_i is the ith signal initial phase; $\theta(t - \tau_i)$ is the ranging-code function.

When generating direct replicas *IP* and *QP*, there occurs accumulation of the signal samples (2.65) in the tracking loop integrator, as a result of which the input effect on the code-delay discriminator looks like:

$$IP = \sum_0^N \alpha_i \cdot \frac{A}{2} \cdot R\left(\tau_i^* - \tau_i\right) cos\left(\varphi_i^* - \varphi_i\right) \tag{2.66}$$

$$QP = \sum_0^N \alpha_i \cdot \frac{A}{2} \cdot R\left(\tau_i^* - \tau_i\right) sin\left(\varphi_i^* - \varphi_i\right), \tag{2.67}$$

where $R\left(\tau_i^* - \tau_i\right)$ is the signal autocorrelation function; τ_i^* and φ_i^* are the estimates for the code pseudo-delay and the carrier phase at the navigational receiver meters' output.

The analytical presentation of signal replicas *IE, IL, QE, QL* is similar, but the auto-correlation function argument additionally involves a lead or a delay of the signal replica by T_d, so that $R(*) = R\left(\tau_i^* - \tau_i \pm T_d\right)$.

Because the kind of the discrimination characteristic depends on the signal auto-correlation function, then the *characteristic for the multipath error envelope will depend on the kind of the phase modulation of the ranging code and on the characteristics of the latter*. Figure 2.16 illustrates this.

As one may ascertain, selecting a modulation type, as well as ranging-code parameters (pulse length and code sequence period duration), may provide a multiple decrease in the multipath error when measuring the code pseudo-range.

Now, let us return to analyzing the ranging error that is caused by the wave propagation medium. The second and third terms (2.62) correspond to the atmospheric

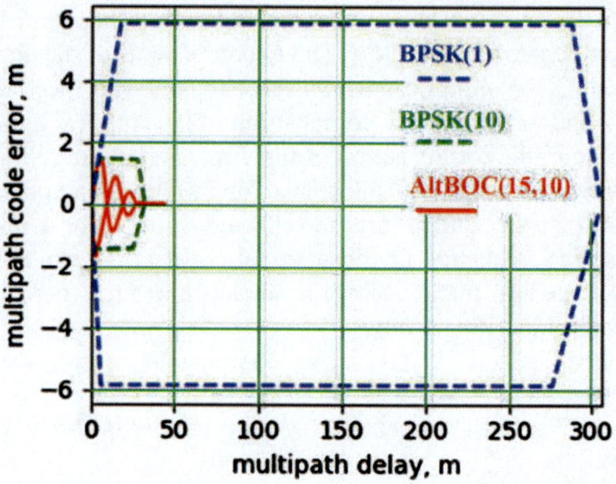

Fig. 2.16 Characteristic for the multipath error when using signals with different modulation types (Padokhin et al. 2021)

refraction effects. The second term (2.62) describes a signal delay due to a change in its propagation velocity. The third term describes a signal delay at its propagation through the paths different from geometrically straight line. The third term value is much less, than that of the second, and further we will neglect it.

One normally detects the ionospheric refraction effect and tropospheric refraction effect from the composition of atmospheric refraction effects, thus, the second term (2.62) may be presented like:

$$\int_s (n-1) \cdot ds = dI_j^i + dT_j^i. \tag{2.68}$$

The character of the ionosphere and the troposphere effects on the navigation signal propagation differs from each other, therefore, we address this problem in greater detail.

The ionosphere as a medium containing free charge particles has a complicated conductance character that depends on the signal frequency. The group (v_G) and the phase (v_P) velocities of wave propagation in the ionosphere differ, and are determined by the group and phase ionospheric refraction coefficients as follows:

$$n_P = c/v_P, n_G = c/v_G. \tag{2.69}$$

The refraction coefficient values, as functions of ionospheric frequency and conductance, may be presented by the following approximation (Yakovlev 1988)

$$n_P = 1 + \frac{a1}{f^2} + \frac{a2}{f^3} + \frac{a3}{f^4} + \ldots \tag{2.70}$$

$$n_G = 1 - \frac{a1}{f^2} - \frac{a2}{f^3} - \frac{a3}{f^4} - \ldots, \tag{2.71}$$

where *al*, *a2*, *a3* are empirical coefficients that depend on the local electron concentration within the site of the group (phase) path through the ionosphere.

Except for some precision problems, one restricts oneself to the first term of expansion (2.70) and (2.71) and neglects the residual terms of the series due to their small contribution (Palamartchouk 2010). With allowance for this, we can write

$$n_P = 1 + \frac{a1}{f^2} \approx 1 - \frac{40.3 \cdot N_e(h)}{f^2} \tag{2.72}$$

$$n_G = 1 - \frac{a1}{f^2} \approx 1 + \frac{40.3 \cdot N_e(h)}{f^2}, \tag{2.73}$$

where $a1 = -40.3 \cdot N_e(h)$ is the first decomposition coefficient; $N_e(h)$ is the local electron concentration at the h height el/m^3.

With allowance for this, one may present the ionospheric ranging error as a refraction gain in the group (phase) path as follows:

$$dI^i_{j,P} = \int_s \left(1 - \frac{40.3 \cdot N_e(h,s)}{f^2}\right) ds - \int_s ds \tag{2.74}$$

$$dI^i_{j,G} = \int_s \left(1 + \frac{40.3 \cdot N_e(h,s)}{f^2}\right) ds - \int_s ds. \tag{2.75}$$

Here, the second summand represents a geometric oblique range between the satellite and the receiver. After conversions, one may write the expressions that enable to directly estimate the group and the phase delays of the signal propagating through ionosphere as follows

$$dI^i_{j,P} = -\frac{40.3}{f^2} \int_s N_e(h,s) ds \tag{2.76}$$

$$dI^i_{j,G} = \frac{40.3}{f^2} \int_s N_e(h,s) ds. \tag{2.77}$$

For convenience of further reasoning, we introduce the notion of *the total electron content* (*TEC*) along the trajectory of the group (phase) path. The *TEC* value shows a full number of electrons that fits the cylinder with a 1-m^2 base cut, whose axis coincides with the "satellite-receiver" direction. This value is measured in a *TEC* Unit (*TECU*), where $1 TECU = 10^{16}$ el/m^2.

$$TEC^i_j = \int_s N_e(h,s) ds. \tag{2.78}$$

By substituting (2.78) to (2.76) and (2.77), we obtain

$$dI^i_{j,P} = -\frac{40.3}{f^2} TEC \tag{2.79}$$

$$dI^i_{j,G} = \frac{40.3}{f^2} TEC. \quad (2.80)$$

Note that the ionospheric ranging error for cases of ranging by code and by phase) has the same absolute value, but an opposite sign. Without effects of the second and higher orders (2.70), (2.71), this error value depends on the carrier frequency as ~ $1/f^2$.

As one can see, the ionospheric ranging error absolute value is the TEC function and does not depend directly on local features of the electron concentration height distribution in the ionosphere. This allows us to formulate the approximation of a spherically stratified ionosphere like a shallow layer at a conditional height of the electron concentration maximum in the ionosphere (*hmF*).

For applied goals, such a presentation enables to simply re-compute the between the *vertical* and *slant* (*oblique)* values of the ionospheric ranging error with a reasonable accuracy, at the set height *hmF* and at the known satellite elevation angle. For this, we use a *scaling function* (*inclination factor*)

$$F_{PP} = \left[1 - (R_E \cos(\gamma)/(R_E + H_{ION}))^2\right]^{-0.5}, \quad (2.81)$$

where R_E is the Earth radius (6378.1363 km); γ is the elevation angle for the satellite over the horizon; $H_{ION} = hmF = 350$ km is a conditional height of the F-layer ionization maximum over the Earth surface.

We re-compute between the vertical and the oblique errors as follows

$$dI^i_{j,SLANT} = F_{PP} \cdot dI^i_{j,VERT}, \quad (2.82)$$

Unlike the ionosphere, the troposphere contains a small number of free charge particles. Therefore, for the signals with < 15 GHz carrier frequency, this medium is not dispersing (Yakovlev 1988). Thereby, the behavior of the GNSS signal group and phase delays in the troposphere does not differ in the sign and does not depend on the frequency. Nevertheless, the tropospheric ranging error value is also a function of the atmosphere refraction coefficient along the group (phase) path and we may write:

$$dT^i_j = \int_s (1 - n_{TR}) ds. \quad (2.83)$$

The troposphere refraction coefficient value depends on the air temperature, on the atmospheric and partial pressure, as well as on the satellite elevation angle, which will be described more scrupulously in Sect. 2.3.2.

To write the code pseudo-range equation in a final form, one should also account for a misfit between the satellite antenna phase center and the satellite mass center, as well as between the receiving antenna phase center and the reference point of its mount.

Let us denote the satellite mass radius-vector as $\mathbf{r}_i$, the reference point radius-vector as $\mathbf{r}_j$, the satellite antenna displacement vector (relative to its mass center) as $\mathbf{dr}_i$, and the receiver antenna displacement vector (relative to its reference point) as $\mathbf{dr}_j$. Then, we may write an improvement for the actual geometric range between the satellite and the receiver:

$$\rho_j^i = \left\| \left(\mathbf{r}_j + \mathbf{dr}_j\right) - \left(\mathbf{r}_i + \mathbf{dr}_i\right) \right\|, \tag{2.84}$$

where $\|\ldots\|$ denotes the geometric range between the satellite and the receiver with allowance for above displacements.

Taking account of the last improvement, we may obtain the final equation for a complete description of the measured code pseudo-range composition:

$$\begin{aligned} \rho_j^i(t) = {} & \left\| \left(r_j\left(t - \tau_j^i\right) + dr_j\left(t - \tau_j^i\right)\right) - \left(r_i(t) + dr_i(t)\right) \right\| + I_j^i + T_j^i \\ & + \mathrm{c}\left(\Delta t_j(t) - \Delta t_i\left(t - \tau_i^j\right)\right) + \mathrm{c}\left(dt_j(t) - dt_i\left(t - \tau_i^j\right)\right) + dm_i^j + e_i^j. \end{aligned} \tag{2.85}$$

The right member of this equation contains (in order):

- $\|\ldots\|$—a geometric distance between the transmitting and the receiving antennas (expressed through the positions of the antenna reference point, of the satellite mass center, and the corresponding displacements);
- I_j^i—a group ionospheric ranging error;
- T_j^i—a tropospheric ranging error;
- $c\left(\Delta t_j(t)\right)$—a receiver timescale deviation error;
- $c\left(\Delta t_i\left(t - \tau_i^j\right)\right)$—a satellite timescale deviation error;
- $c\left(dt_j(t)\right)$—a signal delay in the receiver;
- $c\left(dt_i\left(t - \tau_i^j\right)\right)$—a signal delay in the satellite hardware;
- dm_i^j—a multipath error;
- e_i^j—a measurement noise error.

The positions of the satellite and the receiver, the clock error, and the tropospheric delay do not depend on the signal frequency. All the other components of the observational equation (including the antenna displacement vectors relative to the satellite mass center and the reference point) will be different for different-frequency signals.

The above reasoning is mainly applicable to the error analysis of the phase ranging between the satellite and the receiver. The carrier phase measured in the receiver, equals the difference between the receiver-generated oscillation phase at the signal reception instant and the satellite-generated oscillation phase at its emission instant. In the receiver, only the fractional part of the carrier phase is measured, while the full integer of cycles remains unknown. This number is referred to as the carrier phase measurement ambiguity N_j^i. With allowance for this and by analogy with (2.58), we write

$$\varphi_i^j(t) = \varphi_j(t) - \varphi_i\left(t - \tau_i^j\right) + N_j^i + \epsilon_i^j. \tag{2.86}$$

To present the phase range of the dimensions in meters we multiply the carrier phase by the carrier nominal wavelength $\lambda = c/f_0$, which enables to write the phase observation equation as follows

$$\lambda \cdot \varphi_i^j(t) = \Phi_i^j(t) = c \cdot \tau_i^j + c \cdot \left(\Delta t_j(t) - \Delta t_i\left(t - \tau_i^j\right)\right) \\ + \lambda \cdot (\varphi_j(t_0) - \varphi_i(t_0)) + \lambda \cdot N_j^i + \lambda \cdot \epsilon_i^j, \tag{2.87}$$

where $c \cdot \tau_i^j$ is phase front propagation time; $c \cdot \left(\Delta t_j(t) - \Delta t_i\left(t - \tau_i^j\right)\right)$ is a value of mismatch between the receiver and the satellite timescales; $\lambda \cdot (\varphi_j(t_0) - \varphi_i(t_0))$ is a constant transmitting an inequality to the zero of the satellite and the receiver initial phases; $\lambda \cdot N_j^i$ is a phase ambiguity; $\lambda \cdot \epsilon_i^j$ is the noise of phase ranging.

The phase wavefront propagation time can be described in greater detail by analogy with how it was done above for code pseudo-ranges. This allows one to obtain the final observation equation for phase range (measured in meters)

$$\Phi_i^j(t) = \rho_i^j(t) - I_i^j + T_i^j + \delta m_i^j + c \cdot \left(\Delta t_j(t) - \Delta t_i\left(t - \tau_i^j\right)\right) \\ + c \cdot \left(\delta_j(t) + \delta_i\left(t - \tau_i^j\right)\right) + \lambda \cdot (\varphi_j(t_0) - \varphi_i(t_0)) + \lambda \cdot N_j^i + E_i^j. \tag{2.88}$$

In the left part of Eq. (2.88), the product of the carrier phase and the nominal wavelength is replaced with the $\Phi_i^j(t)$ denoting the carrier phase value in length units. Comparing (2.88) with the corresponding pseudo-range observation (2.85), one note that it contains the same main components. The features of this equation are that:

- the ionospheric ranging error has an opposite sign;
- there are additional terms responsible for the inequality to zero of the initial phases, and a phase ambiguity parameter;
- the multipath error at phase measurements has a lower order, which we analyze further.

In terms of potential accuracy of phase measurements, a multipath error is one of the important factors that is sufficiently difficult to offset at measurements. Under multipath conditions, the superposition signal carrier phase $S_\Sigma(t)$ (2.65) varies randomly relative to the direct signal phase. The receiver reference oscillator is tuned by the signal $S_\Sigma(t)$ phase, generating a corresponding phase measurement error. In the simplest case, when there is one direct signal and one reflected, the multipath-caused phase measurement error may be estimated as follows (Ray 2000)

$$\delta m_i^j = arctan\left(\frac{\alpha_1 \cdot R\left(\tau_S^* - \tau_1\right) \cdot \sin\varphi_1}{R\left(\tau_S^*\right) + \alpha_1 \cdot R\left(\tau_S^* - \tau_1\right) \cdot \cos\varphi_1}\right), \tag{2.89}$$

where $R(\tau_S^*)$ is a direct signal auto-correlation function; $R(\tau_1^* - \tau_1)$ is a cross-correlation function between the direct and the reflected signals; τ_S^* is an estimate for the receiver code pseudo-delay; τ_1 is a reflected signal delay relative to the direct one; α_1 is the reflection coefficient; φ_1 is a reflected signal phase.

Like in the case of code pseudo-delay measurements, the phase multipath error depends on the form of the cross-correlation function $R(\tau_S^* - \tau_1)$, and, hence, is determined by the ranging code parameters and by the form of the phase modulation. We will perform a comparative estimate for the maximal multipath error of the phase measurements when using two ranging codes with different code sequence parameters: C/A (at L1 GPS) and CL (at L5 GPS). The typical values for code-delay estimates are $\tau_S^* = \pm 120 \ldots 150$ m for the C/A-code, and $\tau_S^* = \pm 25 \ldots 30$ m for the CL-code. The phase measurement maximal error may be found through differentiating (2.89) by the parameter φ_1, which yields (Ray 2000)

$$\delta m_{i,MAX}^{j} = cos^{-1}\left(-\frac{\alpha_1 \cdot R(\tau_S^* - \tau_1)}{R(\tau_S^*)}\right). \tag{2.90}$$

Figure 2.17 presents the results for computing the multipath maximal error (expressed in phase range units) for various-level *SMR* (2.64). When computing by Expression (2.90), one uses the code delay spacing $Td = \pm 0.1\ \tau_S$. When modeling correlation functions, supposed was using a coherent pseudo-delay discriminator (Table 2.5) and BPSK-modulation. The solid lines in the figure show the computations for a C/A-code, and the dashed lines denote the same for a CL-code. At the computations, the RF front-end through bandwidth was supposed to be unlimited, and the reflected signal delay varies within $\tau_1 \leq \pm \tau_S$ with a reflected signal phase change φ_1 in 0.1-rad increments within 0 through π.

Apparently, the multipath error is proportional to the relation between the amplitudes of the direct and of the reflected signals. An *SMR* increase leads to the error decrease. As a reflector moves away from a receiving antenna, the multipath error monotonically decreases. The strong multipath area size is determined by the ranging

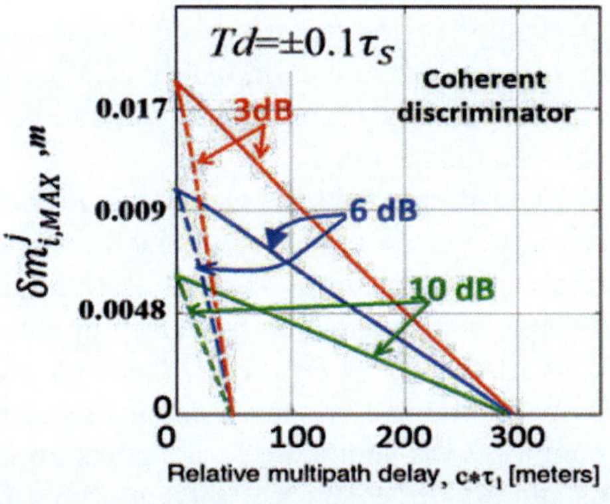

Fig. 2.17 Characteristics of the phase multipath error at L1 when using C/A-code signals (solid lines) and modulated CL-code (dotted lines) with various SMR

code pulse length. When using the CL-code with a short pulse, the size of that area is ~ 40–50 m. A longer C/A-code pulse leads to an increase in the multipath area size up to 300 m. Ray (2000) showed that the maximal phase multipath error hardly depends on the code pseudo-delay discriminator type, as well as on the discriminator spacing *Td*.

The error maximal value in the addressed case was 0.017 m. However, in reality, one should take into account that several reflected beams participate in the multipath formation. Therefore, the error value $\delta m_{i,MAX}^{j}$ may appear substantially higher. This is particularly significant for the signals received from low-elevation satellites.

One may finally write the observation equation for the phase range in terms of the satellite and the receiver positions, and the corresponding antenna displacements, by analogy with Eq. (2.85):

$$\begin{aligned}\Phi_i^j(t) = {} & \left\| \left(r_j\left(t-\tau_j^i\right) + dr_j\left(t-\tau_j^i\right)\right) - \left(r_i(t) + dr_i(t)\right) \right\| - I_i^j + T_i^j \\ & + \delta m_i^j + c \cdot \left(dt_j(t) - dt_i\left(t - \tau_i^j\right)\right) + c \cdot \left(\delta_j(t) - \delta_i\left(t - \tau_i^j\right)\right) \\ & + \lambda \cdot \left(\varphi_j(t_0) - \varphi_i(t_0)\right) + \lambda \cdot N_j^i + E_i^j. \qquad (2.91)\end{aligned}$$

Here, we also used the designations for antenna center displacement vectors, relative to the receiver reference point and to the satellite mass center. Generally, they may differ from those for pseudo-ranges, because the antenna phase center positions differ for these cases.

2.2.2 User Positioning Errors

At a fixed value for range errors to visible satellites, the accuracy and stability of positioning solutions will depend on the features of the chosen method to find the latter, and on the Dilution Of Precision (DOP) geometric factor. Let us address, to what extent each of those factors affects the positioning solution results.

A positioning solution may be done through *one-step* or *filtering* methods. Depending on the ranging error spread and on the extent of the navigation function non-linear system conditionality (2.32), the above methods provide a significantly different accuracy in estimating the state vector parameters (2.30) and the positioning solution robustness.

The one-step methods may be *iterative* or *non-iterative*. In both cases, the position solution is done separately at each instant irrespective of the positioning solution results at the previous instant. The simplest kind of a non-iterative algorithm is a *single-point algorithm* by selecting the set number of satellites from the visible constellation by the *Position Dilution Of Precision* minimum criterion. The approach proper (to build a system of navigation functions by the *PDOP* minimum criterion) minimizes the geometric factor effect on the solution accuracy and robustness. Let us address the essence of this algorithm for a case when using 4 satellites. First, we present the system of navigation functions to determine the object state vector parameters:

$$\begin{cases} (X - x_1)^2 + (Y - y_1)^2 + (Z - z_1)^2 = (\tilde{R}_1 - \Delta R)^2 \\ (X - x_2)^2 + (Y - y_2)^2 + (Z - z_2)^2 = (\tilde{R}_2 - \Delta R)^2 \\ (X - x_3)^2 + (Y - y_3)^2 + (Z - z_3)^2 = (\tilde{R}_3 - \Delta R)^2 \\ (X - x_4)^2 + (Y - y_4)^2 + (Z - z_4)^2 = (\tilde{R}_4 - \Delta R)^2 \end{cases}, \tag{2.92}$$

where x_i, y_i, z_i are the coordinates of the ith satellite in a rectangular reference system; X, Y, Z are a receiver unknown coordinates; $\tilde{R}_i$ is a pseudo-range to the ith satellite; ΔR is a range error caused by a receiver timescale offset.

To solve the system of non-linear Eqs. (2.92), we will linearize it by subtracting one equation from the remaining (the equation selection is not critical). As a result of conversions, we obtain a system of three equations linear relative to the required parameters X, Y, Z and ΔR:

$$\begin{cases} (x_1 - x_2)X + (y_1 - y_2)Y + (z_1 - z_2)Z + (\tilde{R}_2 - \tilde{R}_1)\Delta R = A_1 \\ (x_1 - x_3)X + (y_1 - y_3)Y + (z_1 - z_3)Z + (\tilde{R}_3 - \tilde{R}_1)\Delta R = A_2 \\ (x_1 - x_4)X + (y_1 - y_4)Y + (z_1 - z_4)Z + (\tilde{R}_4 - \tilde{R}_1)\Delta R = A_3 \end{cases}, \tag{2.93}$$

where:

$$A_1 = \tfrac{1}{2}\left(\rho_1^2 - \rho_2^2 - \tilde{R}_1^2 + \tilde{R}_2^2\right); \qquad A_2 = \tfrac{1}{2}\left(\rho_1^2 - \rho_3^2 - \tilde{R}_1^2 + \tilde{R}_3^2\right)$$

$$A_3 = \tfrac{1}{2}\left(\rho_1^2 - \rho_4^2 - \tilde{R}_1^2 + \tilde{R}_4^2\right); \qquad \rho_i^2 = (x_i)^2 + (y_i)^2 + (z_i)^2; i \in [1 - 4]$$

The obtained set of equations describes a 4D straight line produced by crossing of three hyper-planes. To search for the unknowns in the set of Eqs. (2.93), we compose a parametric equation for this straight as follows:

$$\begin{cases} X = x_0 + \alpha_1 \cdot \mu \\ Y = y_0 + \alpha_2 \cdot \mu \\ Y = z_0 + \alpha_3 \cdot \mu \\ \Delta R = s_0 + \alpha_4 \cdot \mu \end{cases}. \tag{2.94}$$

This parametric equation describes a straight line with a parameter μ, that transits through a point with the known coordinates (x_O, y_O, z_O, s_O) in the set direction. This straight orientation in space is specified by the guiding vector $a = \{\alpha_1, \alpha_2, \alpha_3, \alpha_4\}$, where $\alpha_1, \alpha_2, \alpha_3, \alpha_4$ are moduli of the guiding vector projections onto the coordinate axes OX, OY, OZ, and OΔR, respectively.

As an initial point (x_O, y_O, z_O, s_O), one may use any point belonging to the straight (2.94). Assume, for example, $s_O = 0$, and, based on (2.93), we compose a set of equations relative to the required coordinates x_O, y_O, z_O:

$$\begin{cases} (x_1 - x_2)x_0 + (y_1 - y_2)y_0 + (z_1 - z_2)z_0 = A_1 \\ (x_1 - x_3)x_0 + (y_1 - y_3)y_0 + (z_1 - z_3)z_0 = A_2 \\ (x_1 - x_4)x_0 + (y_1 - y_4)y_0 + (z_1 - z_4)z_0 = A_3 \end{cases}. \tag{2.95}$$

Solving this set of equations will be:

$$\begin{aligned} x_0 &= k_{11} \cdot A_1 + k_{12} \cdot A_2 + k_{13} \cdot A_3 \\ y_0 &= k_{21} \cdot A_1 + k_{22} \cdot A_2 + k_{23} \cdot A_3, \\ z_0 &= k_{31} \cdot A_1 + k_{32} \cdot A_2 + k_{33} \cdot A_3 \end{aligned} \tag{2.96}$$

where k_{ij} are elements of the matrix inverse to the equation set coefficient matrix (2.95).

From determining the guiding vector, the following equalities result directly:

$$\begin{aligned} \alpha_1 = {} & (y_1 - y_2)\left[(z_1 - z_3)\left(\overline{R}_4 - \overline{R}_1\right) - (z_1 - z_4)\left(\overline{R}_3 - \overline{R}_1\right)\right] \\ & - (z_1 - z_2)\left[(y_1 - y_3)\left(\overline{R}_4 - \overline{R}_1\right) - (y_1 - y_4)\left(\overline{R}_3 - \overline{R}_1\right)\right] \\ & + \left(\overline{R}_2 - \overline{R}_1\right)[(y_1 - y_3)(z_1 - z_4) - (y_1 - y_4)(z_1 - z_3)] \end{aligned} \tag{2.97}$$

$$\begin{aligned} \alpha_2 = {} & (x_2 - x_1)\left[(z_1 - z_3)(\overline{R}_4 - \overline{R}_1) - (z_1 - z_4)(\overline{R}_3 - \overline{R}_1)\right] \\ & - (z_1 - z_2)\left[(x_1 - x_3)(\overline{R}_4 - \overline{R}_1) - (x_1 - x_4)(\overline{R}_3 - \overline{R}_1)\right] \\ & + (\overline{R}_1 - \overline{R}_2)\left[(x_1 - x_3)(z_1 - z_4) - (x_1 - x_4)(z_1 - z_3)\right] \end{aligned} \tag{2.98}$$

$$\begin{aligned} \alpha_3 = {} & (x_1 - x_2)\left[(y_1 - y_3)\left(\overline{R}_4 - \overline{R}_1\right) - (y_1 - y_4)\left(\overline{R}_3 - \overline{R}_1\right)\right] \\ & - (y_1 - y_2)\left[(x_1 - x_3)\left(\overline{R}_4 - \overline{R}_1\right) - (x_1 - x_4)\left(\overline{R}_3 - \overline{R}_1\right)\right] \\ & + \left(\overline{R}_2 - \overline{R}_1\right)[(x_1 - x_3)(y_1 - y_4) - (x_1 - x_4)(y_1 - y_3)] \end{aligned} \tag{2.99}$$

$$\begin{aligned} \alpha_4 = {} & (x_1 - x_2)[(y_1 - y_3)(z_1 - z_4) - (y_1 - y_4)(z_1 - z_3)] \\ & - (y_2 - y_1)[(x_1 - x_3)(z_1 - z_4) - (x_1 - x_4)(z_1 - z_3)] \\ & + (z_2 - z_1)\left[(x_1 - x_3)(y_1 - y_4) - \left(x_1^1 - x_4\right)(y_1 - y_3)\right] \end{aligned} \tag{2.100}$$

Disposing of these equations, one may find the coordinates of the straight intersection point (2.94) and any of the navigation functions (2.92) (in this case, the navigation functions are spheres). For this purpose, we substitute Expressions (2.94) into any of Eqs. (2.92), for example, into the first of them. Then:

$$(x_0 - x_1 + \alpha_1\mu)^2 + (y_0 - y_1 + \alpha_2\mu)^2 + (z_0 - z_1 + \alpha_3\mu)^2 = (\tilde{R}_1 - \alpha_4\mu)^2 \tag{2.101}$$

This equation solution relative to the parameter μ is ambiguous, and looks like:

$$\mu = -\tfrac{b}{a} \pm \sqrt{\left(\tfrac{b}{a}\right)^2 - \tfrac{c}{a}}, \tag{2.102}$$

$\alpha = \alpha_1^2 + \alpha_2^2 + \alpha_3^2 - \alpha_4^2;$

where: $b = (x_0 - x_1)\alpha_1 + (y_0 - y_1)\alpha_2 + (z_0 - z_1)\alpha_3 + \tilde{R}_1\alpha_4.$

$c = (x_0 - x_1)^2 + (y_0 - y_1)^2 + (z_0 - z_1)^2.$

Selecting the single solution is done empirically while solving the positioning problem. In the case of a true solution, the computed coordinates do not contradict a priori the known location.

As one can see, the presented algorithm is sufficiently simple to implement, and is robust against a dramatic geometric factor dilution. This benefit is implemented, when there is sufficient redundancy of measurements, when one may select another satellite configuration by the *PDOP* minimum. At the same time, this algorithm enables to obtain only coarse direct estimates for coordinates with no possibility of their smoothing due to a posteriori estimation. A limited number of the navigation functions being a part of the equation set (2.92) does not enable to smooth separate components of ranging errors due to their averaging over a great number of instant observations (when satellites are > 4). Another disadvantage of this algorithm is the obtained solution instability, when ranging measurements by signals of separate satellite are not equally accurate. Owing to this, this algorithm has not become widely used and may be recommended only for a priori estimates of the object state vector parameters as an addition to the fundamental positioning algorithm.

As an iterative method to solve a positioning problem, one normally uses the Newton method or one of its modifications (Yarlykov 1980). Provided the initial approximations are set exactly, and the ranging measurements are uniformly accurate, the method provides a fast convergence of the solution and its adequate accuracy. The system of navigation equations for positioning an object represents a system of n non-linear equations like.

$$\tilde{R}_i = \rho_i + \delta R_i = \left[(x_i - X)^2 + (y_i - Y)^2 + (z_i - Z)^2\right]^{0.5} + \delta R_i, \tag{2.103}$$

where δR_i is a ranging error caused by the receiver timescale offset from the GNSS system time and other errors; ρ_i is a geometric range between the receiver and the satellite; x_i, y_i, z_i are the known coordinates of the ith satellite; X, Y, Z are the unknown coordinates of the receiver; n is a number of satellites used for positioning an object.

Solving the set of Eqs. (2.103) through a Newton iterative method assumes obtaining such an estimate for the required parameters X, Y, Z, and δR, at which the following condition is best met:

$$\sum_{i=1}^{n}\left(\left[(x_i - X)^2 + (y_i - Y)^2 + (z_i - Z)^2\right] - \left(\tilde{R}_i - \delta R_i\right)^2\right)^2 \to 0. \tag{2.104}$$

Finding the minimum of function (2.104) requires, thus, solving the system of non-linear equations that we write in a matrix form:

$$\mathbf{F(x)} = 0, \tag{2.105}$$

where: $\mathbf{x} = |X, Y, Z, \delta R|$ is an estimated state vector of an object; $\mathbf{F(x)} = \begin{vmatrix} F_1(X, Y, Z, \delta R_1) \\ F_1(X, Y, Z, \delta R_2) \\ \ldots\ldots\ldots\ldots\ldots\ldots \\ F_n(X, Y, Z, \delta R_n) \end{vmatrix}$ is a vector of navigation functions presented like:

$$F_i(X, Y, Z, \delta R) = \left(\left[(x_i - X)^2 + (y_i - Y)^2 + (z_i - Z)^2\right] - \left(\tilde{R}_i - \delta R_i\right)^2\right)^2.$$

To find a solution for the set of Eqs. (2.105), we linearize the vector $\mathbf{F(x)}$ by its Taylor expansion.

$$\mathbf{F(x}_k) = \mathbf{F(x}_0) + \mathbf{dF(x}_k) \cdot (\mathbf{x}_k - \Delta\mathbf{x}_k) + \mathbf{d}^2\mathbf{F(x}_k) \cdot (\mathbf{x}_k - \Delta\mathbf{x}_k)^2 + \ldots, \tag{2.106}$$

where $\mathbf{F(x_0)}$ is an initial state of the navigation function vector at starting approximations of $\mathbf{x}_0$; $\mathbf{x}_\mathrm{k}$ is an object state vector obtained at the kth iteration; $\boldsymbol{\Delta}\mathbf{x}_\mathrm{k}$ is an incremental vector to the state vector estimates at the kth iteration.

With allowance for (2.106), the solution of the set of Eqs. (2.105) at the kth iteration looks like:

$$\mathbf{x}_k = \mathbf{x}_{k-1} + \Delta\mathbf{x}_k. \tag{2.107}$$

We find the increment $\boldsymbol{\Delta}\mathbf{x}_\mathrm{k}$ at each iteration step as follows:

$$\Delta\mathbf{x}_k = -\mathbf{F}_k^T \cdot \mathbf{dF}_k^{-1} \tag{2.108}$$

where: $\mathbf{F}_k = \begin{vmatrix} \sum_{i=1}^{n} \frac{d}{dX} F_i(X, Y, Z, \delta R_i) \\ \sum_{i=1}^{n} \frac{d}{dY} F_i(X, Y, Z, \delta R_i) \\ \sum_{i=1}^{n} \frac{d}{dZ} F_i(X, Y, Z, \delta R_i) \\ \sum_{i=1}^{n} \frac{d}{d\delta R} F_i(X, Y, Z, \delta R_i) \end{vmatrix}^T$; $\mathbf{dF}_k \begin{vmatrix} \frac{\mathbf{dF}_{X,1}}{dX} & \frac{\mathbf{dF}_{X,1}}{dY} & \frac{\mathbf{dF}_{X,1}}{dZ} & \frac{\mathbf{dF}_{X,1}}{d\delta R_1} \\ \frac{\mathbf{dF}_{X,2}}{dX} & \frac{\mathbf{dF}_{X,2}}{dY} & \frac{\mathbf{dF}_{X,2}}{dZ} & \frac{\mathbf{dF}_{X,2}}{d\delta R_1} \\ \cdot & \cdot & \cdot & \cdot \\ \frac{\mathbf{dF}_{X,n}}{dX} & \frac{\mathbf{dF}_{X,n}}{dY} & \frac{\mathbf{dF}_{X,n}}{dZ} & \frac{\mathbf{dF}_{X,n}}{d\delta R_n} \end{vmatrix}$; and $\mathbf{F}_{\mathrm{k,i}}$ refers to the corresponding element of vector $\mathbf{F}_\mathrm{k}$; the index i indicates the number of the satellite, to which the navigation function F_i $(X, Y, Z, \delta R)$ relates.

Iteration (2.107) continues until the condition $\boldsymbol{\Delta}\mathbf{x}_\mathrm{k} < \varepsilon$ is met; here, ε is the set threshold of an admissible positioning error.

Unlike the previous method, the iterative method to find a positioning solution by an arbitrary number of satellites enables to substantially smooth the effect of uncorrelated components of a ranging error due to a measurement redundancy. However, essential disadvantages of the Newton method (as a minimization gradient method) are as follows:

(1) The solution accuracy directly depends on the accuracy of setting a priori estimates for the state vector $\mathbf{x}_0$. This constraint is particularly significant when positioning dynamic objects;
(2) The solution robustness depends on both how accurate the setting of parameters for the vector $\mathbf{x}_0$ is, and on how uniformly-precise ranging measurement are. In the presence of separate larger-error ranging measurements, the solution robustness and accuracy are not provided. This problem is particularly crucial when using ranging measurements by signals from different-GNSS systems, and when using heterogeneous additional ranging signals from ground- and space-based sources;
(3) The method is not robust to the position dilution of precision factor. This is caused by that PDOP leads to a poor conditionality of the matrix $\mathbf{DF}_\mathrm{k}$, and to a loss of solution stability.

Eliminating the above disadvantages is, mainly, attained by using filtration algorithms for positioning solutions. A typical algorithm is applying a linear discrete Kalman filter. Underlying this filter is a set model for the object state-vector dynamics ($\mathbf{x}_\mathrm{k}$) and a model for dynamics of ranging estimates ($\mathbf{R}_\mathrm{k}$) in the following view (Yarlykov 1980):

$$\mathbf{R}_k = \mathbf{H}_k \cdot \mathbf{x}_k + \mathbf{n}_k \tag{2.109}$$

$$\mathbf{x}_k = \mathbf{F} \cdot \mathbf{x}_{k-1} + \mathbf{n}_{\mathbf{X},k}, \tag{2.110}$$

where: $\mathbf{H}_k = \begin{vmatrix} \frac{dh_{k,1}}{dX} & \frac{dh_{k,1}}{dY} & \frac{dh_{k,1}}{dZ} & \frac{dh_{k,1}}{d\delta R_1} \\ \frac{dh_{k,2}}{dX} & \frac{dh_{k,2}}{dY} & \frac{dh_{k,2}}{dZ} & \frac{dh_{k,2}}{d\delta R_2} \\ \cdot & \cdot & \cdot & \cdot \\ \frac{dh_{k,n}}{dX} & \frac{dh_{k,n}}{dY} & \frac{dh_{k,n}}{dZ} & \frac{dh_{k,n}}{d\delta R_n} \end{vmatrix}$ is a matrix of observations at an instant k, in which: $h_{k,i} = \left[(x_i - X_k)^2 + (y_i - Y_k)^2 + (z_i - Z_k)^2\right]^{0.5} + \delta R_{i,k}$; $\mathbf{n}_k = \left| n_{k,1}\ n_{k,2}\ \ldots\ n_{k,n} \right|^T$ are ranging noise vectors from **Rk** and $\mathbf{n}_{\mathbf{X},k} = \left| n_{X,k,1}\ n_{X,k,2}\ \ldots\ n_{X,k,n} \right|^T$ are object dynamics noise vectors from **xk** that feature white Gaussian noise with a zero ensemble average and a priori set dispersions; $\mathbf{F} = \begin{vmatrix} 1\,0\,0\,0 \\ 0\,1\,0\,0 \\ 0\,0\,1\,0 \\ 0\,0\,0\,1 \end{vmatrix}$ is a transition matrix that can be defined for the case of static or dynamic object.

The algorithm for a consecutive estimate of the object state vector through a discrete linear filter Kalman is implemented in the following sequence.

1. A posteriori estimate of the object ($\tilde{\mathbf{x}}_{\mathrm{k}}$) state vector relative to the previous estimate $\hat{\mathbf{x}}_{\mathrm{k}-1}$ at the instant of next measurement t_k incoming:

$$\tilde{\mathbf{x}}_{\mathrm{k}} = \mathbf{F} \cdot \hat{\mathbf{x}}_{\mathrm{k}-1}; \tag{2.111}$$

2. Computing the predicted state-vector error matrix at t_k:

$$\tilde{\mathbf{D}}_{\mathbf{x},\mathrm{k}} = \mathbf{F} \cdot \mathbf{D}_{\mathbf{x},\mathrm{k}-1} \cdot \mathbf{F}^T + \mathbf{G} \cdot \mathbf{R}_{\mathbf{n},\mathbf{x}} \cdot \mathbf{G}^T, \tag{2.112}$$

where: $\mathbf{G} = \begin{vmatrix} \mathbf{G1} & \mathbf{0} & 0 & 0 \\ 0 & \mathbf{G1} & 0 & 0 \\ 0 & 0 & \mathbf{G1} & 0 \\ 0 & 0 & 0 & \mathbf{G2} \end{vmatrix}$, $\mathbf{G1} = \begin{vmatrix} \mathbf{1} \\ \mathbf{0} \end{vmatrix}$, $\mathbf{G2} = \begin{vmatrix} 1 & 0 \\ 0 & 1 \end{vmatrix}$ is a object dynamics noise forming matrix;

$\mathbf{R}_{\mathbf{n},\mathbf{x}} = \begin{vmatrix} \sigma^2_{n,X} & \mathbf{0} & 0 & 0 \\ 0 & \sigma^2_{n,Y} & 0 & 0 \\ 0 & 0 & \sigma^2_{n,z} & 0 \\ 0 & 0 & 0 & 1 \end{vmatrix}$ is an object dynamics noise correlation matrix;

$\sigma^2_{n,X}, \sigma^2_{n,Y}, \sigma^2_{n,Z}$ are a priori set dispersions for the centered white Gaussian object dynamics noise;

3. Computing the filter amplification constant for an instant t_k:

$$\mathbf{K}_{\mathrm{k}} = -\left[\tilde{\mathbf{D}}_{\mathbf{x},\mathrm{k}} + \mathbf{H}_{\mathrm{k}}^{\mathrm{T}} \cdot \mathbf{R}_{\mathbf{n},\mathbf{x}}^{-1}\right]^{-1} \cdot \mathbf{H}_{\mathrm{k}} \cdot \mathbf{R}_{\mathbf{n},\mathbf{x}}^{-1}; \tag{2.113}$$

4. Correcting the current state vector:

$$\hat{\mathbf{x}}_{\mathrm{k}} = \tilde{\mathbf{x}}_{\mathrm{k}} + K_k \cdot \left[\mathbf{R}_{\mathrm{k}} - \mathbf{H}_{\mathrm{k}} \cdot \tilde{\mathbf{x}}_{\mathrm{k}}\right]; \tag{2.114}$$

5. Correcting the vector-state error matrix

$$\mathbf{D}_{\mathbf{x},\mathrm{k}} = \mathbf{F} \cdot \tilde{\mathbf{D}}_{\mathbf{x},\mathrm{k}} \cdot \mathbf{F}^T + \mathbf{G} \cdot \mathbf{R}_{\mathbf{n},\mathbf{x}} \cdot \mathbf{G}^T; \tag{2.115}$$

6. Transition to the next instant t_{k+1} and repetition of computing items 1…5.

The filtration method may provide obtaining the object coordinate estimates with a sufficiently high accuracy and solution robustness. This is due to:

(1) averaging the estimates for instantaneous ranging errors at each kth step of the filter operation, owing to measurement redundancy;
(2) estimating (a posteriori) the object state vector errors and ranging measurements by predicting these estimates while accumulating the statistics on the noise of measurements ($\mathbf{R}_{\mathrm{n,\,x}}$) and observations ($\mathbf{D}_{\mathrm{x,\,k}}$);
(3) reducing the sensitivity to DOP factor owing to the recurrent computation of the object state vector at each kth step of the filter operation with allowance

for the estimation results at the previous *k-1* step, with consideration of the object dynamics model and the characteristics of observation/measurement noise (accumulated through a posteriori estimation).

Figure 2.18 (panel a) provides a comparative pattern for the variations in the RMS spatial positioning error ($\sigma \Pi$) of a stationary point CENA (65.5°N, 144.7°W) within the IGS network. We use the 2011 Feb 15 measurement data by the GPS C/A-code with no ranging error corrections, in the standalone navigation mode.

The positioning solution was attained by two methods: the Newton iterative (red line) and the Kalman filtering (blue line). The algorithm for both methods is described above. Figure 2.18 (panel b) shows daily variations in the PDOP factor. Figure 2.18 (panel c) presents the ranging RMS obtained from the results of processing of instantaneous errors of all visible satellites at each current measurement epoch.

One can see that the use of the linear Kalman filter gives smoothed positioning errors and provides more accurate positioning. In most cases, an increase in the PDOP value does not lead to a significant synchronous increase in the positioning errors. Variations in the RMS of ranging errors in most cases also do not give a synchronous increase in the error of position determination.

However, the Kalman filtering functions less confidently in the case of a "gain" in the effect of destabilizing factors. One can see that, at simultaneous PDOP and ranging RMS growth (e. g., at 05:00 UT, around 13:00 UT, and 16:00 UT), observed

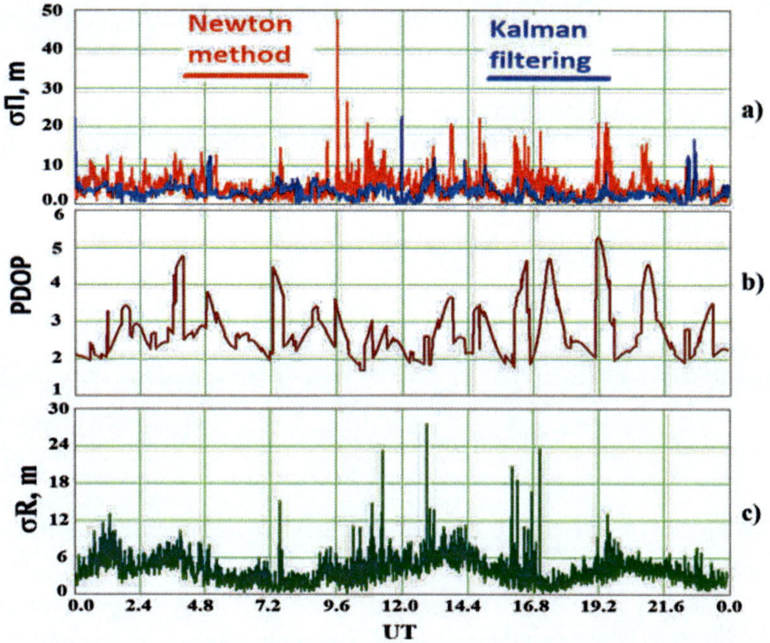

Fig. 2.18 Comparison between the positioning results from the Newton method (red line) and from the Kalman filtering (dark blue line)

is also a $\sigma\Pi$ error increase. Besides, there may emerge $\sigma\Pi$ error surges not related directly to the above destabilizing factors (e. g., at 12:00 UT). One may elucidate this through a filter failure in the case, when a posteriori estimate for the observation and measurement noise mismatches the real situation, or in the case, when the model for the object dynamics and the measurements are incorrect.

The Newton method shows a larger $\sigma\Pi$ dependence on PDOP and σR variations. On average, it features a higher mean $\sigma\Pi$. In particular, this is true in cases of sharp changes in the PDOP values (e.g., at 08:00 UT and between 09:36 and 12:00 UT, as well as at 19:20 UT). Thus, when using the Newton iterative method in positioning solutions, one may expect a greater forcing from PDOP and σR variations on the ultimate positioning accuracy.

2.2.3 Intersystem Biases

When using simultaneously signals from different-GNSS satellites for positioning solutions or for space sounding of the Earth natural shells, there emerge specific measurement errors. These errors are known as *intersystem biases* of the total error in measuring radio-navigation parameters or computing the object coordinates. The main source of the biases is that, between the different-GNSS satellite signal emission and reception instants, there are small random time shifts. The bias phenomenon is a consequence of imperfection in hardware-software tools that generate, broadcast, receive, and process signals, in which these small time shifts while processing different-GNSS signals were not accounted for (Hakansson et al. 2017).

To more explicitly understand the reasons and mechanisms for the above bias emergence, we rewrite the observation equations for the code and phase pseudoranges (2.85), (2.91) as follows:

$$\rho_j^i(t) = \rho_{0,j}^i(t) + c\Big(\Delta t_j(t) - \Delta t_i\Big(t - \tau_i^j\Big) + dt_{j,F}(t) - dt_{i,F}\Big(t - \tau_i^j\Big) + \Delta\tau^{SYS}\Big) \\ + I_{j,DCB}^i + T_j^i + dm_{i,CODE}^j + e_i^j \tag{2.116}$$

$$\Phi_i^j(t) = \rho_{0,i}^j(t) + c\cdot\Big(\Delta t_j(t) - \Delta t_i\Big(t - \tau_i^j\Big) + \delta_{j,F}(t) - \delta_{i,F}\Big(t - \tau_i^j\Big) + \Delta\tau^{SYS}\Big) \\ - I_{i,DCB}^j + T_i^j + \delta m_{i,PHASE}^j + \lambda\cdot(\varphi_j(t_0) - \varphi_i(t_0)) + \lambda\cdot N_j^i + E_i^j. \tag{2.117}$$

By using (2.116 and 2.117), we split the intersystem biases by the emergence source in the following way:

(1) The biases caused onboard the satellite;
(2) The biases originated in the receiver at a simultaneous processing of different-GNSS satellite signals;
(3) The inter-channel biases when using the FDMA (GLONASS only);

(4) The inter-channel biases at a simultaneous processing of different-frequency signals modulated by different ranging codes and different modulation types.

The listed biases may be grouped by their manifestation features as follows.

Bias Group 1 is caused by random fluctuations of the system timescale $\left(\Delta\tau^{SYS}\right)$, of the user receiver $\left(\Delta t_j(t)\right)$ and satellite $\left(\Delta t_i\left(t-\tau_i^j\right)\right)$ reference oscillator frequencies. This bias group manifests itself similarly when measuring code pseudo-range and the carrier phase, and does not depend on the signal frequency, ranging code parameters, and modulation. At the same time, this kind of biases may differ significantly between separate satellites, user receivers, and GNSS systems. Therefore, it requires that the multisystem hardware be calibrated additionally.

Bias Group 2 emerges because of an additional signal delay in a user receiver $(dt_{j,F}(t),\ \delta_{j,F}(t))$ and in a satellite hardware $\left(dt_{i,F}\left(t-\tau_i^j\right),\ \delta_{i,F}\left(t-\tau_i^j\right)\right)$. These biases depend on the type and specific features of the hardware-software part of the user receiver and of the satellite, as well as on the carrier frequency. These factors may also differ significantly among separate satellites, user receivers, and GNSS systems. Besides, this kind of biases differs for measurements of the code pseudo-delay and of the carrier phase.

Bias Group 3 emerges due to multipath noise in a satellite signal reception point $(dm_{i,CODE}^j,\ \delta m_{i,PHASE}^j)$. This kind of biases differs for measurements of the code pseudo-delay and of the phase, and also depends on the carrier frequency. The multipath error was shown above to be determined by type and adjustments of the pseudo-delay and phase discriminator. At a greater extent, this kind of biases and their function domain depend on the ranging code parameters and on the modulation type (see Figs. 2.16 and 2.17). Like in the previous cases, the indicated factors may differ significantly among separate satellites, types of user receivers, and GNSS systems.

When performing a positioning solution in an independent mode, *Bias Group 1* is partially offset by using (within a user receiver) an algorithmic correction based on the model for the satellite timescale drift relative to the GNSS system time (ICD-GPS-200 C 1993). In the differential or relative navigation modes, the components of this bias group are almost completely eliminated by differential correction. The same is attained by using high-precision corrections from analytical centers for processing the network information (PPP mode). The bias caused by an offset of the user receiver timescale is equal for all the instantaneous observations. Therefore, this bias value is included in the navigation function unknowns in (2.32) and is removed directly during a positioning solution.

When sounding the ionosphere by using signals of two and more frequencies, bias group 1 is also removed during the TEC estimation.

Bias Group 2 is much harder to take into account and to offset. To detect this bias at a separate carrier frequency is sufficiently difficult, but not always necessary. Therefore, to analyze these biases, it is more convenient to use the corresponding *DCB* (2.60) computed from the measurements at GNSS two frequencies. The estimates for bias group 2 obtained in this manner and separated by their origins (onboard satellite or in user receiver) may be presented as follows:

$$DCB_{REC} = dt_{j,f1} - dt_{j,f2} \tag{2.118}$$

$$DCB_{SV} = dt_{i,f1} - dt_{i,f2}. \tag{2.119}$$

The characteristics and behavior of these biases feature a great inhomogeneity and are hard to predict. Thus, as per (ICD-GPS-200C 1993), the DCB_{SV} RMS (emerges in a GPS satellite hardware) should be $\leq$ 15 ns (equivalent ranging error being $\leq$ 4.5 m). Real observations show that the DCB_{SV} absolute value may be both positive and negative. However, this bias RMS is sufficiently low (~1.5–3 ns). The DCB_{SV} mean value differs substantially from one satellite to another with mean value scatter within ~ $\pm$ 12 ns (i.e. $\pm$ 3.6 m) (Schaer 2012). Schaer (2012) and Mylnikova et al. (2015) note a slow DCB_{SV} value drift in time, at ~ 0.3–0.5 ns/year.

The DCB_{REC} emerging in a GNSS receiver also have small RMS about 2 ns (0.6 m), but feature a wider mean value scatter. Thus, in (Yasyukevich et al. 2015, Schaer 2012), the DCB_{REC} value was established to vary within ~ $\pm$ 40 ns (i.e. $\pm$ 12 m). Like in the case of the DCB_{SV}, the DCB_{REC} may accept positive and negative values, and have a slow time drift. Besides, the DCB_{REC} absolute values depend on the environment temperature.

A common negative feature for the *DCB* behavior (relates both to satellite hardware and to a user receiver) is a *DCB* unpredictable jump. Figure 2.19 illustrates this phenomenon. Here, panel a presents the time dependence of DCB_{SV} attributed to GPS PRN 03 (gray line) and to GLONASS, the lettered frequency 04 (black line). Panel b in Fig. 2.19 shows the time dependence of the DCB_{REC} measured with the receiver at the IRKJ mid-latitude (52.12°N, 104.26°E) station.

These figures show a significantly different DCB behavior for GPS and GLONASS. Here, one can see an essentially irregular DCB temporal behavior with dramatic changes in the mean levels. Both systems feature a DCB slow time drift. For DCB_{REC}, we may note an explicit dependence of DCB variations on the environment temperature for separate receivers.

Table 2.6 summarizes the information on the above intersystem biases and their offset methods for the main GNSS applications. One can see that, depending on the

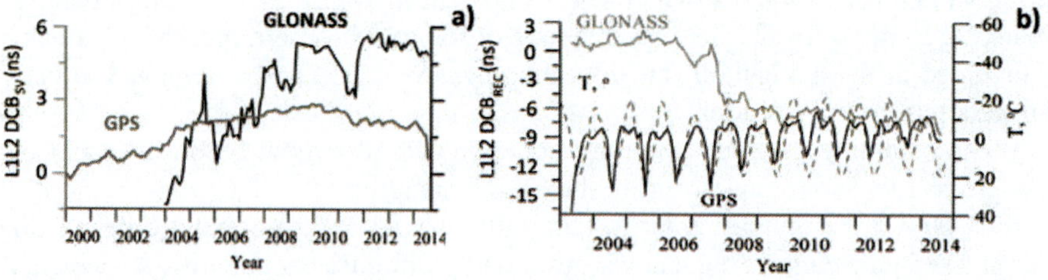

Fig. 2.19 Dynamics and characteristics for the behavior of *DCB* attributed to GPS PRN 03 (panel a, gray line), to GLONASS 04 (panel a, black line), and to the receiver (panel b, GLONASS–gray line, GPS–black line). Dotted line shows temperature time variations within the antenna mount point (Yasyukevich et al. 2015)

source, this or that type of biases causes the emergence of corresponding negative consequences in a GNSS operation. For example, the carrier phase biases introduced onboard the satellite reduce the accuracy of the phase ambiguity solution. The same phenomenon arising in a receiver, when processing signals from several GNSS, leads to deterioration in the positioning solution quality in the PPP and RTK modes (Hakansson et al. 2017).

The inter-channel biases in case of FDMA use cause additional errors in the relative navigation mode, when sharing the GLONASS signals and signals from other GNSS (El-Mowafy 2014). The presence of code pseudo-delay and multipath biases essentially complicate determining the TEC absolute values, while studying the ionosphere from GNSS signals (Yasyukevich et al. 2020).

To offset each of the above biases, one can use various tools and methods, that may be divided into hardware and software. The hardware methods involve:

(1) Upgrading transmission properties of the RF front-end in user receivers and satellite hardware;
(2) Using more advanced modulations (e.g., AltBOC instead of BPSK) and longer code ranging sequences with better cross-correlation characteristics;
(3) Commissioning new reference oscillators with better stability characteristics, including those protected from the external destabilizing factors impact (temperature, mechanical vibrations, etc.);
(4) Applying special antennas providing noise suppression in multipathing and a spatial filtering of GNSS signals.

As software methods to offset intersystem biases, one may consider:

(1) Smoothing code measurements by phase ones at their joint processing;
(2) Using perfect models describing the above bias time drift and accounting for a posteriori statistical characteristic of these biases;
(3) Using redundant measurements to expand the composition of navigation functions or observation equations.

In general, we may state that solving the problem of intersystem biases is one of the crucial goals to be solved in order to achieve the maximal efficiency in using the integrated global GNSS system.

2.3 Differential GNSS

2.3.1 General Notions of Differential Navigation Mode

To increase the accuracy of the object state vector parameter determination (2.30) and the check on the availability of required navigation parameters (RNP), one uses GNSS applications or *augmentation systems*. There are three types of such systems:

– Satellite-Based Augmentation Systems (SBAS);

Table 2.6 GNSS intersystem bias errors

Type (designation)	Offset possibility by GNSS applications				
	Standalone navigation	Differential mode	PPP	RTK	TEC comp
Phase delay in satellite hardware $\left(\delta_{i,F}\left(t-\tau_i^j\right)\right)$	No offset	Partly offset	Offset	Offset	Calibration
Phase delay in user receiver $(\delta_{j,F}(t))$	No offset	Partly offset	Offset	Offset	Calibration
Intersystem biases $(\delta_{j,F,1}(t)-\delta_{j,F,2}(t))$	No offset	Partly offset	Offset	Offset	Calibration
GLONASS inter-channel biases in user receiver $(dt_{j,F1}(t)+kdt_{j,F2}(t))$	Calibration	Calibration	Calibration/Partial offset	Calibration/Partial offset	Calibration
Code delay in satellite hardware $\left(dt_{i,F}\left(t-\tau_i^j\right)\right)$	No offset	Partial offset	Offset	Offset	Calibration
Code delay in user receiver $(dt_{j,F}(t))$	No offset	Partial offset	Offset	Offset	Calibration
Intersystem biases $(dt_{j,F1}(t)-dt_{j,F2}(t))$	Calibration	Calibration	Calibration	Calibration	Calibration
Phase multipath $\left(\delta m_{i,PHASE}^j\right)$	Partial offset	Partial offset	Partial offset	Partial offset	Partial offset
Code multipath $\left(dm_{i,CODE}^j\right)$	Partial offset	Partial offset	Partial offset	Partial offset	Partial offset

- Ground-Based Augmentation Systems (GBAS);
- Local Based (Local Area) Augmentation Systems (LBAS/LAAS).

Wide-area ground-/space-based augmentation systems (WAAS, EGNOS, RSDCM, etc.) and local augmentation systems (to serve the customers that require high navigation characteristics) have become most prevalent.

A notion of the space–time stationarity of navigation parameter determination errors and the user coordinates underlie the operation of such systems. The augmentation system key element is a constantly-operating reference station (WRS). A WRS coordinates are precisely known, and its timescale should be bound to a GNSS system timescale. The difference between the measured and the known WRS coordinates enables to estimate a navigation parameter measurement error or a user current coordinate computation error. These errors persist urgent over a certain time interval (de-correlation time) and within a coverage area (de-correlation radius).

A set of error estimates enables to generate and apply *differential corrections* to a GNSS user state vector determination errors. The accuracy of such corrections will be satisfactory, if their upgrade period and the site of the area, within which they are used (differential correction coverage area), do not exceed the spatial–temporal stationarity radius of the errors subject to corrections.

Further, we will term the mode of navigation-temporal determinations by using differential corrections *a differential mode.* A set of reference stations and facilities to broadcast differential corrections compose an augmentation system *master station* (WMS).

There are two versions of the differential navigation mode: coordinate direct correction mode and navigation parameter correction mode.

Differential mode with direct coordinate correction is based on estimating the difference between the measured and the known WMS coordinates and generates differential corrections that enable to correct the user state vector as follows:

$$\Delta\Pi = \Pi_{ref} - \hat{\Pi}, \tag{2.120}$$

$$\Pi_{us} = \hat{\Pi}_{us} + \Delta\Pi, \tag{2.121}$$

where $\hat{\Pi}$, Π_{ref} are the reference and the measured (during current measurements) WMS state vectors. Π_{us}, $\hat{\Pi}_{us}$ are the estimated and the object (user) corrected state vectors.

This method may be used in local augmentation systems only if determining the differential correction $\Delta\Pi$ and the user state vector $\hat{\Pi}_{us}$ in the user receiver is done by the same GPS constellation (Perov and Kharisov 2010). At a change in the observed satellite constellation at the WMS, it is compulsory that the composition of the satellite constellations used in the user receiver be changed, too. Because the type and parameters for adjusting the navigation receiver also affect the positioning accuracy, it is desirable that the user and the WMS have the agreed adjustments in such a mode of differential correction.

An advantage of the differential mode with direct correction coordinate of is an insignificant scope of the broadcast correcting information and sufficiently simple algorithm for applying a differential correction. However, in most cases, the above requirements are difficult to implement in practice, particularly, at mobile objects or when using the mode for large groups of users equipped with non-standard hardware. Therefore, the differential mode with direct coordinate correction has not been applied widely.

The differential mode with navigation parameter correction is based on generating differential corrections to navigation parameters separately for each satellite. This version is widespread, and it is this mode that we will mean further. The measured navigation parameters are compared with their WMS-pre-computed reference values, based on the known ephemeris and timescales offset. The comparative estimate for the measured and the pre-computed navigation parameter for each satellite enables to generate the corresponding correction for this parameter measurement error. Herewith, in a navigation receiver, an individual correction for the measured navigation parameter is done for each visible satellite.

Depending on the used type of navigation parameters, one may indicate three kinds of the differential mode. In the simplest case, *raw* measurements of the code pseudo-range without smoothing noise are used. An improvement of this differential mode kind is to use measurements of the code pseudo-range and of the carrier pseudo-phase simultaneously. Pseudo-phase measurements are used to smooth measurements of pseudo-range to suppress the high-frequency noise (first of all, the thermal noise introduced by user hardware and by multipath noise).

Smoothing the measurements of the code pseudo-range $\rho(t)$ by the carrier phase synchronous measurements $\varphi(t)$ are performed by using range increments (i.e., be object velocity increments $V(t)$) with accumulating the measurements over the T interval. The functional association between the above navigation parameters has the following form (Perov and Kharisov 2010):

$$V(t) = \frac{d\rho(t)}{dt} = \frac{\lambda}{2} \cdot \frac{d\varphi(t)}{dt} \tag{2.122}$$

The pseudo-range smoothed estimate at an instant i with the k measurement accumulation over the interval T may be determined as follows (Perov and Kharisov 2010):

$$\rho_{i,k} = \frac{1}{k+1} Y_{\rho,k} + \frac{k}{k+1} \left(\rho_{i-1,k-1} + Y_{V,i-1}\right) \tag{2.123}$$

$$Y_{\rho,k} = \rho_0 + k \cdot T \cdot V_0 + \eta_{\rho,i} \tag{2.124}$$

$$Y_{V,i} = V_0 + \eta_{V,i}, \tag{2.125}$$

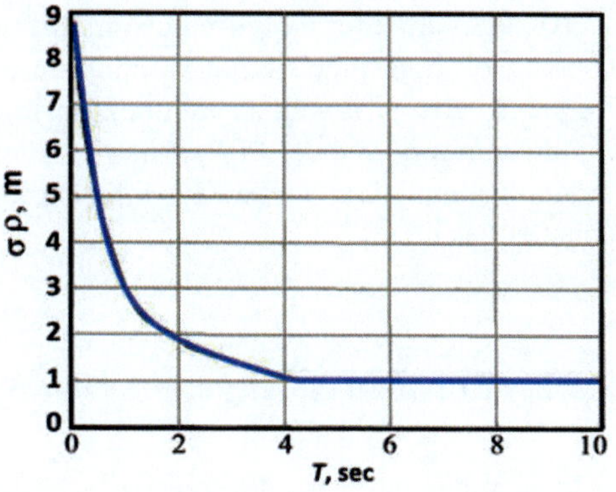

Fig. 2.20 RMS for the code pseudo-range smoothed estimate with measurement accumulation over the interval *T*. Re-produced, modified and translated from Russian figure based on figure from the published book (Perov and Kharisov 2010)

where ρ_0 and V_0 are zero (a priori) estimates for the object pseudo-range and pseudo-velocity; $\eta_{\rho,i}$ and $\eta_{V,i}$ are the noise of pseudo-range and pseudo-velocity measurements.

Apparently, as the time *T* increases, the first summand (2.124) decreases. Due to this, the pseudo-range estimate starts generating, primarily, owing to the range increments (i.e., owing to the phase measurements). Because the phase measurement noise is considerably lower, than the pseudo-ranging noise, this results in an essential decrease in the noise component of the error when measuring the main navigation parameter (pseudo-range), which Fig. 2.20 illustrates explicitly.

Thus, the differential correction quality may be increased and effectively used to correct the errors with a stationarity time $\gg T$.

Both in the first and in the second case, the differential correction mode is referred to as *a standard differential mode with pseudo-range correction.* Owing to a higher quality of the smoothed differential corrections robust to high-frequency measurement noise, one applies more often the second kind of this differential mode.

To increase the accuracy of positioning solutions (e.g., for geodetic purposes), one may use the third kind of the differential mode: *a high-precision differential mode.* In this case, the carrier phase is the radio-navigation parameter that is generated at the WMS and is corrected in the user receiver.

In case of using the carrier phase, one measures the radio-navigation parameter upon solving the phase ambiguity within a complete cycle of the phase variations (0…0.2 π). A potential phase measurement accuracy does not exceed the wavelength, and, at the experimental estimate, equals $\sigma_\varphi \approx 0.01 \cdot \lambda$, which corresponds to the ~ 0.2 cm equivalent ranging error at the *f1* GPS. This is significantly higher, than the measurement potential accuracy when using the ranging code (7.5 m for the C/A-code and 0.75 m for the P-code, see Sect. 2.1.3). At the same time, in this differential correction mode, one attains the maximal positioning solution accuracy by accumulating and post processing the measurements to remove the phase ambiguity and the errors with a set stationarity time.

An alternative for a high-precision differential correction mode is using external sources of high-precision correcting information on the ephemeris and time–frequency errors, as well as on ionospheric and tropospheric ranging errors (*Precise Point Positioning* mode, *PPP*). Such data have been already freely provided now by numerous analytical centers for information processing from measuring networks (see Sect. 2.1.1).

2.3.2 *Local-Area Augmentation System (LAAS)*

When implementing the differential mode based on one ground-based station (*Local-Area Augmentation System*), *scalar* differential corrections are generated. A scalar correction for the ranging error to each of visible satellites is estimated at the reference station and involves the sum of the ranging error components with a common variance equal to (GPS-WAAS-PS 2008):

$$\sigma_{Di}^2 = \sigma_{i,e}^2 + \sigma_{i,ion}^2 + \sigma_{i,tr}^2 + \sigma_{i,mp}^2 + \sigma_{i,rec}^2, \tag{2.126}$$

where $\sigma_{i,e}^2$, $\sigma_{i,ion}^2$, $\sigma_{i,tr}^2$ are variances of the errors caused by inaccuracy in ephemeris prediction, by ionospheric and tropospheric effects; $\sigma_{i,mp}^2$ is a multipath noise variance; $\sigma_{i,rec}^2$ is an inner noise variance of a user receiver.

The errors caused by multipath ($\sigma_{i,mp}^2$) and inner noise of user receivers ($\sigma_{i,rec}^2$) and of a reference station are not correlated. Therefore, the navigation differential mode cannot offset the latter, and one should take extra measures to eliminate them (e.g., by smoothing the code measurements by phase ones).

There are four main types of errors in measuring navigation parameters, that are *correlated* and may be offset by using differential correction:

(1) satellite Reference Oscillator errors;
(2) satellite ephemerides determination errors;
(3) tropospheric errors;
(4) ionospheric errors.

The quality of these differential corrections deteriorates as the latter become obsolete and as the distance between a user receiver and a reference station increases.

Let us introduce a notion of *spatial and temporal de-correlating factors* that determine a necessary time for updating differential corrections and of an LAAS coverage area radius. For this purpose, we present the differential correction like a Taylor expansion in the surroundings of a user receiver location with the coordinates $\mathbf{x}_U$ at the instant t_U (Perov and Kharisov 2010):

$$\Delta\rho(t_0, \mathbf{x}_0) \approx \Delta\rho(t_U, \mathbf{x}_U) + \frac{\partial \Delta\rho(t_U, \mathbf{x}_U)}{\partial \mathbf{x}_U}(\mathbf{x}_0 - \mathbf{x}_U) + \frac{\partial \Delta\rho(t_U, \mathbf{x}_U)}{\partial t}(t_0 - t_U), \tag{2.127}$$

where $\Delta\rho(t_0, \mathbf{x}_0)$ is a differential correction for a ranging error, generated at a reference station with the coordinates $\mathbf{x}_0$ at the instant t_0; $\Delta\rho(t_U, \mathbf{x}_U)$ is a differential correction used in the user receiver with the coordinates $\mathbf{x}_U$ at the instant t_U by the timescale of the latter.

The second and third summands in (2.127) correspond to the *spatial and temporal de-correlating factors* of the differential correction. They determine an extent of a correction deviation from its reference value as a user moves away from a base station and as the time after a previous correction increases.

The ranging error correlated components are split into *slow* and *fast*, depending on the extent of the effect of time de-correlating factor. Group 1 includes the errors of satellite ephemeris prediction (including the action of relativistic and gravity effects) and the errors of satellite Reference Oscillators. Group 2 involves ionospheric and tropospheric errors. Below, we will show that error Group 2 determines the time necessary to update differential corrections and the LAAS coverage area radius.

When addressing the satellite Reference Oscillator errors, the base station and the user receiver are supposed to use the same navigation message bound to the same reference issue-of-data (*IOD*) time. Thus, generating a correction for this kind of error at the reference station and in the user receiver will be uniform for the same system time (ICD GPS-200C 1993). Therefore, the Reference Oscillator error systematic part is practically offset completely during differential correction and weakly depends on the effect from temporal and spatial de-correlation. The residual uncompensated part of this error is mostly determined by random variations in the satellite Reference Oscillator (see Subsect. 2.1.2).

Similar reasoning can be applicable to *the satellite orbital position prediction error*. The reference station and the user estimate the satellite velocity and its orbital position based on the same navigation message from that satellite. Therefore, the time de-correlating factor effect here will be negligible, if one takes into account the known regularity of the satellite radial velocity variation and the range to the satellite relative to the user receiver. The spatial de-correlating factor may affect the residual error more significantly. This is caused by that, at the reference station and in the user receiver, the same satellite is observed at different angles and has different radial velocities. Owing to this, the predictions of the satellite orbital position relative to the reference station and relative to the user receiver slightly differ. This effect leads to the emergence of the ~ 0.05-m/100-km error spatial de-correlating factor of the ephemeris prediction error (Perov and Kharisov 2010).

Thus, like in the previous case, using a differential correction almost completely offsets the systematic part of the ranging error related to the satellite orbital position prediction error. The residual part of this error will be determined by unaccounted variations in the satellite orbital position (see Subsect. 2.1.2).

A tropospheric error depends substantially on the spatial and time de-correlating factors. The reason for this error emergence is a regular and irregular refraction of a satellite navigation signal in the lower atmosphere, including the troposphere, as well as the stratosphere, mesosphere, and thermosphere (Monteiro et al. 2005). The tropospheric error may be decomposed into the *dry* (80–90% of the entire error) and the *wet* components (Kaplan 1996):

$$dt_{tr} = \frac{10^{-6}}{5}(n_{d,0}H_d + n_{w,0}H_w), \tag{2.128}$$

where $n_{d,0}$ and $n_{w,0}$ are initial values for the troposphere refraction coefficients within its *dry* and *wet* layers at the layer lower boundary; $H_d \approx 40$ km and $H_w \approx 10$ km are the heights for the upper boundaries of the *dry* and *wet* tropospheric layers above sea level.

Within the *dry* tropospheric layer, the water vapor concentration is low and does not affect essentially the value of and variations in the medium refraction coefficient for that layer. Therefore, the refraction coefficient in the layer depends primarily on the *dry* component partial pressure of the air and the temperature of the latter. At the same time, the refraction coefficient in the *wet* layer also depends on the water vapor partial pressure. Hence, we have (Kaplan 1996):

$$n_{d,0} \approx a_1 \frac{P_0}{T_0} \tag{2.129}$$

$$n_{w,0} \approx a_2 \frac{e_0}{T_0} + a_3 \frac{e_0}{T_0^2}, \tag{2.130}$$

where P_0 and T_0 are the partial pressure and the temperature of the dry air component at the sea level (mbar, K); e_0 is the partial water–vapor pressure at the sea level (mbar); $a_1 = 77.624$ K/mbar, $a_2 = -12.92$ K/mbar, $a_3 = 371.9$ K^2/mbar are empirical constants.

The tropospheric error may vary due to a spatial–temporal variability in the above meteorological parameters. However, the variability of almost all these parameters (except for the partial pressure *e*), is insignificant within a day (Brunner and Welsch, 1993). On the contrary, the parameter *e* has a high spatial–temporal variability, particularly with a height variation, and it is related to climatic phenomena in the troposphere. But this parameter determines only the *wet* component n_w that is 10–20% of the tropospheric error combined value. Therefore, dynamic climate phenomena in the troposphere result in insignificant variations in the absolute value of the tropospheric error in general. Iwabuchi et al. (2003) showed that the spatial de-correlating factor for the tropospheric error varies within ~ 1–6 mm/km.

The tropospheric error time de-correlating factor is determined by the size and velocity of the tropospheric inhomogeneity drift relative to the satellite line-of-sight. In (Iwabuchi et al. 2003), the authors established that one may neglect the effect of the tropospheric error time de-correlating factor over the < 10–15-min interval. Herewith, the effect of the spatial and of the time de-correlating factors depends significantly on the user height over the Earth surface (see the comments further).

The ionospheric error in measuring navigation parameters, as a rule, exceeds the other errors by the absolute value, and, in particular, by the extent of the spatial–temporal variability. Therefore, the offset of this error through a differential correction is a complicated problem.

The main contribution to the ionospheric ranging error variations is the *TEC* daily ionospheric variations. In this case, the error maximum is reached at a local midday, with the minimum at night. The variations in the ionospheric error absolute maxima and minima have a positive correlation with the 11-yr solar activity cycle (Akasofu and Chapman 1972). Geomagnetic variations and storms impact on the above cyclic TEC changes. These phenomena cause the emergence of irregular TEC variations with an amplitude of up to 50% of the background level and with a wide spatial–temporal spectrum of TEC periods and disturbance wavelengths: from 1 to 2 thousand km (≤1-h period) to tens of km (≤1–2 min period) (Afraimovich and Perevalova 2006).

A significant variability of the ionosphere, particularly, in the equatorial anomaly region and with an increase in the geomagnetic activity, leads to a substantial growth in the spatial and temporal de-correlating factors. Konno et al. (2005) estimate the spatial gradient for the ionospheric ranging error at mid-latitudes to be more than 3…5 mm/km. But, at lower latitudes, intensive small-scale turbulences of the electron density result in a sharp transient increase in the gradient to 500–550 mm/km (Saito et al. 2017). Konno et al. (2005) proposed to introduce a threshold to detect a *dangerous gradient* of the ionospheric ranging error *G* into the algorithm for local differential system operation. The threshold is measured on the line-of-sight of each of the visible satellites, with allowance for the elevation angle in a point of a user receiver location:

$$\frac{\partial \Delta\rho_{ION}(t_U, \mathbf{x}_U)}{\partial \mathbf{x}_U} = \left(\begin{array}{c|c} G_i > 80\text{ mm/km} & \gamma > 60^\circ \\ G_i > 100\text{ mm/km} & \gamma \leq 60^\circ \end{array} \right). \tag{2.131}$$

Excessing that threshold is supposed to signal about a dangerous growth in the spatial de-correlating factor of the ionospheric ranging error $\Delta\rho_{ION}(t_U, \mathbf{x}_U)$, which may violate the satisfactory quality of differential correction.

The time de-correlating factor of the ionospheric ranging error, according to the estimates from Jakowski and Hoque (2019), is ~ 0.1–0.3 mm/s. However, by the estimates from the same authors, this value may grow up to 5–6 mm/s.

When implementing practically an LAAS, one can determine the coverage area radius by setting an admissible value for the residual ranging error that is caused by an action of each of the above factors.

The residual ephemeris prediction error $\Delta\varepsilon_E$ caused by a satellite motion relative to the user equals (Kaplan 1996):

$$\Delta\varepsilon_E = \left(\frac{\varepsilon_E}{\rho}\right) \cdot \Delta R \cdot sin^2(\gamma), \tag{2.132}$$

where ρ is the measured range from the reference station to the satellite; ε_E is an error of the satellite orbital position prediction; ΔR is a distance between the user and the reference station; γ is a satellite elevation angle in the user location.

Admitting, by example, $\varepsilon_E = 1\text{M}$, $\rho = 100$ km and $\gamma_i = 10^\circ$, we will obtain the value for an offset error of ephemeris prediction $\Delta\varepsilon_E = 0.005$ m. Such a small value corroborates that the ephemeris prediction error will not impact on the LAAS coverage area maximal size.

The residual pseudo-ranging tropospheric error depends on a distance between a reference station and a user. However, it depends, even to a greater extent, on a height difference between a user and a reference station. In case of the ground user, the residual tropospheric ranging error after the differential correction equals (Kaplan 1996):

$$\Delta\varepsilon_{TR} \approx \frac{\cos(\gamma_{RS})}{d} \cdot \Delta R \cdot (1.4588 + 0.0029611 \cdot N) \\ - 0.3048 \cdot \left(0.00586 \cdot (N - 360)^2 + 294\right) \cdot \left(\gamma_{RS}^{-2.3} - \gamma^{-2.3}\right), \quad (2.133)$$

where d is the satellite normal height over the Earth surface; N is a tropospheric refraction coefficient at the Earth surface (typical mean value being $\langle N \rangle = 360$); ΔR is a distance between a reference station and a user; γ, γ_{RS} are elevation angles of satellites in the user and in the reference station locations, respectively.

Due to a fast pressure and air temperature change with height, an increase in the navigation receiver height relative to the reference station height leads to a fast growth in a residual tropospheric error under the exponential law (Kaplan 1996):

$$\Delta\varepsilon_{TR} \approx \varepsilon_{TR} \cdot exp\left[-\left((0.002 \cdot N + 0.07) \cdot h + \left(\frac{0.83}{N} - 0.0017\right) \cdot h^2\right)\right], \quad (2.134)$$

where ε_{TR} is a tropospheric pseudo-ranging error determined in the reference station location; h is a height of the user receiver above the LAAS reference station.

One can compute that, at a user is horizontally 100 km away from a reference station, and at $\varepsilon_{TR} = 1$ m, $\gamma_{RS} = 45°$ and $N = \langle N \rangle = 360$, the residual tropospheric error will be $\Delta\varepsilon_{TR} \approx 0.01$ m. Such a value is not significant for most practical applications. However, if a user is at a 1 km height, then, with the same parameters ($\varepsilon_{TR} = 1$ m and $\gamma_{RS} = 45°$), the residual error will be $\Delta\varepsilon_{TR} \approx 0.45$ m. Thus, the user height increase by 1 km will cause the emergence of essential non-offset error, whose value equals 55% relative to the initial value ε_{TR}.

The non-offset ionospheric ranging error will depend on the distance ΔR showing how far a user is from a reference station, and on the difference in the TEC values along a satellite signal propagation path to a reference station and to a navigation receiver. Its value may be estimated in an undisturbed spherically-layered ionosphere approximation as follows (Kaplan 1996):

$$\Delta\varepsilon_{ION} \approx \frac{\Delta R}{\rho} \cdot \left(\frac{\Delta R}{\rho} - cos(\theta)\right) \cdot \frac{40.3}{f^2} TEC, \quad (2.135)$$

where θ is a satellite elevation angle measured at the ionospheric F-layer ionization maximum height in the intersection point of the line-of-sight and the F-layer relative

Table 2.7 Characteristics of the ranging error main components

Differential correction component	Spatial de-correlating factor, mm/km	Time de-correlating factor, mm/s	Typical values
Ephemeris error	~ 0.5	–	0.5–1 m
Ionospheric error	3…850	0.1…6	1–15 m
Tropospheric error	1…6	0.05…2	0.1–2 m

to a reference station; *TEC* is a total electron content along the *reference station—satellite* path; f is a satellite signal carrier frequency; ρ is a slant range between a reference station and a satellite.

For example, at the $TEC = 50\ TECU$ and $\theta = 45°$, for the GPS *f1* frequency and the distance between the user and the reference station $\Delta R = 100$ km, the residual ionospheric ranging error will equal $\Delta\varepsilon_{ION} \approx 0.03$ m.

Table 2.7 generalizes the estimates for the spatial and time de-correlating factors, for the main components of the ranging error, obtained by generalizing the addressed sources.

Table 2.7 shows that the main component of the navigation parameter measurement error is the ionospheric error. This error, as a rule, has the greatest absolute value and the spatial–temporal de-correlating factor. The characteristics of this error depend on the geomagnetic activity level and on the geomagnetic observational latitude. Owing to this, the LAAS coverage area radius is restricted by the influence of the ionospheric error proper, and varies from 30 to 100 km. Herewith, the recommended time to update differential corrections should be ≥ 1 s (MOPS GPS/GLONASS/ABAS 2014).

2.3.3 Wide-Area Augmentation Systems

The basic disadvantage of Local-Area Augmentation Systems is their small radius of action. Thereby, LAASs have specialized applications and are non-available to a common use. The differential mode with a pseudo-range correction may be also used and for larger areas with ~ 5000 km coverage area. Such systems are referred to as Wide-Area Augmentation Systems (WAAS). Figure 2.21 shows the WAAS overview.

The basis for wide-area augmentation systems is a network of ground-based reference stations. These stations are arranged so that they provide a relatively uniform coverage of an extended territory. The WAAS master station monitors continuously the state of a GNSS satellite segment, and estimates ranging errors. The *fast components* of those errors feature a significant spatial de-correlating factor and cannot be used farther than 50–100 km from the master station. Therefore, measurements from the reference station network are periodically broadcast to the WAAS master station. There, the reference station measurements are converted into a set of mixed

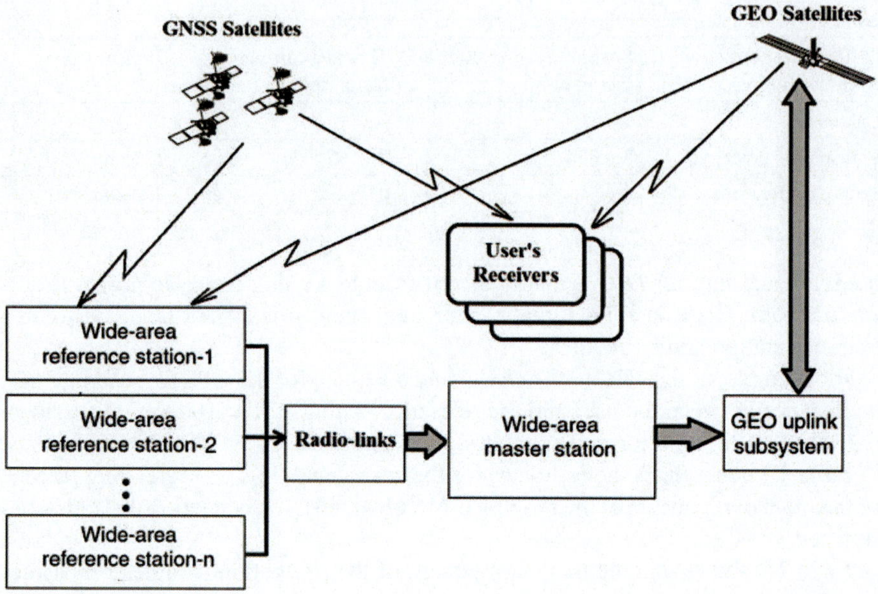

Fig. 2.21 WAAS overview

scalar-vector corrections uniform for the entire WAAS working zone. The set may be used by users within the zone over certain time. The *scalar-vector* corrections enable to uniformly apply a differential correction within the entire coverage area, unlike the LAAS scalar corrections.

Because it is required that the scalar-vector corrections be broadcast at large distances, there are geostationary (GEO) satellites within a WAAS. Signals from those satellites are also used as additional navigation signals. Due to this, one may improve the PDOP factor, in case of signal tracking failures at separate GNSS satellites. Table 2.8 shows the configuration and main characteristics for the basic WAASs.

The set of scalar-vector corrections is determined for each GNSS satellite like $\{\Delta\mathbf{X}_i, \Delta\mathbf{V}_i, \Delta\rho_i, \Delta v_i,\}$, where: $\Delta\mathbf{X}_i = |\Delta x_i, \Delta y_i, \Delta z_i|^T$ and $\Delta\mathbf{V}_i = \left|\Delta V_{x,i}, \Delta V_{y,i}, \Delta V_{z,i}\right|^T$ are the vectors of corrections to positioning estimate errors for each (ith) satellite in the orbit and to its velocity vector components; $\Delta\rho_i$ and Δv_i is a complex scalar correction for a pseudo-range and pseudo-velocity estimates for the ith satellite-receiver pair.

Because the spatial–temporal de-correlating factor weakly affects the vector corrections $\Delta\mathbf{X}$, $\Delta\mathbf{V}$, then the differential correction of the satellite position prediction errors, and of the positioning errors is computed for the entire coverage area. The fundamental difficulty in a wide-area differential correction is in generating and applying (uniformly) scalar corrections to pseudo-ranges (and to corresponding pseudo-velocities) for each pair "satellite- user's receiver".

Table 2.8 WAASs main characteristics

WAAS name (main GNSS)	Coverage area	WRS/MS	GEO satellites' name	Longitude/GEO Sat PRN
Wide area Augmentation System–WAAS (GPS)	North America, including maritime belt, Hawaii	38/3	Inmarsat-3-F3/POR Inmarsat-3-F4/AOR-W Intelsat galaxy XV TeleSat F1R	178°E/134 142°W/122 133°W/135 107.3°W/138
European geostationary navigation overlay system–EGNOS (Galileo)	Western and Eastern Europe, Mediterranean sea, North Africa	34/4	Inmarsat-3-F2/AOR-E ESA Artemis Inmarsat-3-F5/IOR-W	15.5°W/120 21.5°E/124 25°E/126
Russian system of differential correction and monitoring–RSDCM (GLONASS)	Russia, including maritime belt and Eastern Europe	(46 + 7)/3	Luch-5B Luch-5C Luch-4	16°W/125 95°E/140 167°E/141

Applying WAAS differential corrections to pseudo-range at an instant t for each of the satellites may be presented as follows (GPS WAAS PS 2008):

$$\rho(t)_{COR} = \rho(t)_0 + FC + RRC(t - t_0) + \Delta\rho_{ION} + \Delta\rho_{TR}, \tag{2.136}$$

where $\rho(t)_0$ is a measured pseudo-range to the satellite; FC is a *fast* differential correction for non-correlated errors; $RRC(t - t_0)$ is a correction for range rate of change owing to non-correlated error effect; $\Delta\rho_{ION}$ is a correction for an ionospheric ranging error; $\Delta\rho_{TR}$ is a correction for a tropospheric ranging error; t_0 is a reference time corresponding to the instant of broadcasting the latest *fast* correction (always coincides with the onset of a regular 1-s epoch of applying a differential correction); t is a GNSS current system time. All the above corrections relate to a particular satellite.

A correction for non-correlated errors FC for a separate satellite equals:

$$FC(t) = \rho(t)_0 - \Delta r_E(t) - \Delta\rho_{ION} - \Delta\rho_{TR}, \tag{2.137}$$

where $\Delta r_E(t)$ is a correction for a satellite orbital position prediction error (ephemeris error).

A range variation rate that originates owing to the effect of non-correlated errors *RRC* at an instant t_k is estimated relative to the measured range in the previous instant $t_{k\text{-}1}$ as follows:

$$RRC(t_k) = \frac{FC(t_k) - FC(t_{k-1})}{t_k - t_{k-1}}. \tag{2.138}$$

A set of differential corrections *FC* and *FFC* serves as a convenient tool to detect abnormal non-correlated errors caused by hardware failure in a separate satellite, as well as by strong multipath phenomena when receiving signals from that satellite.

A correction for an ionospheric ranging error is represented like a reference network of ionospheric grid points (IGPs). This network nods are determined as a result of interpolations of the current ionospheric ranging measurement error at all the WAAS reference stations. These nods are distributed evenly by longitude and by latitude (at 5° and 2.5° step for the WAAS network) and cover the entire WAAS coverage area. Geostationary satellites (Table 2.8) broadcast the current values for the ionospheric error at IGP-network nodes (τ_{PP}) within its coverage area. Based on the τ_{PP} values and on the known location in a navigation receiver, one computes the current "oblique" ionospheric ranging error $\Delta\rho_{ION}$ for each satellite relative to the user location by the following algorithm:

$$\Delta\rho_{ION} = -F_{PP} \cdot \tau_U, \tag{2.139}$$

where τ_U is an interpolation-computed IGP error value for a chosen satellite and a user location; F_{PP} is the oblique factor relating the values for the vertical and oblique ionospheric pseudo-ranging errors (2.82).

A vertical ionospheric error value in a user location τ_U is computed by bilinear interpolation within an IGP-network cell, in which a user is. For this, one uses values for an ionospheric correction τ_{PP} at the nods of this cell:

$$\tau_U = \sum_{k=1}^{4} W_k \tau_{PP,k}, \tag{2.140}$$

where W_k are the weight coefficients that determine the contribution of the corresponding correction $\tau_{PP,k}$ to the total estimate τ_U depending on the distance between a user and k-th reference point within an IGP-network.

The correction for the ionospheric ranging error determined by WAAS facilities will differ from real values. The correction used in a user receiver location (2.139) is a product of double interpolation. First, one interpolates the real ionospheric error measurements taken at a reference station network in order to convert them to a uniform IGP-network. Further, the user estimates the correction value, by using interpolation (2.140) again. Such an approach to determining corrections may result in a crucial deterioration of differential correction during geomagnetic disturbances (Conker et al. 2003; Demyanov et al. 2019).

When generating a correction for a tropospheric ranging error within the extended working zone, one neglects the spatial–temporal de-correlating factor. A WAAS GEO satellite broadcasts a common differential correction for this kind of error. The correction is presented as a current mean index of refraction in the troposphere N_R and the tropospheric layer thickness h_0. The tropospheric error value also depends substantially on the height of a user location relative to sea-level Δh, and on a satellite elevation angle γ. Therefore, these parameters are also accounted for when computing a correction for a tropospheric ranging error $\Delta\rho_{TR}$ through the following expression:

$$\Delta\rho_{TR} = N_R h_0 \frac{10^{-6}}{\sqrt{0.02 + sin^2(\gamma)}}\left(1 - \exp\left(-\frac{\Delta h}{h_0}\right)\right). \quad (2.141)$$

Upon applying all above corrections, the corrected pseudo-range to a satellite will contain corresponding residual errors. The total residual error for a pseudo-range estimate to each satellite is described by the corresponding RMS value as follows

$$\sigma^2 = \sigma_{flt}^2 + \sigma_{UIRE}^2 + \sigma_{air}^2 + \sigma_{TR}^2, \quad (2.142)$$

where σ_{flt} is a User Differential Range Error (*UDRE*) upon applying corrections for the satellite orbital position prediction errors and smoothing code ranging; σ_{UIRE} is a residual User Ionosphere Range Error (*UIRE*); $\sigma_{i,air}$ is ranging error RMS caused by multiple satellite-signal re-reflectances that emerge for low satellites along an aircraft motion trajectory (this error component is accounted for when organizing air traffic in an airport zone); σ_{TR} is a residual tropospheric error RMS.

A full application of ranging error differential correction is possible only in a complex with a *system integrity check* and *availability of navigation standards in a coverage area.*

During generating and using differential corrections, one should solve a problem of checking their integrity (i.e., timely detect an anomalous correction value or impossibility of its usage). For this, applied are *degradation factors* of differential corrections dependent on the time from the generation to the actual use of those corrections. The degradation factors are broadcast to the user as a part of the system service message and provide *a short-time* prediction of RMS of the corresponding components of the residual ranging error for each of satellites (2.142).

If the current error component degradation factor x for the satellite i exceeds a threshold value (see Table 2.9), then a negative prediction for the residual ranging error component is generated in the user receiver. Here, a residual error negative prediction is thought to be an event (Annex 10 2006):

$$\sigma_{i,x}^2 > \sigma_{x,MAX}^2, \quad (2.143)$$

where $\sigma_{x,MAX}^2$ is a maximal threshold value for the component x of the residual ranging error, determined from Table 2.9.

Table 2.9 Limiting variances of residual pseudo-ranging errors

Type of error $\sigma_{xx,MAX}^2$	Limiting value, m^2
$\sigma_{flt,MAX}^2$(*UDRE*)	> 2078.695
$\sigma_{UIRE,MAX}^2$(*GIVE*)	> 187.0826

In this case, the user receiver does not apply the corresponding differential correction until its next update in the service message. Over this time, a coarse prediction for the corresponding residual ranging error will be used in the user receiver instead of the correction. Below, we will discuss this.

The *UDRE* RMS prediction in the user receiver with allowance for degradation factors looks like (Annex 10 2016):

$$\sigma_{flt}^2 = (\sigma_{UDRE} + \varepsilon_{FC} + \varepsilon_{RRC} + \varepsilon_{LTC} + \varepsilon_{ER})^2, \tag{2.144}$$

where σ_{UDRE} is a *UDRE* RMS upon applying error corrections for ephemeris prediction and smoothing the code range by phase measurements for a selected satellite; ε_{FC} is a correction degradation parameter; ε_{RRC} is a parameter for degradation of a correction change rate estimate; ε_{LTC} is a parameter for a long-time degradation of the correction in the absence of periodic update; ε_{ER} is a parameter for a correction degradation during non-categorized approach (for aviation users).

The above degradation factors are broadcast in the system service message and are updated every 5 min. If the degradation factors are unknown, the prediction for a variance of this residual error is coarsened and looks like (Annex 10 2016):

$$\sigma_{flt}^2 = (\sigma_{UDRE} + 8)^2. \tag{2.145}$$

In the worst case, when smoothing of the code pseudo-range and its correction for an ephemeris prediction error are not done, then:

$$\sigma_{flt}^2 = (60)^2. \tag{2.146}$$

The residual ionospheric ranging error RMS prediction is provided not only with allowance for the "aging" of the corresponding differential correction, but also accounting for the IGP-network interpolation error (Annex 10 2016) as follows:

$$\sigma_{UIRE}^2 = F_{PP}^2 \cdot \sigma_{UIVE}^2 \tag{2.147}$$

$$\sigma_{UIVE}^2 = \sum_{k=1}^{4} W_k \sigma_{k,IGP}^2 \tag{2.148}$$

$$\sigma_{k,IGP}^2 = (\sigma_{GIVE} + \varepsilon_{ION})^2, \tag{2.149}$$

where σ_{UIVE} is an ionospheric "vertical" error RMS in the user location with allowance for interpolation inaccuracy within an IGP-network cell; $\sigma_{k,IGP}^2$ is an ionospheric "vertical" ranging error at a kth node of an IGP-network; σ_{GIVE} is a residual ionospheric ranging error RMS obtained upon applying the correction (2.139–2.141); ε_{ION} is a spatial–temporal degradation factor for estimates of an ionospheric "vertical" error at IGP-network nodes.

In the case, when the ionospheric correction (2.139) cannot be applied, one uses a coarsened prediction of an expected residual ionospheric error in the form:

$$\sigma_{UIRE}^2 = MAX\left\{\left(\frac{\Delta R_{MOD}}{5}\right)^2, (F_{PP} \cdot \Delta R_{VR})^2\right\}, \tag{2.150}$$

where ΔR_{MOD} is an ionospheric ranging error for a particular satellite, determined by a user-selected model (for example Klobuchar model—for GPS users); ΔR_{VR} is a coarse averaged estimate for the navigation signal vertical refraction ionospheric delay determined with allowance for the user geographic latitude φ as follows

$$\Delta R_{VR} = \left\{\begin{array}{l} 9\text{ m}, 0^\circ \leq |\varphi| \leq 20^\circ \\ 4,5\text{ m}, 22^\circ \leq |\varphi| \leq 55^\circ \\ 6\text{ m}, 55^\circ < |\varphi| \end{array}\right\}. \tag{2.151}$$

When predicting a residual tropospheric error, one uses the signature of presence/absence of the necessary correcting information within (2.141). In this case, the prediction for the tropospheric error residual RMS σ_{TR} depends on a satellite elevation angle (γ) and has the following form:

$$\sigma_{TR} = \frac{0.12}{\sqrt{0.002 + sin^2(\gamma)}}. \tag{2.152}$$

If a differential navigation object is an aircraft at an approach stage, there are stricter requirements for the accuracy of its guidance in the horizontal and vertical planes. In this case, one should account for the residual ranging error RMS prediction caused by multiple re-reflectances of navigation signals from infrastructure objects located along a glide path. This error depends on satellite elevation angles as follows:

$$\sigma_{air}(\gamma) \leq 0.15 + 0.43 \cdot exp(-\gamma/\gamma_0), \tag{2.153}$$

where $\gamma_0 \geq 0.5^\circ$ is an aircraft elevation angle relative to a runway level; $\gamma \geq 6.9^\circ$ is a satellite elevation angle relative to a runway level.

Thus, the WAAS differential correction integrity check is in the following. If the RMS prediction has not exceeded the limits indicated in Table 2.9, the user receiver applies a regular differential correction to ranging to the given satellite and uses those measurements for positioning solutions. If a differential correction by the signals of this or that satellite is not applied, the total complex RMS of the residual ranging error (2.142) for a satellite is made of coarsened predictions (2.146), (2.150) and estimates (2.152) and (2.153). The total variance of the predicted ranging error is applied to check the measurement integrity at each particular satellite. In the case, if for the total variance condition (2.143) is fulfilled, ranging by this satellite is excluded from the positioning solution.

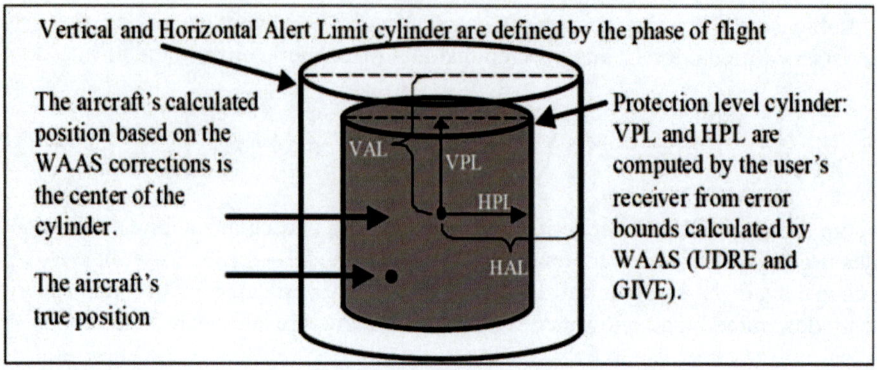

Fig. 2.22 Illustrating the principle of checking the system integrity and availability of required navigation standards to a WAAS user (GPS-WAAS-PS 2008. https://www.gps.gov/technical/ps/2008-WAAS-performance-standard.pdf)

After a differential correction and determining the composition of satellites, one performs *a check of availability of required navigation parameters* to solve a positioning problem. The algorithm of such a check is performed separately in the horizontal and the vertical planes and is based on the *safety cylinder* concept. A safety cylinder is understood a domain, in which one expects the availability of navigation standards. The domain center corresponds to the user location coordinates. The boundaries of that domain are determined by the limiting levels of a safe error in determining the coordinates in the horizontal (Horizontal Alert Limit–*HAL*) and in the vertical (Vertical Alert Limit–*VAL*) planes. The user actual location may vary randomly within an area restricted by horizontal and vertical protection levels: Horizontal Protection Level (*HPL*) and Vertical Protection Level (*VPL*). The RNP are regarded available at a current instant, if the area restricted by the *HPL* and *VPL* boundaries has not exceeded beyond the safety cylinder limited by the *HAL* and *VAL* boundaries. Figure 2.22 illustrates the introduced concepts (GPS-WAAS-PS 2008).

The quantitative value for the *HAL*, *VAL*, *HPL*, and *VPL* bounds depends on a user's category and on the corresponding safety requirements for that category. A description of typical GNSS users and required navigation parameters are specified by State Radio-navigation Plans, and are discussed in Sect. 2.4.

The process of checking the availability of required navigation parameters to a WAAS user comes to computing the current position of the *HPL* and *VPL* bounds and checking the fulfillment of the $HPL \leq HAL$ and $VPL \leq VAL$ conditions. Different WAAS algorithms may imply different approaches to the solution of this problem. Let us address one of those algorithms used in a WAAS system. Computing the current *HPL* and *VPL* protective level values for a given category of users is performed as follows (Annex 10 2006):

$$HPL = \begin{cases} K_{HNPA} \cdot D_{mjr} \\ \text{or} \\ K_{HPA} \cdot D_{mjr} \end{cases}; \quad VPL = K_{VPA} \cdot D_V$$

$$D_{mjr} = \sqrt{\frac{D_X^2 + D_Y^2}{2} + \sqrt{\left(\frac{D_X^2 - D_Y^2}{2}\right)^2 + D_{XY}^2}}$$

$$D_X^2 = \sum_{i=1}^{N} S_{X,i}^2 \cdot \sigma_i^2; \quad D_Y^2 = \sum_{i=1}^{N} S_{Y,i}^2 \cdot \sigma_i^2; \quad D_V^2 = \sum_{i=1}^{N} S_{V,i}^2 \cdot \sigma_i^2$$

$$D_{XY} = \sum_{i=1}^{N} S_{X,i} \cdot S_{Y,i} \cdot \sigma_i^2$$

$$\mathbf{S} = \begin{vmatrix} S_{X,1} & S_{X,2} & \ldots\ldots & S_{X,N} \\ S_{Y,1} & S_{Y,2} & \ldots\ldots & S_{Y,N} \\ S_{V,1} & S_{V,2} & \ldots\ldots & S_{V,N} \\ S_{t,1} & S_{t,2} & \ldots\ldots & S_{t,N} \end{vmatrix} = \left(\mathbf{G}^T \cdot \mathbf{W} \cdot \mathbf{G}\right)^{-1} \cdot \mathbf{G}^T \cdot \mathbf{W} \tag{2.154}$$

where $K_{HNPA} = 6.18$ is a confidence probability coefficient for a domain of errors in positioning solution in a horizontal plane (for non-precise applications); KHA $= 6.0$ is the same for precise applications; K_{VPA} is a confidence probability coefficient for a domain of errors in positioning solution in a vertical plane; **G** is a matrix for direction cosines of the observed navigation constellation; **W** is a diagonal matrix for ranging error weight coefficients; σ_i is a total residual ranging error (2.142) for an *i*th satellite; *N* is a number of satellites within the observed navigation constellation.

The efficiency of checking the WAAS integrity and availability of required navigation standards is wholly determined by a correspondence degree between a real status of "fast" pseudo-ranging errors and their prediction generated by the system. The ionospheric ranging error features a greatest spatial–temporal variability. Therefore, one may expect a decrease in the quality of the corresponding differential corrections under the conditions, when there emerge intensive ionospheric inhomogeneities and scintillations (Conker et al. 2003; Walter et al. 2004; Demyanov et al. 2019). The principal result of this is an increase in the probability of issuing false alerts for system integrity violation and unavailability of required navigation standards (Tossaint et al. 2007, WAAS Report #53 2015).

2.3.4 Real Time Kinematic and Precise Point Positioning

There are two varieties of the differential mode: Precise Point Positioning (*PPP*) and Real Time Kinematic (*RTK*). They are both supposed to apply reference station networks, phase ranging or code measurements smoothed by phase ones (2.123–2.125).

The *PPP* is based on using high-precision differential corrections to ephemerides, and to deviation of a satellite Reference Oscillator. These corrections are determined by separate facilities, such as analytical data processing centers of the local or global measurement networks (IGS, CORS, etc.). Therefore, to practically implement the *PPP*, one should provide a user with a quick access to corresponding data storage servers through a communication channel.

Upon accounting for the above high-precision differential corrections, one uses synchronous carrier-phase measurements at two GNSS operation frequencies to remove ionospheric ranging error in the user receiver. After performing this operation, the corrected phase range to an *i-th* satellite (L_i) will have the form

$$L_i = \frac{L_{1,i} f_1^2}{f_1^2 - f_2^2} - \frac{L_{2,i} f_2^2}{f_1^2 - f_2^2} = \rho_i + c\Delta t + \Delta\rho_{TR,i} + \left(\frac{\lambda_1 N_{1,i} f_1^2}{f_1^2 - f_2^2} - \frac{\lambda_2 N_{2,i} f_2^2}{f_1^2 - f_2^2} \right), \tag{2.155}$$

where ρ_i is a geometric range between a user and a *i*-satellite; L_1 and L_2 are the phase range measured at f_1 and f_2, and reduced to meters; $c\Delta t$ is a user receiver timescale deviation error; $\Delta\rho_{TR}$ is a tropospheric ranging error; N_1 and N_2 are phase ambiguity parameters at f_1 and f_2, respectively.

For a positioning solution in the *PPP*, there is accumulation of measurements in a user receiver (2.155). Due to accumulating and post-processing, one provides a phase ambiguity resolution, a decrease in the noise non-correlated errors and in a tropospheric ranging error. Such post-processing requires measurement accumulation within several minutes to achieve high positioning accuracy. Thus, the *PPP* may not be addressed as a real-time navigation.

The real-time kinematic mode removes phase measurement ambiguity, ephemeris prediction errors and Reference Oscillator deviation, main part of the ionospheric and tropospheric errors in real time. This is attained by using constantly operating reference stations with a 10–100 km baseline length between the latter. Also, at each instant t_k, processed are the differences in the measured ranges $L_{A,k}$, $L_{B,k}$ on each A-B baseline relative to the same satellite at a reference station and in a user receiver, as Fig. 2.23 illustrates this.

The range to a satellite measured at the initial instant (t_0) at a reference station *A* is:

$$L_{A,0} = \rho_{A,0} + \Delta\rho_{A,0}^S + \Delta R_0^S + \Delta R_0^r, \tag{2.156}$$

where $\rho_{A,0}$ is a pre-computed geometric range from a reference station *A* to a satellite; $\Delta\rho_{A,0}^S$ is a total ranging error to a satellite, caused by the ionosphere and atmosphere effect; ΔR_0^S is a satellite timescale deviation error and satellite position prediction inaccuracy; ΔR_0^r is an error caused by peculiarities of signal receiving and processing at a reference station (receiver timescale deviation, multipathing).

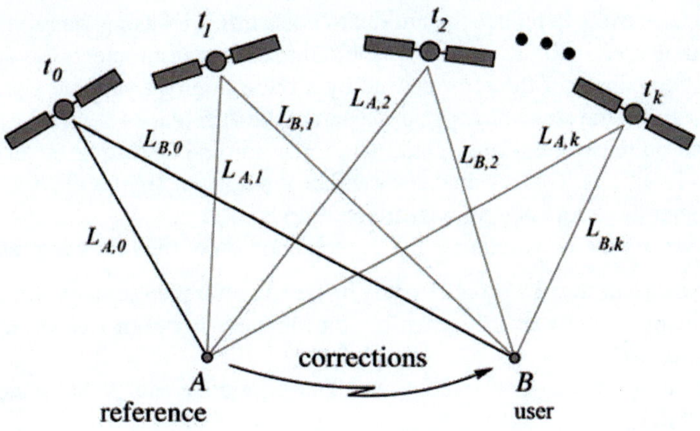

Fig. 2.23 *RTK* implementation principle

The corresponding differential correction for a ranging error, generated at an instant t_0, will be equal to:

$$RC_{A,0} = \rho_{A,0} - L_{A,0} = -\Delta\rho_{A0}^{S} - \Delta R_0^{S} - \Delta R_0^{r}. \tag{2.157}$$

Because this correction will be applied with a delay at an instant *t*, its application requires accounting for time correction degradation. The degradation is accounted for by the Range Rate Corrections (*RRC*) as follows:

$$RC_A(t) = RC_{A,0} + RRC(t - t_0). \tag{2.158}$$

At an instant *t*, one may present the measured range to a satellite in a user location (*B*) as follows:

$$L_B(t) = \rho_B(t) + \Delta\rho_B^S(t) + \Delta R^S(t) + \Delta R^r(t), \tag{2.159}$$

where $\rho_B(t)$ is a pre-computed geometric range from point *B* to a satellite.

Upon applying corrections (2.157–2.158), the corrected range between a satellite and a user location will be equal to:

$$\begin{aligned} L_{B,COR}(t) = \rho_B(t) + RC_A(t) = \rho_B(t) + \left(\Delta\rho_B^S(t) - \Delta\rho_A^S(t)\right) \\ + \left(\Delta R_B^S(t) - \Delta R_A^S(t)\right) + \left(\Delta R_B^r(t) - \Delta R_A^r(t)\right). \end{aligned} \tag{2.160}$$

Owing to measuring by the same satellite, the component $\left(\Delta R_B^S(t) - \Delta R_A^S(t)\right) = 0$. In the case of using the same type of a receiver and of satellite signal processing, one may assume that $\left(\Delta R_B^r(t) - \Delta R_A^r(t)\right) \to min$. However, this assumption is true only in the case, if one may neglect a multipath noise difference in points *A* and *B*. When using phase range measurement, this requirement is, as a rule, fulfilled.

Besides, it is possible to reduce the multipath noise effect by using the satellites with elevation angles > 20–30°, as well as by smoothing code measurements through phase ones. The component of the error caused by a refraction of a satellite signal in the ionosphere and in the atmosphere, that is a part of the difference $\left(\Delta\rho_B^S(t) - \Delta\rho_A^S(t)\right)$, will depend on the value of the spatial and temporal de-correlating factor of those errors (see Table 2.7). This residual error will be the less, the shorter the A-B baseline and the higher the frequency of measurement update.

Implementing the RTK requires the implementation of the following measures:

(1) deploying a network of continuously operating reference stations (CORS) with a sufficient density of filling the region (distance between stations should be from 10 to 100 km);
(2) conjugating signals and intersystem biases of different GNSS systems (see Sect. 2.2.3);
(3) checking the measurement integrity;
(4) providing reliability, performance, and noise immunity of data transmission lines within an RTK system.

To increase the RTK operation quality and decrease the costs for the above measures, nowadays, widely used is integration of high-precision navigation modes within a unified RTK-PPP strategy (Rizos 2003, 2007).

2.4 Common and Specific Requirements for GNSS Positioning

Solving an essentially wide range of problems requires object positioning solution, determining its motion parameters and its spatial-angular orientation. Particular requirements for GNSS users' navigation provision are determined proceeding from problems to solve. These problems are subdivided into the following main groups:

- traffic problems, i.e., the problems related to motion of mobile objects;
- special (including military) problems;
- surveying.

The last group of problems may be attributed to a separate category that features essentially high requirements for the accuracy of determining spatial coordinates, relative travel velocities, and objects' angular orientation. Such positioning solutions are possible only due to measurements of the carrier phase as a fundamental radio-navigation parameter. Herewith, as a rule, one uses the RTK, PPP modes, or a differential mode with measurement accumulating and smoothing. A number of special problems also requires a high-precision determination of coordinates, velocity, and the angular position of objects.

Figure 2.24 shows the structure of problems, to solve which one should involve GNSS, as a fundamental navigation facility (RNP RF 2019). Based on what class of problems needs solving, certain *Required Navigation Parameters* standards are

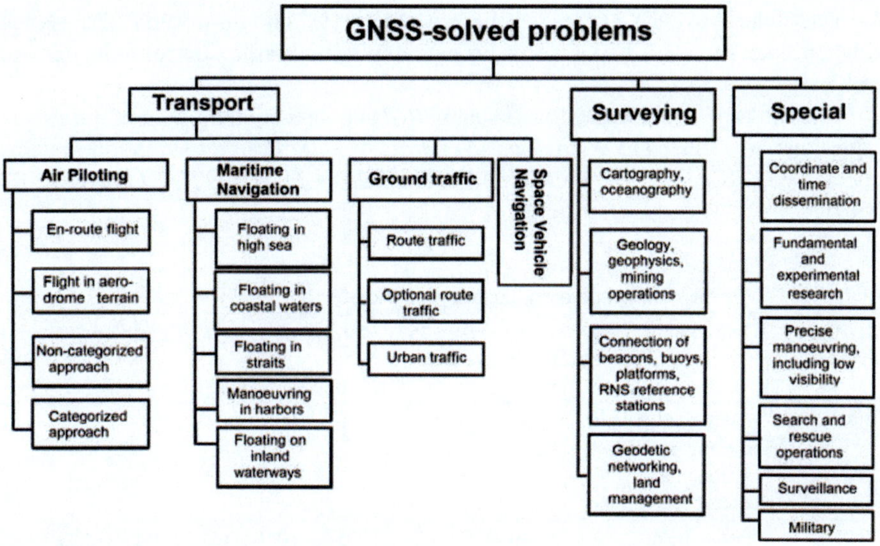

Fig. 2.24 Structure of GNSS-solved problems. Re-produced, modified and translated from Russian based on figure from the open assess source (https://docs.cntd.ru/document/456034328)

established. Each RNP type determines a marginal error for navigation standards, that may be attained due to using all available combinations of navigation information data units (including GNSS hardware) within a set space. RNP may be established for an en-route, a number of en-routes, a region, a volume of aerial or of water space (to plan the motion of air or water traffic). The established RNPs will also determine the requirements for a user navigation receiver and a navigation space infrastructure (parameters for an air flight level, a vehicles' succession time, etc.).

As example of a RNP system arrangement, one may address the problems of air traffic control, that feature the strictest requirements for navigation standards. For such problems, six RNP types are specified. They differ depending on the navigation standard admissible error with a 95% confidence level (DOT-VNTSC-OST-R-15-01 2019):

- *RNP1* is supposed to provide flights with precise aging of the set en-routes of an Air Traffic Flow Management (ATFM) for flights from an aerodrome zone, required en-route and back.
- *RNP4* and *RNP5* are intended for AFTM en-routes beyond aerodrome zones within an inland air space.
- *RNP6*, *RNP10*, and *RNP12* provide a limited en-route optimization in a region with a lower level of navigation provisions, in any controllable air space.
- *RNP20* is a minimal level of requirements, that should be provided by navigation hardware of any aircraft in any controllable air space at any time.

Each of the above RNP types is split into subclasses, with improved *RNPs* as they toughen, like Table 2.10 shows this by example of air traffic control management problem.

From the Table one can see that, in each *RNP* subclass contains a set of numerical values that are referred to as *navigation characteristics* of radio-navigation systems (including GNSS and its applications) (RNP RF 2019, DOT-VNTSC-OST-R-15-01 2019):

Table 2.10 *RNPs* for landing approach, landing, and takeoff (RNP 2019)

RNP type	Operation	Accuracy, m. 95% (lateral/vertical planes)	Integrity/T_{ALARM}	Continuity/T_{MAX}
RNP 1	En-route flight and transition to approach and takeoff stage	± 1853	–	–
RNP 0,5	Initial site of approach, takeoff	± 926	$1.0–1 \cdot 10^{-5}$/1 h	$1.0–1 \cdot 10^{-4}$/1 h
RNP 0,3	Initial or intermediate approach segment, missed approach, takeoff	± 556	$1.0–1 \cdot 10^{-5}$/1 h	$1.0–1 \cdot 10^{-4}$/1 h
RNP 0,3	Instrument approach with guidance in the vertical plain	± 556/38	$1.0–1 \cdot 10^{-5}$/1 h	$1.0–1 \cdot 10^{-4}$/1 h
RNP 0,03	Accurate approach to a runway 100 m (350 feet) by ICAO Category I	± 56/15	$1.0–3.5 \cdot 10^{-7}$/6 s	$1–1 \cdot 10^{-5}$/15 s
RNP 0,02	Accurate approach to a runway 60 m (200 feet) by ICAO Category I	± 37/12	$1.0–3.5 \cdot 10^{-7}$/6 s	$1.0–1 \cdot 10^{-5}$/15 s
RNP 0,01	Accurate approach to a runway 30 m (100 feet) by ICAO Category II	± 19/5	$1.0–2.5 \cdot 10^{-9}$/1 s	$1.0–6.0 \cdot 10^{-5}$/15 s
RNP 0,003	Accurate approach to a runway under 30 m (100 feet), landing and takeoff, as well as guidance at a run by ICAO category III	± 6	$1.0–2.0 \cdot 10^{-9}$/1 s	$1.0–6.0 \cdot 10^{-5}$/30 s

- working zone (coverage area) size;
- positioning accuracy;
- availability;
- integrity;
- service continuity;
- positioning discreteness;
- system capacity.

A working zone (coverage area) is an area of space, in which the required navigation standards are provided for any category of users in any point.

The up-to-date GNSS provide a global working zone, however, the level of navigation characteristics in it is not equal everywhere. So, above, the PDOP factor was shown to increase with a geographic latitude growth. Local effects of multipathing, electromagnetic interference and ionosphere geomagnetic disturbances may also cause disruptions in the global radio navigation field, where the set RNPs will not be ensured. In this case, as well as when solving high-RNP problems, one uses GNSS differential subsystems that produce local working zones of certain size (see Sect. 2.3).

Positioning accuracy is a degree of correspondence between the object measured coordinates and their true value at the same instant. The accuracy quantitative measure is the absolute difference between the measured and the true coordinates:

$$\Delta x_k = X_0 - x_k; \quad \Delta y_k = Y_0 - y_k; \quad \Delta z_k = Z_0 - z_k, \tag{2.161}$$

where X_O, Y_O, Z_O are true, while x_k, y_k, z_k are the measured (computed) at a kth instant the object coordinates in a rectangular geocentric coordinate system.

Normally, the absolute error distribution (2.161) is considered to obey the normal distribution law (Perov and Kharisov 2010). When defining an RNP, the positioning accuracy is set in the form of a confidence interval at the set probability (95% and higher).

Availability defines the ability of a system to achieve a positioning solution. The availability numerical characteristic is the probability for a user to obtain the information on the user's location over a certain time interval and with the required accuracy.

The requirements for availability are defined proceeding from providing the set safety level when solving a specific problem. For example, the highest requirements for availability are produced by air users at an approach and at landing by ICAO categories, by marine and river customers at harbor maneuvering and moving down inland waterways, as well as at automated management of traffic flows and separate vehicles transporting dangerous or valuable cargo.

Integrity specifies the ability of a system to detect its failure and to exclude the possibility of its use in the case, when the system specifications go beyond the tolerance limits. Numerically, a system integrity is estimated by a probability of alerting customers to the system operation breakdown over a time-to-alert time (T_{ALARM}).

The requirements for a GNSS integrity also defined proceeding from a required safety level and with allowance for the dynamics of a particular problem solution. These requirements will be the highest for piloting an aircraft at the landing stage, for solving the problems of automated high-speed ground transportation, and when fulfilling some military problems.

To timely detect and alert to dangerous failures within GNSS segments and their applications, the integrity check is implemented by:

(1) control segment facilities (detection of satellite segment abnormal status);
(2) receiver autonomous integrity monitoring (*RAIM*);
(3) outer integrity check with using differential systems (see Sect. 2.3) and external independent sensors of navigation information.

Service continuity is an ability of a system to function with no breaks with the set performance standards over a certain time interval. The service continuity represents a probability of the system no-failure operation over most important time intervals of a problem solution. Numerically, it is estimated as a threshold time (T_{MAX}) with the set probability of that, during $\Delta t \leq T_{\text{MAX}}$, a system will provide the required navigation characteristics for the given category of users.

Positioning discreteness features a time interval, after which a new positioning by signals of the same GNSS with satisfactory navigation standards is possible.

Capacity features a number of radio-navigation system users, who can be served simultaneously. Due to a push character of radio-navigation parameter measurements, the up-to-date GNSS have an unlimited capacity in the global working zone.

The particular required navigation characteristics to solve various problems are stated in national Radio-navigation Plans of the GNSS countries-operators (RNP 2019, DOT-VNTSC-OST-R-15-01 2019 et al.) Reviewing these requirements shows the presence of a great number of traffic and special problems, to solve which one should provide the accuracy of coordinate determination with an admissible error no worse than 1–10 m (within a 95% confidence interval) in the global working zone. Herewith, the availability should be no worse than 0.999 … 0.9999, and the integrity at 0.999999 … … 0.9999999995, at the admissible time-to-alert $T_{\text{ALARM}} \leq 10$ s.

2.5 Main Problems and Trends in Up-To-Date GNSS

Like above stated, over the last 20 years, there has occurred a full-scale deployment of a number of GNSS, LAASs and WAASs, high-precision navigation network systems, measuring networks, and analytical data processing centers. Apparently, now, we evidence a «quantity-to-quality transition». This implies the formation of an integrated World Wide GNSS based on the above ground and space components. The World Wide GNSS facilities enable to reach an unprecedented accuracy, continuity, and integrity of positioning solutions for any mobile and stationary objects in the global working zone. Along with this, one observes a sharp price reduction for the corresponding services and hardware in the GNSS user segment. By using a set of the above tools enables (already today) to attain a millimeter positioning accuracy when applying the PPP and RTK modes. In the autonomous navigation mode within the global working zone, one reaches the positioning error of about meters at availability of more than 99% coordinate measurements.

Nowadays, one can note the following possibilities to improve the GNSS as per the next trends (Demyanov et al. 2018):

1. expanding GNSS orbital constellations, updating and renovating the satellite fleet;
2. upgrading the frequency-time provision based on new optical frequency standards with a 10^{-15}–10^{-16} relative frequency instability (ground segment) and with a 10^{-14} one for reference oscillators onboard a satellite;
3. upgrading ephemeris provision due to expanding the ground networks of reference stations to perform passive radio and active laser measurements, as well as by intersatellite passive relative measurements via an additional radio channel (Ka-band);
4. optimizing the frequency plan and selecting operational frequencies for future generations of orbital SVs with allowance for increasing the noise immunity and the power potential of RF links;
5. selecting spectrally effective modulations of GNSS signals, that ensure the best noise immunity of a system at an optimal use of the frequency resource;
6. deploying and updating national WAAS, and LAAS systems;
7. expanding reference ground measuring networks and increasing a number of analytical centers to process the measuring information on the frequency-time and ephemeris provisions, ionospheric and tropospheric errors;
8. introducing the e-GNSS, PPP, RTK + PPP technologies with possibilities for users to access free analytical center servers containing actual corrections for GNSS measurement errors through Internet and through open radio channels;
9. improving stability for the operation of GNSS and their applications under the conditions of extreme factors from space weather (this problem will be addressed more explicitly in Chap. 7).

In terms of the World Wide GNSS users, it is important to define economic feasibility and most perspective fields, where, first of all, one should introduce high-precision navigation technologies (PPP, RTK, e-GNSS). First of all, it is recommended to estimate the expediency and potentials from introducing high-precision navigation in such fields as complex systems of public safety; smart transport systems; agriculture, and water resources management; urban infrastructure management and monitoring its current status; sync of national time-and- frequency standards.

Because reference ground networks are the key element that, in the future, may provide a "global" navigation with a millimeter error level, then, it is important to solve two issues. First, what, ultimately, an optimal density of reference station arrangement should be within the integrated global reference network. Secondly, is it expedient to place reference networks and stations throughout the entire territory of the Earth, or is there an optimal configuration of the reference network and stations, at which the concept of the "global" high-precision navigation may be best implemented at optimal costs?

To provide a high-precision mode of differential or relative navigation, as well as precise determination of necessary geodetic and geo-information parameters, one should use a network of reference stations with 10–100 km spacing. To cover the entire Earth surface with such a dense network of reference stations, almost unacceptable number of reference stations in terms of operational costs for their deployment and maintenance. In general, solving this issue has remained open so far.

The best solution is, obviously, in a further expansion of "private" CORS networks with solving the issues of creating global shared high-precision database and implementing (based on this) high-precision navigation technologies via Internet (e-GNSS). But here, also, one should solve a number of accompanying issues, the main of which are:

- "property right" to products of each "private network";
- balance between "national" and "global" interests;
- complication of measurement algorithms and models when processing multi-system and multi-network measurements;
- further toughening of requirements for the frequency-time and ephemeris provision;
- necessity to unify corresponding technical and structural characteristics for separate "national" satellite radio navigation systems.

One of the World Wide GNSS relevant trends is further development of applications (WAAS and LAAS) with integrating their possibilities within the world-wide GNSS. Initially, those applications were produced for aviation (navigation within an aerodrome terrain, precise landing approach and landing by the ICAO categories). Under the ICAO requirements, a positioning error is to be about tenths of meter. Nowadays, there is a specific «blank spot» within a peak-to-peak error that are provided by involving differential systems (decimeters) and by high-precision navigation (millimeters).

The first question to answer is: can one provide centimetric and millimeter accuracy by using the WAAS navigation? The answer largely depends on how successfully one will be able to solve the problem of the space weather effects on the WAAS operations. The second question arises from the first: what is a circle of potential users of such a service and will it be economically expedient to further develop WAASs? The third question is: is there an expediency to produce new national WAASs in the countries that have not possessed the latter at present? In general, solving the above issues also has remained open awhile.

Possible alternatives to WAASs and, even, to LAASs are:

- using CORS networks as reference differential networks;
- using e-GNSS facilities and databases;
- developing and promoting onto the market low-cost two- and multi-frequency user receivers, as well as midget phased arrays.

Implementing these measures enables to efficiently offset most significant ionospheric ranging errors and multipath noise in "urban canyons" and on a rugged terrain.

Another important problem that needs solving efficiently is in producing the algorithms for the "inner" and "outer" GNSS integrity check. Here, there are difficulties in offsetting positioning errors related to intersystem biases (see Subsect. 2.2.3). This problem may arise at synchronous measurements by different-GNSS satellite signals within a GNSS user the multi-system receiver. To solve this problem, one should unify the performance specifications and calibration procedures for multi-frequency and multi-system user receivers. Besides, it is necessary to develop algorithms for the inner integrity check within a user receiver. The algorithms are to efficiently detect:

- "smooth" anomalies in the navigation function system, that are caused by a monotonous ranging error accuracy divergence among measurements by different-GNSS signals;
- essential ranging unequality in the navigation function system, that emerges owing to the multipath effect when using signals with different modulations and frequencies.

One more problem is a divergence of national mobile geodetic coordinate systems (WGS-84, PZ-90). The divergence is caused by a regular continental motions and the polar shift. Annual systems corrections for the national coordinate system parameters result in additional costs that may be avoided at a transition to the concept of the World Wide GNSS. There emerges an important international problem related to creating or selecting a world universal coordinate system. Apparently, the problem will be solved by addressing the International Terrestrial Reference Frame (ITRF). This coordinate system is maintained by the international community based on the measurements at the world net of reference stations by involving Satellite Laser Ranging (SLR), Very Long Baseline Interferometry (VLBI), existing GNSS, analytical data processing centers and International Earth Rotation and Reference Systems Service (IERS).

Note separately the highlights related to the program of updating and developing the American GPS, as a "fashion-maker" in designing and building all the remaining GNSS segments and systems. GPS is traditionally regarded as a leading system in implementing the future World Wide GNSS. Therefore, the USA administration and the corresponding departments are interested in the uttermost integration of GPS and its applications into the World Wide GNSS. All the plans in the further GPS and WAAS development are made with allowance for the above. The other participants of the world navigation systems are encouraged to integrate in it. The main trends in further GPS updating (that will underlie the World Wide GNSS development) for the near future will, apparently, be as follows:

1. developing new generation of user receivers that should provide robust receiving and processing the signals from all national GNSS in operation (GLONASS, Galileo, BeiDou, QZSS, NavIC);

2. matching the characteristics for separate national GNSS in terms of rational use of frequency bands, suppression of inter-frequency and inter-system noise, positioning solution robustness under the conditions of unequal-accuracy measurements and inter-system biases;
3. developing further orbital constellations by commissioning new-generation orbital blocks broadcasting signals at the frequencies and having such a structure that ensures their easy integration into the World Wide GNSS;
4. solving cybersecurity problems, because more and more *e*-GNSS technologies are planned to be applied;
5. intensifying and deepening R&D to improve the frequency-time and ephemeris provisions;
6. involving facilities to monitor the Sun and the near-Earth space conditions in order to generate actual alerts to violations of the GNSS integrity and their differential applications;
7. involving the data space weather services to generate correcting information from analytical centers that process ground reference network measurements.

Apparently, one should expect further development of national GNSS segments and their merging into a worldwide navigation system within the above concept.

References

Aarons J (1982) Global morphology of ionospheric scintillations. Proc IEEE 70(4):360–378. https://doi.org/10.1109/PROC.1982.12314

Afraimovich EL, Perevalova NP (2006) GPS—monitoring of the Earth's upper atmosphere. Irkutsk, GU NC RVH VSNC SO RAMN, p 480. ISBN 5-98277-033-7 (in Russian)

Akasofu SI, Chapman S (1972) Solar-terrestrial physics. Chapter 2. Oxford University Press, p 901

Annex 10 to the Convention on International Civil Aviation (2006) Aeronautical telecommunications. Radio navigation aids, vol I, 6th edn. ICAO. https://www.pilot18.com/wp-content/uploads/2017/10/Pilot18.com-ICAO-Annex-10-Volume-1-Radio-Navigation-Aids.pdf

BeiDou navigation satellite system signal in space interface control document (2020) Precise point positioning service signal PPP-B2b (version 1.0). China satellite navigation office, 02.2019. BDS-SIS-ICD-PPP-B2b-1.0

BeiDou navigation satellite system signal in space interface control document (2017) Open service signal B2a (version 1.0). China satellite navigation office, 02.2019. BDS-SIS-ICD-B2a-1.0

BeiDou navigation satellite system signal in space interface control document (2017) Open service signal B1C (version 1.0). China satellite navigation office, 02.2019. BDS-SIS-ICD-B1C-1.0

BeiDou navigation satellite system signal in space interface control document (2018) Open service signal B3I (version 1.0). China satellite navigation office, 02.2019. BDS-SIS-ICD-B3I-1.0

BeiDou navigation satellite system signal in space interface control document (2019) Open service signal B1I (version 3.0). China satellite navigation office, 02.2019. BDS-SIS-ICD-B1I-3.0

Brunner F, Welsch W (1993) Effect of the troposphere on GPS measurements. GPS World Mag 4(1)

Bruyninx C, Habrich H, Söhne W, Kenyeres A, Stangl G, Völksen C (2012) Enhancement of the EUREF permanent network services and products. Geodesy for planet earth. IAG symposia series, vol 136, pp 27–35. https://doi.org/10.1007/978-3-642-20338-1_4

Chen Z, Gao Y, Liu Z (2005) Evaluation of solar radio bursts' effect on GPS receiver signal tracking within international GPS service network. Radio Sci 40:RS3012. https://doi.org/10.1029/2004RS003066

Conker RS, El-Arini MB, Hegarty CJ, Hsiao T (2003) Modeling the effects of ionospheric scintillation on GPS/satellite-based augmentation system availability. Radio Sci 38(1):1001. https://doi.org/10.1029/2000RS002604

Crane RK (1977) Ionospheric scintillation. Proc IEEE 65(2):180–199

Demyanov V, Zhang X, Lu X (2019) Moderate geomagnetic storm condition, WAAS alerts and real GPS positioning quality. J Atmos Sci Res 2(1):10–23. https://doi.org/10.30564/jasr.v2i1.343

Demyanov VV, Konuyshkin GY, Khlyzova IV (2010) Systematic changing in positioning errors for GNSS users in different latitudes. Aviakosmicheskoye Priborostroenie. 7:9–16. (in Russian)

Demyanov VV, Imarova OB, Garmysheva ES, Yasyukevich YV (2018) Trends in the development of GNSS technologies and directions of their application in transport. Mod Technol. Syst Anal. Model. https://doi.org/10.26731/1813-9108.2018.2(58).82-90

Dow JM, Neilan RE, Rizos C (2009) The international GNSS service in a changing landscape of global navigation satellite systems. J Geodesy 83:191–198. https://doi.org/10.1007/s0019000803003

EGNOS safety of life service (SoL) SDD, issue 3.1 (2016) European GNSS agency. https://doi.org/10.2878/181522

El-Mowafy A (2014) GNSS multi-frequency receiver single-satellite measurement validation method. GPS Solutions 18(4):553–561. https://doi.org/10.1007/s10291-013-0352-6

European GNSS (GALILEO) open service signal in space interface control document (2021) Issue 2.0. OS SIS ICD. European GNSS agency

Exertier P, Belli A, Samain E, Meng W, Zhang H, Tang K, Schlicht A, Schreiber U, Hugentobler U, Prochàzka I, Sun X, McGarry JF, Mao D, Neumann A (2018) Time and laser ranging: a window of opportunity for geodesy, navigation, and metrology. J Geodesy. https://doi.org/10.1007/s00190-018-1173-8

Federal Radionavigation Plan (2019) DOT-VNTSC-OST-R-15-01. Springfield, Virginia

Freymueller JT, AEIC (Kodiak) (2014) The GAGE facility operated by UNAVCO, Inc., GPS/GNSS observations dataset (2017). https://doi.org/10.7283/T5X928QS

Global Navigation Satellite System GLONASS (2002) Interface control document: ICD GLONASS. 5th edn. Russian Space Systems. (in Russian)

Global Navigation Satellite System GLONASS (2016) Interface control document. Code division multiple access service navigation signal in L1 frequency band. ICD GLONASS CDMA L1, 1 edn

Global Navigation Satellite System GLONASS (2016) Interface control document. Code division multiple access service navigation signal in L3 frequency band. ICD GLONASS CDMA L3, 1 edn

Global Navigation Satellite System GLONASS (2016) Interface control document. General description of code division multiple access signal system. ICD GLONASS CDMA general description, 1 edn

Global Positioning System Wing (GPSW) (2010–06) Systems engineering and integration interface specification. IS-GPS-705 A

Global positioning systems directorate systems engineering and integration interface specification. IS-GPS-200 J

GNSS data assembly center for the global sea level observing system—SONEL. https://www.sonel.org/

Gonorovsky IS (1986) Radio chains and signals. Moskow, Telecommunications, p 446 (in Russian)

Global positioning system wide area augmentation system (WAAS) performance standard. GPS-WAAS-PS (2008–10)

Gurtner W, Estey L (2007) RINEX: the receiver independent exchange format version 2.11. Bern, Astronomical Institute, University of Bern. ftp://igs.org/pub/data/format/rinex211.txt

Gurtner W, Estey L (2009) RINEX: the receiver independent exchange format version 3.01. Bern, Astronomical Institute, University of Bern. https://files.igs.org/pub/data/format/rinex301.pdf

Hakansson M, Jensen ABO, Horemuz M, Hedling G (2017) Review of code and phase biases in multi-GNSS positioning. GPS Solut 21:849–860. https://doi.org/10.1007/s10291-016-0572-7

Herring T, Melbourne T, Szeliga W, Murray M and Phillips DA (2017) UNAVCO GAGE GPS Data Analysis Plan and Products. Updated 12 Sept 2017. https://www.unavco.org/data/gps-gnss/derived-products/docs/GAGE_GPS_Analysis_Plan_20170912.pdf

Interface Control Document (1993) ICD-GPS-200 C

IS-QZSS (5 Nov 2018) Quasi-Zenith satellite system interface specification centimeter level augmentation service (IS-QZSS-L6-001). Available at https://qzss.go.jp/en/technical/download/pdf/ps-is-qzss/is-qzss-l6-001.pdf

ISRO-IRNSS-ICD-SPS-1.1. IRNSS SIS ICD for standard positioning service. V. 1.1. August 2017. Available at https://www.isro.gov.in/sites/default/files/irnss_sps_icd_version1.1-2017.pdf

Iwabuchi T, Miyazaki S, Heki K, Naito I, Hatanaka Y (2003) An impact of estimating tropospheric delay gradients on tropospheric delay estimations in the summer using the Japanese nationwide GPS array. J Geophys Res 108(D10):4315. https://doi.org/10.1029/2002JD002214

Jakowski N, Hoque MM (2019) Estimation of spatial gradients and temporal variations of the total electron content using ground-based GNSS measurements. Space Weather 17:339–356. https://doi.org/10.1029/2018SW002119

Jayachandran PT, Langley RB, MacDougall JW, Mushini SC, Pokhotelov D, Hamza AM, Mann IR, Milling DK, Kale ZC, Chadwick R, Kelly T, Danskin DW, Carrano CS (2009) Canadian high arctic ionospheric network (Chain). Radio Sci 44(1):RS0A03. https://doi.org/10.1029/2008RS004046

Kaplan ED (1996) Understanding GPS: principles and applications. Artech House Publisher, Boston, London, p 556

Klobuchar JA, Kunches JM, Van Dierendonck AJ (1999) Eye on the ionosphere: potential solar radio burst effects on GPS signal to noise. GPS Solutions 3(2):69–71

Konno H, Pullen S, Luo M, Enge P (Jan 2005) Analysis of ionosphere gradient using Japan GEONET data. In: Proceedings ION-NTM-2005, institute of navigation, San Diego, CA, pp 1118–1129

Linty N. (2010) Codeless tracking algorithms for GNSS software receivers: Tesi di Laurea Magistrale, N. Linty.—Torino. https://nanopdf.com/queue/pdf-positioning-location-and-navigation-group_pdf?queue_id=-1&x=1663588338&z=MTk0LjE0My4xNDguMjc=

Minimum Operational Performance Standards for GPS/GLONASS/ABAS airborne equipment (Draft 2) (2014) MOPS GPS/GLONASS/ABAS

Monteiro LS, Moore T, Hill C (2005) What is the accuracy of DGPS? J Navig 58:207–225. https://doi.org/10.1017/S037346330500322X. The Royal Institute of Navigation

Mylnikova AA, Yasyukevich YV, Kunitsyn VE, Padokhin AM (2015) Variability of GPS/GLONASS differential code biases. Results Phys 5:9–10. https://doi.org/10.1016/j.rinp.2014.11.002

Navstar GPS space segment/navigation user interfaces. ICD-GPS-200 C.1993-10

Padokhin AM, Mylnikova AA, Yasyukevich YV, Morozov YV, Kurbatov GA, Vesnin AM (2021) Galileo E5 AltBOC signals: application for single-frequency total electron content estimations. Remote Sens 13:3973. https://doi.org/10.3390/rs13193973

Palamartchouk K (2010) Apparent geocenter oscillations in Global navigation satellite systems solutions caused by the ionospheric effect of second order. J Geophys Res 115:B03415. https://doi.org/10.1029/2008JB006099

Perov AI, Kharisov VN et al (2010) GLONASS: principles of the construction and functioning, 4th edn. Moscow, Radiotechnika, p 800. (in Russian)

Povalyaev AA (2008) Satellite radio-navigation systems: time, clock, measurements and relative positioning. Moscow, IPRZhR, p 328. (in Russian)

Psiaki ML, Powell SP, Jung H, Kintner Jr PM (Sept 2003) Design and practical implementation of multi-frequency RF front ends using direct RF sampling. In: Proceedings of the 16th international

technical meeting of the satellite division of the institute of navigation (ION GPS/GNSS 2003), Portland, pp 90–102
Radio-navigation plan of Russian Federation. RNP RF. Moskow (2019). https://rntfnd.org/wp-content/uploads/CIS-Russia-Radionav-Plan-2019-2024.pdf
Ray JK (2000) Mitigation of GPS code and carrier phase multipath effects using a multi-antenna system. Calgary, Alberta, p 286
Report C#337/13 evaluation of the standard ephemerides and time support of the global positioning systems (2013) Analytical center of central institute of the machinery production. Korolev city. (in Russian)
Riehle F (2017) Optical clock networks. Published online: 3 Jan 2017. https://doi.org/10.1038/nphoton.2016.235
Rizos C (2003) GNSS interoperability: future or fantasy? GPS World 14:24–25
Rizos C (2007) Alternatives to current GPS-RTK services and some implications for CORS infrastructure and operations. GPS Solutions 11:151–158. https://doi.org/10.1007/s10291-007-0056-x
Saito S, Sunda S, Lee J, Pullen S, Supriadi S, Yoshihara T, Terkildsen M, Lecat F (2017) Ionospheric delay gradient model for GBAS in the Asia-Pacific region. GPS Solutions 21:1937–1947. https://doi.org/10.1007/s10291-017-0662-1
Schaer S (2012) Overview of GNSS biases. International GNSS service. Workshop on GNSS biases. [Electronic resource]. http://www.biasws2012.unibe.ch/pdf/bws12_1.3.1.pdf
The System of Differential Correction and Monitoring (SDCM) (2011) Interface control document: ICD SDCM, 1st edn. Russian Space Systems (in Russian)
Teunissen PJG, Kleusberg A (1998) GPS for geodesy. Springer, p 650
Tossaint M, Samson J, Toran F, Ventura-Traveset J, Hernandez-Pajares M, Juan JM, Sanz J, Ramos-Bosch P (2007) The stanford—ESA integrity diagram: a new tool for the user domain SBAS integrity assessment. Navig, Inst Navig 54(2):153–162. https://doi.org/10.1002/j.2161-4296.2007.tb00401.x
Vdovin VS, Dvorkin VV, Karlik AP et al (2018) Problems and perspectives of the Russian active geodetic networks improvement and their integration into ITRF. Vestnik SGUGiT, vol 23. # 2., pp 6–27. (in Russian)
Villiger A, Dach R (eds) (2021) International GNSS service technical report 2020 (IGS annual report). Bern Open Publishing, IGS Central Bureau and University of Bern. https://doi.org/10.48350/156425
Walter T, Blanch J, Rife J (2004) Treatment of biased error distributions in SBAS. Presented at the international symposium on GNSS/GPS, Dec, 2004, Sydney, Australia [Electronic resource].—Stanford, 2004. http://web.stanford.edu/group/scpnt/gpslab/pubs/papers/Walter_ISGNSS_2004.pdf
Warren DLM, Raquet JF (2003) Broadcast versus precise GPS ephemerides: a historical perspective. GPS Solutions 7:151–156. https://doi.org/10.1007/s10291-003-0065-3
Wide-area augmentation system performance analysis report, Report #53. Reporting period: April 1 to June 30, 2015. Available from: http://www.nstb.tc.faa.gov/DisplayArchive.htm
Xu G (2007) GPS theory, algorithms and applications. 2nd edn. Springer Berlin Heidelberg, New York, p 354. ISBN 978-3-540-72714-9
Yakovlev OI (1988) Space radio-physics. Voronezh, Science Book Ed, p 432. (in Russian)
Yarlykov MS (1980) Aviation radio-navigation devises and systems. Moskow, N.E. Zhukovsky Military Aviation Engineering Academy, p 235. (in Russian)
Yasyukevich Y, Mylnikova A, Vesnin A (2020) GNSS-based non-negative absolute ionosphere total electron content, its spatial gradients, time derivatives and differential code biases: bounded-variable least-squares and taylor series. Sensors 20(19):5702. https://doi.org/10.3390/s20195702
Yasyukevich YV, Mylnikova AA, Kunitsyn VE, Padokhin AM (2015) Influence of GPS/GLONASS differential code biases on the determination accuracy of the absolute total electron content in the ionosphere. Geomagnetism and Aeronomy 55(6) 763-769 https://doi.org/10.1134/S001679321506016X

Chapter 3
Space Weather General Concepts

Maria A. Sergeeva

3.1 Brief Introduction to Phenomena Related to Space Weather

Space Weather or Space Meteorology is a concept that unites a combination of phenomena in the near-Earth space which can have an effect on modern technology. It includes time-variable conditions and dynamic processes on the Sun, in the solar wind and interplanetary medium, in the Earth's magnetosphere, atmosphere, thermosphere and ionosphere (Koskinen et al. 2001; Wright et al. 1997). This chapter provides a brief introduction to the subject of Space Weather; several concepts are discussed further in the book in more detail.

Let us briefly introduce some phenomena important in regard to Space Weather. The *Sun* (Ambastha 2020; Jenkins 2013; Hanslmeier 2007) continuously emits flows of electromagnetic radiation and charged particles. Magnetic fields of the Sun are also transported by solar wind filling the interplanetary space. A *solar wind* is a turbulent outflow of ionized gas (plasma) that is continuously extending in all directions from the Sun into interplanetary space (Baker 1996). The solar wind consists of ions and electrons. Its parameters can significantly vary depending on the phenomena on the Sun (especially during eruptive events). Its average velocity is 450 km/s, but can vary in a wide range. The interaction of the solar wind with the geomagnetic field forms the magnetosphere. The direction of the interplanetary magnetic field (IMF) is an important parameter: if the Bz-component of IMF is southward, the IMF force lines reconnect with Earth magnetic field lines letting the solar wind and particles enter the magnetosphere and cause geomagnetic disturbances. The reader is referred to works (Campbell 1997; Nagatsuma 2002; Borovsky and Valdivia 2018) for the detailed description of the processes.

Dark spots on the Sun's surface, which contain very strong magnetic fields, are called *sunspots*. They are colder than the surrounding plasma. Sunspots tend to appear in groups and can live from some days to several Sun rotations (Davies 1965; Vaquero 2007 and references therein). The more sunspots on the Sun, the greater the probability of intense solar events. The sunspot number (SSN) is an important

V. Demyanov et al., *Space Weather Impact on GNSS Performance*,
https://doi.org/10.1007/978-3-031-15874-2_3

parameter whose variation has a periodicity of about 11 years, though some individual cycles may be shorter or longer (Ivanov 2021). *SSN* is obtained from observations of the Sun by optical instruments (telescopes). Its variation defines the quasi-periodic change in the solar activity manifestations called *solar cycle* during which the low solar activity is followed by the high solar activity and vice versa (Hathaway 2015; Obridko et al. 2020). The average annual SSN is mostly used as an indicator of solar cycle evolution. Extended areas on the Sun characterized by the intense magnetic fields that are seen as local brightening in the ultraviolet (UV) and X-ray frequencies are referred to as *active regions* (Bruzek 1972; Jaeggli and Norton 2016; Zhang et al. 2010 and references therein). Four digits numbers are consecutively assigned to active regions as they are observed on the solar disk since 1972 by the National Oceanic and Atmospheric Administration, U.S. (NOAA) (previously by different observatories individually). As the 10,000 value had been reached in 2002, the first digit has been omitted from then. If an active region survives more than one solar rotation and reappear, it is numbered again. An active region is a set of changing structural formations in a certain limited area of the solar atmosphere and the major source of various solar activities mentioned further. Sunspots develop in active regions. Figure 3.1 illustrates two sunspots observed in the solar disk on October 3–4, 2017. H-Alpha is the strongest spectral line of the solar spectrum, one of those used for monitoring of the Sun.

A *solar flare* is a sudden large explosion on the Sun with a release of energy characterized by the emission of electromagnetic radiation at all wavelengths and

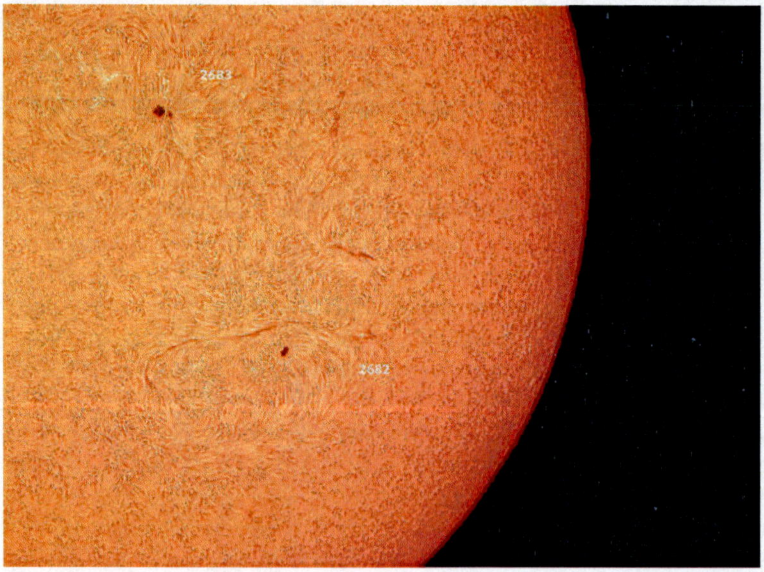

Fig. 3.1 Image of the solar chromosphere in H-Alpha (6562.8 Å) provided by the Laboratory of Geo-Spatial Sciences (LACIGE), Morelia, Mexico (Courtesy of Dr. M. Rodríguez-Martínez)

Fig. 3.2 An example of the Sun image in H-Alpha (6562.8 Å) provided by LACIGE (Courtesy of Dr. M. Rodríguez-Martínez). The solar flare is observed at the east edge of the solar disk

a rapid increase of brightness of part of the Sun's surface. An example is given in Fig. 3.2. A large amount of charged particles and significantly increased electromagnetic radiation (X-rays, extreme UV, gamma rays, infrared, visible light, microwaves, radiowaves) are erupted from a small region on the Sun. Such bursts significantly affect the state of solar wind and eventually impact the Earth. Solar flares occur in active regions and are associated with strong localized magnetic fields in sunspots (Cander 2019). However, not every active region leads to the formation of flares. The sunspots shown in Fig. 3.1 did not cause any strong solar flares.

Flares are classified according to their intensity (e.g. (Bai and Sturrock 1989)) and impact on Earth. The X-ray brightness wavelength (1 to 8 Angstroms) defines five categories of flares (from the weakest to the strongest): A, B, C, M, X (Hanslmeier 2007, Schrijver 2009). These categories are subdivided using the numbers from 1 to 9. The scale is logarithmic. For instance, M7 is 10 times weaker than X7. The most intense X-class flares have no upper limit of intensity.

Solar flares are usually detected through X-rays observations. The X-ray and Extreme UV sensors installed on GOES satellites (Thiemann et al. 2019; Machol et al. 2020) provide us with this information. Figure 3.3 shows the example of the solar X-ray flux variation with two intense flares (right part of the Fig.): M8.1 flare at 7:40 UT on September 8, 2017, and X8.2 class flare at 15:35 UT on September 10, 2017

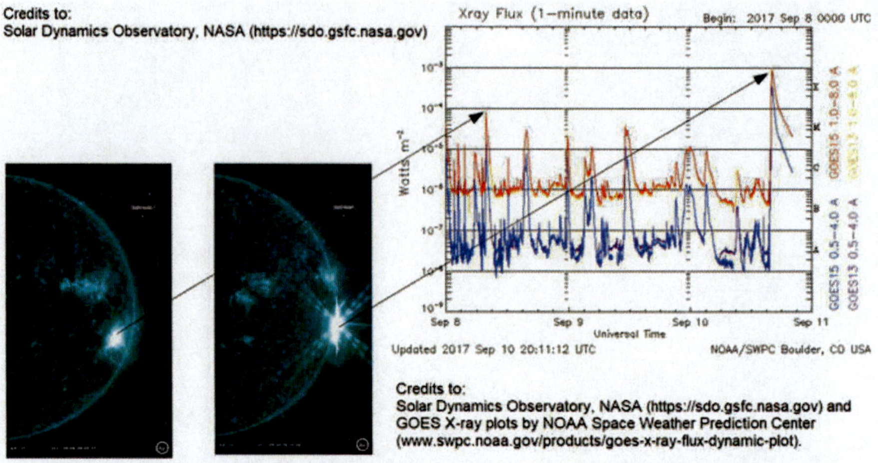

Fig. 3.3 Solar X-ray flux detected using GOES satellites. (Modified from Mexican Space Weather Service (SCiESMEX) weekly report of Space Weather conditions freely available at www.sciesmex.unam.mx/static/media/uploads/reporte-clima-espacial-2017-09-14.pdf)

(Courtesy of GOES X-ray plots by NOAA Space Weather Prediction Center freely-available by following the link: www.swpc.noaa.gov/products/goes-x-ray-flux-dynamic-plot). Their corresponding flare-associated active regions can be seen in the SDO satellite images of the Sun (left part of the Fig.) taken in a wavelength of extreme UV light (Courtesy of NASA/SDO and the AIA, EVE, and HMI science teams).

Under favourable observation conditions (the corresponding active region on the Sun is facing the Earth), an intense flare (mostly X-class, sometimes M-class) can cause significant ionospheric disturbances on the day-side (sunlit face) of Earth.

Flares are more likely during periods of high solar activity. Often they are accompanied by *coronal mass ejections* (CME). These are the significant releases of plasma and magnetic fields from the solar corona. This material perturbs the solar wind. When plasma and magnetic fields arrive at Earth's space environment, they can interact with the geomagnetic field disturbing the stability of the Earth's magnetosphere (Temmer 2021).

Solar radio bursts are enhanced radio emissions from the Sun that can interfere with communication, radar, and navigation systems (Giersch et al. 2017 and references therein). These "noise storms" are associated with strong solar flares and certain geophysical disturbances. They can last for hours or days (Davies 1965). The issue is discussed in more detail in the next section.

A *coronal hole* is a region of the solar corona with a locally open solar magnetic field. It can be seen as a dark area (meaning cooler and less dense) in X-rays and UV light. An example is given in Fig. 3.4: the large coronal hole is observed at the South Pole extending to mid-latitudes and facing towards the Earth. The magnetic field lines from this hole expand in the solar wind and do not return back as in

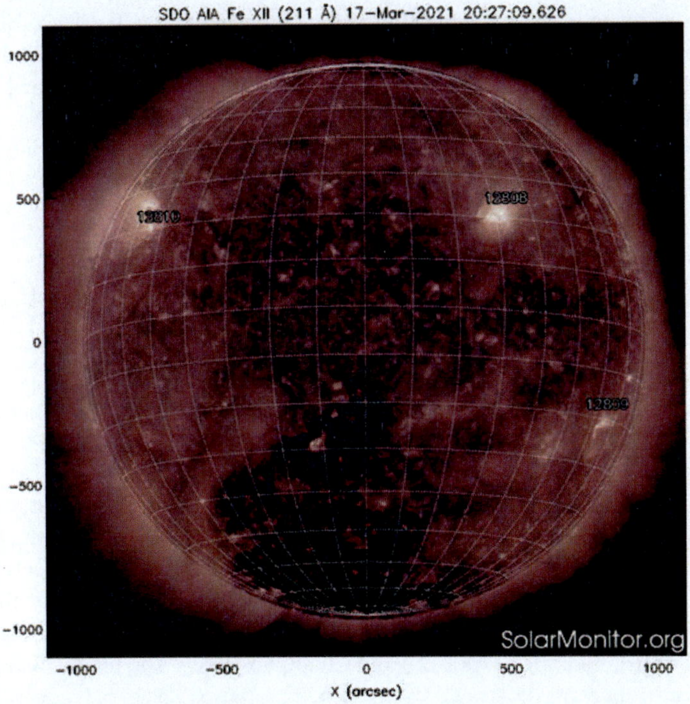

Fig. 3.4 The Sun in soft X-rays (211 Å) on March 17, 2021. Data supplied courtesy of solarmonitor.org (Gallagher et al. 2002) and SDO/HMI and SDO/AIA consortia

other regions. Coronal holes are the source of fast solar wind streams. *Corotating Interaction Regions* (CIR) are the regions of compression that are formed when the high speed solar wind catches up with the slower stream and their interaction results in plasma compression and magnetic field intensification (Kunow 2001). The slower solar wind originates in CME and the faster wind in coronal holes. The concepts of CME and CIR and their development are very complex; please refer to (Richardson 2018; Jian et al. 2006; Howard 2014; Andreeva and Malashchuk 2020). Both can trigger Space Weather disturbed conditions.

There is always a background flux of galactic cosmic rays (mostly, protons and ions). Sometimes energetic particle events occur. *Solar energetic particles* (SEP) are the particles (mainly protons, neutrons, electrons and alpha particles) of high energies (10^3 eV–10^{10} eV) coming from the Sun. They are caused in some intense flare events (accelerated by magnetic reconnection) and shock waves from CMEs. Usually, it takes them from ~ 15 min to a few hours to reach the Earth's orbit. The high-latitude regions of the Earth are more vulnerable to particle precipitations due to geomagnetic field force lines configuration.

The list of books and articles that discuss the solar-terrestrial relationships and the near-Earth environment in regard to Space Weather is very long. Let us mention a few works to which the reader may refer (Hargreaves 1979, 1992; McNamara 1994; Kamide and Chian 2007; Prölss 2011, 2012; Carovillano and Forbes 1983; Singh et al. 2021; Crosby et al. 2006; Cander 2019; Pulkkinen 2007; Khazanov 2016; Tsurutani et al. 2020; Moldwin 2008b; Schwenn 2006).

Further, we cover the topic of solar radio bursts in a little more detail as it is important for the subject of this book.

3.2 Introduction to the Sources and Main Characteristics of Solar Radio Bursts

The Sun is a source of a continuous radiation in a wide range of frequencies from units of MHz to 30 GHz and higher. The intensity of this radio emission can increase drastically during solar flare events and active region presence. These increases are referred as Solar Radio Bursts. Different physical mechanisms can be responsible for solar radio emission: accelerated and non-uniform motion of solar plasma electrons, collision of electrons with each other and with ions or spiral motion of electrons under the action of the force lines of the magnetic field of the Sun. Consequently, the sources for solar radio bursts are thermal radio emission of plasma, plasma oscillations and gyro-synchrotron radiation (Akasofu and Chapman 1972). The mechanisms of these sources occurrence are illustrated in Fig. 3.5.

Thermal radio radiation is excited in plasma within the process of solar plasma electron and proton collisions within the active regions on the Sun which have caused a solar flare. Another mechanism of radio emission is related to the high-energy electrons rising from the surface of the Sun. In the last case, the energy of Cherenkov plasma waves is dissipated in the plasma cloud which escaped from the flare due to the reconnection of magnetic field lines.

Plasma oscillations occur as a consequence of shock wave propagation. This wave forms during a solar flare and propagates through the Sun's magnetosphere filled with plasma.

Another plasma oscillations source is a prominence propagation during the process of its eruption. The wave-like oscillations of the plasma, and therefore of the electrons, can act as a traveling source of radio emission.

Gyro-synchrotron emission is attributable to the fact that the high-energy electrons from plasma emissions during solar flares are trapped by the magnetic field of sunspots. These electrons spiral movement along the magnetic field force lines (Fig. 3.5) generates the gyro-synchrotron emission. The polarization of such emission due to the wave excitation mechanism is generally circular. Due to the unequal influence of the solar magnetized plasma on the propagation of the ordinary and extraordinary components of the wave, the polarization sign and the dominant wave mode depend on the direction of rotation of the electrons with reference to the magnetic field line.

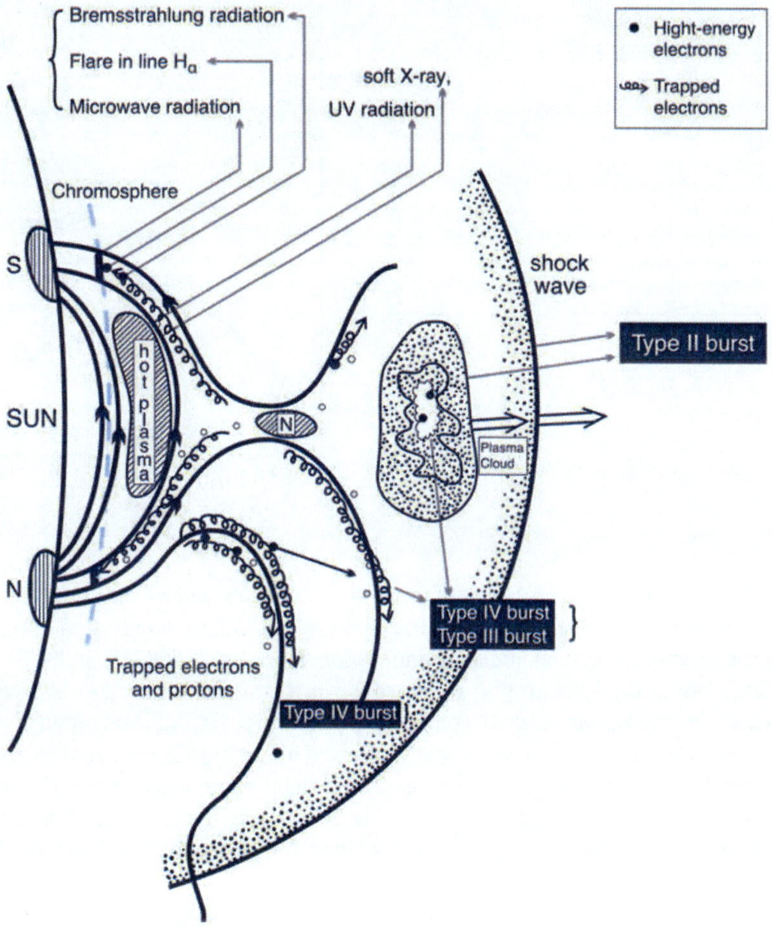

Fig. 3.5 Schematic representation of processes in the Sun during the typical solar storm. Reproduced and modified based on the figure published in the book (Akasofu and Chapman 1972)

The Sun's surface outside the active regions emits a background radio emission, the source of which is the bremsstrahlung radiation of thermal electrons of the solar plasma (thermal radio emission). The frequency of such emission $F_{SF} \geq f_p$ (f_p is plasma frequency). As plasma frequency differs significantly at different regions in the Sun, thermal radio emissions of different wave frequencies is generated. Meter waves of background radio emission are generated in the solar corona, decimeter waves in the transition region, and centimeter waves in the solar chromosphere. The intensity of radio emission depends on the equivalent brightness temperature of the emitting region.

According to (Bruzek 1972), the level of the background thermal radio emission of the Sun within the operating frequency range of GNSS is rather low ~ 50 s.f.u. (s.f.u. is a unit of solar radio emission density flux 1 s.f.u. = 10^{-22} W/(m^2•Hz)) (Fig. 3.6).

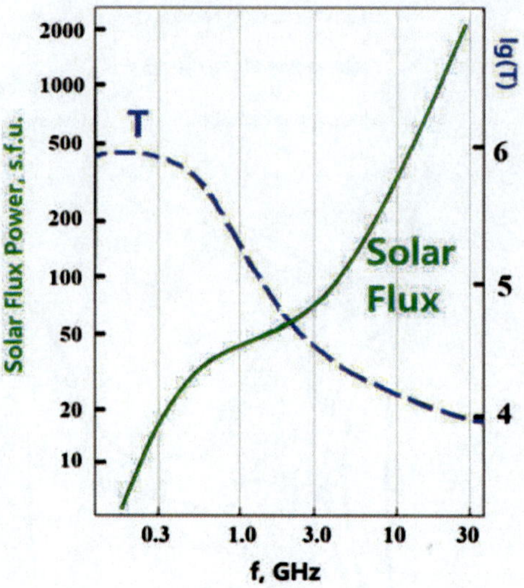

Fig. 3.6 Spectrum of the quiet Sun radio emission and equivalent brightness temperature. Re-produced from (Bruzek 1972))

The intensity of this radio emission of the Sun is characterized by the strong positive correlation with SSN. The intensity of the background radiation emission of the Sun during the years of maximal sunspot number increases by 25–60%.

In addition to the background radio emission from the Sun, there is always a *slowly varying* component, or *S-component*. Its source is the intensified thermal plasma radiation from the Sun's active regions. This component of radio emission is generated within the range of centimeter and decimeter waves (meaning that it includes the GNSS operating frequencies). The S-component radio emission intensity is also characterized by a strong correlation with the sunspot number. In the years of SSN maximum the S-component magnitude is comparable to the level of the background radio emission of the Sun.

The bursts of radio emission comprise a special class of events during which the level of radio emissions from the Sun is sharply increased during the short period of time from fractions of a second to ~ 10 min. The energy of the burst can be concentrated in one or in several consequent bursts and in a narrow frequency range or cover several frequency ranges.

During the process of radio burst, a drift of the average frequencies of the radiation is possible. Types I–V of bursts of radio emission from the Sun are distinguished (Dulk 1985). The origin of some of the bursts is non-thermal, related to the gyro-synchrotron mechanism of emission. Another part of the bursts are generated due to the dissipation of the energy of high-energy electrons and by waves in plasma.

Type I solar radio burst comprises multiple (from 2–3 to 100) short-lived bursts of radio emission lasting during (0.1–1) s, which are observed mostly within the range of metric waves. The energy of the bursts is usually concentrated in the narrow band of frequencies within ~ (5–15) MHz. Sometimes, the drift of the average frequency of type I burst occurs towards the lower frequencies. The intensity of these bursts varies in a wide range: from fractions of the background radio emission of the Sun to the level exceeding the background emission by many times. The well-pronounced circular polarization of the waves in these bursts is frequently observed. Unpolarized radiation waves of type I bursts occur less often. Their sources are located near the solar limb.

Type II solar radio bursts are the separate large disturbances of radio emission in meter, decameter and kilometer wavelength ranges with duration of ~ 10 min (they can also last about 24 h). The characteristic frequencies of this type of emission drift in the region of low frequencies with the velocity of ~ 0.25 MHz/s. Type II bursts are related to the intense solar flares (in the optical and X-ray range) and manifest themselves approximately 10 min after the flash phase of the flare. It is suggested that the source of these bursts is the plasma oscillations that are excited in front and behind the shock wave propagating from the flare region (Fig. 3.5). They can be also produced by CME-associated shocks. That is why they can last for many hours.

Type III solar radio bursts are the group bursts of radio emission with the separate burst duration of several seconds. These groups of bursts manifest themselves in the meter and decameter ranges and are characterized by the fast drift of the average frequencies towards the low-frequency region with the drift velocity of $\sim -0.01 \bullet f^{1.84}$. Usually, the burst intensity exceeds the value of 100 s.f.u. In the majority of cases, the wave polarization is absent, but sometimes the circular or linear polarization is observed. The main source of the type III bursts is the scatter of Cherenkov plasma waves that are generated by the high-energy electron fluxes with the speed of $\geq 0.3 \bullet s$ (~1/3 c) in plasma.

Type IV solar radio burst is also related to solar flares. It represents the powerful radio emission with the complex amplitude-frequency form and the overall duration of several minutes. A type IV radio burst radiation can cover a wide frequency range (from millimeter to decameter waves) depending on its generation mechanism. Therefore, this type of bursts has a complex nature of origin and formation mechanism.

Microwave type IV burst (IVμ) has a gyro-synchrotron mechanism and generates extraordinary wave of circular polarization. *Decimeter type IV burst (IVdm)* covers the frequency range within (200–2000) MHz and generates the ordinary wave of circular polarization. The supposed source of decimeter bursts is the dispersion of Cherenkov plasma waves. Type IV burst in the meter range can be of two categories: *moving Type IV burst (IVmB)* and *stationary Type IV burst (IVmA)*. The moving type IV burst is caused by gyro-synchrotron radiation that is generated in the plasma ejections moving from the Sun (eruptive prominences, expanding arcs or shock front in plasma, Fig. 3.5).

Type V solar radio burst manifests itself as a broadband continuous radiation at long meter waves which occurs after the type III burst. The duration of this burst is ~ 1 min. This type of radio bursts is suggested to be generated by energetic electrons of Cherenkov plasma waves which are temporally captured by the magnetic force lines in the vicinity of the type III burst formation.

Microwave burst of solar radio emission manifests itself in a wide range of wavelengths from millimeters to decimeters. The burst power flux maxima are observed approximately simultaneously at different frequencies. In many cases, a gradual increase of radio emission intensity to a single maximum is followed by its decrease. The total duration of such bursts is (15–60) min, and the maximum intensity reaches 50 s.f.u. Single impulse bursts with a sudden onset and a sharp increase of the intensity of radio emission can be observed. In this case, the radiation maximum is reached some seconds after the onset and the whole burst duration is of units of minutes.

Another type of microwave burst is called a *complex burst*. It has several radiation maxima.

A small part of microwave bursts can reach the radiation flux density > 500 s.f.u. Such bursts are usually complex and are called *great bursts*. In the majority of cases, this type of bursts represents the microwave part of type IV radio burst. According to the shape of the spectrum, the characteristic categories of complex microwave bursts with the following features may be distinguished:

(1) continuous increase of burst intensity with frequency;
(2) continuous decrease of burst intensity with frequency;
(3) a single maximum in the spectrum (usually in the centimeter range);
(4) U-shaped spectrum with a minimum intensity in the frequency range 500–2000 MHz.

In the majority of cases (except for the weak lasting bursts), the microwave bursts are generated by gyro-synchrotron radiation (Fig. 3.5).

The brief review of solar radio bursts given above does not cover the whole variety of solar bursts. At the same time, even this summary shows a wide variety of generation mechanisms of solar radio emission and radiation characteristics. Table 3.1 (Modified from: (Dulk 1985)) presents the main characteristics of the considered types of radio bursts. Here, o-mode means the dominance of the ordinary wave of circular polarization and x-mode means the dominance of the extraordinary wave.

According to Table 3.1, not all the categories of solar radio emission pose a potential hazard for satellite broadband navigation and communication system operations. In this regard, the type IV solar radio bursts and microwave bursts draw our attention. These events often generate radiation in the range of GNSS operation frequencies. At the same time, the large part of radiation has the right-handed circular polarization of the wave which coincides with the polarization of antennas and GNSS signals. Further in this book, we consider these types of solar radio emissions to be the potential danger for jam-resistant operation of GNSS equipment.

3.3 Space Weather Conditions

It is worth noting that the term "Space Weather" is mostly related to the applied problems and used in terms of operational tasks. Ultimately, it is also widely used by research scientists as a concise and comprehensive term that embraces a wide range of phenomena. Space Weather (SW) is a multidisciplinary issue that is based on the knowledge of and is related to different fields of research including but not limited to solar-terrestrial physics, geophysics, heliophysics, atmospheric physics,

Table 3.1 Major characteristics of solar radio bursts

Type	Duration	Polarization	Frequency range	Association	Generation mechanism
I	≤ 1 s	50–100% circular	50–300 MHz	a number of solar sunspots	Thermal emission of plasma
II	≥ 10 min	none	1–200 MHz	flare, shock wave	Thermal emission/waves in plasma
III	≥ 1 s	o-mode 10–30%	1–200 MHz	high energy electron flux	Waves in plasma
IVμ	≥ 10 min	x-mode circular	1–30 GHz	Intense flares (X-rays and UV)	Gyro-synchrotron radiation
IVdm	≥ 10 min	o-mode circular	200–2000 MHz	Intense flares (X-rays and UV)	Waves in plasma
IVmB	$\approx$30 min	x-mode sign-variable	10–200 MHz	Weak flares	Gyro-synchrotron radiation
IVmA	$\approx$20 min	o-mode circular 0–40%	10–200 MHz	Moderate to intense flares	Thermal emission of plasma
V	≥ 1 min	x-mode circular < 10%	10–100 MHz	Type III burst	Waves in plasma
Microwave, impulse	≥ 1 min	x-mode circular ~ 30%	3–30 GHz	flares, hard X-ray flares	Gyro-synchrotron radiation
Microwave, complex	$\approx$10 min	x-mode circular ~ 10%	1–30 GHz	Intense flares (X-rays and UV)	Gyro-synchrotron radiation

Earth science, radio propagation studies and other related disciplines. Within the SW concept, the harmful impacts on modern technology, whether on Earth or in space (Cannon 2013), and even the effects on human health and other living beings are considered.

Why the term "weather" is applied to describe the processes in the space environment? This is because there are all the attributes of weather in the near-Earth space. There are storms (magnetic and ionospheric), clouds (magnetic, mesospheric), wind (solar wind) and even rain (particle precipitations, especially at the high-latitude ionosphere) (http://ipg.geospace.ru/what-is-space-weather.html). Moreover, quiet (calm) periods (e.g. during low solar activity intervals) are followed by periods of sudden environmental changes (e.g. during high solar activity) and vice versa. Bothmer and

Daglis (2007) discussed the analogies between SW and terrestrial weather in more detail.

The *quiet SW conditions* are defined as regular behaviour of SW parameters which basically implies their systematic variations. The main known regular variations are diurnal variation, day-to-day variability, monthly, seasonal, annual (and semi-annual) variations and solar-cycle variation (e.g., Richards 2001; Fuller-Rowell et al. 1996; Kurazhkovskaya 2020; Ratovsky et al. 2015). There are also more short-term variations that can be controlled by regional/local processes and more long-term variations controlled by global natural and probably anthropogenic processes (e.g., Danilov 1998; Danilov and Konstantinova 2020; Sergeenko 2020). The regular variations numbered above are the result of change of day and night as well as the change of season, dynamic atmospheric processes (winds, atmospheric waves, tides), Earth and solar rotations and other natural factors.

SW disturbances are the periods characterized by the SW parameter behaviour that is beyond its regular systematic variation. There are different types of SW disturbances related to the phenomena mentioned above. The origin of SW disturbances is the Sun (Rycroft 2004; Bothmer and Zhukov 2007). That is why SW disturbances of notable intensity are called solar storms in common language. In general, the following solar phenomena are the most geoeffective (effective to cause disturbances) and important in terms of SW effects on the ionosphere: solar X-ray flares, Solar Energetic Particles (SEPs), Coronal Mass Ejections (CMEs), high speed streams from coronal holes and associated Corotating Interaction Regions (CIRs) (Cander 2019). Figure 3.7 gives the idea of time scales needed for different types of emission to reach the Earth's environment.

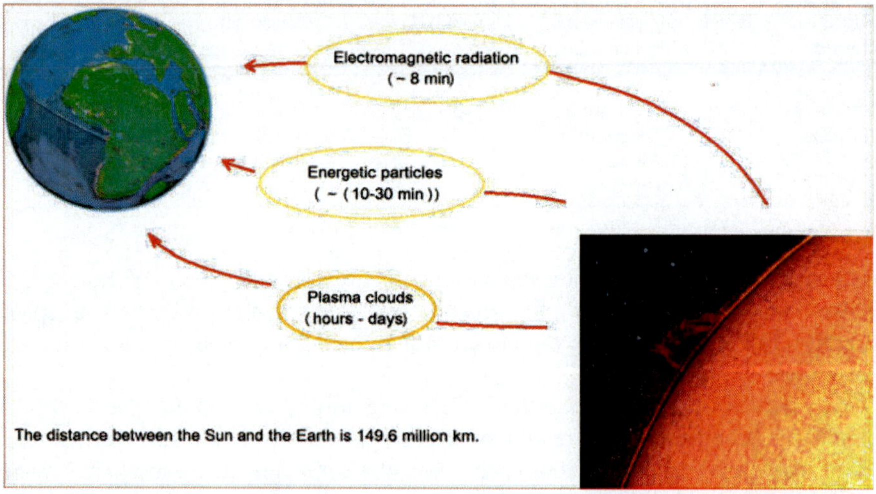

Fig. 3.7 Three main types of phenomena that have the potential of negative SW effects (not to scale). Image of the Sun in H-Alpha (6562.8 Å) provided by LACIGE is used for the schematic representation (Courtesy of Dr. M. Rodríguez-Martínez)

From the SW perspective, every day is considered unique even under quiet conditions. The identical SW storms are even less possible (Lanzerotti 1992; Cannon 2013). This highlights the importance of case studies of SW events to benefit our understanding of physical processes in the near-Earth space. For example, relevant case studies presented in the following works (though this list may be much longer): (Mannucci et al. 2005; Jin et al. 2008; Nava et al. 2016; Sergeeva et al. 2021b; Blagoveshchensky et al. 2019; Blagoveshchensky and Sergeeva 2018; Hayakawa et al. 2020; Trivedi et al. 2007; Sahai et al. 2011; Joshua et al. 2014; Zolotukhina et al. 2017). At the same time, there are many successful attempts to reveal interesting features by searching similarities and differences within the set of the similar/different SW disturbances or in some particular region (Zolotukhina et al. 2018; Wang et al. 2010; Szuszczewicz et al. 1998; Blagoveshchensky et al. 2021; Blagoveshchensky 2018; Shagimuratov et al. 2009; Klimenko et al. 2017). The statistical studies fall into another category of the research works (Kouris and Fotiadis 2002; Wang et al. 2010; Vijaya Lekshmi et al. 2011; Blagoveshchensky et al. 2018; Deminov et al. 2011; Wang et al. 2019; Jacobsen and Dähnn 2014; Chen et al. 2019, Oinats et al. 2016; Zhao et al. 2005b; Gordienko et al. 2001; Ding et al. 2008; Song et al. 2013; Tsugawa et al. 2004; Borovik and Zhdanov 2018 and many others).

3.4 Space Weather Parameters

There are different parameters and characteristics (functional dependence of one parameter on another) that characterize SW conditions. Figure 3.8 shows an example of variations of some of the most commonly used SW parameters.

F10.7-index is a solar radio noise flux. It shows the density of the energy flux of solar radio emission at a wavelength of 10.7 cm. F10.7-index characterizes the overall solar activity on equal footing with SSN and, as a consequence, serves to describe the solar cycle development (Fig. 3.9). Interplanetary Magnetic Field (IMF) magnitude and its Bz-component describe the processes in the interplanetary medium and in the solar wind and characterize solar wind interaction with Earth's magnetosphere. Solar wind proton temperature, solar wind speed (V) and proton density (Np) in the solar wind plasma are parameters of the solar wind.

Usually, to estimate geomagnetic conditions, different global geomagnetic indices are used (three lower panels of Fig. 3.8). It should be noted that inspite of being called global, every geomagnetic index has its particular meaning. Different indices are not interchangeable. In regard to the SW state, geomagnetic disturbance, in general, and geomagnetic storm and substorm, in particular, are among the most important phenomena. Therefore, we enlarge upon the point in a bit more detail.

Geomagnetic conditions are defined mostly in terms of the horizontal (H-/X-) component of the geomagnetic field variation that is measured at different stations (a regular variation during the quiet time or a depression followed by a gradual recovery during storms). Data from various networks of magnetic observatories serves as a basis for the calculation of different global geomagnetic indices. These indices allow

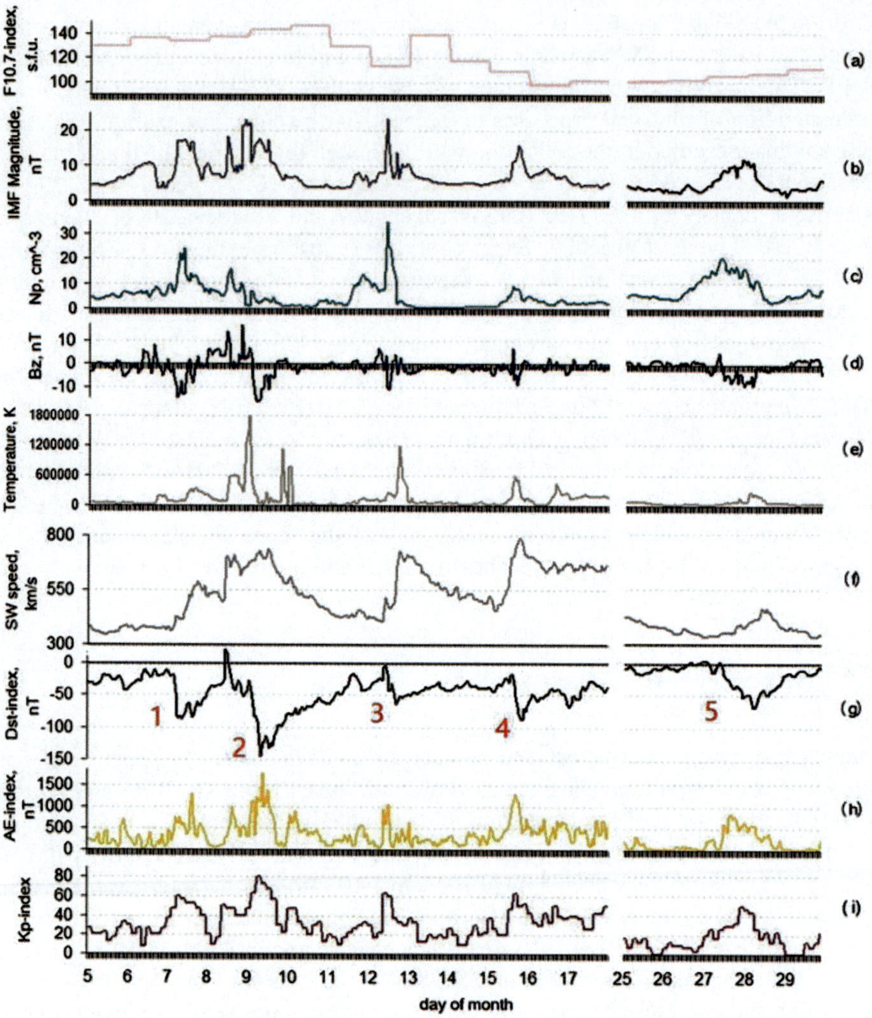

Fig. 3.8 SW parameter variations during March 7–17 and 25–29, 2012: **a** F10.7-index; **b** averaged magnitude of IMF; **c** proton density; **d** Bz-component of IMF; **e** solar wind proton temperature; **f** solar wind speed; **g** Dst-index; **h** AE-index; **i** Kp-index. Geomagnetic storms are numbered 1–5. (Taken from (Sergeeva et al. 2021b))

us to quantify the intensity of geomagnetic disturbances (Campbell 1997; Borovsky and Valdivia 2018; Hunsucker and Hargreaves 2003; Gonzalez et al. 1994) and characterize storms on a global scale. Let us mention the most popular of them. Dst-index (Disturbance storm-time index) represents geomagnetic field decreases mostly due to the equatorial current system in the magnetosphere (ring current) and also the magnetosphere compression from the increases of solar wind pressure (Sugiura 1964; Sugiura and Kamei 1991; Nagatsuma 2002). There are three versions

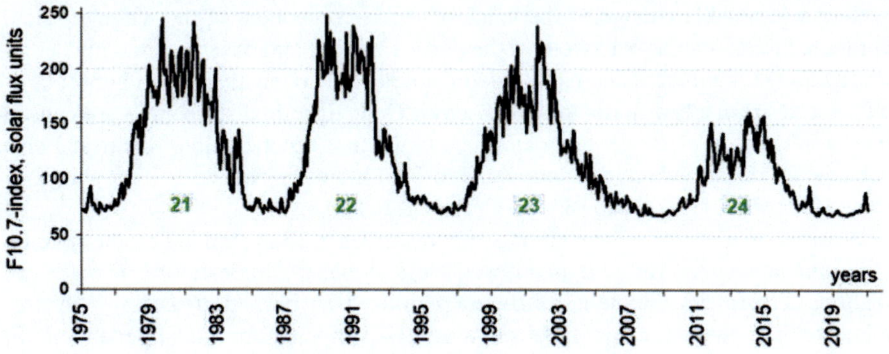

Fig. 3.9 Solar cycles 21–24 by F10.7 data (27-day averaged values)

of the index: real-time (after the initial checking of magnetic data), provisional (after second quality checking) and final (definitive information). Dst is calculated with a 1-h time resolution based on data from four low latitude stations. SYM-H is essentially the analog of the Dst-index but with a 1-min time resolution based on data from six stations evenly distributed in longitude (Iyemori et al. 2010). AE-index (Auroral Electrojet index) serves to estimate the global electrojet activity in the auroral zone (Davis and Sugiura 1966). It has 1-min resolution and is calculated with data from 12 stations in the Northern auroral zone. AE = (AU–AL), where AU and AL are the envelopes of the superposed plot of H-component curves from all the stations. SML and SMU, in essence, are the AL and AU indices respectively but based on data from 100 or more stations (Newell and Gjerloev 2011; Gjerloev 2012). Kp-index is based on the largest difference between the highest and the lowest values of X- and Y- components of geomagnetic field (Bartels 1949, 1957; Bartels and Veldkamp 1954). It is calculated with data from eleven stations in the Northern hemisphere and two in the Southern hemisphere situated at mid- and high-latitudes (44°–60°) and has a three-hour resolution. PC-index is an indicator of solar wind energy entry into the magnetosphere. It characterizes magnetic activity in the polar cap and is derived from data of two stations in the near-pole regions: one in the Northern hemisphere (PCN) and another in the Southern (PCS) (Troshichev et al. 2006, 2014; Troshichev and Janzhura 2012).

The preference of which index to use depends on the particular task and the analyzed region on Earth.

3.5 Geomagnetic Disturbances

In general, Earth's magnetic field variations are resulted from different regular and irregular processes on the Sun and in the solar wind that impacts the Earth's magnetosphere and ionosphere. In essence, a *geomagnetic disturbance* is a temporary irregular variation (a notable change of the value of the H-component of the magnetic field compared to its regular behaviour at the particular hour and observation point) characterized by a significant decrease of geomagnetic field intensity (tens or hundreds of nT) lasting from a few hours to several days. Geomagnetic disturbances are caused by the interaction of the geomagnetic field with the magnetic field of the near-Earth solar wind, during which the geomagnetic field intensity decreases on a planetary scale (Campbell 1997; Borovsky and Valdivia 2018).

A *geomagnetic storm* is a geomagnetic disturbance characterized by the specific H-component curve form. It manifests itself globally demonstrating the general features of storm development in different regions. It can be identified using measurements of an individual magnetic observatory (H-component or X-component curve) or by the Dst-index curve. Further, we discuss Dst (and its analog SYM-H) variations as it is used to estimate the geomagnetic disturbance development on a global scale. During the quiet conditions, Dst-index mostly varies near the zero value or is slightly lower than this baseline. The Dst values range is approximately between + 100 nT and − 600 nT (Gonzalez et al. 1994). Figure 3.10a shows typical Dst-index variation during a storm. The circle on the Fig. 3.10a shows the part of variation presented by the SYM-H index as Dst-index temporal resolution was not enough to identify

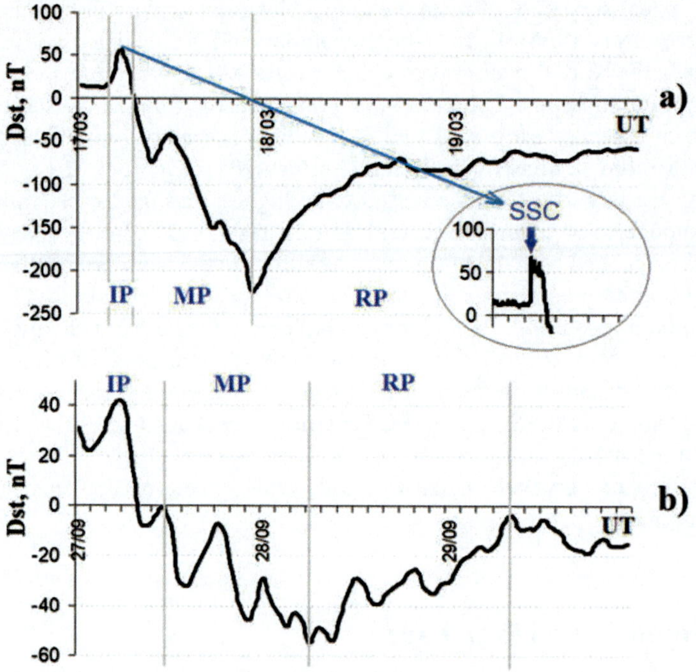

Fig. 3.10 Dst-index variations during the geomagnetic storms occurred on March 17–21, 2015 (last days not shown) (panel a) and September 27–29, 2017 (panel b)

clearly some features. The last two days of the storm of March 17–21, 2015 are not shown as they do not provide essential information.

A classical geomagnetic storm on Fig. 3.10 consists of the following phases:

- A sharp increase of the magnetic field is referred to as a Storm Sudden Commencement (SSC). Usually, it has duration of some minutes and magnitude of a 20–30 nT (though it can reach up to 100 nT). Physically, SSC is a signature of the interplanetary shock impacts and has a short duration. Dst-index does not always show SSC presence due to its low temporal resolution, but it is clearly seen with SYM-H (shown in the circle on the Fig. 3.10a).
- SSC is followed by the initial phase of the storm (IP) characterized by the increase of the H-component/Dst-index value and usually lasts from half an hour to several hours. During the IP, the Earth plunges into the plasma and magnetic field behind the front of the shock wave.
- IP is followed by the main phase (MP) of the storm that is a result of the development of the ring current and the intensification of the magnetotail currents (Borovsky and Valdivia 2018). Significant deformation of the magnetosphere structure takes place. MP is characterized by the gradual notable decrease of the H-component/Dst-index value, which can last several hours (sometimes days).
- The last is the recovery phase (RP) during which the ring current decay leads to recovery of the magnetic field. RP consists of the slow return of the H-component/Dst-index to its pre-storm level or sometimes to a level slightly lower than baseline. It can last from several hours to several days.

Please note that sometimes, no storm is initiated after SSC, in which case the sharp increase of the magnetic field is called a Sudden Impulse (SI) instead of SSC. In some cases, SI can be negative. It is also important to emphasize that some storms begin without IP and/or SSC. In contrast, MP and RP are the obligatory steps of any magnetic storm. The development of the magnetic storm and the physics of the processes are widely described in the literature. Please refer to (Campbell 1997; Tsurutani et al. 2020; Borovsky and Valdivia 2018 and references therein).

Geomagnetic storms can be caused by solar flares often accompanied by a CME or by high speed solar wind streams from coronal holes. The storms triggered by CME can reach larger intensities than those triggered by flows from coronal holes. Dst-index curve during the storm triggered by the high speed streams tends to be more rugged. Figure 3.10 provides examples. Figure 3.10a presents the CME-driven storm of March 17–21, 2015. Figure 3.10b presents the storm originated from the coronal hole on September 27–29, 2017. In the first case, SSC was caused by an interplanetary shock wave and the arrival of the material of the plasma-sheath at Earth during the first hours of March 17, 2015. Few hours later, the plasma-sheath material showed the Bz-south polarity, a condition that marked the storm's IP. The Bz-component of IMF in the magnetic cloud of CME turned southward and triggered the MP development. CME left the Earth's orbit on March 18th, 2015. In the second case (Fig. 3.10b), the fast solar wind with southern Bz-component of IMF originated from the equatorial coronal hole on the Sun, reached the terrestrial environment and caused the geomagnetic storm of September 27–29, 2017.

Sometimes an active region responsible for a storm can live more than 27 days (~ solar rotation period) on the Sun and consequently cause storms more than once each time an active region facing the Earth. Geomagnetic storms depend on the solar activity phase: they are more frequent during the high solar activity periods as more active regions appear. According to (Kurazhkovskaya 2020), the maximum number of geomagnetically disturbed days occurs during the descending phase of solar activity.

There are different *classifications of storms*. According to (Gonzalez et al. 1994), the storms with a Dst peak value between −30 nT and −50 nT are weak (small). Those with Dst peak value between −50 nT and −100 nT are moderate. The intense (great) storms are characterized by Dst peak value of −100 nT or less. Loewe and Prölss (1997) suggested the following classification: weak (< −30 nT), moderate (< −50 nT), strong (< −100 nT), severe (< −200 nT), and great (< −350 nT) magnetic storms. The geomagnetic storm magnitude can also be described with Kp-index. For the operational tasks, it is commonly used the Kp-based classification developed by the Space Weather Prediction Center of NOAA, U.S. (Poppe 2000) which is freely-available at www.swpc.noaa.gov/sites/default/files/images/NOAAscales.pdf. These scales are a popular tool as they characterize the probability of occurrence of SW disturbances and list their possible impacts. It is a good reference for users, but it should be taken with some reserve because SW negative consequences in the technological sectors are often not reported publicly, which means a certain lack of information (especially on the consequences of not extreme events). In addition, users should recall that the consequences at different latitudes and longitudes can be rather different.

It is a known fact that geomagnetic disturbances can cause ionospheric disturbances (Davies 1965; Kelley 2009; Hargreaves 1979; Prölss 2011, 2012; Tsurutani 2000; Förster and Jakowski 2000; Afraimovich et al. 2013; Mendillo 2006; Mandrikova et al. 2018; Blagoveshchenskii 2013; Ratovsky et al. 2020; Pei et al. 2021; Wang et al. 2010). The next section provides a brief introduction to the ionosphere principles.

3.6 General Concepts of the Ionosphere Principles

The atmosphere is the gaseous environment that surrounds our planet. The *ionosphere* is a part of the upper atmosphere. It begins at heights of (50–80) km above the ground (depending on the latitude of the observation point). The air in this region is very ionized: large numbers of free electrons (within the limits of 10^3–10^6 per 1 cm^3) and positive ions are present. In general, the medium is electrically neutral.

The presence of free electrons makes possible radio wave reflection in the ionosphere for waves with wavelengths greater than 10 m. Due to this fact, radio waves can propagate over very long distances by series of reflections between the layers of the ionosphere and the ground (and even can circumnavigate the globe if conditions are favorable) making possible communications. The waves that propagate by cause of reflection in the ionosphere are called *ionospheric waves.* Variations of electron density in the ionosphere directly affect radio propagation. It is also true for the waves that pass through the ionosphere without being reflected. For instance, the signals emitted by satellites and received by ground receivers travel through all the atmospheric regions including the ionosphere. It should be noted that the effects on radio propagation depend on the frequency at which the signal is emitted. In modern life, knowing the current state of the ionosphere is one of the most important tasks for the stable and reliable operation of various technology-based systems. SW phenomena cause changes in the state of the Earth's ionosphere. In turn, these changes can affect the operation of telecommunications, navigation systems, radars, electrical power distribution networks, oil and gas pipelines, the serviceability of satellites, global positioning systems, control systems, computers, radio communication with ships, airplanes, etc.

The main ionospheric parameter is the electron density (the number of electrons per unit volume). The main characteristic of the ionosphere is its height distribution called vertical electron-density profile (*N(h)-profile*). This characteristic indicates the presence of several maxima of electron density in the ionosphere including the main most pronounced maximum. Examples of profiles are shown in Fig. 3.11a, b. According to the height distribution of electron density, the ionosphere may be divided into two larger parts: above the main maximum (topside) and below it (bottomside ionosphere). The first is characterized by the gradual electron density decrease with height. The second has a more complex structure and is subdivided into several regions different by their atmospheric composition, dynamic processes and ionizing spectrum. The lowest is the ionospheric D region found below 90 km. The E region manifests itself approximately between 90 and 140 km. The F region is above 140 km and contains the main electron density (Ne) maximum. Sometimes, during the daylight hours it subdivides into F1 and F2 regions. At night, the D and Fl regions disappear and the E region weakens significantly. Please note that the height limits of the regions are relative and inconstant. A part of the region that contains a relative Ne maximum and/or characterized by a drastic change in concentration is called a layer. In general, four regular layers can be distinguished: D, E, F1 and F2. Some irregular layers are possible.

Radio signal reflection from the particular ionospheric layer depends on the wave frequency and the ionospheric state. The higher Ne, the higher the frequency of waves that can be reflected. It is generally considered that the waves with frequencies higher than 100 MHz pass through the ionosphere without reflection (but are still affected by its state). Therefore, on the one hand, the ionosphere is a "mirror" that reflects radio waves (mostly, HF range) allowing the signal propagation over very long distances. On the other hand, it is a means that adds error to satellite signals (e.g., to GNSS signals that experience delay in the ionosphere in accordance to its Ne). Thus, the ionosphere plays an important role, defining conditions of radio propagation both for space- and ground-based systems for various applications (Goodman 2005). For the detailed description of radio propagation in the ionosphere, the reader is referred to (Davies 1965; Kelley 2009; McNamara 1994).

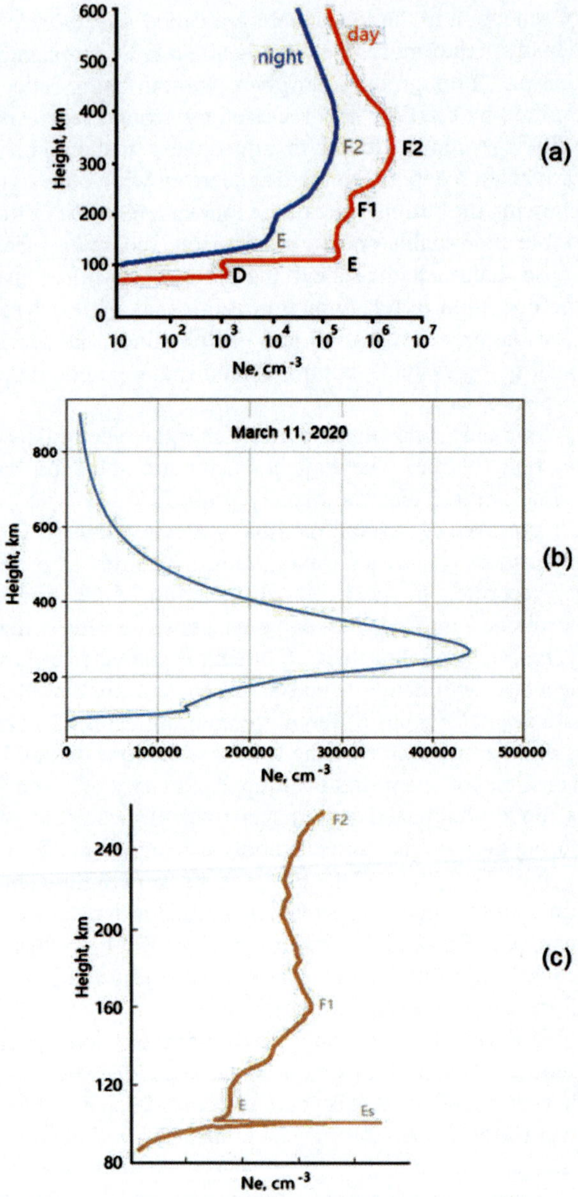

Fig. 3.11 **a** Typical electron-density profiles at day and night (Re-produced from the published book (Izyumov and Linde 1976)); **b** Ionospheric profile reconstructed with the use of IRI2016-model (Bilitza et al. 2017) and local daytime experimental data from Mexico; **c** Example of the intense Es layer (Re-produced, modified and translated from Russian based on figure from the published book (Alpert 1960))

As mentioned above, in addition to regular ionospheric layers, several specific *irregular structures* can be formed causing effects on radio propagation (Davies 1965). Let us mention just two of them (though the list is much longer). A thin and dense sporadic layer that occurs in the E region (*Es-layer*) is a remarkable feature

(Haldoupis 2011). Figure 3.11c shows a schematic illustration of the intense Es-layer that was observed during the rocket launch experiment described in (Alpert 1960). The dotted line stands for the results measured at the ascent of the rocket, the solid line during its fall. Es-layer physical causes can be related to the wind-shear mechanism or meteor impacts; attributed to ionization by precipitating particles or being a result of the restructure of the ionosphere during geomagnetic storms (Yeh et al. 2014 and references therein). At the same time, it is not as sporadic as its name suggests (Haldoupis 2011). In some regions, its occurrence can reveal some systematic features depending, for example, on local time or season. If significantly dense, the Es-layer "reflects" HF signals that are usually reflected in the F-region or even not reflected at all in the ionosphere (VHF signals). To add, it can totally prevent normal propagation or be semitransparent for radio waves. Hence, Es-layer diagnostics is important for radio propagation tasks and is a subject for extensive study (e.g., (Kirkwood and Nilsson 2000; Sun et al. 2018; Maeda and Heki 2015) and many others).

Another type of irregular structures is *Travelling Ionospheric Disturbance* (TID). It is a wave-like quasi-periodic Ne variation driven by atmospheric gravity waves, which can travel over long distances in the ionosphere. TIDs are classified by their spatiotemporal parameters and dynamics: time scale, horizontal size, relative Ne change and velocity of movement (Hocke and Schlegel 1996; Afraimovich and Perevalova 2006). These parameters are determined by triggering phenomena. Usually, large-scale TIDs (~1000 km) are formed in the auroral regions during geomagnetic storms and move from high to low latitudes. Medium-scale TIDs can be caused by a range of phenomena including meteorological phenomena, solar terminator passage and intense earthquakes. They can move in different directions and be tilted. Small-scale TIDs (100–1000 m) are mostly attributed to meteorological processes.

Electron density in the ionosphere is subject to both (a) regular/systematic changes (daily, seasonal, annual/semi-annual and solar cycle) and (b) irregular (disturbances). SW events of a different kind (solar flares, CMEs, changes in solar wind parameters, particle precipitation, etc.) cause alterations in the complex structure of the ionosphere. It is important to detect systematic variations and the presence of specific features/anomalies in each region on the globe, as well as to understand better the physical mechanisms that cause them, since they serve as a "quiet level" against which the disturbances are revealed. Based on this knowledge, the level of possible effects on modern technology related to the ionosphere state changes can be estimated.

In a first approximation, *ionospheric disturbances* can be divided into two general types: positive (Ne higher than under quiet conditions at the particular time moment and location) and negative (Ne lower than under quiet conditions) (Danilov 2013). Figure 3.12 provides the example of the ionospheric response over the Mexican region to the moderate geomagnetic storm that occurred on June 22–23, 2015 during the descending part of solar cycle 24. The state of the ionosphere at the F2-layer heights is represented by the parameter "Total Electron Content" which is discussed further in the text. In simplistic terms, it may be said that Ne is represented here by Total Electron Content whose distribution is indicated by color, from the blue for the lowest values to the red for the highest (see the color bar in the right part of the Fig. 3.12). The picture of regular electron density at 20 h of Universal Time (UT), or 14 h of local time (LT), is given in the Fig. 3.12a. The Ne distribution is non-homogeneous. The lower the latitude, the higher the density. The structure of the enhanced density is seen near the southern coast of Mexico (discussed further in Sect. 3.8). The geomagnetic storm led to changes in this picture. First, the general Ne increase (in comparison to the quiet conditions) can be observed in the Fig. 3.12b designating the positive disturbance. The ionospheric structures became more pronounced; the zone of the increased Ne enhanced its size and was shifted in latitude. A general Ne decrease (in comparison to the regular picture) is observed on the next day (3.12c). The structure of the enhanced Ne at the south did not manifest itself.

The intensity of disturbances (positive or negative) and the significance of their consequences may be estimated by comparing of disturbed and quiet conditions. That is why it is important to perform a continuous monitoring of ionospheric behaviour on a regional basis. We recall that the state of the ionosphere is a subject to regular daily, monthly, and seasonal changes, as well as the changes during the 11-year cycle of solar activity. The morphology of the ionosphere also depends on the observation point coordinates (latitude/longitude). Therefore, a continuous regional/local diagnostics is needed.

The same change in the ionosphere can provoke the contradictory effects for different applications. For example, in terms of radio propagation, F2-layer is the most important part of the ionosphere. First, the main propagation modes for the reliable HF communication are via reflections in this layer. Furthermore, the satellite signals are dependent on its state as well. In a broad summary, it can be assumed that the negative disturbances in the F-region present risks for the HF communication, especially at nighttime and during solar minimum when the operational frequencies of signals propagating through F2-reflections are very low due to their regular variations. The frequency range gets narrow significantly. In contrast, the positive ionospheric disturbances if intense can be dangerous for the satellite signals especially for those used for global positioning and precise timing tasks. This is because the higher the electron concentration in the F-region, the more the signals delay when passing through this layer and the larger is a positioning error. To sum up, both significant Ne increases and decreases concerning its quiet value can cause negative effects. The physics of ionospheric perturbations and their positive and negative phases are discussed in (Buonsanto 1999; Lastovicka 2002; Prölss 1995; Mendillo 2006; Danilov 2013 and references therein).

Apart from the temporal systematic changes there are also *spatial variations*: the ionosphere morphology and behaviour have their peculiarities at different latitudes and longitudes as their controlling processes at high-, mid- and low- (including equatorial) latitudes differ (Hargreaves 1992; Kelley 2009; Prölss 2011). Without going into details, we only mention that the configuration of the Earth's magnetic field lines plays an important role in the spatial inhomogeneity of the ionosphere. The idealized picture of these lines is shown in Fig. 3.13a. In reality, the solar wind modifies their form, especially on the night side of our planet (Fig. 3.13b). Both schematic representations prove that the atmosphere's protection from solar wind material by the

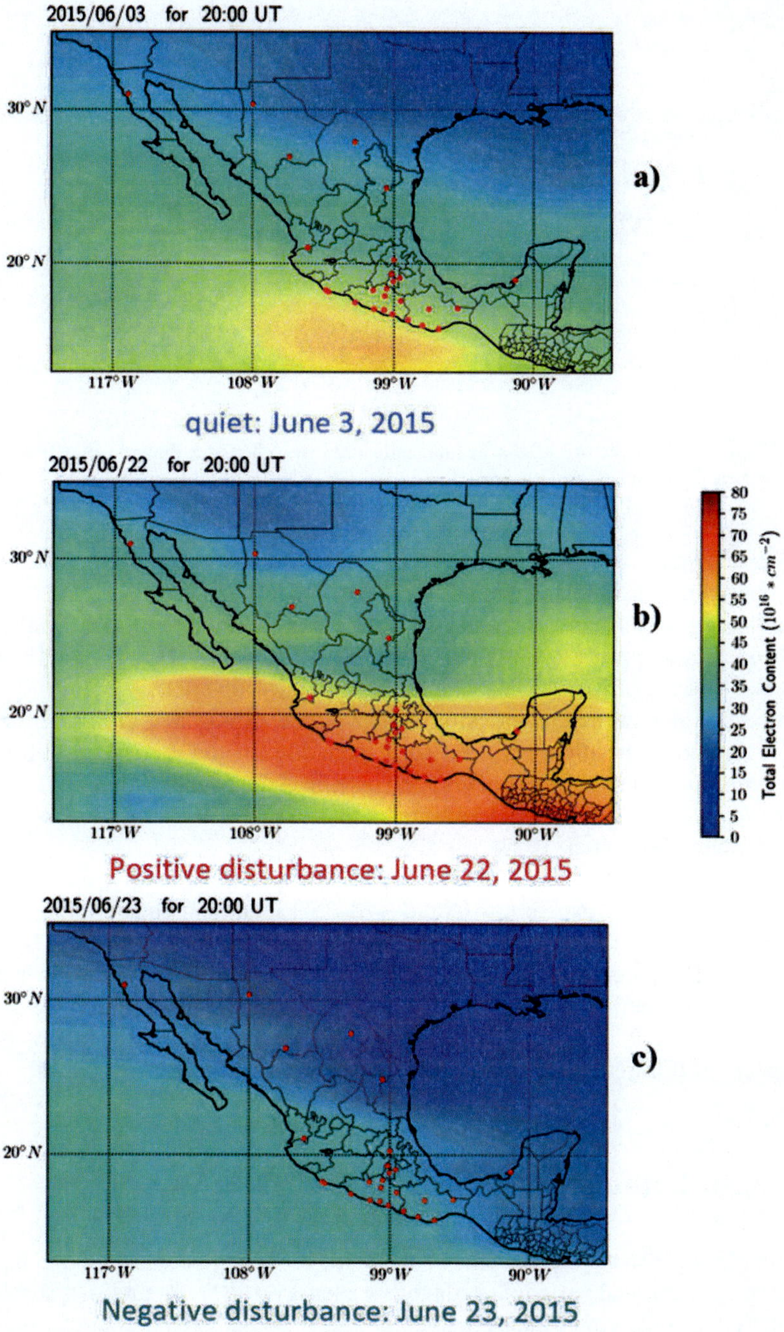

Fig. 3.12 Total Electron Content distribution over Mexico at the hour of its diurnal maximum (reconstructed from GNSS data). **a** June 3, **b** June 22 and **c** June 23, 2015. (Provided by Mexican Space Weather Service and LACIGE)

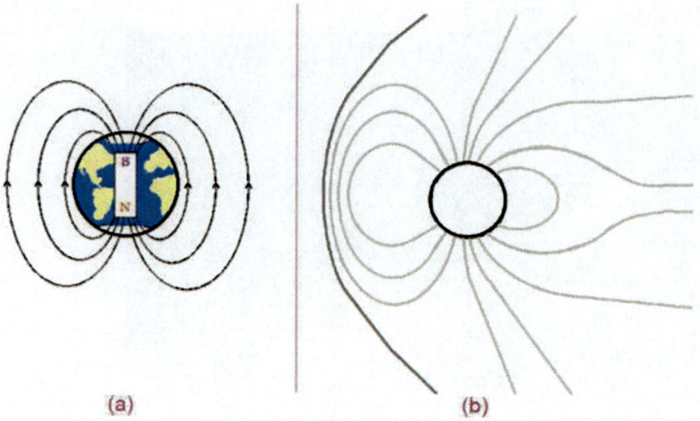

Fig. 3.13 **a** Idealized picture of the geomagnetic field. **b** Schematic drawing of Earth and its magnetic field in the solar-wind plasma

geomagnetic field varies with latitude: more protected near the geomagnetic equator (field lines almost parallel to the Earth's surface) and less at the northern and southern near-polar regions. To add, there is also a large region in the Southern Hemisphere which is characterized by the geomagnetic field anomaly called South American Magnetic Anomaly (Mandea and Chambodut 2020; Abdu et al. 2005).

The difference in the dynamical processes in the atmosphere of different zones and local/regional peculiarities are also responsible for the ionosphere difference at different latitudes and longitudes. For instance, the Weddell Sea and Yakutsk Anomalies may be mentioned as examples of such peculiarities (Klimenko et al. 2015; Zakharenkova et al. 2017).

Thuswise, in regard to the ionospheric structure and variability, the following regions can be distinguished: low latitudes (0–35°) including equatorial region (0–20°), mid-latitudes (35–55°), subauroral ionosphere (55–65°) and high-latitudes (above 65°). Different impacts prevail in each region resulting in the specific features of regular ionospheric behavior. Let us briefly mention the main spatial features of different regions. For the detailed description of the characteristic features of the ionosphere of different latitudes the reader is referred to (Hargreaves 1992).

3.7 High-Latitude Ionosphere

The high-latitude ionosphere is fairly considered the less predictable due to its very complex structure and variations (Blagoveshchensky 1981). Figure 3.14 shows the schematic representation of this complexity.

The *subauroral zone* main feature is the presence of the depleted ionization region called Main Ionospheric Trough (MIT) observed during night hours. It is a natural

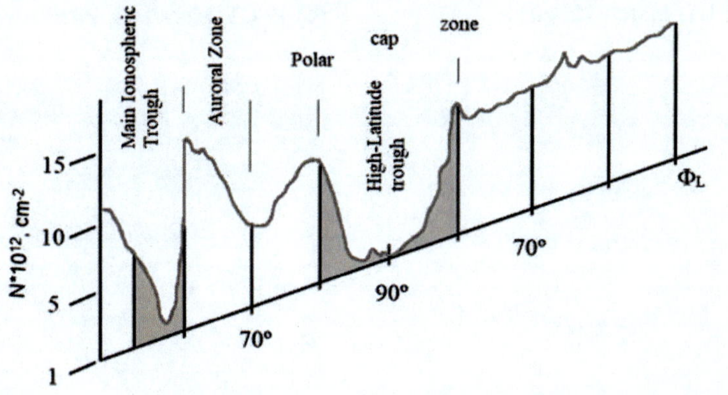

Fig. 3.14 Schematic representation of typical large-scale structures in the high-latitude ionosphere during wintertime for 0–12 LT. N characterizes latitudinal distribution of integral content of electrons (from the ground to 1000 km). Φ_L are the invariant latitudes. Modified from (Re-produced, modified and translated from Russian based on figure from the published book (Blagoveshchensky 1981))

boundary between the mid-latitude ionosphere formed mainly at the expense of solar ultraviolet emission and the auroral ionosphere where the major source of ionization is the intense particle fluxes.

Auroral zone is a zone of the form of a ring shifted in latitude to the night side (relative to the magnetic pole), over which auroras are frequently observed. It is not circular but oval. Usually, the auroral oval has its maximum near the geomagnetic latitude of 67° at local midnight and 77° at local noon. The position, width and form of MIT and auroral oval substantially depend on magnetic activity and local time.

Polar cap zone or near-pole ionosphere is located in a quasi-circular area centered at the geomagnetic pole. It is often called the ionosphere connected with open magnetosphere. Here, the rather chaotic corpuscular invasions are observed, magnetic field disturbances are weaker than in the auroral zone and Ne diurnal variation is much less pronounced.

During geomagnetic disturbances the high-latitude structures tend to expand to lower latitudes changing radically radio propagation conditions (Hunsucker and Hargreaves 2003). To provide an example, the auroral oval position was estimated using the Auroral Precipitation Model during the MP of the storm and at the same hour of the day for the quiet geomagnetic conditions. The model is freely available at http://apm.pgia.ru/webtool/frontend/about (Vorobjev and Yagodkina 2005, 2007). The results are shown in Fig. 3.15.

The subject of high-latitude ionosphere is very complex even under quiet conditions not to mention during SW disturbances. The monograph (Hunsucker and Hargreaves 2003) is recommended as further reading.

3.8 Low-Latitude Ionosphere

The most prominent feature of the low-latitude ionosphere is the *Equatorial Ionization Anomaly* (EIA) (Appleton 1946; Rastogi 1966; Anderson 1973; Moffett 1979

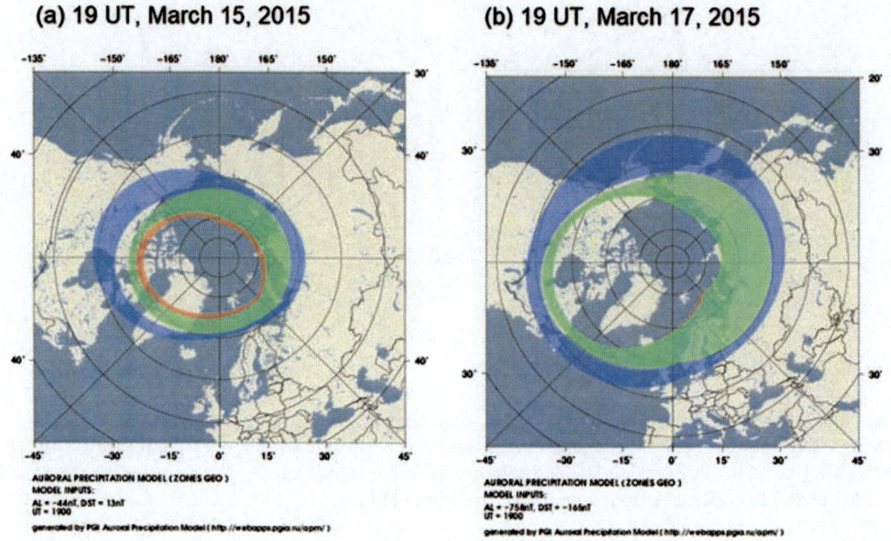

Fig. 3.15 Modeling results **a** for the quiet and **b** disturbed day. Blue color stands for the diffuse auroral zone equatorward from the auroral oval, green for the auroral oval precipitation region and orange for the soft diffuse precipitation

Abdu et al. 1991; Balan et al. 2018). Two ionization maxima (crests) are formed on both sides of geomagnetic equator approximately within the geomagnetic latitudes (10º–20º) and a trough is formed over the geomagnetic equator. This phenomenon is also known as Appleton anomaly. Figure 3.16 illustrates two anomaly crests on either side of equator. Here, Total Electron Content again should be regarded as a representative for Ne at F2 heights. The structure of the enhanced Ne in Fig. 3.12 discussed earlier (in Sect. 3.6) is a Northern crest of EIA. Physically, EIA is attributed to the so-called "fountain" effect explained in Fig. 3.17a. This plasma fountain is due to the combination of the directions of the electric field E (eastward during the day) and the magnetic field B (northward) over the equator which creates E $\times$ B plasma drift in the direction transverse to both fields meaning upward at heights of F-region (Kelley 2009). Therefore, the ionospheric plasma rises. Then, as geomagnetic field line dip changes northward and southward from the equator (increases), plasma gradually slides down the magnetic field lines, assisted by gravity (Kelley 2009). Large density gradients are developed in the horizontal plane (between the latitudes covered with crests and those that are not). Sometimes, the northern and southern crests are non-identical (Hargreaves 1992; Walker 1981; Huang et al. 2013). The regional features of EIA at different longitudes and its role during SW events are actively studied (e.g. (Walker 1981; Mo et al. 2018; Hanson and Moffett 1966; Venkatesh et al. 2015; Kumar et al. 2016; Bolaji et al. 2022; Karpachev 2021)).

The irregularity characteristic for the low-latitudes is *plasma bubbles* or *equatorial Spread F* phenomenon (Ezquer and Radicella 2008; Li et al. 2021; Oliveira

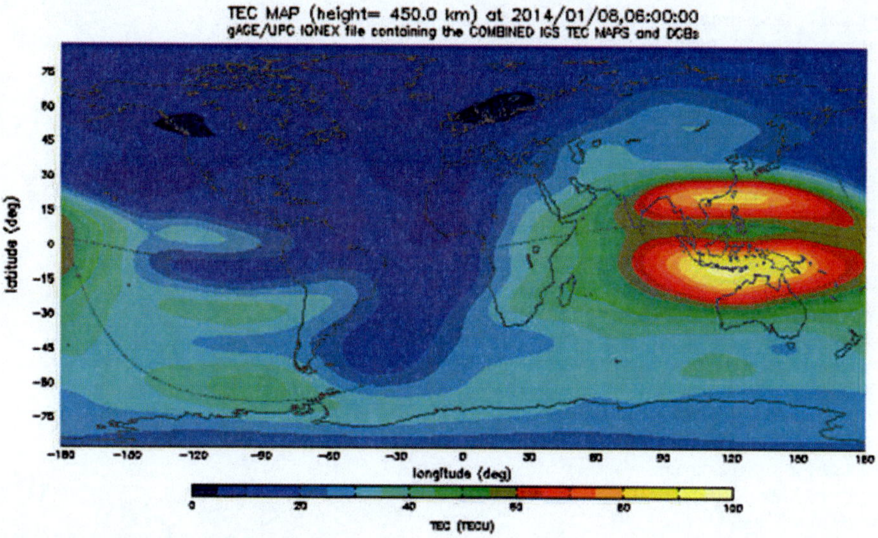

Fig. 3.16 GIM Total Electron Content map (Hernandez-Pajares et al. 2009) constructed for 06 UT January 8, 2014 (solar maximum conditions)

et al. 2020). The bubbles are plasma depletions with respect to the background ionosphere. Equatorial plasma bubbles (EPBs) are developed when low-density structures with complex shapes move vertically (upward) through the higher density medium (Woodman and La Hoz 1976; Kil 2015; Yokoyama et al 2015; Hickey et al. 2020 and references therein). The process is illustrated schematically in Fig. 3.17b.

Spread-F is defined in terms of radio propagation and consists in the scattering of the radio signal from electron density irregularities embedded in the in F-region of the ionosphere which is the reflecting layer (Davies 1965; Glover 1960). This results in the diffuseness near the critical frequency of the F2-layer (the highest frequency reflected vertically from the F2-layer) or the range diffuseness of the signal trace in ionograms (graphical representation of the height-frequency characteristic measured by ionosondes) (McNamara 1994)). Spread-F has been extensively studied (e.g. (Kelley et al. 2003; Abdu et al. 1981; Cohen and Bowles 1961; Wang et al. 2019; Gonzalez 2021; Thammavongsy et al. 2020)), though its spatiotemporal variability is still not fully understood.

All-sky imagers with a narrowband filter at 630.0 nm capture airglow emissions generated at ionospheric heights. When plasma irregularities are present they are detected as airglow depletions or dark bands indicating electron density depletions (e.g. Martinis et al. 2019, 2015). An example of plasma/airglow depletions associated with equatorial spread-F is presented in Fig. 3.17c. During extreme SW events these depletions can reach magnetic latitudes close to 40° but it is not always so.

Low-latitude irregularities can have a significant effect on transionospheric radio propagation, satellite communications, navigation and geodesic systems (Kil 2015; Karan et al. 2020). Therefore, they are recognized as important ionospheric features

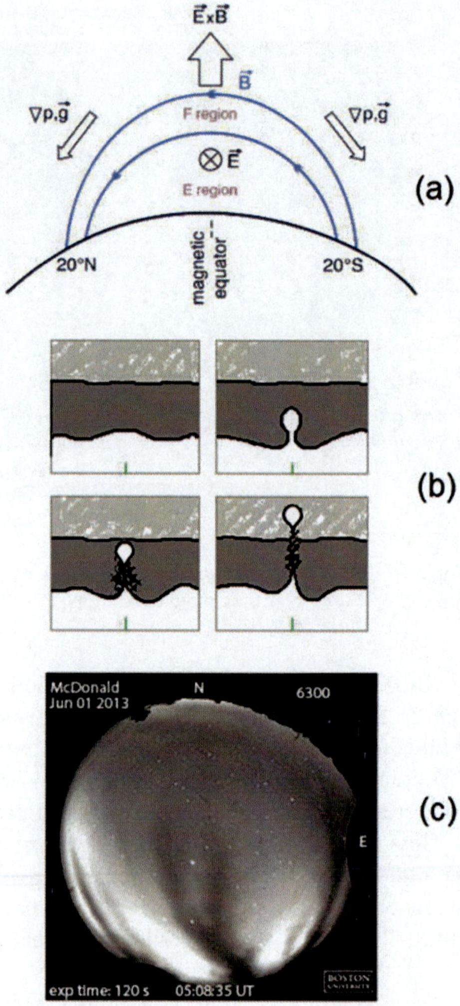

Fig. 3.17 **a** Schematic explanation of the fountain effect (Re-produced and modified from (Kelley 2009)) where g means gravity and p pressure gradients. **b** Schematic explanation of plasma bubble phenomenon (Re-produced and modified from (Woodman and La Hoz, 1976)). **c** 630.0 nm image from all-sky imager at the McDonald Observatory. Large equatorial spread-F structures are observed reaching zenith. The field of view of the imager covers Northern Mexico and South West United States (courtesy of Dr. Carlos Martinis).

in regard to SW. For further detail on the dynamics and processes in the equatorial zone the reader is referred to (Kelley 2009).

To sum up, different Ne maxima are present in the ionosphere. They are subject to daily, seasonal, annual/semi-annual and solar cycle changes. In addition, SW disturbances of different types (solar flares, coronal mass ejections, changes in solar wind parameters, particle precipitation, etc.) cause alterations in the complex ionospheric structure. It is important to detect systematic variations and the presence of anomalies in each region of the globe, as well as to better understand the physical mechanisms that cause them. Based on this knowledge, the level of possible damages in modern technology due to alterations in the state of the ionosphere can be estimated. This field of studies is connected with the fields of electrodynamics, electromagnetism and radio propagation.

In conclusion, it is worth noting that the ionosphere is a very sensitive and exposed to a variety of effects environment. It responds not only to SW events, but also to the slightest changes of other nature (Afraimovich et al. 2013; Ding et al. 2014; Polekh et al. 2019; Xiao et al. 2007; Song et al. 2013; Mikhalev et al. 2019; Shpynev et al. 2017; Edemskiy and Yasyukevich 2018; Chernogor 2020, Shagimuratov et al. 2020). First of all, it should be mentioned the influences "from below" (Lastovicka 2006, Kazimirovsky et al. 2003, Afraimovich and Perevalova 2006). Formation of plasma irregularities can be a response to earthquakes of high magnitude, strong volcanic erruptions, explosions of different nature (e.g. meteor explosion) and tropical cyclons. Ionospheric parameters reveal signatures of solar terminator passage, industrial noise (for example, power transmission lines), meteoroid flight and destruction in the atmosphere, etc.

More information on ionospheric physics and basics of the ionospheric studies can be found in (Beynon and Brown 1958; Kelley 2009; Prölss 1995; Mendillo 2006; Buonsanto 1999; Kazimirovsky et al. 2006; Friedrich and Rapp 2009; Lastovicka 2002; Reinisch and Galkin 2011; Rishbeth 1997; Ratovsky et al. 2018; Maltseva and Nikitenko 2019; Rishbeth and Mendillo 2001 and references in these works).

3.9 Overview on the Ionospheric Indices

Different ionospheric parameters are used to estimate Ne and its regular (systematic) and irregular changes (perturbations) of different spatial and temporal scales. Historically, the most commonly used parameters are the critical frequencies of the F2 and Es ionospheric layers (foF2 and foEs) that directly represent the maximum Ne of these layers and their virtual heights (h'F2 and h'Es) (McNamara 1994). These parameters are obtained by vertical ionospheric sounding (VIS) performed using ionosondes of various types. The oblique ionospheric sounding (OIS) technique is applied to obtain the Maximum and Lowest Observed Frequency (MOF and LOF) values along the particular radio paths. There is a large family of other important parameters that can be obtained via ionospheric sounding experiments, but we do not labour this point, as it does not concern the main subject of this book.

GNSS signals received by ground stations can provide us with much information about the ionosphere state (for instance, (Davies and Hartmann 1997)). The most popular parameter derived from GNSS data is the already mentioned Total Electron Content (TEC). It is defined as the number of electrons in a column of unit cross-section (Davies and Hartmann 1997, Afraimovich and Perevalova 2006) and serves both for research and operational tasks. It would be no exaggeration to say that in recent years TEC has become the most popular parameter for ionosphere diagnostics due to the availability of GNSS signals at any time and in all weather conditions. Moreover, GNSS stations have almost global coverage unlike other ground-based instruments such as radars and ionosondes (Blagoveshchensky et al. 2018). TEC calculation is discussed further in the book in detail. Usually, it is based on data of phase and code delays of GNSS satellite signals received by dual-frequency ground receivers. The absolute TEC value can differ depending on the method of its calculation (Arikan et al. 2003). Though TEC is an integral parameter (electron content from the satellite to the ground), it is assumed that it characterizes the F region of the ionosphere as this layer provides the main contribution to it. In general, it is correctly reasoned that the TEC value is mostly affected by changes in the ionospheric electron density approximately at the height of the F2-layer density maximum; consequently TEC is used as an ionospheric parameter. At the same time, plasmaspheric and tropospheric phenomena also can impact TEC. We urge the reader to consider this fact when applying simplification and approximation to the concept of TEC for each particular task.

As in the case of solar and geomagnetic condition studies, scientists seek simple and convenient indices to describe the ionosphere state for SW tasks. It can be stated, with some reserve, that the ionosphere morphology and state during disturbances depend on the coordinates significantly more than the geomagnetic field variations. Despite the possibility of anomalies and peculiarities of Earth's magnetic field in different regions (Campbell 1997), such global phenomena as geomagnetic storms may be described with some global geomagnetic index that provides us with the overall picture of the event on a global scale (Dst, Kp, etc.). That is impossible for ionospheric conditions whose variability highly depends on the local conditions. For instance, (Blagoveshchensky et al. 2018) provides examples of geomagnetic storms that caused different types of ionospheric response (positive/negative; different structures; effect presence/absence) in different regions. There are different helpful indices, but none of them is used to characterize the ionospheric conditions on a global scale. Some of the widely used ionospheric indices are mentioned below.

T-index characterizes HF propagation conditions (Turner 1968; Caruana 1989; Wilkinson 1995; www.sws.bom.gov.au/Educational/5/2/1). To obtain its value, a large amount of historical data on maximum usable frequency (MUF) is needed (during different solar cycles). MUF is the highest frequency at which a radio wave can propagate trough reflection in the ionosphere between two terminals of the particular path. MUF is measured using ionosondes. Its values are higher during the high solar activity (meaning the wider bandwidth for HF communications use) and vice versa. A linear relationship is fitted between MUF and SSN (separately for each hour of the day, each month and each observation point). When based on recent MUF observations, this relationship serves as a "predictor" for the ionospheric conditions and is called the "equivalent SSN" or T-index. It is regularly calculated by the SW Services of the Australian Bureau of Meteorology using global foF2 data converted to MUF.

W-index is the ionospheric weather index that allows us to reveal the ionospheric disturbances and estimate their intensity and sign (Gulyaeva et al. 2013; Stanislawska and Gulyaeva 2015). It is defined by the level of the logarithmic deviation of the observed TEC value from its median of the 27 days prior to the day of observation: DTEC = log(TEC/TECmed). W-index has nine values: intense positive W + storm (W = 4, DTEC > 0.301), moderate W + storm or substorm (W = 3, 0.155 < DTEC

≤ 0.301), moderate W + disturbance (W = 2, 0.046 < DTEC $\leq$ 0.155), quiet W + state (W = 1, 0.0 < DTEC $\leq$ 0.046), reference quiet state (W = 0, DTEC = 0), quiet W- state (W = −1, −0.046 $\leq$ DTEC < 0.0), moderate W- disturbance (W = −2, −0.155 $\leq$ DTEC < 0.046), moderate W- storm or substorm (W = −3, −0.301 $\leq$ DTEC < −0.155) and intense negative W- storm (W = −4, DTEC < −0.301). DfoF2, calculated similarly to DTEC, can be used instead of DTEC as a basis for the index.

WTEC is the index of TEC vertical variation intensity which shows a regional TEC disturbance that is equal to the mean TEC variation (with periods less than the specified ones) amplitude for a single GPS/GLONASS site (Voeykov et al. 2016, 2018).

The disturbance ionosphere index (DIX) (Jakowski et al. 2012; Wilken et al. 2018) also describes the disturbance level of the ionosphere while avoiding being related to the exact propagation conditions at the measurement site. It indicates the probability of a potential impact on radio systems used in communication, navigation, and remote sensing based on GNSS measurements (Jakowski et al. 2012). The approach for DIX calculation may depend on the user's task. More details can be found in the cited works.

The global electron content (GEC) is also one of the important ionospheric indices (Afraimovich et al. 2008; She et al. 2008; Gulyaeva and Veselovsky 2014). It is the total number of electrons in the near-Earth space environment within the GPS orbital altitude of about 20,200 km. It discards the peculiarities of local/regional ionospheric features and serves to estimate the ionosphere dynamics. Basically, it reflects the dynamics of the solar cycle as it is "modulated" by solar activity.

The Along Arc TEC Rate (AATR) allows us to identify the disturbances significant for GNSS performance at a regional scale (Juan et al. 2018; Berngardt et al. 2020). It is well correlated with ionospheric and geomagnetic activity. AATR index is used for the International Civil Aviation Organization studies in Satellite-based Augmentation System (SBAS) (Juan et al. 2018).

It is worth noting a large family of TEC-based indices. The most popular of them are the rate of TEC change (ROT), Rate of TEC changes Index (ROTI) and its derivative DROTI (Pi et al. 1997, Bhattacharryya et al. 2000, Gogie 2021). They serve to detect different ionospheric irregularities. The definition of these indices is provided in Chap. 4 (Table 4.1). Please find the details on each index in the cited papers and references therein.

The most often used scintillation indices are S4 and σ_ϕ. Scintillation is the phenomenon of rapid fluctuation of the radio wave amplitude and/or phase caused by irregular ionospheric structures of different scales and origins (Aarons 1982; Yeh and Liu 1982; Doherty et al. 2003; Salles et al. 2021; Priyadarshi 2015). Scintillations have a negative impact on radio systems performance (Kintner et al. 2007). S4 and σ_ϕ are the standard deviations of amplitude and phase of GNSS signal, respectively (Briggs and Parkin 1963; Yeh and Liu 1982). The threshold values of S4 and σ_ϕ that indicate the intensity of scintillations are defined arbitrary. For example, according to (Dubei et al. 2005), the thresholds S4 > 0.6, 0.4 < S4 $\leq$ 0.6 and S4 $\leq$ 0.4 indicate strong, moderate and weak amplitude scintillations. $0.5 < \sigma_\phi \leq 0.6$, $0.25 < \sigma_\phi \leq$

0.5, $0.1 < \sigma_{\phi} \leq 0.25$ (in radians) are the thresholds for the corresponding levels of phase scintillations (Dubei et al. 2005; Prikryl et al. 2016). Please find more details on these indices in Chap. 4.3.2.

Another interesting index that can provide us with information about the small-scale electron density irregularities in the ionosphere is D2fi. It is calculated as the second-order derivative of the carrier phase of the GPS signal (Demyanov et al. 2019a, Demyanov et al. 2019b). Please find more details in Chap. 4.3.3.

The list of indices mentioned above is not comprehensive as there are more without a doubt. More details on different ionospheric indices can be found in the work (Borries et al. 2020) and references therein.

Scintillations, TIDs, plusma bubbles, high-latitude ionospheric phenomena, etc. can be considered as subjects of independent areas of research within the ionospheric studies. In some cases, special dedicated tools are needed to detect certain phenomena (especially the short-lived small-scale structures). For instance, SIMURG (Yasyukevich et al. 2020) provides many useful instruments to study TIDs and their spatiotemporal scale.

In general, the choise of the appropriate ionospheric index or parameter depends on the task to be solved and data availability (similarly to geomagnetic index). It also depends on the type and scale of disturbances to be detected and the region under analysis. There are active discussions in the scientific community on the possibility of the use of some ionospheric index globally (though based on local data) and on how to move from the ionospheric indices towards the standardized activity scales for SW Services (for instance, (Wilken et al. 2018)). The task is very comprehensive due to the fact that the ionosphere and its responses are very different in different zones to be characterized with the same measure that could be adopted internationally.

3.10 SW Disturbance Consequences

Why the role of the SW state (especially during disturbances) is so important in present life? Every year society becomes more and more dependent on fine modern technology that in turn is vulnerable to SW perturbations inclusive ionospheric disturbances. The possible consequences of SW intense events are widely discussed in the literature. The reader is referred to the following works and references therein (Cannon 2013; Baker 2002, 2004; Bothmer and Daglis 2007; Gary and Keller 2004; Kappenman 2012; Moldwin 2008a; Echer et al. 2005; Koskinen et al. 2001). Here, we briefly mention the main negative SW effects.

A *solar flare* causes increased radiation in the far ultraviolet and soft X-rays which penetrate the lower ionosphere (D region and lower E region) resulting in a sudden Ne enhancement. In the worst case, this can lead to the total absorption of energy of radio waves that propagate by reflection in the ionosphere. The waves of the HF band, especially of lower frequencies of this range are most vulnerable (both voice communications and data links). The regions on Earth with a smaller solar zenith angle are more affected as they receive more energy from the flare. The most pronounced effect is at the sub-solar point (local noon). HF blackouts (the whole or a part of the HF spectrum unusable) can last from several minutes to hours (Daglis 2005). Even if there is no blackout, HF communication can be degraded. The low-frequency range also is vulnerable (Hayes et al. 2021). Communication links for the marine and aviation sectors can be affected. Solar flare radio noise can affect the performance of mobile networks. Increased satellite navigation errors can be observed (Tsurutani et al. 2009).

Solar Energetic particle events can result in increased radiation, HF signal enhanced absorption, navigation problems, dielectric charging, malfunctioning or failure of the components of satellites and ships. The discharge can affect the electronics (inclusive microelectronics), causing variations in data, false commands and even damage to the components. Particles with energy > 40 meV are a risk to life in space (to cosmonauts and astronauts at the international space station and even to crew and passengers of commercial aviation of high-latitude routes). (Lee et al. 2014; Clement and Sasaki 2013). There are no conclusive studies on effects on human health within the atmosphere (at the ground level).

Geomagnetic storms have an impact on a variety of applications (Baker 2002). Most of the effects mentioned below are related to drastic changes in the ionosphere, for example (Goodman 2005; Demyanov and Yasyukevich 2021). There can be negative impacts on operation of telecommunications systems, navigation systems, over-the-horizon radar, electrical power generation and distribution networks as well as oil and gas pipelines, railway lines and underwater communication cables (Eroshenko et al. 2010; Thaduri et al. 2020; Viljanen and Pirjola 1994; Viljanen 1997), satellite services (solar cell damages, control problems including space traffic management, orientation and tracking errors, charging of surfaces), low altitude satellite drag, global positioning systems and their augmentation segments (positioning precision degradation or, in the worst case, even collapse), radio communications for land, sea and air services (signal absorption, attenuation, interference, amplitude and phase scintillations, change of operation frequency range (wavelengths at which communication is possible) including unavailability of the higher frequencies of the HF range, etc.).

High-latitude HF radio paths are particularly sensitive to SW phenomena (e.g. Warrington et al. 2004; Blagoveshchensky et al. 2019; Blagoveshchensky et al. 2016; Hunsucker and Hargreaves 2003). Signal distortions, fadeouts and the total absence of HF communication are not rare in this region. Rapid vertical movements of the ionospheric layer heights during geomagnetic storms can introduce errors in operation of different radio systems, affect the accuracy of marine navigation as well as cause errors in the positioning of objects with the use of GNSS signals (Goodman 2005; Daglis 2005; Blagoveshchensky and Sergeeva 2018; Ding et al. 2008). The effects exerted on GNSS systems hold a special place in this list and are the focus of this book. GNSS radio signals emitted at the frequencies of L-band pass through the atmosphere, including its ionospheric part, and can experience delays of the modulated carrier signal (group delay), resulting from signal interaction with the free electrons of the ionosphere (Soicher 1977; Klobuchar 1983; Davies and Hartmann 1997), the phase advance and Doppler shift of the radio frequency carrier signal, Faraday rotation of the plane of polarization of linearly polarized waves, angular refraction or bending of the path of the radio wave as it passes through the ionosphere; signal waveform distortion; and scintillations (Klobuchar 1983; Aarons et al. 1996). During geomagnetic storms, some satellites come to be out of the magnetosphere lacking protection and are more exposable to solar wind's drastic influence (Lanzerotti 2001; Jachhia 1959). Different control systems can be affected significantly.

To summarize, the intense SW storms can seriously affect five strategic sectors of the national economy (Fig. 3.18): power generation and distribution systems (Marshall et al., 2013; Pilipenko 2021; Caraballo et al. 2020), aviation (e.g. Jones 2004; Lucke et al. 2013), satellites (e.g. Horne and Pitchford 2015; Rycroft 2013), global positioning systems (e.g. Zhbankov et al. 2022; Petrie et al. 2011) and communications (e.g. Redmon et al. 2018; McNamara 1994; Tulunay and Bradley 2004; Lefeuvre and Tanzi 2014). Indeed, the major part of modern technological systems, especially those that rely on GNSS timing and positioning data and communication links (both ground and satellite), are under significant risks related to SW impacts. To add, the effects of the induced currents can accumulate and then manifest themselves even during the weak geomagnetic disturbances (Barbosa et al. 2015). In this book, the emphasis is placed on the effects on different GNSS system performance.

To mitigate these deleterious effects on technology systems, a reliable diagnostics of the near-Earth space, particularly the state of the ionosphere, is critical. Therefore, SW has become the subject of national and international programs involving significant efforts of researchers, planning and operating entities.

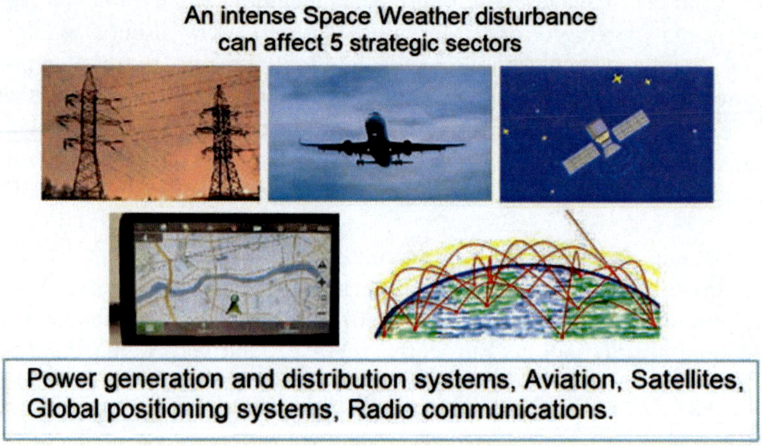

Fig. 3.18 Strategic sectors of national economies vulnerable to SW events

3.11 Space Weather as a Natural Hazard

In regard to SW, the diagnostics of conditions in near-Earth space is one of the main tasks for the continuous safe and reliable operation of these systems. There is historical evidence of critical infrastructures vulnerability to SW. It is noteworthy, that modern technology (e.g. satellite systems) is rather young in terms of SW threats. Indeed, the space era began with the first artificial satellite launched in 1957 under the leadership of the founder of practical cosmonautics, academician Sergei Korolev and the first manned flight by Yuri Gagarin in 1961. Many other technologies highly dependent on SW were developed only recently compared to the estimated average period between severe SW events.

Up to now, the most intense SW event known is the so-called "*Carrington event*" which occurred on September 1–2, 1859 (Carrington 1859). It was discussed widely in a number of works (Cliver 2006, Hayakawa et al. 2019; Keika et al. 2015; Morina et al. 2019; Gonzalez-Esparza and Cuevas-Cardona 2018; Muller 2014; Lakhina et al. 2005). During geomagnetic perturbation within this event the Dst-index by different estimates reached the values within the range of ~ (− 850 ÷ − 1760) nT (Lakhina et al. 2005; Siscoe 2006; Love et al. 2015; Ngwira et al. 2014 and references therein). Though the estimated value varies in a wide range (Siscoe 2006), any of the suggested scenarios seems extreme.

According to (Cannon 2013), a superstorm like the Carrington event can occur once in ~ 100 years. Nonetheless, it should be taken into account that our statistics are rather poor to provide a reliable estimate. Had there been strong storms in history before 1859? Most probably, yes. However, there were neither contemporary technologies that could have been drastically affected nor registered SW parameter deviations. For instance, the geomagnetic field data can be found at longest during the last ~ 180 years, not to mention much less period of satellite measurements (Cannon 2013). In 1859 the most noticeable reported effect was the disruption of telegraph services (Boteler 2006). The more recent superstorm events occurred during *October–November 2003* (most famous as "Halloween storms", Dstmin = − 422 nT). This period attracted much attention, which resulted in a long list of publications focused on different aspects of these events, for instance (Gulyaeva et al. 2013; Gulyaeva and Stanislawska 2013; Gordienko et al. 2005; Mannucci et al. 2005; Poppe et al. 2004; Blagoveshchensky et al. 2006; Zhao et al. 2005a; Lakhina et al. 2005; Borries et al. 2017; Jin et al. 2008) and many others. The *Quebec Power Outage of 1989* due to SW disturbance can also be mentioned (Dstmin = −589 nT) (Lakhina et al. 2005 and references therein, Béland and Small 2004). The more recent events in Mexico are worth mentioning. In this case, the SW conditions were not extreme like those made known above. At the same time, it is an interesting example in terms of civil protection management, as several natural hazards almost coincided in time in the region (see Fig. 3.19). Gonzalez-Esparza et al. (2018) discussed these events in detail.

During *September 4–10, 2017*, multiple X-class intense solar flares occurred and the CME interaction with Earth's magnetosphere resulted in the geomagnetic storm

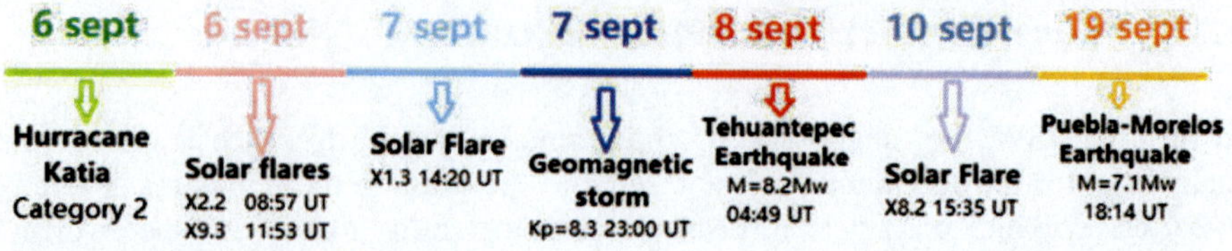

Fig. 3.19 Timeline of events

(Dstmin = −124 nT). At the same time, three hurricanes were causing destructions in the Gulf of Mexico and the Caribbean. In particular, Katia (category 2) affected at least 53 municipalities in the country and Irma (category 5) produced significant damages in Antigua and Barbuda, Saint Martin, the Bahamas, Cuba and the United States. Moreover, two major earthquakes in Mexico on September 8th and 19th with magnitudes of 8.2 Mw and 7.1 Mw correspondingly caused severe damages and took lives. Such a combination of almost simultaneous events required significant efforts from civil protection organizations and rescue services. It was particularly remarkable that the SW disturbances affected radio communications for the emergency management operating in the Caribbean region at that moment (Redmon et al. 2018; Gonzalez-Esparza et al. 2018). Fortunately, the flares occurred early in the morning and thus did not cause major blackouts and the geomagnetic storm was not severe. Otherwise the rescue communications could have been more affected. However, these events are an illustration of how simultaneous SW and other natural hazards can represent a significant danger.

Aiming to understand the risks, *attempts are made to estimate* the occurrence probability and the level of the consequences of SW disturbance comparable by its intensity to the Carrington event if it happened now, for example (Townsend et al. 2006; Ngwira et al. 2014; Bala et al. 2015; Ebihara et al. 2021; Baker et al. 2008; Hapgood 2011; Keika et al. 2015; Yermolaev et al. 2013; Love et al. 2015; Odenwald and Green 2007; Curto 2020; WMO 2011; Ishkov 2021) and others. An extensive analysis of risks is made in the work freely-available at www.oecd.org/gov/risk/46891645.pdf (last access: February 25, 2022).

It must not escape our attention that the SW extreme storms can provoke a *cascading effect*. Imagine the power supplies affected. As an effect, ATM and banking operations stop. Water reservoir substations, work of hospitals, transport operations would be interrupted without electricity. Not to mention transport being affected by disrupted communications and errors of navigation applications. In turn, transport problems might result in delays in food supplies delivery. An interesting survey of customers of SW information was reported in (Schrijver and Rabanal 2013). According to the authors, about 40% of the respondents expected serious SW consequences if not prepare for it or without adequate information on SW conditions. As infrastructures are dependent on each other, a deleterious impact can be widespread. To provide examples, during the Halloween storms there were reported the power system failure in Sweden, high current levels in transmission lines of Wisconsin and New York, high-latitude GPS receiver outages and Wide Area Augmentation System

(WAAS) service interruptions. Many satellites and spacecrafts experienced spontaneous computer resets, solar panels rotated, contact loss, etc. Many missions required preventive actions from operators. Airlines rerouted polar flights because of bad HF/VHF communication. Trans-Atlantic routes moved south. Astronauts onboard the International Space Station went into the service module for radiation protection (Barbieri and Mahmot 2004; Doherty et al. 2004; Webb and Allen 2004; Cannon 2013).

It should be emphasized that apart from the complex technological problems, the economic aspect should be considered because the economic losses may be significant (Oughton 2018; Eastwood et al. 2017). Furthermore, there is also a social aspect of the events. For instance, what a drastic change of plans, expenses increase and stress would experience people if the majority of flights were redirected and cancelled. In addition, apart from technical and socio-economical impacts on modern society, ethical issues are also discussed (Ritter et al. 2020).

3.12 Space Weather Initiatives

3.12.1 What Can Be Done?

Many studies are devoted to the development of complex theoretical and empirical models that help both to understand what happens under some scenario and to forecast SW conditions. Making predictions and forecasts is a sophisticated line of research. The already known climatological features of ionospheric, atmospheric, thermospheric and magnetospheric behaviour help us to elaborate forecasts (mostly, short-term) during the periods of quiet SW conditions. During disturbances, the complexity of underlying physical processes and the sudden (unexpected) occurrence of some SW manifestations makes the task of short-term forecasting much more difficult (Daglis et al. 2021). The probability of solar flares can be estimated by monitoring active regions on the Sun, but we can not be certain if a flare is going to happen in the next moment and of what category as it is a sudden eruptive event and it takes only a few minutes for radiation to reach the Earth (Fig. 3.7). At the same time, the scientific community makes advances in this field, for example (Gallagher et al. 2002; Wheatland 2005; Park et al. 2018; Junior et al. 2018; Barkhatov et al. 2020; Abramov-Maximov and Bakunina 2020; Bakunina et al. 2020; Abed et al. 2021) and other scientific articles. Forecasting of magnetic cloud arrival from CME or interaction region seems more promising as CME material and compression region need more time to reach the Earth and cause magnetic storms. ENLIL (Pizzo et al. 2011) is one of the most popular models used to make such predictions. Modelling based on European Heliospheric Forecasting Information Asset (EUHFORIA) (Pomoell and Poedts

2018), SUSANOO (Shiota et al. 2014) and an early-alert system SPArToS (Corona-Romero et al. 2017; Corona-Romero and Riley 2020; www.sciesmex.unam.mx/productos-y-servicios/experimentales/spartos) are also worth mentioning. However, our estimations are still far from being precise. Moreover, the ionospheric response to geomagnetic storms can be rather different in each case (e.g., Zolotukhina et al. 2018). Here, we do not go further into details. The medium- and long-term forecasts can be even more difficult. In regard to operational risks, the estimates of the probability of different types of events and their intensity may be helpful. For instance, such estimations may be derived from the NOAA SW scales mentioned earlier. Cannon (2013) provided an interesting assessment of the likelihood of experiencing a severe event against system lifetimes (12% risk per decade). According to this work, if a system is planned to be operational for more than 8.25 years then it should be designed with allowances made for the severe event occurrence.

The relevance of the SW problem also comes from the potentially catastrophic effects it can have on economics, the life of society and each person. This is because an extreme event could temporally or permanently disable critical communication infrastructure, electric power generation and distribution systems in large regions of the planet. Some of the technical damages can be unrecoverable. However, even in the case of a non-catastrophic but rather intense solar storm, the impacts on technology products can be significant (Lanzerotti 1992), prevent the users of different applications from using them (as in the Mexican case discussed above), reduce the useful life of technologies which often are not rapidly replaceable, and eventually change our everyday life. The demand for SW monitoring and understanding was urged in the middle of the twentieth century because of the development of space-based technology and communications and the growing society's dependence on technological systems that are vulnerable to the negative effects of solar activity. We recall that SW perturbations have repercussions on a global scale but different regions on Earth can be affected to a different extent or in a different way. Hence, SW programs have been promoted in many countries studying the underlying physics (research), providing the historical data and real-time information about the SW conditions (monitoring services) and elaborating of specific action plans for risk mitigation (government agencies).

3.12.2 National Space Weather Services

SW service is an organization that provides users with SW related products. Usually, it is an interdisciplinary group that includes researchers, experts and technicians. A national SW service mission is to provide information on current conditions, estimate the possibility of affectations on the national territory predicated on basic and applied research in heliophysics and solar-terrestrial physics. To accomplish this mission, a Service usually operates scientific instruments (optical and radio telescopes, radars, GNSS receivers, ionosondes, riometers, magnetometers and many kinds of other sensors (Zherebtsov 2020)), collects, processes and interpretes a variety of data.

The SW service initiatives are discussed in a number of works, e.g. (Simpson 2003; Belehaki et al. 2009; Onsager 2012). Their main objective is to offer the users the products that meet their requirements. Let us mention some tasks addressed by SW services.

(a) Provision of information about the SW state.

The key issues for SW service operation are the monitoring in the quasi real-time (usually, there is a some-minutes-delay as the data from instruments are downloaded and processed before being shown), current SW diagnostics based on this monitoring, short-term forecasts and a client-oriented analysis of specific events (on demand). A continuous monitoring of current conditions may combine both automatic features and operator work 24/7. The products may include nowcasting and short-term forecasting (to the extent possible) of SW conditions, early-warnings and alerts of solar, magnetospheric, geomagnetic and ionospheric conditions. When a strong event happens, a SW service is expected to answer such questions as: "What is happening?", "What can happen in 12, 24, 48 and 72 h?", "Is it dangerous? If yes, for what kind of applications and technological sector?", "What are the most possibly affected areas on Earth/in the country"? A large part of the work is dedicated to data collection, pre-processing, processing and classification followed by data analysis and data exchange with other services. SW service operation relies on scientific research of SW physical mechanisms (case studies and statistical studies).

(b) Transition of scientific results into operational solutions.

Essentially, it results in "translation" of the scientific SW knowledge into the semaphore lights and indices more convenient for users who are not specialists in the field of solar-terrestrial interactions. Usually, SW events are categorized with assigning to each category a specific color that for the particular user would mean "normal" conditions (e.g., represented by green light), "attention" (e.g., yellow light) or "alert" (e.g., red light). The last two may require some actions from operators. There can be more than three colors. This "translation" is not an easy task. Sometimes, the light/index interpretation is rather tricky. A semaphore should be designed for each user sector separately because the same physical phenomena can represent danger for one economy sector and make no (or different) effect on another. To provide an example, we recall that the positive ionospheric disturbance can cause an increased delay of satellite signals and at the same time extend a working frequency range for HF communications.

(c) Preparedness for an extreme event.

This topic is widely discussed during scientific and operational conferences, assemblies and other meetings of SW community that includes engineers, scientists, researchers, model and software developers, data providers, industry specialists and students (Bingham et al. 2019). One of the main questions to be answered is in what case the authorities should be warned and how. To answer it, the studies of worst-case scenarios and risk assessments are developed.

(d) Standardization and automatization.

Many recognized Services work on standardization of their data and products, automatization of the majority of processes within the service operations (including automatic alerts sent to colleagues and users) and quality management. The last includes product/service certification according to national and international standards, for instance International Organization for Standardization (ISO) norms. These activities aim to provide the products of the best possible quality and certitude complying with the delivery times and to address promptly and effectively the user needs and requests. The next two items may be also a subject for standardization.

(e) The development of regional hazard levels and their determination based on observational data from regional/local and international space- and ground-based tools and modeling.
(f) Development of communication protocols between the Service employees as well as with governmental agencies and other users.
(g) Assisting the public in understanding of SW.

SW service is a "guide" in the scientific word not only for different kind of users but also for general public. Each year the societal interest for this topic and awareness of SW adverse effects on critical infrastructure grows. SW services are expected to actively participate in public outreach activities.

The potential users of SW service products are the operators and decision-making responsibles in the sectors of the economy affected by the SW (Fig. 3.18), researchers, the mass media providing the public with information, government agencies and society in general. One of the common and freely-available products of SW services is the weekly report of SW conditions which represents a brief description and a preliminary analysis of what had happened during the last week. These reports are usually freely-available. Examples are given below (last access to the links: February 28, 2022).

www.stce.be/newsletter/newsletter.php (by the Solar-Terrestrial Centre of Excellence in Belgium).

www2.inpe.br/climaespacial/portal/weekly-summary/ (by National Institute of Space Research in Brazil).

www.sciesmex.unam.mx/blog/category/reporte-semanal-de-clima-espacial/ (by Mexican Space Weather Service).

www.sws.bom.gov.au/Geophysical/3/2 (by Australian Bureau of Meteorology, Space Weather Services).

Please note that the list of provided products and the particular tasks can vary significantly between the services.

3.12.3 Collaborative Efforts

National SW services work to address SW challenges and eventually to develop the national action strategies to protect space-based and ground-based technological systems *(as far as possible)* under the scenario of an extreme SW event. The coordination between different groups is done by the International Space Environment Service (ISES) (www.spaceweather.org). ISES members are the Regional Warning Centers (national SW services) and collaborative expert centers which provide SW forecasts, warnings, alerts, and environmental data to government and private-industry users around the globe. Each member has to correspond to the specific requirements to be accepted to ISES community. The mission of ISES is to deliver, coordinate and improve operational SW services. This is accomplished through the rapid exchange of information; sharing of best experiences for data analysis and product development; and the open dissemination of products and services. Regional Warning Centers must be endorsed by national governments as ISES SW service provider.

There are also other international entities that cooperate scientific and operational efforts of different working groups. These organisms define product requirements, establish standards (for instance, the exchanging file-formats), promote research collaborations and are involved in other activities including public outreach. Let us mention some of them (Fig. 3.20).

The United Nations Office for the Peaceful Uses of Outer Space (*UNCOPUOS*) plays an important role in SW-related activities. UNCOPUS experts work to enhance global resilience to SW (United Nations General Assembly 2021; Committee on the Peaceful Uses of Outer Space 2022). UNCOPUS provides a unique platform at the global level to monitor and discuss space exploration and related developments, international cooperation in space and the use of space technology applications. Information-sharing and cooperation are among the priorities. The international collaboration on SW is determined as one of the main demands of the 2030 agenda towards sustainable development, which promote the United Nations. Please see (Mann et al. 2018 and references therein) for more details. The International Space

Fig. 3.20 Logos of international organizations that unite efforts of different SW groups

Weather Initiative (*ISWI*) of the UNCOPUOS is a program that focuses on SW instrumentation, data analysis, modeling, education, training and public outreach.

The Committee on Space Research (*COSPAR*) of the International Council for Science (https://cosparhq.cnes.fr) promotes the scientific research, discussions and publications in the space field. COSPAR strategic action plan 2019–2023 is published at its official web-page (https://cosparhq.cnes.fr/about/cospar-strategy-statement). The International Living With a Star (*ILWS*) program (Opgenoorth et al. 2003; Fichtner and Liu 2011) has a mission is to stimulate, strengthen and coordinate space research. COSPAR and ILWS global roadmap for 2015–2025 is given in detail in the paper (Schrijver et al. 2015). This article provides an extensive guide to future goals and gives the specific recommendations grouped by physical domain. It highlights the need to increase investment and development of scientific infrastructure for SW observations, analysis and prediction, advances in data assimilation and modeling understanding the physics of the underlying processes, developing specifications. Such issues as the priorities for research to benefit the services, the need for better forecasts, an important role of collaboration between agencies/communities and information-sharing as well as the end-user requirements are addressed. The insufficient coordination between international and national organizations for SW is discussed in (Kuznetsova 2018). Let us make a short digression.

Indeed, a better communication between science and industry is needed to provide relevant, reliable and usable information to operators for decision making (Krausmann et al. 2016). Probably, there are many unknown cases of damages caused by SW effects on technology which are not reported by the corresponding organisms as most of them form part of critical infrastructures and their information is not public. More feedback researchers and forecasters have, more reliable will be their estimations, predictions and early warnings. That is why, nowadays, different initiatives of collaborative work are developed between scientists and infrastructure operators. For example, different panel discussions are organized aiming to unite the efforts in a manner that both research and industry may benefit from bringing them together (Bingham et al. 2019; Lilensten et al. 2005). Information on the current collaborative work can be found at International Space Weather Action Teams web-site (freely-available at http://ccmc.gsfc.nasa.gov/iswat). What is on the agenda of scientists, SW services and users (SW vulnerable applications) and how much interaction with various users (representatives of strategic sectors) is necessary can be understood from the review (Bingham et al. 2019).

The Scientific Committee on Solar-Terrestrial Physics (*SCOSTEP*) is a thematic body of the International Science Council (ISC) (https://scostep.org). Its focus is on awakening the interest for the solar-terrestrial topics among the students, promote data exchange in this research field and run long-term international interdisciplinary scientific programs.

The World Meteorological Organization (*WMO*) Space Programme (https://community.wmo.int/activity-areas/wmo-space-programme-wsp) aims to archive the maximal benefits from satellite data and space-related activities including for SW products and services. A four-year plan for WMO activities related to SW 2016–2019 and a four-year plan for WMO coordination of SW activities 2020–2023 are available at WMO web-site (https://public.wmo.int). WMO Inter-Programme Team on Space Weather Information, Systems and Services (IPT SWeISS) is a group of experts working on SW issues.

As discussed above, SW impact on operations in aviation sector may be drastic (Jones et al. 2005). The concerns are about the aircraft avionics (reboots and anomalies), communications (HF voice and HF data link, L-band SATCOM performance problems), degraded performance of navigation and surveillance that rely on GNSS and the radiation doses for crews and passengers. More details about risks in this sector can be found in International Civil Aviation Organization (*ICAO*) documentation (ICAO 2019b). Therefore, to avoid the risks posed to flight safety and to maintain the high safety levels, the provision of SW advisory information is needed. Operators for pre-flight planning and flight crew members should be supplied with information before departure (ICAO 2018). The manual published by ICAO (ICAO 2019b) specifies which parameters to be monitored by SW centers (Knipp and Hapgood 2019) and provides also thresholds of advisory issuances for two intensity levels: moderate and severe. SW advisories are issued by the designated by ICAO centers following WMO recommendations (ICAO 2019a). SW information provider should meet specific criteria. The requirements for emitted warnings to aviation are defined by ICAO (2018). Operational products are delivered to aviation decision-makers.

In 2019, *a 24/7 operational service* was launched to provide the real-time and worldwide SW information. Some of the established SW regional warning centers (see the list of ISES members) join in consortia to raise their chances to be elected as SW information provider for International Air Navigation needs. Consortium members fulfill SW monitoring (usually from different local time zones which contributes well the work) and forecasting providing products useful for aviation; make verification, validation and mutual sharing of information; work on quality management and data backups (e.g., ICAO 2019c).

At present, ICAO-WMO SW global centers are operated by: (a) ACFJ consortium (comprising Services of Australia, Canada, France and Japan) (ACFJ services for Aviation), (b) PECASUS consortium (comprising Services of Austria, Belgium, Cyprus, Finland, Germany, Italy, Netherlands, Poland and the United Kingdom) (Kauristie et al. 2021), (c) Space Weather Prediction Center (USA) (e.g. (Bain et al. 2021; Crown 2012)) and (d) consortium of People's Republic of China and the Russian Federation (CRC) (Kholodkov et al. 2021; Aleshin et al. 2021). The responsibilities are rotated between the global centers each two weeks. South African National Space Agency (SANSA) (McKinnell et al. 2021) is a regional SW detection center (with potential to convert to advisory center in the future). Recently, in December 2021, it was signed a Letter of Agreement to enhance the collaboration in monitoring and SW studying in the Latin American region. The aim is the creation of a consortium comprising Services of Latin American countries. The agreement was signed by three ISES regional warning centers (Brazil, Mexico and Argentine) (Denardini et al. 2016; Gonzalez-Esparza et al. 2017; Lanabere et al. 2020) and the University of Chile. The idea is to include more countries in the near future. The requirements for SW centers and their obligations are defined in the ICAO documentation (ICAO 2018). The optimal number of global and regional SW information providers is planned to be reassessed at the latest in 2027 by ICAO and WMO (https://www.icao.int/SAM/Documents).

To summarize, all these initiatives are aimed at being ready for strong and especially extreme SW event occurrence. Some countries have recognized the threat of extreme SW and included it in their strategic national risk assessment. They prepare the official documents that define the degree of risk and possible consequences in each particular region. To the best of the authors' knowledge, *national action protocols* have been developed in U.K., U.S., Republic of Korea, Mexico and other countries. They concern the risk management and responsibilities of ministries and organizations in the face of SW events. The importance of regional programs is due to SW impact dependence on coordinates (both geographical and geomagnetic latitudes and longitudes), local time and other factors (e.g., the presence of anomalies in the regular behavior of the magnetic field or the ionosphere). Certainly, the type of impact should be considered as well. For example, a powerful CME and subsequent intense geomagnetic storm would have a global impact. In contrast, a powerful solar flare or an intense particle precipitation would affect the illuminated side of the Earth and high latitudes (more vulnerable to precipitations), respectively. SW is a recognized global threat and a top challenge for any country. International collaborations are critical for advancing in SW research (Onsager 2012). An additional point to emphasize is that apart from scientific and instrumentation resources, SW research and services require substantial financial resources, the issue which is easier to address within the international collaborations.

In conclusion, it is important to note that *threats other than SW hazards* are present in near-Earth space. In some cases, those threats, such as those resulting from satellite failure or space debris solar-wind drag, can be provoked by SW events. SW impact in these cases, besides possible asset losses, may increase collision risk between anthropogenic space objects or cause ahead of time reentry of these objects into Earth's atmosphere. Those, in turn, can cause local disturbances of the ionosphere or, if reach the Earth's surface, can result in property damage and casualties. The following civil projects on tracking and registration of anthropogenic space objects can be mentioned: EU SST (Hermoso et al. 2021), LeoLabs (Rowland et al. 2021), ISON (Mokhnatkin et al. 2021). The Planetary Defense Coordination Office of NASA is currently responsible for the majority of discoveries and observations of near-Earth objects. The near-Earth space is also under the risks of complex meteoroid-meteor-meteorite phenomenon. In case of large objects entering Earth's atmosphere, apart from their obvious destructive consequences, they cause detectable local disturbances of the ionosphere (Chernogor 2014, 2020; Ruzhin et al. 2014; Sergeeva et al. 2021a). Essentially, these disturbances are related to the effects triggered in the object's wake in the atmosphere and waves excitation consequences (especially shock waves (Silber et al. 2018)). Notable examples include the famous Chelyabinsk meteoroid explosion in 2013 (Perevalova et al. 2015; Yang et al. 2014) and Tunguska explosion in 1908 (Chyba et al. 1993) and much less powerful Romanian meteoroid in 2015 (Borovicka et al. 2017) and Morelian meteoroid in 2020 (Sergeeva et al. 2021a).

Acknowledgements LANCE acknowledges support from CONACYT-AEM 292684, CONACYT LN 315829. The OMNI data (F10.7-index, solar wind parameters and geomagnetic indices) were obtained from the GSFC/SPDF OMNIWeb interface at https://omniweb.gsfc.nasa.gov. The SYM-H index used in this paper was provided by the WDC for Geomagnetism, Kyoto (http://wdc.kugi.kyoto-u.ac.jp/wdc/Sec3.html). The authors also thank NOAA's National Centers for Environmental Information for the opportunity to use GOES satellite data available at https://satdat.ngdc.noaa.gov/sem/goes. We thank Laboratory of Geo-Spatial Sciences (LACIGE), UNAM, Mexico and Dr. M. Rodríguez-Martínez in particular for solar H-Alpha images (grants: CONACyT INFR 253691, PAPIIT IN118119 and PAPIME PE103419). We thank Dr. Carlos Martinis for providing the image from the Boston University all-sky imager at the McDonald Observatory. We gratefully acknowledge the service SolarMonitor.org for the Sun image in soft X-rays. We thank the Auroral Precipitation Model developers for the opportunity of using it online at http://apm.pgia.ru/webtool. Regional TEC maps are a joint SCiESMEX-LACIGE product. The gratitude is expressed to the NOAA Space Weather Prediction Center for providing the analysis software used for TEC maps production. We also express our gratitude to the personnel of SSN, TLALOCNet and SSN-TLALOCNet networks (station maintenance, data acquisition, IT support and data distribution) whose GNSS data were used for TEC calculations. We also express our gratitude to the IGS services and IGS Ionosphere Working Group for the opportunity of downloading global ionospheric maps via Internet (https://igsiono.uwm.edu.pl).

References

Aarons J, Mendillo M, Yantosca R, Kudeki E (1996) GPS phase fluctuations in the equatorial region during the MISETA 1994 campaign. J Geophys Res Sp Phys 101:26851–26862

Aarons J (1982) Global morphology of ionospheric scintillations. Proc IEEE 70(4):360

Abdu MA, Batista IS, Carrasco AJ, Brum CGM (2005) South Atlantic magnetic anomaly ionization: a review and a new focus on electrodynamic effects in the equatorial ionosphere. J Atmos Solar Terr Phys 67:1643–1657. https://doi.org/10.1016/j.jastp.2005.01.014

Abdu MA, Bittencourt JA, Batista IS (1981) Magnetic declination control of the equatorial F-region dynamo electric field development and spread-F. J Geophys Res 86. https://doi.org/10.1029/JA086iA13p11443

Abdu MA, Sobral JHA, de Paula ER, Batista IS (1991) Magnetospheric disturbance effects on the equatorial ionization anomaly: an overview. J Atm Terr Phys 53(8):757–771

Abed AK, Qahwaji R, Abed A (2021) The automated prediction of solar flares from SDO images using deep learning. Adv Space Res 67(8):2544–2557. https://doi.org/10.1016/j.asr.2021.01.042

Abramov-Maximov VE, Bakunina IA (2020) Solar-flare precursors in the microwave range. Geomagn Aeron 60:846–852. https://doi.org/10.1134/S0016793220070038

ACFJ services for Aviation (available at www.swpc.noaa.gov/sites/default/files/images/u59/05%20David%20Boteler%20Official.pdf)

Afraimovich EL, Astafyeva EI, Oinats AV, Yasukevich YuV, Zhivetiev IV (2008) Global electron content: a new conception to track solar activity. Ann Geophys 26:335–344. https://doi.org/10.5194/angeo-26-335-2008

Afraimovich EL, Perevalova NP (2006) GPS-monitoring of earth upper atmosphere, Irkutsk, Russian academy of sciences, Siberian Branch, p 460. ISBN 5-98277-033-7

Afraimovich EL, Astafyeva EI, Demyanov VV, Edemskiy EK, Gavrilyuk NS, Ishin AB, Kosogorov EA, Leonovich LA, Lesyuta OS, Palamartchouk KS, Perevalova NP, Polyakova AS, Smolkov GI, Voeykov SV, Yasyukevich YV, Zhivetiev IV (2013) A review of GPS/GLONASS studies of the ionospheric response to natural and anthropogenic processes and phenomena. J Space Weather Space Clim 3. https://doi.org/10.1051/swsc/2013049

Akasofu SI, Chapman S (1972) Solar-terrestrial physics

Aleshin IM, Arakelov AS, Bruevich EA, Burov VA, Ivanov SD, Ochelkov YuP, Repin AY, Kholodkov KI (2021) Methods for monitoring strong space weather disturbances to support international air navigation. Russ Meteorol Hydrol 46(3):205–211. https://doi.org/10.3103/S1068373921030109

Alpert JL (1960) Propagation of radio waves and ionosphere, Moscow, p 479

Ambastha A (2020) Physics of the invisible Sun: instrumentation, observations, and inferences. CRC Press

Anderson DN (1973) A theoretical study of the ionospheric F region equatorial anomaly—I. Theor, Planet Space Sci 21(3):409–419. https://doi.org/10.1016/0032-0633(73)90040-8

Andreeva OA, Malashchuk VM (2020) Sunspot activity and coronal holes during the declining phase of cycle 24. Geomagn Aeron 60:1093–1100. https://doi.org/10.1134/S0016793220080022

Appleton EV (1946) Nature 157(3995):691–691

Arikan F, Erol CB, Arikan O (2003) Regularized estimation of vertical total electron content from global positioning system data. J Geophys Res 108:1469. https://doi.org/10.1029/2002JA009605

Bai T, Sturrock PA (1989) Classification of solar flares. Annu Rev Astron Astrophys 27:421–467

Bain HM, Steenburgh RA, Onsager TG, Stitely EM (2021) A summary of national oceanic and atmospheric administration space weather prediction center proton event forecast performance and skill. Space Weather 19. https://doi.org/10.1029/2020SW002670

Baker DN (1996) Solar wind—magnetosphere drivers of space weather. J Atmos Terr Phys 58(11):1509–1526

Baker DN (2002) How to cope with space weather. Science 297(5586):1486–1487. https://doi.org/10.1126/science.1074956

Baker DN, Bodeau JM, Cameron E, Fennell JF, Fisher GM, Forbes KF, Kintner PL, Leffler LG, Lewis WS, Reagan JB, Small AA III, Stansell TA, Strachan L Jr, Graham SJ, Fisher TM, Swisher V, Gruber CA (2008) Severe space weather events—understanding societal and economic impacts. The National Academy Press, Washington, D.C., pp 1–144

Baker DN (2004) Introduction to space weather, In: Scherer K, Fichtner H, Heber B, Mall U (eds) Space weather, lecture notes in physics, 656, Springer. https://doi.org/10.1007/978-3-540-31534-6_1

Bakunina IA, Melnikov VF, Morgachev AS (2020) Signs of preflare situation in solar ultraviolet and microwave emission. Geomagn Aeron 60:853–859. https://doi.org/10.1134/S001679322007004X

Bala R, Reiff P, Russell CT (2015) Testing the estimated hypothetical response of a major CME impact on earth and its implications to space weather. J Geophys Res Space Physics 120:3432–3443. https://doi.org/10.1002/2014JA020739

Balan N, Souza J, Bailey GJ (2018) Recent developments in the understanding of equatorial ionization anomaly: a review. J Atmos Solar Terr Phys 171:3–11. https://doi.org/10.1016/j.jastp.2017.06.020

Barbieri LP, Mahmot RE (2004) October–November 2003's space weather and operations lessons learned. Space Weather 2. https://doi.org/10.1029/2004SW000064

Barbosa C, Alves L, Caraballo R, Hatrmann GA, Papa ARR, Pirjola RJ (2015) Analysis of geomagnetically induced currents at a low-latitude region over the solar cycles 23 and 24: comparison between measurements and calculations. J Space Weather Clim 5. https://doi.org/10.1051/swsc/2015036

Barkhatov NA, Revunov SE, Barkhatova OM, Revunova EA, Kosolapova NV (2020) Algorithm for detecting an extreme solar flare of september 6, 2017 from its geomagnetic precursors. Radiophys Quantum Electron 62(10):651–656

Bartels J (1957) The geomagnetic measures for the time-variations of solar corpuscular radiation, described for use in correlation studies in other geophysical fields. Ann Intern Geophys 4:227–236

Bartels J, Veldkamp J (1954) International data on magnetic disturbances, fourth quarter. J Geophys Res 59:297–302. https://doi.org/10.1029/JZ059i002p00297

Bartels J (1949) The standardized index Ks and the planetary index Kp, IATME Bulletin 12b, 97

Béland J, Small K (2004) Space weather effects on power transmission systems: the cases of hydro-québec and transpower New Zealand Ltd. In: Daglis IA (eds) Effects of space weather on technology infrastructure, vol 176. Springer. https://doi.org/10.1007/1-4020-2754-0_15

Belehaki A, Watermann J, Lilensten J, Glover A, Hapgood M, Messerotti M, Van der Linden R, Lundstedt H (2009) Renewed support dawns in Europe: an action to develop space weather products and services. Space Weather 7. https://doi.org/10.1029/2008SW000451

Berngardt OI, Voeykov SV, Perevalova NP (2020) Comparison of the TEC-based ionospheric disturbance indices AATR and WTEC. J Atmos Solar-Terr Phys 203. https://doi.org/10.1016/j.jastp.2020.105254

Beynon WJG, Brown GM (1958) The Ionosphere. Pergamon Press, Annals of the International Geophysical Year

Bhattacharrya A, Beach TL, Basu S, Kintner PM (2000) Nighttime equatorial ionosphere: GPS scintillations and differential carrier phase fluctuations. Radio Sci 35(1):209–224

Bilitza D, Altadill D, Truhlik V, Shubin V, Galkin I, Reinisch B, Huang X (2017) International reference Ionosphere 2016: from ionospheric climate to real-time weather predictions. Space Weather 15:418–429. https://doi.org/10.1002/2016SW001593

Bingham S, Murray SA, Guerrero A, Glover A, Thorn P (2019) Summary of the plenary sessions at European space weather week 15: space weather users and service providers working together now and in the future. J Space Weather Space Clim 9. https://doi.org/10.1051/swsc/2019031

Blagoveshchenskii DV (2013) Effect of geomagnetic storms (substorms) on the ionosphere: 1. A review. Geomagn Aeron 53:275–290. https://doi.org/10.1134/S0016793213030031

Blagoveshchensky DV (1981) Propagation of decameter radio waves at high latitudes. Nauka Publishers, Moscow (in Russian), p 180

Blagoveshchensky DV (2018) Comparison of ionospheric parameters during similar geomagnetic storms. Geomagn Aeron 58(2):207–213. https://doi.org/10.1134/S0016793218020032

Blagoveshchensky DV, MacDougall JW, Piatkova AV (2006) Ionospheric effects preceding the october 2003 Halloween storm. J Atmos Solar Terr Phys 68(7):821–831. https://doi.org/10.1016/j.jastp.2005.10.017

Blagoveshchensky DV, Maltseva OA, Anishin MM, Sergeeva MA, Rogov DD (2016) Impact of the magnetic superstorm on march 17–19, 2015 on subpolar HF radio paths: experiment and modeling. Adv Space Res 58(6):835–846. https://doi.org/10.1016/j.asr.2016.05.027

Blagoveshchensky DV, Maltseva OA, Sergeeva MA (2018) Impact of magnetic storms on the global TEC distribution. Ann Geophys 36:1–15. https://doi.org/10.5194/angeo-36-1057-2018

Blagoveshchensky DV, Sergeeva MA, Corona-Romero P (2019) Features of the magnetic disturbance on september 7–8, 2017 by geophysical data. Adv Space Res 64(1):171–182. https://doi.org/10.1016/j.asr.2019.03.037.1

Blagoveshchensky DV, Sergeeva MA (2018) Impact of geomagnetic storm of september 7–8, 2017 on ionosphere and HF propagation: a multi-instrument study. Adv Space Res. https://doi.org/10.1016/j.asr.2018.07.016.9

Blagoveshchensky DV, Sergeeva M, Raita T (2021) Riometer absorption during four similar storms. Adv Space Res. https://doi.org/10.1016/j.asr.2021.10.001

Bolaji OS, Adekoya BJ, Adebiyi SJ, Adebesin BO, Ikubanni SO (2022) The African equatorial ionization anomaly response to the St. Patrick's day storms of march 2013 and 2015. Astrophys Space Sci 367:1. https://doi.org/10.1007/s10509-021-04022-5

Borovicka J, Spurny P, Grigore VI, Svoren J (2017) The january 7, 2015, superbolide over Romania and structural diversity of meter-sized asteroids. Planet Space Sci 143:147–158. https://doi.org/10.1016/j.pss.2017.02.006

Borovik AV, Zhdanov AA (2018) Statistical studies of duration of low-power solar flares. Solar-Terr Phys 4(2):8–16. https://doi.org/10.12737/stp-42201803

Borovsky JE, Valdivia JA (2018) The earth's magnetosphere: a systems science overview and assessment. Surv Geophys 39:817–859. https://doi.org/10.1007/s10712-018-9487-x

Borries C, Jakowski N, Kauristie K, Amm O, Mielich J, Kouba D (2017) On the dynamics of large-scale traveling ionospheric disturbances over Europe on 20 november 2003. J Geophys Res Space Phys 122:1199–1211. https://doi.org/10.1002/2016JA023050

Borries C, Wilken V, Jacobsen KS, García-Rigo A, Dziak-Jankowska B, Kervalishvili G, Jakowski N, Tsagouri I, Hernández-Pajares M, Ferreira AA, Hoque MM (2020) Assessment of the capabilities and applicability of ionospheric perturbation indices provided in Europe. Adv Space Res 66(3):546–562. https://doi.org/10.1016/j.asr.2020.04.013

Boteler DH (2006) The super storms of august/september 1859 and their effects on the telegraph system. Adv Space Res 38:159–172. https://doi.org/10.1016/j.asr.2006.01.013

Bothmer V, Daglis IA (2007) Space weather—physics and effects. Springer. IBSN 10-3-540-23907-3

Bothmer V, Zhukov A (2007) The Sun as the prime source of space weather. In: Space weather—physics and effects. Springer. https://doi.org/10.1007/978-3-540-34578-7-3

Briggs B, Parkin I (1963) On the variation of radio star and satellite scintillations with zenith angle. J Atmos Terr Phys 25:339–366. https://doi.org/10.1016/00219169(63)90150-8

Bruzek A (1972) Properties of solar active regions. In: de Jager C, Dyer ER (eds) Solar-terrestrial physics, astrophysics and space science library, vol 29. Springer. https://doi.org/10.1007/978-94-010-3126-4-5

Buonsanto MJ (1999) Ionospheric storms—a review. Space Sci Rev 88:563–601

Campbell WH (1997) Introduction to geomagnetic fields, Cambridge University Press, p 209. ISBN 571936

Cander LR (2019) Ionospheric space weather. Springer Geophys. https://doi.org/10.1007/978-3-319-99331-7

Cannon P (2013) Extreme space weather: impacts on engineered systems and infrastructure, royal academy of engineering, prince Philip house. ISBN 1-903496-95-0. Available online at: www.raeng.org.uk/spaceweather

Caraballo R, Gonzalez-Esparza JA, Sergeeva MA, Pacheco CR (2020) First GIC estimates for the Mexican power grid. Space Weather 18. https://doi.org/10.1029/2019SW002260

Carovillano RL, Forbes JM (1983) Solar-terrestrial physics, principles and theoretical foundations, D. Reidel Publishing Company, e-ISBN-I3: 978-94-009-7194-3. https://doi.org/10.1007/978-94-009-7194-3

Carrington RC (1859) Description of a singular appearance seen in the Sun on september 1, 1859. Mon Not R Astron Soc 20:13–151859

Caruana J (1990) The IPS monthly T index, solar-terrestrial predictions. In: Proceedings of a workshop at Leura, Australia October 16–20, 1989, vol 2, pp 257–263

Chen G, Zhou C, Liu Y, Zhao J, Tang Q, Wang X, Zhao Z (2019) A statistical analysis of medium-scale traveling ionospheric disturbances during 2014–2017 using the Hong Kong CORS network. Earth Planets Space 71:52. https://doi.org/10.1186/s40623-019-1031-9

Chernogor LF (2014) Geomagnetic field effects of the Chelyabinsk meteoroid. Geomagn Aeron 54:613–624. https://doi.org/10.1134/S001679321405003X

Chernogor LF (2020) Effects of the Lipetsk meteoroid in the geomagnetic field. Geomagn Aeron 60:355–372. https://doi.org/10.1134/S0016793220030032

Chyba C, Thomas P, Zahnle K (1993) The 1908 Tunguska explosion: atmospheric disruption of a stony asteroid. Nature 361:40–44

Clement CH, Sasaki M (2013) Assessment of radiation exposure of astronauts in space, ICRP publication, 123. https://doi.org/10.1016/j.icrp.2013.05.004

Cliver EW (2006) The 1859 space weather event: then and now. Adv Space Res 38:119–129. https://doi.org/10.1016/j.asr.2005.07.077

Cohen R, Bowles KL (1961) On the nature of equatorial spread F. J Geophys Res. https://doi.org/10.1029/JZ066i004p01081

Committee on the peaceful uses of outer space, non-consensus paper of the expert group on space weather on the survey of the state of member state preparedness, and current and future activities and needs for space weather impact mitigation, Report A/AC.105/C.1/2022/CRP.10 (2022)

Corona-Romero P, Riley P (2020) Development of a formalism for computing in situ transits of earth-directed CMEs—part 2: towards a forecasting tool. Ann Geophys 38:657–681. https://doi.org/10.5194/angeo-38-657-2020

Corona-Romero P, Gonzalez-Esparza JA, Perez-Alanis CA, Aguilar-Rodriguez E, De la Luz V, Mejia-Ambriz JC (2017) Calculating travel times andarrival speeds of CMEs to earth: ananalytic tool for space weather fore-casting. Space Weather 15:464–483. https://doi.org/10.1002/2016SW001489

Crosby NB, Scherer K, Fichtner H, Heber B, Mall U (eds) (2006) Space weather: the physics behind a slogan, lecture notes in physics. Surv Geophys 27:489–490. https://doi.org/10.1007/s10712-006-9005-4

Crown MD (2012) Validation of the NOAA space weather prediction center's solar flare forecasting look-up table and forecaster-issued probabilities. Space Weather 10. https://doi.org/10.1029/2011SW000760

Curto JJ (2020) Geomagnetic solar flare effects: a review. J Space Weather Space Clim 10. https://doi.org/10.1051/swsc/2020027

Daglis ID, Chang LC, Dasso S, Gopalswamy N, Khabarova OV, Kilpua E, Lopez R, Marsh D, Matthes K, Nandy D, Seppälä A, Shiokawa K, Thiéblemont R, Zong Q (2021) Predictability of variable solar–terrestrial coupling. Ann Geophys 39:1013–1035. https://doi.org/10.5194/angeo-39-1013-2021

Daglis IA (2005) Effects of space weather on technology infrastructure. Kluwer Academic Publishers, p 334

Danilov AD (1998) Review of long-term trends in the upper mesosphere, thermosphere and ionosphere. Adv Space Res 22(6):907–915. https://doi.org/10.1016/S0273-1177(98)00122-7

Danilov AD (2013) Ionospheric F-region response to geomagnetic disturbances. Adv Space Res 53(3):343–366. https://doi.org/10.1016/j.asr.2013.04.019

Danilov AD, Konstantinova AV (2020) Long-term variations in the parameters of the middle and upper atmosphere and ionosphere (review). Geomagn Aeron 60:397–420. https://doi.org/10.1134/S0016793220040040

Davies K, Hartmann GK (1997) Studying the ionosphere with the global positioning system. Radio Sci 32(4):1695–1703. https://doi.org/10.1029/97RS00451

Davies K (1965) Ionospheric radio propagation, national bureau of standards, monograph, vol 80, p 487

Davis TN, Sugiura M (1966) Auroral electroject activity index AE and its universal time variations. J Geophys Res 71:785–801. https://doi.org/10.1029/JZ071i003p00785

Deminov MG, Deminova GF, Zherebtsov GA, Deminov MG, Deminova GF, Zherebtsov GA, Pirog OM, Polekh NM (2011) Variability of parameters of the F2-layer maximum in the quiet midlatitude ionosphere under low solar activity: 1. Statistical properties. Geomagn Aeron 51:348–355. https://doi.org/10.1134/S0016793211020058

Demyanov V, Yasyukevich Y (2021) Space weather: risk factors for global navigation satellite systems. Solar-Terr Phys 7(2):28–47. https://doi.org/10.12737/stp-72202104

Demyanov VV, Sergeeva MA, Yasyukevich AS (2019a) GNSS high-rate data and the efficiency of ionospheric scintillation indices, book chapter in "Ionospheric and atmospheric threats for GNSS and satellite telecommunications". IntechOpen. https://doi.org/10.5772/intechopen.90078

Demyanov VV, Yasyukevich YV, Jin S, Sergeeva MA (2019a) The second-order derivative of GPS carrier phase as a promising means for ionospheric scintillation research. Pure Appl Geophys 1–19. https://doi.org/10.1007/s00024-019-02281-6

Denardini CM, Dasso S, Gonzalez-Esparza JA (2016) Review on space weather in Latin America. 2. The research networks ready for space weather. Adv Space Res 58:1940–1959. https://doi.org/10.1016/j.asr.2016.03.013

Ding F, Wan WX, Mao T, Wang M, Ning BQ, Zhao BQ, Xiong B (2014) Ionospheric response to the shock and acoustic waves excited by the launch of the Shenzhou 10 spacecraft. Geophys Res Lett 41(10):3351–3358. https://doi.org/10.1002/2014GL060107

Ding F, Wan W, Liu L, Afraimovich EL, Voeykov SV, Perevalova NP (2008) A statistical study of large-scaletraveling ionospheric disturbances observed by GPS TEC during major magnetic storms over the years 2003–2005 J Geophys Res 113. https://doi.org/10.1029/2008JA013037

Doherty PH, Delay SH, Valladares CE, Klobuchar JA (2003) Ionospheric scintillation effects in the equatorial and auroral regions, Navig. J Inst Navig 50(4):235–245. https://doi.org/10.1002/j.2161-4296.2003.tb00332.x

Doherty P, Coster AJ, Murtagh W (2004) Space weather effects of october–november 2003. GPS Solutions 8:267–271. https://doi.org/10.1007/s10291-004-0109-3

Dubei S, Wahi R, Mingkhwan E, Gwal AK (2005) Study of amplitude and phase scintillation at GPS frequency. Indian J Radio Space Phys 34:402–407

Dulk GA (1985) Radio emission from the Sun and stars. Ann Rev Astron Astrophys 23:169–224

Eastwood JP, Biffis E, Hapgood MA, Green L, Bisi MM, Bentley RD, Wicks R, McKinnell L-A, Gibbs M, Burnett C (2017) The economic impact of space weather: where do we stand? Risk Anal 37(I2):206–218. https://doi.org/10.1111/risa.12765

Ebihara Y, Watari S, Kumar S (2021) Prediction of geomagnetically induced currents (GICs) flowing in Japanese power grid for carrington-class magnetic storms. In: Ebihara et al. Earth, planet space, vol 73, p 163. https://doi.org/10.1186/s40623-021-01493-2

Echer E, Gonzalez WD, Guarnieri FL, Dal Lago A, Vieira LEA (2005) Introduction to space weather. Adv Space Res 35:855–865. https://doi.org/10.1016/j.asr.2005.02.098

Edemskiy IK, Yasyukevich AS (2018) Observing wave packets generated by solar terminator in tec during typhoons. Solar-Terr Phys 4(2):33–40. https://doi.org/10.12737/stp-42201806

Eroshenko EA, Belov AV, Boteler D, Gaidash SP, Lobkov SL, Pirjola R, Trichtchenko L (2010) Effects of strong geomagnetic storms on Northern railways in Russia. Adv Space Res 46(9):1102–1110

Ezquer RG, Radicella SM (2008) Ionosfera de bajas latitudes: efectos sobre señales de satélites recibidas en Tucumán, Física de la Tierra, pp 61–81

Fichtner H, Liu WW (2011) Advances in coordinated Sun–Earth system science through interdisciplinary initiatives and international programs. In: Miralles M, Sánchez Almeida J (eds) The Sun, the solar wind, and the heliosphere. IAGA special sopron book series, vol 4, Springer. https://doi.org/10.1007/978-90-481-9787-3_24

Förster M, Jakowski N (2000) Geomagnetic storm effects on the topside ionosphere and plasmasphere: a compact tutorial and new results. Surv Geophys 21:47–87. https://doi.org/10.1023/A:1006775125220

Friedrich M, Rapp M (2009) News from the lower ionosphere: a review of recent developments. Surv Geophys 30:525–559. https://doi.org/10.1007/s10712-009-9074-2

Fuller-Rowell TJ, Codrescu MV, Rishbeth H, Moffett RJ, Quegan S (1996) On the seasonal response of the thermosphere and ionosphere to geomagnetic storms. J Geophys Res: Space Phys 101:2343–2353. https://doi.org/10.1029/95JA01614

Gallagher PT, Moon YJ, Wang H (2002) Active-region monitoring and flare forecasting—I. Data Process First Results, Solar Phys 209:171–183. https://doi.org/10.1023/A:1020950221179

Gary DE, Keller CU (2004) Solar and space weather radiophysics. Springer. ISBN 140202813X

Giersch OD, Kennewel LJ, Lynch M (2017) Solar radio burst statistics and implications for space weather effects. Space Weather 15:1511–1522. https://doi.org/10.1002/2017SW001658

Gjerloev JW (2012) The superMAG data processing technique. J Geophys Res 117. https://doi.org/10.1029/2012JA017683

Glover FN (1960) A study of spread F, NBS technical note, boulder laboratories, vol 82, p 67

Gogie TK (2021) The rate of ionospheric total electron content index (ROTI) as a proxy for nighttime ionospheric irregularity using ethiopian low-latitude GPS data. Geomagn Aeron 61:464–475. https://doi.org/10.1134/S0016793221030051

Gonzalez WD, Joselyn JA, Kamide D, Kroehl HW, Rostoker G, Tsurutani BT, Vasyliunas VM (1994) What is a geomagnetic storm? J Geophys Res 99(6):5771–5792

Gonzalez-Esparza JA, De la Luz V, Corona-Romero P, Mejia-Ambriz JC, Gonzalez LX, Sergeeva MA, Romero-Hernandez E, Aguilar-Rodriguez E (2017) Mexican space weather service (SCIESMEX). Space Weather 15(1):3–11. https://doi.org/10.1002/2016SW001496

Gonzalez-Esparza JA, Sergeeva MA, Corona-Romero P, Mejia-Ambriz JC, Gonzalez LX, De la Luz V et al (2018) Space weather events, hurricanes, and earthquakes in Mexico in September 2017. Space Weather 16:2038–2051. https://doi.org/10.1029/2018SW001995

Gonzalez-Esparza JA, Cuevas-Cardona MC (2018) Observations of low-latitude redaurora in Mexico during the 1859 carrington geomagnetic storm. Space Weather 16. https://doi.org/10.1029/2017SW001789

Goodman JM (2005) Operational communication systems and relationships to the ionosphere and space weather. Adv Space Res 36:2241–2252

Gordienko GI, Zachateiskiy DE, Kaliev MZ, Mukasheva SN, Vodyannikov VV (2001) Nighttime increases in ionosphere electron content (a statistical and experimental study). J Atmos Solar Terr Phys 63(6):617–625. https://doi.org/10.1016/S1364-6826(01)00002-5

Gordienko GI, Vodyannikov VV, Yakovets AF (2005) Ionospheric disturbances over Alma-Ata during the october–november 2003 magnetic storms. J Geophys Res 110. https://doi.org/10.1029/2004JA010945

Gulyaeva T, Stanislawska I (2013) Deformation of the ionosphere structure during the space weather events on October–November 2003. Adv Space Res 51:683–690. https://doi.org/10.1016/j.asr.2011.09.022

Gulyaeva TL, Veselovsky IS (2014) Imaging global electron content backwards in time more than 160 years ago. Adv Space Res 53(3):403–411. https://doi.org/10.1016/j.asr.2013.11.036

Gulyaeva TL, Arikan F, Hernandez-Pajares M, Stanislawska I (2013) GIM-TEC adaptive ionospheric weather assessment and forecast system. J Atmosph Solar-Terr Phys 102:329–340. https://doi.org/10.1016/j.jastp.2013.06.011

Haldoupis C (2011) A tutorial review on sporadic E layers. In: Abdu M, Pancheva D (eds) Aeronomy of the earth's atmosphere and ionosphere. IAGA special sopron book series, vol 2. Springer, Dordrecht. https://doi.org/10.1007/978-94-007-0326-1_29

Hanslmeier A (2007) The Sun and space weather, 2nd edn. Springer, Netherlands, p 315

Hanson WB, Moffett RJ (1966) Ionization transport effects in the equatorial F region. J Geophys Res 71:5559–5572. https://doi.org/10.1029/JZ071i023p05559

Hapgood MA (2011) Towards a scientific understanding of the risk from extreme space weather. Adv Space Res 47(12):2059–2072. https://doi.org/10.1016/j.asr.2010.02.007

Hargreaves JK (1979) The upper atmosphere and solar-terrestrial relations—an introduction to the aerospace environment. New York, Van Nostrand Reinhold Co., p 312

Hargreaves JK (1992) The solar-terrestrial environment. Cambridge University Press. ISBN 9780511628924. https://doi.org/10.1017/CBO9780511628924

Hathaway DH (2015) The solar cycle. Living Rev Sol Phys 12:4. https://doi.org/10.1007/lrsp-2015-4

Hayakawa H, Ebihara Y, Willis DM, Toriumi S, Iju T, Hattori K, Wild MN, Oliveira DM, Ermolli I, Ribeiro JR, Correia AP, Ribeiro AI, Knipp DJ (2019) Temporal and spatial evolutions of a large sunspot group and great auroral storms around the Carrington event in 1859. Space Weather 17(11):1553–1569. https://doi.org/10.1029/2019sw002269

Hayakawa H, Ribeiro P, Vaquero JM, Gallego MC, Knipp DJ, Mekhaldi F, Bhaskar A, Oliveira DM, Notsu Y, Carrasco VMS, Caccavari A, Veenadhari B, Mukherjee S, Ebihara Y (2020) The extreme space weather event in 1903 october/november: an outburst from the quiet Sun. Astrophys J Lett 897. https://doi.org/10.3847/2041-8213/ab6a18

Hayes LA, O'Hara OSD, Murray SA, Gallagher PT (2021) Solar flare effects on the earth's lower ionosphere. Sol Phys 296:157. https://doi.org/10.1007/s11207-021-01898-y

Hermoso HM, Morand V, Gelhaus J, Vellutini E, Poenaru V, Freitas J, Krużyński M, Garcia-Yarnoz D, Guijarro N (2021) System approach to analyze the performance of the EU space surveillance and tracking system. AMOS conference proceedings

Hernandez-Pajares M, Juan J, Sanz J, Orus R, Garcia-Rigo A, Feltens J, Komjathy A, Schaer S, Krankowski A (2009) The IGS vTEC maps: a reliable source of ionospheric information since 1998. J Geodesy 83:263–275

Hickey DA, Sau S, Narayanan VL, Gurubaran S (2020) A possible explanation of interhemispheric asymmetry of equatorial plasma bubbles in airglow images. JGR Space Phys. https://doi.org/10.1029/2019JA027592

Hocke K, Schlegel K (1996) A review of atmospheric gravity waves and travelling ionospheric disturbances 1982–1995. Ann Geophys 14(5):917–940

Horne RB, Pitchford D (2015) Space weather concerns for all-electric propulsion satellites. Space Weather 13:430–433. https://doi.org/10.1002/2015SW001198

Howard T (2014) Space weather and coronal mass ejections. Springer, New York

Huang L, Huang J, Wang J, Jiang Y, Deng D, Zhao K, Lin G (2013) Analysis of the north–south asymmetry of the equatorial ionization anomaly around 110°E longitude. J Atmos Solar Terr Phys 102:354–361. https://doi.org/10.1016/j.jastp.2013.06.010

Hunsucker RD, Hargreaves JK (2003) The high-latitude ionosphere and its effects on radio propagation. Cambridge University Press

ICAO (2019a) Dissemination of space weather advisories. www.icao.int/airnavigation/METP/Panel%20Documents/SWXA_Additional%20Information.pdf

ICAO (2019b) Manual on space weather information in support of international air navigation, ICAO Doc 10100, 1st edn, p 35 https://store.icao.int/en/manual-on-space-weather-information-in-support-of-international-air-navigation-doc-10100

ICAO (2019c) Assembly—40th session, technical commission A40-Wp/336 Te/140. https://www.icao.int/Meetings/a40/Documents/WP/wp_336_en.pdf

ICAO (2018) Annex 3: meteorological service for international air navigation, 20th edn

Ishkov VN (2021) Extreme space weather events during the first cycles of epochs with decreased solar activity. Geomagn Aeron 61:801–809. https://doi.org/10.1134/S0016793221060074

Ivanov VG (2021) Two links between parameters of 11-year cycle of solar activity. Geomagn Aeron 61:1029–1034. https://doi.org/10.1134/S0016793221070069

Iyemori T, Takeda M, Nose M, Odagi Y, Toh H (2010) Mid-latitude geomagnetic indices ASY and SYM for 2009 (Provisional), internal report of data analysis center for geomagnetism and space magnetism. Kyoto University, Japan

Izyumov N, Linde D (1976) Fundamentals of radio. Mir Publishers, Moscow

Jachhia L (1959) Two atmospheric effects in the orbital acceleration of artificial satellites. Nature 183:526–527. https://doi.org/10.1038/183526a0

Jacobsen KS, Dähnn M (2014) Statistics of ionospheric disturbances and their correlation with GNSS positioning errors at high latitudes. J Space Weather Space Clim 4. https://doi.org/10.1051/swsc/2014024

Jaeggli SA, Norton AA (2016) The magnetic classification of solar active regions 1992–2015. Astrophys J Lett

Jakowski N, Borries C, Wilken V (2012) Introducing a disturbance ionosphere index (DIX). Radio Sci 47. https://doi.org/10.1029/2011RS004939

Jenkins JJ (2013) Observing the Sun, a pocket field guide. Springer. https://doi.org/10.1007/978-1-4614-8015-0

Jian L, Russell CT, Luhmann JG, Skoug R (2006) Properties of stream interactions at one AU during 1995–2004. Sol Phys 239(1):337–392

Jin SG, Luo OF, Park P (2008) GPS observations of the ionospheric F2-layer behavior during the 20th november 2003 geomagnetic storm over South Korea. J Geod 82(12):883–892. https://doi.org/10.1007/s00190-008-0217-x

Jones JBL, Bentley RD, Hunter R, Iles RHA, Taylor GC, Thomas DJ (2005) Space weather and commercial airlines. Adv Space Res 36:2258–2267. https://doi.org/10.1016/j.asr.2004.04.017

Joshua BW, Adeniyi JO, Adimula IA, Oladipo OA, Olawepo OA, Adebiyi SJ (2014) Ionospheric response to the storm-time disturbance of 29 may, 2010. Adv Space Res 53(2):219–225. https://doi.org/10.1016/j.asr.2013.11.003

Juan JM, Sanz J, Rovira-Garcia A, González-Casado G, Ibáñez D, Perez RO (2018) AATR an ionospheric activity indicator specifically based on GNSS measurements. J Space Weather Space Clim 8. https://doi.org/10.1051/swsc/2017044

Junior SLD, Cecatto JR, Fernandes MM, Ribeiro MX (2018) An optimized data mining method to support solar flare forecast. In: Latifi S (eds) Information technology—new generations, Advances in intelligent systems and computing, vol 558. Springer. https://doi.org/10.1007/978-3-319-54978-1_60

Kamide Y, Chian AC-L (2007) Handbook of the solar-terrestrial environment. Springer, Science & Business Media

Kappenman JG (2012) A perfect storm of planetary proportions. IEEE Spectr 49:26–31

Karan DK, Daniell RE, England SL, Martinis CR, Eastes RW, Burns AG, McClintock WE (2020) First zonal drift velocity measurement of equatorial plasma bubbles (EPBs) from a geostationary orbit using gold data. JGR Space Phys. https://doi.org/10.1029/2020JA028173

Karpachev AT (2021) Diurnal and longitudinal variations in the equatorial anomaly for winter solstice according to interkosmos-19 satellite data. Geomagn Aeron 61:29–43. https://doi.org/10.1134/S0016793221010060

Kauristie K, Andries J, Beck P, Berdermann J, Berghmans D, Cesaroni C, De Donder E, de Patoul J, Dierckxsens M, Doornbos E et al (2021) Space weather services for civil aviation—challenges and solutions. Remote Sens 13. https://doi.org/10.3390/rs13183685

Kazimirovsky E, Herraiz M, De La Morena BA (2003) Effects on the ionosphere due to phenomena occurring below it. Surv Geophys 24:139–184. https://doi.org/10.1023/A:1023206426746

Kazimirovsky ES, Kokourov VD, Vergasova GV (2006) Dynamical climatology of the upper mesosphere, lower thermosphere and ionosphere. Surv Geophys 27:211–255. https://doi.org/10.1007/s10712-005-3819-3

Keika K, Ebihara Y, Kataoka R (2015) What caused the rapid recovery of the Carrington storm? Earth Planets Space 67(1):1–12. https://doi.org/10.1186/s40623-015-0234-y

Kelley MC, Makela JJ, Paxton LJ, Kamalabadi F, Comberiate JM, Kil H (2003) The first coordinated ground- and space-based optical observations of equatorial plasma bubbles. Geophys Res Lett. https://doi.org/10.1029/2003GL017301

Kelley MC (2009) The earth's ionosphere: plasma physics and electrodynamics, 2nd edn. International geophysics series; vol 96. Elsevier, ISBN 978-0-12-088425-4

Khazanov GV (2016) Space weather fundamentals. ISBN 978-1-4987-4907-7

Kholodkov K, Aleshin I, Arakelov A, Burov V, Vasiliev A, Ivanov S (2021) Sino-Russian space weather effort for global air navigation safety. Russ J Earth Sci 21. https://doi.org/10.2205/2021ES000077

Kil H (2015) The morphology of equatorial plasma bubbles—a review. J Astron Space Sci 32(1):13–19. https://doi.org/10.5140/JASS.2015.32.1.13

Kintner PM, Ledvina BM, De Paula ER (2007) GPS and ionospheric scintillations. Space Weather 5. https://doi.org/10.1029/2006SW000260

Kirkwood S, Nilsson H (2000) High-latitude sporadic-E and other thin layers—the role of magnetospheric electric fields. Space Sci Rev 91:579–613

Klimenko MV, Klimenko VV, Karpachev A, Ratovsky KG, Stepanovm AE (2015) Spatial features of Weddell sea and Yakutsk anomalies in foF2 diurnal variations during high solar activity periods: interkosmos-19 satellite and ground-based ionosonde observations. IRI Reprod GSM TIP Model Simul, Adv Space Res 55:8. https://doi.org/10.1016/j.asr.2014.12.032

Klimenko MV, Klimenko VV, Zakharenkova IE, Ratovsky KG, Korenkova NA, Yasyukevich YV, Mylnikova AA, Cherniak IV (2017) Similarity and differences in morphology and mechanisms of the foF2 and TEC disturbances during the geomagnetic storms on 26–30 September 2011. Ann Geophys 35:923–938. https://doi.org/10.5194/angeo-35-923-2017

Klobuchar JA (1983) Ionospheric effects on earth-space propagation. Ionos. effects earth-sp. propagation, p 31

Knipp DJ, Hapgood M (2019) Space weather aviation forecasting on a global scale. Eos 100. https://doi.org/10.1029/2019EO135277

Koskinen H, Tanskanen E, Pirjola AER, Pulkkinen A, Dyer C, Rodgers D, Cannon P, Mandeville J-C, Boscher D (2001) Space weather effects catalogue. ESWS-FMI-RP-0001 (2.2). http://www.spaceweather.org/ISES/swxeff/1.pdf

Kouris SS, Fotiadis DN (2002) Ionospheric variability: a comparative statistical study. Adv Space Res 29(6):977–985. https://doi.org/10.1016/S0273-1177(02)00045-5

Krausmann E, Andersson E, Gibbs M, Murtagh W (2016) Space weather and critical infrastructures: findings and outlook. https://doi.org/10.2788/15287

Kumar S, Patel K, Singh AK (2016) TEC variation over an equatorial and anomaly crest region in India during 2012 and 2013. GPS Solut 20:617–626. https://doi.org/10.1007/s10291-015-0470-4

Kunow H (2001) Solar wind: corotating interaction regions. Encyclopedia of astronomy and astrophysics. CRC Press, pp 1–5

Kurazhkovskaya NA (2020) Global disturbance of earth's magnetosphere and its connection with space weather. Solar-Terr Phys 6(1):41–49. https://doi.org/10.12737/stp-61202005

Kuznetsova M (2018) COSPAR panel on space weather: en route to a global space weather forum—establishing the coordinated research initiative targeting the improvements of space weather operational services, 42nd COSPAR scientific assembly, Abstract id. PSW.5-27-18

Lakhina GS, Alex S, Tsurutani BT, Gonzalez WD (2005) Research on historical records of geomagnetic storms. Proceedings IAU symposium no 226. https://doi.org/10.1017/S1743921305000074

Lanabere V, Dasso S, Gulisano AM, Lopez VE, Niemela-Celeda AE (2020) Space weather service activities and initiatives at LAMP (Argentinean space weather laboratory group). Adv Space Res 65:2223–2234. https://doi.org/10.1016/j.asr.2019.08.016

Lanzerotti LJ (1992) Comment on "Great magnetic storms" by Tsurutani et al. Geophys Res Lett 19. https://doi.org/10.1029/92gl02238

Lanzerotti LJ (2001) Space weather effects on technologies, space weather. Geophysics monographs series 125. https://doi.org/10.1029/GM125p0011

Lastovicka J (2002) Monitoring and forecasting of ionospheric space weather—effects of geomagnetic storms. J Atmos Solar Terr Phys 64:697–705

Lastovicka J (2006) Forcing of the ionosphere by waves from below. J Atmos Solar-Terr Phys 68:479–497. https://doi.org/10.1016/j.jastp.2005.01.018

Lee J, Park I, Kim J, Lee J, Hwang J-A, Kim YC (2014) Measurement of the space radiation dose for the flight aircrew at high-altitude. J Astron Space Sci 31(1):33–39. https://doi.org/10.5140/JASS.2014.31.1.33

Lefeuvre F, Tanzi T (2014) Disaster risk reduction—availability of radio communications services at the time of space weather events. IEEE Publications, pp 1–4. https://doi.org/10.1109/CAMA.2014.7003443

Li G, Ning B, Otsuka Y, Li G, Ning B, Otsuka Y, Abdu MA, Abadi P, Liu Z, Spogli L, Wan W (2021) Challenges to equatorial plasma bubble and ionospheric scintillation short-term forecasting and future aspects in east and Southeast Asia. Surv Geophys 42:201–238. https://doi.org/10.1007/s10712-020-09613-5

Lilensten J, Glover A, Hilgers A, Belehaki A, Lefeuvre F, Opgenoorth H, Zolesi B, Fuellekrug M (2005) An introduction to the 1st European space weather week (ESWW). Ann Geophys 23:2967–2968. SRef-ID: 1432-0576/ag/2005-23-2967

Loewe CA, Prölss GW (1997) Classification and mean behavior of magnetic storms. J Geophys Res 102:14209–14213. https://doi.org/10.1029/96JA04020

Love JJ, Rigler EJ, Pulkkinen A, Riley P (2015) On the log-normality of historical magneticstorm intensity statistics: implica-tions for extreme-event probabilities. Geophys Res Lett 42:6544–6553. https://doi.org/10.1002/2015GL064842

Lucke O, Fazli EH, Gaubert J, Gallois R, Forster C, Tafferner A (2013) Planet-2: providing in-flight weather services and observations to and from business and regional aircraft. IEEE Publications, J4, conference paper, p. 12

Machol JL, Eparvier FG, Viereck RA, Woodraska DL, Snow M, Thiemann E, Woods TN, McClintock WE, Mueller S, Eden Jr. TD, Meisner R, Codrescu S, Bouwer SD, Reinard AA (2020)

GOES-R series solar X-ray and ultraviolet irradiance the GOES-R series. Elsevier Inc. https://doi.org/10.1016/B978-0-12-814327-8.00019-6

Maeda J, Heki K (2015) Morphology and dynamics of daytime mid-latitude sporadic-E patches revealed by GPS total electron content observations in Japan Earth. Planet Space 67:89. https://doi.org/10.1186/s40623-015-0257

Maltseva OA, Nikitenko TV (2019) Features of the behavior of ionospheric parameters B0 and B1 in the high-latitude zone. Geomagn Aeron 59:681–695. https://doi.org/10.1134/S0016793219060082

Mandea M, Chambodut A (2020) Geomagnetic field processes and their implications for space weather. Surv Geophys 41:1611–1627. https://doi.org/10.1007/s10712-020-09598-1

Mandrikova O, Polozov Y, Fetisova N, Zalyaev T (2018) Analysis of the dynamics of ionospheric parameters during periods of increased solar activity and magnetic storms. J Atmos Solar-Terr Phys 181 B:116–126. https://doi.org/10.1016/j.jastp.2018.10.019

Mann IR, Di Pippo S, Opgenoorth HJ, Kuznetsova M, Kendall DJ (2018) International collaboration within the United Nations committee on the peaceful uses of outer space: framework for international space weather services (2018–2030). Space Weather 16:428–433. https://doi.org/10.1029/2018SW001815

Mannucci AJ, Tsurutani BT, Iijima BA, Saito AKA, Gonzalez WD, Guarnieri FL, Kozyra JU, Skoug R (2005) Dayside global ionospheric response to the major interplanetary events of October 29–30, 2003 "Halloween Storms". Geophys Res Lett 32. https://doi.org/10.1029/2004GL021467

Marshall RA, Gorniak H, Van Der Walt T, Waters CL, Sciffer MD, Miller M, Dalzell M, Daly T, Pouferis G, Hesse G, Wilkinson P (2013) Observations of geomagnetically induced currents in the Australian power network. Space Weather 11:6–16. https://doi.org/10.1029/2012SW000849

Martinis C, Baumgardner J, Mendillo M, Wroten J, Coster A, Paxton L (2015) The night when the auroral and equatorial ionospheres converged. J Geophys Res Space Phys 120:8085–8095. https://doi.org/10.1002/2015JA021555

Martinis C, Yokoyama T, Nishioka M (2019) All-sky imaging observations and modeling of bright 630-nm airglow structures associated with MSTIDs. J Geophys Res Space Phys 124:7332–7340. https://doi.org/10.1029/2019JA026935

McKinnell L-A, Tshisaphungo MT, Nndanganeni RR, Matamba TM, Cilliers P, Witten J (2021) Improving operational space weather services in the African region, 43rd COSPAR scientific assembly, abstract PSW.1-0003-21, id. 2351

McNamara LF (1994) Radio amateurs guide to ionosphere. Krieger Publishing. IBSN 0-89464-804-7

Mendillo M (2006) Storms in the ionosphere: patterns and processes for total electron content. Rev Geophys 47. https://doi.org/10.1029/2005RG000193

Mikhalev AV, Beletsk AB, Vasilyev RV, Eselevich MV, Ivanov KI, Komarova ES, Podlesnyi AV, Podlesny SV, Syrenova TE (2019) Long-lived meteor trails. Solar-Terr Phys 5(3):109–116. https://doi.org/10.12737/stp-53201913

Mo XH, Zhang DH, Liu J, Hao YQ, Ye JF, Qin JS, Xiao Z (2018) Morphological characteristics of equatorial ionization anomaly crest over Nanning region. Radio Sci 53:37–47. https://doi.org/10.1002/2017RS006386

Moffett RJ (1979) The equatorial anomaly in the electron distribution of the terrestrial F region. Fundam Cosm Phys 4:313–391

Mokhnatkin A, Voropaev V, Schmalz S, Streltsov A, Zakhvatkin M (2021) KIAM RAS/ISON optical facilities and applications of their data, 8th European conference on space debris proceedings

Moldwin M (2008a) The technological impacts of space storms. In: An introduction to space weather. Cambridge University Press, pp 79–94. https://doi.org/10.1017/CBO9780511801365.007

Moldwin M (2008b) An introduction to space weather. Cambridge University Press. https://doi.org/10.1017/CBO978051180136

Morina D, Serra I, Puig P, Corral A (2019) Probability estimation of a carrington like geomagnetic storm. Sci Rep 9(1):2393. https://doi.org/10.1038/s41598-019-38918-8

Muller C (2014) The carrington solar flares of 1859: consequences on life, origins of life and evolution of the biosphere. 44:185–195. https://doi.org/10.1007/s11084-014-9368-3

Nagatsuma T (2002) Geomagnetic storms. J Commun Res Lab 49(3):139–154

Nava B, Rodríguez-Zuluaga J, Alazo Cuartas K, Kashcheyev A, Migoya-Orué Y, Radicella SM, Amory-Mazaudier C, Fleury R (2016) Middle- and low-latitude ionosphere response to 2015 St. Patrick's day geomagnetic storm. J Geophys Res Space Phys https://doi.org/10.1002/2015JA022299

Newell PT, Gjerloev JW (2011) Evaluation of superMAG auroral electrojet indices as indicators of substorms and auroral power. J Geophys Res 116. https://doi.org/10.1029/2011JA016779

Ngwira CM, Pulkkinen A, Kuznetsova MM, Glocer A (2014) Modeling extreme "Carrington-type" space weather events using three-dimensional global MHD simulations. J Geophys Res Space Phys 119. https://doi.org/10.1002/2013JA019661

Obridko VN, Abunin AA, Georgieva K, Kirov B, Shelting BD, Livshits IM (2020) Some features of the present-day transition period in solar activity. Geomagn Aeron 60:1007–1016. https://doi.org/10.1134/S0016793220080186

Odenwald SF, Green JL (2007) Forecasting the impact of an 1859-caliber superstorm on geosynchronous earth-orbiting satellites: transponder resources. Space Weather 5(6):1–16

Oinats AV, Nishitani N, Ponomarenko P, Berngardt OI, Ratovsky KG (2016) Statistical characteristics of medium-scale traveling ionospheric disturbances revealed from the Hokkaido East and Ekaterinburg HF radar data. Earth Planet Sp. 68:8. https://doi.org/10.1186/s40623-016-0390-8

Oliveira CB, Espejo TMS, Moraes A et al (2020) Analysis of plasma bubble signatures in total electron content maps of the low-latitude ionosphere: a simplified methodology. Surv Geophys 41:897–931. https://doi.org/10.1007/s10712-020-09584-7

Onsager T (2012) Advancing space weather services through international coordination. Space Weather 10. https://doi.org/10.1029/2011SW000763

Opgenoorth HJ, Guhathakurta M, Liu W, Kosugi T, Zelenyi L (2003) International living with a star (ILWS), a new collaborative space program in solar. Heliospheric Solar-Terr Phys EGS—AGU—EUG Joint Assembly, abstract id.5360

Oughton EJ (2018) The economic impact of critical national infrastructure failure due to space weather, Oxford Research Encyclopedia of Natural Hazard Science, Oxford University Press USA. https://doi.org/10.1093/acrefore/9780199389407.013.315

Park E, Moon Y-J, Shin S, Yi K, Lim D, Lee H, Shin G (2018) Application of the deep convolutional neural network to the forecast of solar flare occurrence using full-disk solar magnetograms. Astrophys J 869:91. https://doi.org/10.3847/1538-4357/aaed40

Pei N, Wu Y, Su R, Li J, Wang Y, Li X, Wu Z (2021) Global-scale coupling characteristics between geomagnetic storms and ionospheric disturbances. Geomagn Aeron 61:632–646. https://doi.org/10.1134/S0016793221040101

Perevalova NP, Shestakov NV, Voeykov SV, Takahashi H, Guojie M (2015) Ionospheric disturbances in the vicinity of the Chelyabinsk meteoroid explosive disruption as inferred from dense GPS observations. Geophys Res Lett 42:6535–6543. https://doi.org/10.1002/2015GL064792

Petrie EJ, Hernández-Pajares M, Spalla P, Moore P, King MA (2011) A review of higher order ionospheric refraction effects on dual frequency GPS. Surv Geophys 32:197–253. https://doi.org/10.1007/s10712-010-9105-z

Pi X, Mannucci AJ, Lindqwister UJ, Ho CM (1997) Monitoring of global ionospheric irregularities using the worldwide GPS-network. Geophys Res Lett 24:2283–2286. https://doi.org/10.1029/97GL02273

Pilipenko VA (2021) Space weather impact on ground-based technological systems. Solar-Terr Phys 7(3):68–104. https://doi.org/10.12737/stp-73202106

Pizzo V, Millward G, Parsons A, Biesecker D, Hill S, Odstrcil D (2011) Wang-sheeley-arge-enlil cone model transitions to operations. Space Weather 9(3). https://doi.org/10.1029/2011SW000663

Polekh NM, Chernigovskaya MA, Yakovleva OE (2019) On the formation of the f1 layer during sudden stratospheric warming events. Solar-Terr Phys 5(3):117–127. https://doi.org/10.12737/stp-53201914

Pomoell J, Poedts S (2018) EUHFORIA: European heliospheric forecasting information asset. J Space Weather Space Clim 8:A35. https://doi.org/10.1051/swsc/2018020

Poppe B, Onsager TG, Murtagh W (2004) Halloween storms star in space weather week 2004 meeting. Space Weather. https://doi.org/10.1029/2004sw000099

Poppe BB (2000) New scales help public, technicians understand space weather. EOS, AGU 81(29):322–328. https://doi.org/10.1029/00EO00244

Prikryl P, Ghoddousi-Fard R, Weygand JM, Viljanen A, Connors M, Danskin DW, Jayachandran PT, Jacobsen KS, Andalsvik YL, Thomas EG, Ruohoniemi JM, Durgonics T, Oksavik K, Zhang Y, Spanswick E, Aquino M, Sreeja V (2016) GPS phase scintillation at high latitudes during thegeomagnetic storm of 17–18 March 2015. J Geophys Res Space Phys 121:10448–10465. https://doi.org/10.1002/2016JA023171

Priyadarshi S (2015) A review of ionospheric scintillation models. Surv Geophys 36:295–324. https://doi.org/10.1007/s10712-015-9319-1

Prölss GW (2011) Density perturbations in the upper atmosphere caused by the dissipation of solar wind energy. Surv Geophys 32:101–195. https://doi.org/10.1007/s10712-010-9104-0

Prölss GW (1995) Ionospheric F-region storms, Handbook of atmospheric electrodynamics, 2nd edn, pp 195–248

Prölss GW (2012) Physics of the earth's space environment, an introduction. Springer

Pulkkinen T (2007) Space weather: terrestrial perspective. Living Rev Sol Phys 4(1). https://doi.org/10.12942/lrsp-2007-1

Rastogi RG (1966) The equatorial anomaly in the F2 region of the ionosphere. IETE J Res 12(4):245–256. https://doi.org/10.1080/03772063.1966.11485300

Ratovsky KG, Medvedev AV, Tolstikov MV (2015) Diurnal, seasonal and solar activity pattern of ionospheric variability from Irkutsk Digisonde data. Adv Space Res 55(8):2041–2047

Ratovsky KG, Klimenko MV, Klimenko VV, Chirik NV, Korenkova NA, Kotova DS (2018) After-effects of geomagnetic storms: statistical analysis and theoretical explanation. Solar-Terr Phys 4:26–32. https://doi.org/10.12737/stp-44201804

Ratovsky KG, Klimenko MV, Yasyukevich Y, Vesnin MA, Klimenko VV (2020) Statistical analysis of the ionospheric response to geomagnetic storms based on the data from global ionospheric maps. Russ J Phys Chem B. https://doi.org/10.1134/S1990793120050243

Redmon RJ, Seaton DB, Steenburgh R, He J, Rodriguez JV (2018) September 2017's geoeffective space weather andimpacts to Caribbean radiocommunications during hurricaneresponse. Space Weather 16:1190–1201. https://doi.org/10.1029/2018SW001897

Reinisch BW, Galkin IA (2011) Global ionospheric radio observatory (GIRO). Earth, Planet Space 63:377–381. https://doi.org/10.5047/eps.2011.03.001

Richards PG (2001) Seasonal and solar cycle variations of the ionospheric peak electron density: comparison of measurement and models. J Geophys Res: Space Phys 106:12803–12819. https://doi.org/10.1029/2000JA000365

Richardson IG (2018) Solar wind stream interaction regions throughout the heliosphere. Living Rev Sol Phys 15(1):1–95

Rishbeth H (1997) The ionospheric E-layer and F-layer dynamos—a tutorial review. J Atmos Solar Terr Phys 59(15):1873–1880. https://doi.org/10.1016/S1364-6826(97)00005-9

Rishbeth H, Mendillo M (2001) Patterns of F2-layer variability. J Atmos Sol Terr Phys 63:1661–1680

Ritter S, Nawal A, Diggewadi A (2020) International legal and ethical issues of a future carrington event: existing frameworks. Shortcomings, Recommendations, New Space 8:1. https://doi.org/10.1089/space.2019.0026

Rowland J, McKnight D, Prado Pino B, Reihs B, Stevenson M (2021) A worldwide network of radars for space domain awareness in low earth orbit, AMOS conference proceedings

Ruzhin YY, Kuznetsov VD, Smirnov VM (2014) The ionosphere effects of the chelyabinsk meteoroid explosion. Int J Electron 1(2):39–60

Rycroft M (2004) The Sun: an introduction, 2nd edn; Surveys Geophys 25(2):183–184. https://doi.org/10.1023/B:GEOP.0000028254.65371.8b

Rycroft MJ (2013) Space weather and hazards to application satellites. Handbook of Satellite Applications pp 1175–1193

Sahai Y, Fagundes PR, de Jesus R, de Abreu AJ, Crowley G, Kikuchi T, Huang C-S, Pillat VG, Guarnieri FL, Abalde JR, Bittencourt JA (2011) Studies of ionospheric F-region response in the Latin American sector during the geomagnetic storm of 21–22 January 2005. Ann Geophys 29:919–929. https://doi.org/10.5194/angeo-29-919-2011

Salles LA, Vani BC, Moraes A, Costa E, De Paula ER (2021) Investigating ionospheric scintillation effects on multifrequency GPS signals. Surv Geophys 42:999–1025. https://doi.org/10.1007/s10712-021-09643-7

Schrijver CJ, Rabanal JP (2013) A survey of customers of space weather information. Space Weather 11:529–541. https://doi.org/10.1002/swe.20092

Schrijver CJ, Kauristie K, Aylward AD, Denardini CM, Gibson SE, Glover A, Gopalswamy N, Grande M, Hapgood M, Heynderickx D, Jakowski N, Kalegaev VV, Lapenta G, Linker JA, Liu S, Mandrini CH, Mann IR, Nagatsuma T, Nandy D, Obara T, O'Brien TP, Onsager T, Hermann J, Opgenoorth HJ, Terkildsen M, Valladares CE, Vilmer N (2015) Understanding space weather to shield society: a global road map for 2015–2025 commissioned by COSPAR and ILWS. Adv Space Res 55:2745–2807. https://doi.org/10.1016/j.asr.2015.03.023

Schrijver CJ (2009) Driving major solar flares and eruptions: a review. Adv Space Res 43(5):739–755. https://doi.org/10.1016/j.asr.2008.11.004

Schwenn R (2006) Space weather: the solar perspective. Living Rev Sol Phys 3:2. https://doi.org/10.12942/lrsp-2006-2

Sergeenko NP (2020) Long-term evolution of the parameters of ionospheric disturbances of the F2 layer. Geomagn Aeron 60:318–324

Sergeeva MA, Demyanov VV, Maltseva OA, Mokhnatkin A, Rodriguez-Martinez M, Gutierrez R, Vesnin AM, Gatica-Acevedo VJ, Gonzalez-Esparza JA, Fedorov ME, Ishina TV, Pazos M, Gonzalez LX, Corona-Romero P, Mejia-Ambriz JC, Gonzalez-Aviles JJ, Aguilar-Rodriguez E, Cabral-Cano E, Mendoza B, Romero-Hernandez E, Caraballo R, Orrala-Legorreta ID (2021a) Assessment of morelian meteoroid impact on Mexican environment. Atmosphere 12:185. https://doi.org/10.3390/atmos12020185

Sergeeva M, Maltseva O, Caraballo R, Gonzalez-Esparza JA, Corona-Romero P (2021b) Latitudinal dependence of the ionospheric slab thickness for estimation of ionospheric response to geomagnetic storms. Atmosphere 12:164. https://doi.org/10.3390/atmos12020164

Shagimuratov II, Zakharenkova IE, Tepenitsyna NY et al (2020) Features of the ionospheric total electronic content response to the annular solar eclipse of June 21, 2020. Geomagn Aeron 61:756–762. https://doi.org/10.1134/S001679322105011X

Shagimuratov I, Ephishov I, Tepenitsyna N, Koltunenko L (2009) Similarities and differences of storm time occurrence of GPS phase fluctuations at northern and southern hemispheres, 3rd European conference on antennas and propagation, pp 1432–1436

She C, Wan W, Xu G (2008) Climatological analysis and modeling of the ionospheric global electron content. Chin Sci Bull 53:282–288. https://doi.org/10.1007/s11434-007-0519-z

Shiota D, Kataoka R, Miyoshi Y, Hara T, Tao C, Masunaga K, Futaana Y, Terada N (2014) Inner heliosphere MHD modeling system applicable to space weather forecasting for the other planets. Space Weather 12:187–204. https://doi.org/10.1002/2013SW000989

Shpynev BG, Alsatkin SS, Khakhinov VV, Lebed VP (2017) Investigating the ionosphere response to exhaust products of PROGRESS cargo spacecraft engines on the basis of Irkutsk incoherent scatter radar data. Solar-Terr Phys 3(1):114–127. https://doi.org/10.12737/article_58f9722a233f33.557381

Silber EA, Boslough M, Hocking WK, Gritsevich M, Whitaker RW (2018) Physics of meteor generated shock waves in the earth's atmosphere—a review. Adv Space Res 62(3):489–532. https://doi.org/10.1016/j.asr.2018.05.010

Simpson S (2003) Europe to develop new space weather services. Space Weather 1:1008. https://doi.org/10.1029/2003SW000032

Singh AK, Bhargawa A, Siingh D, Singh RP (2021) Physics of space weather phenomena: a review. Geosciences 11:286. https://doi.org/10.3390/geosciences11070286

Siscoe G, Crooker NU, Clauer CR (2006) Dst of the carrington storm of 1859. Adv Space Res 38(2):173–179. https://doi.org/10.1016/j.asr.2005.02.102

Soicher H (1977) Ionospheric and plasmaspheric effects in satellite navigation systems. IEEE Trans Antennas Propag 25(5):705–708

Song Q, Ding F, Wan W, Ning B, Liu L, Zhao B, Li Q, Zhang R (2013) Statistical study of large-scaletraveling ionospheric disturbances generated by the solar terminator over China. J Geophys Res Space Physics 118:4583–4593. https://doi.org/10.1002/jgra.50423

Stanislawska I, Gulyaeva T (2015) Satellite positioning. IntechOpen, chapter ionospheric W index based on GNSS TEC in the operational use for navigation systems

Sugiura M, Kamei T (1991) Equatorial Dst index 1957–1986, IAGA Bull., IUGG, Paris, p 40

Sugiura M (1964) Hourly values of equatorial Dst for the IGY, Annual international geophysics year, vol 35. Pergamon Press, Oxford, p 9

Sun W, Ning B, Yue X, Li G, Hu L, Chang S et al (2018) Strong sporadic E occurrence detected by ground-based GNSS. J Geophys Res Space Phys 123:3050–3062. https://doi.org/10.1002/2017JA025133

Szuszczewicz EP, Lester M, Wilkinson P, Blanchard P, Abdu M, Hanbaba R, Igarashi K, Pulinets S, Reddy BM (1998) A comparative study of global ionospheric responses to intense magnetic storm conditions. J Geophys Res 103:11665–11684

Temmer M (2021) Space weather: the solar perspective. Living Rev Sol Phys 18:4. https://doi.org/10.1007/s41116-021-00030-3

Thaduri A, Galar D, Kumar U (2020) Space weather climate impacts on railway infrastructure. Int J Syst Assur Eng Manag https://doi.org/10.1007/s13198-020-01003-9

Thammavongsy P, Supnithi P, Phakphisut W, Hozumi K, Tsugawa T (2020) Spread-F prediction model for the equatorial Chumphon station, Thailand. Adv Space Res 65:152–162. https://doi.org/10.1016/j.asr.2019.09.040

Thiemann EMB, Eparvier FG, Woodraska D, Chamberlin PC, Machol J, Eden T, Jones AR, Meisner R, Mueller S, Snow M, Viereck R, Woods TN (2019) The GOES-R EUVS model for EUV irradiance variability. J Space Weather Space Clim 9. https://doi.org/10.1051/swsc/2019041

Townsend LW, Stephens DL Jr, Hoff JL, Zapp EN, Moussa HM, Miller TM, Campbell CE, Nichols TF (2006) The Carrington event: possible doses to crews in space from a comparable event. Adv Space Res 38:226–231. https://doi.org/10.1016/j.asr.2005.01.111

Trivedi NB, Vitorello I, Kabata W, Dutr SLG, Padilha AL, Bologna MS, De Padua MB, Soares PS, Luz GS, De Pinto AF, Pirjola R, Viljanen A (2007) Geomagnetically induced currents in an electric power transmission system at lowlatitudes in Brazil: a case study. Space Weather 5. https://doi.org/10.1029/2006SW000282

Troshichev OA, Podorozhkina NA, Sormakov DA, Janzhura AS (2014) PC index as a proxy of the solar wind energy that entered into the magnetosphere: development of magnetic substorms. J Geophys Res Space Physics 119:6521–6540. https://doi.org/10.1002/2014JA019940

Troshichev O, Janzhura A (2012) Space weather monitoring by ground-based means: PC index. Springer Praxis Books, p 287. https://doi.org/10.1007/978-3-642-16803-1

Troshichev O, Janzhura A, Stauning P (2006) Unified PCN and PCS indices: method of calculation, physical sense, and dependence on the IMF azimuthal and northward components. J Geophys Res 111. https://doi.org/10.1029/2005JA011402

Tsugawa T, Saito A, Otsuka Y (2004) A statistical study of large-scale traveling ionospheric disturbances using the GPS network in Japan. J Geophys Res 109. https://doi.org/10.1029/2003JA010302

Tsurutani BT, Verkhoglyadova OP, Mannucci AJ, Lakhina GS, Li G, Zank GP (2009) A brief review of "solar flare effects" on the ionosphere. Radio Sci 4:1. https://doi.org/10.1029/2008RS004029

Tsurutani BT, Lakhina GS, Hajra R (2020) The physics of space weather/solar-terrestrial physics (STP): what we know now and what the current and future challenges are, Nonlin. Process Geophys 27:75–119. https://doi.org/10.5194/npg-27-75-2020

Tsurutani BT (2000) Solar/Interplanetary plasma phenomena causing geomagnetic activity at earth. In: Coppi B, Ferrari A, Sindoni E (eds) Proceedings of the international school of physics 'Enrico Fermi'. IOS Press, Amsterdam, pp 273–286

Tulunay YK, Bradley PA (2004) The impact of space weather on communication. Ann Geophys Supplement to 47(2/3)

Turner JF (1968) The development of the ionospheric index T. IPS Series R Report, R11

United Nations General Assembly (2021) Draft final report of the expert group on space weather: towards improved international coordination for space weather services. https://undocs.org/en/A/AC.105/C.1/L.401

Vaquero JM (2007) Historical sunspot observations: a review. Adv Space Res 40(7):929–941. https://doi.org/10.1016/j.asr.2007.01.087

Venkatesh K, Fagundes PR, Prasad DSVVD, Denardini CM, De Abreu AJ, De Jesus R, Gende M (2015) Day-to-day variability of equatorial electrojet and its role on the day-to-day characteristics of the equatorial ionization anomaly over the Indian and Brazilian sectors. J Geophys Res Space Phys 120:9117–9131. https://doi.org/10.1002/2015JA021307

Vijaya Lekshmi D, Balan N, Tulasi RS, Liu JY (2011) Statistics of geomagnetic storms and ionospheric storms at low and mid latitudes in two solar cycles. J Geophys Res: Space Phys 116. https://doi.org/10.1029/2011JA017042

Viljanen A (1997) The relation between geomagnetic variations and their time derivatives and implications for estimation of induction risks. Geophys Res Let 24:631–634

Viljanen A, Pirjola R (1994) Geomagnetically induced currents in the finnish high voltage power system. Surv Geophys 15:383–408

Voeykov SV, Berngardt OI, Shestakov NV (2016) Geomag Aeron 56(2):219–228. https://doi.org/10.1134/S0016793216020122

Voeykov SV, Yasyukevich AS, Edemskiy IK, Perevalova NP, Yasyukevich YuV (2018) WTEC: a new index to estimate the intensity of ionospheric disturbances. Results Phys 11:1056–1057. https://doi.org/10.1016/j.rinp.2018.11.023

Vorobjev VG, Yagodkina OI (2005) Effect of magnetic activity on the global distribution of auroral precipitation zone. Geomag Aeron 45(49):438–444

Vorobjev VG, Yagodkina OI (2007) Auroral precipitation dynamics during strong magnetic storms. Geomag Aeron 47(2):185–192

Walker GO (1981) Longitudinal structure of the F-region equatorial anomaly—a review. J Atmos Terr Phys 43(8):763–774. https://doi.org/10.1016/0021-9169(81)90052-0

Wang N, Guo L, Ding Z, Zhao Z, Xu Z, Xu T, Hu Y (2019) Longitudinal differences in the statistical characteristics of ionospheric spread-F occurrences at midlatitude in Eastern Asia. Earth, Planet Space 71:47. https://doi.org/10.1186/s40623-019-1026-6

Wang W, Lei J, Burns AG, Solomon SC, Wiltberger M, Xu J, Zhang Y, Paxton L, Coster A (2010) Ionospheric response to the initial phase of geomagnetic storms: common features. J Geophys Res 115. https://doi.org/10.1029/2009JA014461

Warrington EM, Stocker AJ, Zaalov NY, Siddle DR, Nasyrov IA (2004) Propagation of HF radio waves over northerly paths: measurements, simulation and systems aspects. Ann Geophys 47(2/3):1161–1177. https://doi.org/10.4401/ag-3292

Webb DF, Allen JH (2004) Spacecraft and ground anomalies related to the October–November 2003 solar activity. Space Weather 2:3. https://doi.org/10.1029/2004SW000075

Wheatland MS (2005) A statistical solar flare forecast method. Space Weather 3. https://doi.org/10.1029/2004SW000131

Wilken V, Kriegel M, Jakowski N, Berdermann J (2018) An ionospheric index suitable for estimating the degree of ionospheric perturbations. J Space Weather Space Clim 8. https://doi.org/10.1051/swsc/2018008

Wilkinson PJ (1995) Predictability of ionospheric variations for quiet and disturbed conditions. J Atmosph Terr Phys 57:1469–1481

WMO (2011) Global preparedness for space weather hazards, Abridged final report of the sixteenth world meteorological congress, Annex 4, p 404. www.wmo.int/pages/prog/sat/documents/SAT-GEN_ST-14-Space-Weather-statement-Cg16-Annex4.pdf

Woodman RF, La Hoz C (1976) Radar observations of F region equatorial irregularities. JGR 81:5447–5466. https://doi.org/10.1029/JA081i031p05447

Wright JM, Lewis FP, Corell RW et al (1997) U.S. National space weather program, the implementation plan, FCM-P31-1997. Washington, D.C

Xiao Z, Xiao S, Hao Y, Zhang D (2007) Morphological features of ionospheric response to typhoon. J Geophys Res 112. https://doi.org/10.1029/2006JA01167

Yang Y-M, Komjathy A, Langley RB, Vergados P, Butala MD, Mannucci AJ (2014) The 2013 Chelyabinsk meteor ionospheric impact studied using GPS measurements. Radio Sci 49(5):341–350. https://doi.org/10.1002/2013RS005344

Yasyukevich YuV, Kiselev AV, Zhivetiev IV, Edemskiy IK, Syrovatskii SV, Maletckii BM, Vesnin AM (2020) SIMuRG: system for ionosphere monitoring and research from GNSS. GPS Solut 24:69. https://doi.org/10.1007/s10291-020-00983-2

Yeh CK, Liu CH (1982) Radio wave scintillations in the ionosphere. Proc IEEE 70(4):324–360. https://doi.org/10.1109/PROC.1982.1231

Yeh W-H, Liu J-Y, Huang C-Y, Chen S-P (2014) Explanation of the sporadic-elayer formation by comparing FORMOSAT-3/COSMIC data with meteorand wind shear information. J Geophys Res Atmos 119:4568–4579. https://doi.org/10.1002/2013JD020798

Yermolaev YI, Lodkina IG, Nikolaeva NS, Yermolaev MY (2013) Occurrence rate of extreme magnetic storms. J Geophys Res Space Physics 118:4760–4765. https://doi.org/10.1002/jgra.50467

Yokoyama T, Jin H, Shinagawa H (2015) West wall structuring of equatorial plasma bubbles simulated by three-dimensional HIRB model. JGR Space Phys. https://doi.org/10.1002/2015JA021799

Zakharenkova I, Cherniak IV, Shagimuratov I (2017) Observations of the weddell sea anomaly in the ground-based and space-borne TEC measurements. J Atmos Solar-Terr Phys 161. https://doi.org/10.1016/j.jastp.2017.06.014

Zhang J, Wang Y, Liu Y (2010) Statistical properties of solar active regions obtained from an automatic detection system and the computational biases. Astrophys J 723:1006–1018. https://doi.org/10.1088/0004-637X/723/2/1006

Zhao B, Wan W, Liu L (2005a) Responses of equatorial anomaly to the October–November 2003 superstorms. Ann Geophys 23:693–706

Zhao B, Wan W, Liu L, Yue X, Venkatraman S (2005b) Statistical characteristics of the total ion density in the topside ionosphere during the period 1996–2004 using empirical orthogonal function (EOF) analysis. Ann Geophys 23:3615–3631. https://doi.org/10.5194/angeo-23-3615-2005

Zhbankov GA, Danilkin NP, Maltseva OA (2022) Influence of the ionosphere on the accuracy of the satellite navigation system. Acta Astronaut 190:194–201. https://doi.org/10.1016/j.actaastro.2021.10.004

Zherebtsov GA (2020) Complex of heliogeophysical instruments of new generation. Solar-Terr Phys 6(2):3–13. https://doi.org/10.12737/stp-62202001

Zolotukhina N, Polekh N, Kurkin V, Rogov D, Romanova E, Chelpanov M (2017) Ionospheric effects of St. Patrick's storm over Asian Russia: 17–19 march 2015. J Geophys Res Space Phys 122:2484–2504. https://doi.org/10.1002/2016JA023180

Zolotukhina NA, Kurkin VI, Polekh NM (2018) Ionospheric disturbances over East Asia during intense december magnetic storms of 2006 and 2015: similarities and differences. Solar-Terr Phys 4(3):28–42. https://doi.org/10.12737/stp-43201805

Chapter 4
Methodology for GNSS Capability Analysis

Yury Yasyukevich and Vladislav Demyanov

4.1 Total Electron Content in the Ionosphere

In a cold plasma approximation, the refraction index in the ionosphere is described by the Appleton equation (Appleton-Hartree, Appleton-Lassen-Hartree) (Davies 1969):

$$n^2 = 1 - \frac{X}{1 - iZ - \frac{Y_T^2}{2(1-X-iZ)} \pm \sqrt{\frac{Y_T^4}{4(1-X-iZ)^2} + Y_T^2}} \tag{4.1}$$

where $X = (f_N/f)^2$ is a square of the relation of plasma frequency f_N to a signal frequency f; $Y = f_{\mathrm{H}}/f = eB_0/2\pi mf$ is a relation of a gyrofrequency f_H to a signal frequency (e being electron charge, B_0 being magnetic field strength, m being electron mass), $Y_{\mathrm{L}} = Y \cdot \cos\theta$, $Y_{\mathrm{T}} = Y \cdot \sin\theta$, θ is an angle between a signal propagation path and a geomagnetic field line, $Z = \upsilon/2\pi f$ is a relation of electron collision frequency in the ionospheric plasma υ to a signal frequency f.

At high frequencies of a transionospheric signal ($f \gg f_{\mathrm{H}}, f \gg f_{\mathrm{N}}, f \gg \upsilon$), the Appleton formula is considerably simplified

$$n \approx \sqrt{1 - X} \approx 1 - \frac{40.308N_e}{f^2} \tag{4.2}$$

In Chap. 1, we showed (see Expressions 1.72–1.80) that a gain in the transionospheric signal phase track is proportional to the integral distribution of the ionosphere refraction coefficient along a signal propagation path. Therefore, the carrier phase measured at each of the two frequencies may be recorded as:

$$\varphi_{1,2} = \frac{2\pi f_{1,2}}{c} \int_0^D n_{1,2} ds + \varphi_0 \tag{4.3}$$

where D is a distance between a satellite and a receiver (integration is performed along all the signal s propagation path); φ_0 is an unknown initial phase.

V. Demyanov et al., *Space Weather Impact on GNSS Performance*,
https://doi.org/10.1007/978-3-031-15874-2_4

Owing to the ionosphere refraction index dependence on frequency, the difference in the phase track gains at two different frequencies will be proportional to the TEC value along the phase wavefront propagation path. Therefore, for a dual-frequency combination of phase measurements, one can reconstruct the TEC value as follows:

$$I_{\varphi} = \frac{1}{40.308} \frac{f_1^2 f_2^2}{f_1^2 - f_2^2} [(L_1 \lambda_1 - L_2 \lambda_2) + K + \sigma L] \tag{4.4}$$

where f_1, f_2 are GNSS signal frequencies; $L = \varphi/2\pi$ is whole phase cycles (including the fractional part); λ_1, λ_2 is a signal wavelength at L1 and L2; K is an ambiguity of phase measurements, σL is a summand related to errors and noise of phase measurement.

Similarly, by combining measurements of a satellite signal code delay at two operational frequencies, one may reconstruct the TEC value by group measurements:

$$I_P = \frac{1}{40.308} \frac{f_1^2 f_2^2}{f_1^2 - f_2^2} [(P_2 - P_1) + c\Delta t + \sigma P] \tag{4.5}$$

where P_2, P_1 is a pseudo-ranges at the second and first frequencies; $c\Delta t$ are related to the differential code biases; c is the speed of light in vacuum; σP is related to errors and noise of code pseudo-range measurement.

Different sign of ionospheric effects for phase and pseudo-range allow single-frequency (for example at L1) estimating the TEC:

$$I_{P\varphi} = \frac{1}{2} \frac{f_1^2}{40.308} [(P_1 - L_1 \lambda_1) + PK + \sigma LP] \tag{4.6}$$

where PK is a constant related to both the phase ambiguity and to a DCB in a user receiver and in a satellite hardware (formula 2.60); σLP is the summand related to phase and pseudo-delay measurement noise (because the pseudo-delay noise is much higher than the phase measurement noise, the latter summand is determined by the pseudo-delay measurement noise, including multipath).

Due to high pseudo-range noise (up to 100% and more of the measured TEC value), to investigate ionospheric irregularities, one uses TEC phase measurements (4.4) (Afraimovich et al. 2013). Figure 4.1 shows a typical example for the TEC series computed by different measurement combinations of phase and code pseudo-delay (formula 4.4–4.6).

Apparently, the TEC values computed by the P1P2 combination have the greatest measurement noise. This is caused by larger multipath noise of code measurements (see Fig. 2.14). Another reason is that estimating the code pseudo-delay in the P2 measurements requires using a semi-codeless measuring technique that causes an essential CNR decrease, and, as a consequence, this parameter estimation RMS increase (see (2.35) and (2.36)).

TEC measurements through a combination of code and phase measurements at one frequency is essentially lower. This is caused by that: (1) strong measurement

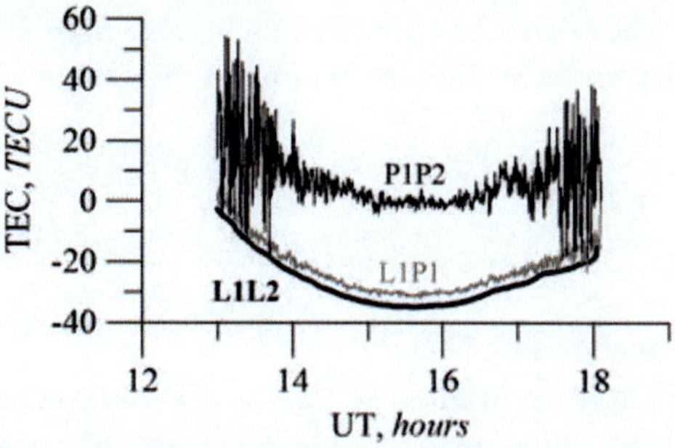

Fig. 4.1 TEC dynamics obtained by the dual-frequency L1L2 combination (lower black curve), single-frequency L1P1 combination (middle gray curve), dual-frequency P1P2 combination (upper black curve). The example is for GPS PRN29 signal processing. The data were obtained on 2015 Jun 22 at the ISTP station (ISTP SB RAS, Irkutsk) that is a part of the SibNet measuring network

noise caused by multipath propagation are present only in one channel; (2) used are code range measurements by open code, therefore, the measurement channel CNR is substantially higher, and the measurement noise RMS is lower.

The TEC measurements computed by the combination of phase measurements have the least noise level, which is caused by a small noise level of phase measurements as compared with the code pseudo-delay measurements (see Formulas 2.89, 2.90).

The TEC values measured by the code pseudo-range P1P2 difference is inexpedient to use for ionospheric investigations, except for a qualitative estimate of general trends in the ionosphere behavior. In all the other cases, the TEC estimates obtained by the combination of code measurements are subjected to preliminary smoothing by using synchronous phase measurements at the same frequencies and with the subsequent reconstruction of TEC values.

The most typical procedure of processing the combination of code and phase measurements is in leveling the phase TEC measurements (4.4) to pseudo-range TEC measurements (4.5), which enables to reduce ambiguities of phase TEC measurements to the accuracy within a DCB value. To perform the latter procedure, one defines the phase ambiguity constant as a mean for the difference between phase TEC and pseudo-range TEC with some statistical weights:

$$K^j = \frac{1}{\sum_{i=1}^{N^j} \omega_i^j} \sum_{i=1}^{N^j} \omega_i^j \left(I_P^j - I_\varphi^j \right)_i \tag{4.7}$$

where ω is a statistical weight; j is the number of a continuous interval (i.e., the measurement interval, over which there are no cycle slips not eliminated during the preliminary processing); i is the number of a phase record in a measurement series; N^j is the number of measurements for a j-th continuous interval, in which there are no measurement anomalies.

The statistical weight ω may be chosen as an inverse mapping function *Fpp*. So, we will have following phase ambiguity parameter (Yasyukevich et al. 2020a):

$$K^j = \frac{1}{\sum_{i=1}^{N^j} \frac{1}{Fpp_i^j}} \sum_{i=1}^{N^j} \frac{1}{Fpp_i^j} \left(I_P^j - I_\varphi^j \right)_i \tag{4.8}$$

*F*pp is a mapping function (2.81).

To summarize the narrative in Sects. 2.1.3, 2.2.1 and the considerations here, we may indicate that TEC estimate noise will differ essentially depending on a particular combination of code measurement types (see Fig. 2.16), operational frequencies, and features of satellite and receiver hardware applied in different GNSSs (see Sect. 2.2.3). This should be accounted for both when arranging measurements and when interpreting their results.

4.2 GNSS Data Processing

4.2.1 GNSS Receiver Networks

At present, GNSS receiver networks develop actively. They are one of the main sources to measure radionavigation parameters for performing global geophysical studies of the ionosphere, atmosphere, and the Earth crust. First of all, this is related to applying base station networks for real-time kinematic (RTK, and RTK Network). The largest structures are created to provide and maintain coordinate systems. Below, we present a number of large international network structures that provide access to measuring radionavigation parameters and GNSS measurement error components.

(1) International GNSS Service, or IGS (Dow et al. 2009). This structure involves more than 200 organizations of more than 100 countries (http://www.igs.org). The IGS data are available at the Crustal Dynamics Data Information System NASA website https://cddis.nasa.gov/archive/gnss/data/daily.

(2) UNAVCO (Smith 2000). A major provider for the GNSS measurement data, representing a non-commercial educational and geophysical consortium. The UNAVCO main goal is geodesy development. The UNAVCO archives are available at ftp://data-out.unavco.org.

(3) EUREF (Bruyninx et al. 2019). European association of more than 100 agencies, universities, and research organizations under the administration of the International Association of Geodesy. The key goal of the EUREF is maintaining the European Terrestrial Reference System 89 (ETRS89). The data are available at ftp://igs.bkg.bund.de/EUREF/obs/.

(4) SONEL (Gravelle et al. 2019), or Système d'Observation du Niveau des Eaux Littorales. The association of more than 200 organizations from 67 countries is

a part of the Global Sea Level Observing System (GLOSS). The measurement data are available at ftp://ftp.sonel.org/.

(5) SIRGAS (Sánchez and Kehm 2021), or Geocentric Reference System for the Americas. A Pan-American inter-departmental association of organizations, whose goal is to determine and describe the continental geocentric reference system and the geoid gravimetric model. The organizations composing the association store their data at their resources at (https://www.sirgas.org/en/).

(6) Geoscience Australia (Owen et al. 2018). This governmental association involves about 100 permanently operating stations being parts of several measuring GNSS networks. Among those networks, the main are: Australian Regional GNSS Network (ARGN), South Pacific Regional GNSS Network (SPRGN), AuScope Network (ftp://ftp.ga.gov.au).

(7) AFREF (Combrinck 2008). The African Geodetic Reference Frame. AFREF unifies the many geodetic reference frames of Africa using GNSS data as the primary data source for the implementation of a uniform reference frame (ftp://ftp.afrefdata.org).

Besides the above organizations, we can provide additional information on a number of other national/departmental networks and GNSS measurement data analysis centers. Those networks, in some cases, may not be parts of the above international organizations.

(8) BEV: Austrian data center are maintained by Federal Office of Metrology and Surveying. Austrian POsitioning Service (APOS) based on BEV realizes, distributes and maintains ETRS89 (ftp://gnss.bev.gv.at).

(9) BKG (Thorandt and Wojdziak 2011), or Bundesamt für Kartographie und Geodäsie Data Center. BKG provides high-accuracy observational and navigational GNSS data and products. The data come from national German GREF network as well as from a European and a world-wide network of continously operated reference stations. It serves as Regional Data Center for Europe (ftp://igs.bkg.bund.de).

(10) CASC (Duval et al. 1997), or Canadian Active Control System. Each CACS antennas anchored to bedrock to continuously record high-quality GNSS data. Some stations are also equipped with meteorological sensors measuring atmospheric pressure, temperature and humidity (https://webapp.geod.nrcan.gc.ca/geod/data-donnees/cacs-scca.php?locale=en).

(11) CHAIN (Jayachandran et al. 2009), or Canadian High Arctic Ionospheric Network. A Canadian high-latitude network of stations involving diverse ground-based equipment for monitoring the polar cap processes, including 25 specialized GNSS receivers-ionospheric monitors with a possibility to record scintillation indices and high temporal-resolution (50–100 Hz) data (ftp://chain.physics.unb.ca).

(12) DPGA, or Dutch Permanent GNSS Array consist of 18 receivers in the Netherlands. It aims to assist international programs such as IGS, EPN, E-GVAP et al. (ftp://gnss1.tudelft.nl).

(13) ERGNSS (Morin et al. 2000). Network in Spain contains more than 120 receivers. It is maintained by Instituto Geográfico Nacional (ftp://ftp.geodesia.ign.es).
(14) FReDNet (Battaglia et al. 2003). The Seismological Research Centre of OGS (Italian National Institute for Oceanography and Experimental Geophysics) monitors crustal deformation along northeast boundary of the Adriatic microplate (ftp://frednet.crs.inogs.it/).
(15) GeoDAF (Ferraro et al. 2022), or Geodetic Data Archiving Facility aims to provide the geodetic data and products stored by Italian space agency (ftp://geodaf.mt.asi.it).
(16) GSI (former GEONET) (Tsuji and Yamaguch 2013). One of the densest regional networks of GNSS receivers in the world. It involves more than 1300 receivers in the territory of Japan, with an ~ 20 km mean distance between them. The network is used to study earth crust deformations and to solve geodesy problems (http://terras.gsi.go.jp/).
(17) IGN: An analytical center within the Institute Geographique National (France) that administers its own measuring network, as well as acquires the world network data. ftp://igs.ensg.ign.fr.
(18) ITACyL: Network of GNSS measuring stations in Spain under the administration of Instituto Tecnológico Agrario de Castilla y León. Intended to provide high-precision RTK navigation and support of the ETRS89 coordinate system (ftp://ftp.itacyl.es).
(19) KASI (Jo et al. 2006), or Korea Astronomy and Space Institute. Regional center of the IGS data in Asia. The main goals of the center are to estimate small-scale changes in the Earth shape and its rotation parameters. Along with the IGS data, the data archives of this center includes measurements at its own network of stations (ftp://nfs.kasi.re.kr).
(20) New Zealand GeoNet (Gentle et al. 2016). New Zealand GeoNet network functions under the joint management of the Earthquake Commission (EQC), GNS Science geophysical service, the government agency on property and geodesy (Land Information New Zealand, LINZ). The main goal is monitoring the geological threats in New Zealand (ftp://ftp.geonet.org.nz).
(21) NCEDC (2014), or Northern California Earthquake Data Center. The network main goal is monitoring the earth crust deformations along the discontinuity between the Pacific and North American plates and in the San Francisco area with a view of the early alert to earthquakes (www.ncedc.org).
(22) NOANET (Chousianitis et al. 2021) is a GNSS network in Greece whose primary role is to enhance and support geophysical research. The network operates under the administration from the National Observatory of Athens (https://www.gein.noa.gr/services/GPSData/).
(23) PANGA (1996), or Pacific Northwest Geodetic Array comprises about 220 CORS. Receivers located throughout the Pacific Northwest and greater Cascadia Subduction Zone (www.geodesy.cwu.edu < http://www.geodesy.cwu.edu >).

(24) RAMSAC (Pinon et al. 2018) is National terrestrial reference fame of Argentina maintained by Argentinian National Geographic Institute (<https://www.ign.gob.ar/NuestrasActividades/Geodesia/Ramsac/DescargaRinex>).
(25) RBMC, or Rede Brasileira de Monitoramento Contínuo dos Sistemas GNSS (former LPIM) (Fortes et al. 1998). The Brazilian network of continuous monitoring under the administration from Instituto Brasileiro de Geografia e Estatística (ftp://geoftp.ibge.gov.br).
(26) RENAG (RESIF 2017), or REseau NAtional GPS permanent. GNSS monitoring and observation services in the northern French Alps (ftp://renag.resif.fr).
(27) RING (INGV 2016), or Rete Integrata Nazionale GPS. RING installs GPS receivers with broad-band seismometers and accelerometers to detect all the frequency content of the earthquake process (ftp://bancadati2.gm.ingv.it).
(28) ROB (Bruyninx and Defraigne 2018), or Royal Observatory of Belgium. Data are included in EUREF and are used to maintain the ETRS89 (ftp://ftp.epncb.oma.be).
(29) SatRef (Chan and Li 2007). Hong Kong Satellite Positioning Reference Station Network (ftp://ftp.geodetic.gov.hk).
(30) SOPAC (Jamason et al. 2004), or Scripps Orbit and Permanent Array Center. A center to acquire and analyze geodesic and seismic data in the interests of studying earthquakes, tsunami, tectonic plate motions, and earth crust deformations (ftp://garner.ucsd.edu, ftp://lox.ucsd.edu).
(31) State GPS network of the Republic of Bulgaria (BNCGG 2011). The network aims to realize, distribute, and maintain the European terrestrial reference system ETRS89 (ftp://195.96.249.3).
(32) SuGAr (Mcloughlin et al. 2011) is the Sumatran GPS Array monitors plate boundary between Indo-Australian and Asian tectonic plates. It includes about 50 stations (ftp://ftp.earthobservatory.sg).
(33) TLALOCNet (Cabral-Cano et al. 2018) is the GNSS network in Mexico aims to study climate, atmospheric processes, the earthquake cycle, and tectonic processes (ftp://tlalocnet.udg.mx).
(34) TrigNet (Hedling et al. 2000) is a South African network, that includes more than 50 GNSS stations spaced by 80–300 km (ftp://trignet.co.za).
(35) WCDA (Chen et al. 1996), or the Western Canada Deformation Array. Stations monitor crustal dynamics along the Cascadia subduction zone (ftp://wcda.pgc.nrcan.gc.ca).
(36) Wuhan University IGS analysis center implements backup of the IGS archives data (ftp://igs.gnsswhu.cn).

Within the Russian Federation, there exist a number of institutional GNSS networks: RosGidroMet (Federal Hydrometereology and Environmental Monitoring Service) network (Aleshin et al. 2014), network for geodynamic observations at the Far East Branch of the Russian Academy of Sciences (Sorokin et al. 2017), SibNet (Siberian network of ISTP SB RAS receivers) (Yasyukevich et al. 2018), as well as several commercial networks that require registration or subscription:

(1) EFT-CORS (https://eft-cors.ru/);
(2) HIVE, Industrial geodesic systems (https://hive.geosystems.aero/);
(3) SmartNet (http://smartnet-ru.com/);
(4) RTKNet—«Geodetika» (http://rtknet.ru/).

One of the largest GNSS-measurement coordinators in Russia is the Federal Center of Navigation Data (fcnd.ru).

4.2.2 RINEX: GNSS Data Storage Format

A great number of receiver manufacturers use their own unique data formats to measure radio navigation parameters, time, and service information in order to interface with other hardware and to store data. The GNSS measurement data at the points of measuring networks (see Sect. 4.3) may be presented in diverse formats. This substantially impedes the data exchange process among the communities concerned and the process to automate processing and analyzing great volumes of measurements at GNSS networks.

In the late 1980s, a number of organizations concerned in acquiring GNSS data produced 4 unified formats for data exchange (Gurtner 1994). Among those formats, the Astronomical Institute of the University of Berne proposed the RINEX format (Receiver Independent Exchange Format). The IGS recommended the format that gained the greatest spreading as a unitized format for data exchange to all the international scientific organizations concerned.

Among the RINEX versions during the existence of two GNSSs (GPS plus GLONASS), and with the Galileo under development, most data providers used RINEX 2.11 (Gurtner and Estey 2007). Commissioning new GNSSs (Galileo and BeiDou) required that the format be significantly updated, with allowance for a possibility to write a larger number of measurements into a file. In particular, including BeiDou signal measurements led to RINEX version 2.12. Except for that, RINEX 2 had an 80-character limitation on the string length. For RINEX 3, a similar limitation remains only for the title (for inverse compatibility), which enables to automate batch processing for great volumes of data presented in different RINEX versions.

Due to the further development of GNSS systems and introduction of new operational frequencies and signals, there emerged a need for a cardinal change in the RINEX format implemented in RINEX 3 (Gurtner and Estey 2009). In 2021 Dec, the IGS community proposed the newest version, RINEX 4 (Romero 2021) in which the main changes affected the navigation message storage format. A concerned reader may be referred to https://files.igs.org/pub/data/format/ for more details on the RINEX versions and their evolution.

RINEX is textual (ASCII), and, besides the data structure, it establishes the standard for a file name. There are three types of data files presented in all the RINEX versions:

1. Observation data file;

2. Navigation message (broadcast) file;
3. Meteorological data file.

Radionavigation parameter measurements are stored in an Observation (OBS) file. Those measurements include: the carrier phase, the code pseudo-range, and the Doppler frequency shift. Besides, additional measurements are stored here: signal strength; measurement integrity signatures; current system time by the satellite clocks; information concerning the station (station name, its rectangular geocentric coordinates, receiver and antenna types, data recording period, etc.). Also, newer formats store the measurements obtained by various ranging codes (e.g., the GPS L2 phase may be measured and stored in RINEX 3.00 files in ten modes, including those by the C/A-code, by the full L2C-code, as a result of codeless processing, etc.).

The IGS recommended a special archival format to better arrange a storage for a vast data archive, Compact RINEX (Hatanaka 2008).

In a RINEX OBS-file (ver. 3 and higher), one uses a three-character notation for separate types of radio navigation parameters (e.g., C1P or S2X), whereas, in RINEX 2 files, the third character is not indicated (similar parameters look like C1 or S2). The structure of those designations assumes the following:

(1) The first character denotes a measurement type: *C* stands for pseudo-range, *L* for phase, *D* for Doppler frequency shift, *S* for signal strength.
(2) The second character indicates a frequency subband, and, in some cases, a ranging code type. For GPS, 1, 2, 5 corresponds to the L1, L2, L5 subbands. For the other systems, the correspondence is indicated in Appendix 4.1. For example, for Galileo, C8 signifies a measurement within the E5 range by using E5a and E5b signals (see Fig. 2.5).
(3) The third character signifies a signal processing technique for the encrypted and open access. C signifies using an open civil code (C/A), P being an encrypted code (e.g., P(Y) GPS), W stands for the z-tracking technology, X stands for the processing methods using two orthogonal codes (e.g., Galileo E5a and E5b signals).

Appendix 4.1 shows the designations accepted in a RINEX OBS-file (ver. 3.04). For example, L2W signifies the L2 (1575.42 MHz) carrier phase estimated at a semi-codeless processing through the z-tracking technology. The same designation for BeiDou signals shows the carrier phase at B1-2 (1561.098 MHz), estimated at a code processing by using I + Q signals simultaneously. The S2P signal strength record shows the strength by P-code at L2 for GLONASS and GPS signals. A more detailed description of the data components recorded in RINEX 3.0 and higher can be found in (Gurtner and Estey 2009).

Figures 4.2 (RINEX 2.12) and 4.3 (RINEX 3.04) show the examples for the structure and the content of an OBS-file. Figures 4.2 and 4.3 show the same types of measurements presented in different formats. Colors in both figures mark the same data types. In RINEX 3 files, besides the measured parameter (C1, L1, S1), it is indicated directly by which code the measurement is taken. For example, C1W, L1P, S3X, etc., where last character (W, P, X, etc.) indicates the ranging code type, by which the measurement is performed. Appendix 4.1 provides decoding for those parameters (indices) in the RINEX specification and duplicates it for RINEX 3.04.

Color emphasis in the figures corresponds to the same measured parameters (in rectangular frames: red–L1W, green–L2W, pink–S2P, yellow–L5I, dark blue–C2I; blue oval marks the format version). In the example presented in Fig. 4.2, GPS L1 measurements correspond to the L1W designation (not, for example, L1C), but, for GLONASS, L1 corresponds to L1P. When generating files, some programs/manufacturers may mix measurements. Wanninger (2018) established that about 2.7% of ~ 2600 files (GPS-receivers) mix the L2P(Y)/L2C measurements in a RINEX 2 file, which is impossible to check without special processing.

The navigation data file format (NAV-files) differs for different systems. So, for GPS, Galileo, and BeiDou, the NAV-file contains elements of SV orbits (ephemeris), by which a user can compute the satellite coordinates in the rectangular geocentric coordinate system at any current instant. A GLONASS and SBAS NAV-file

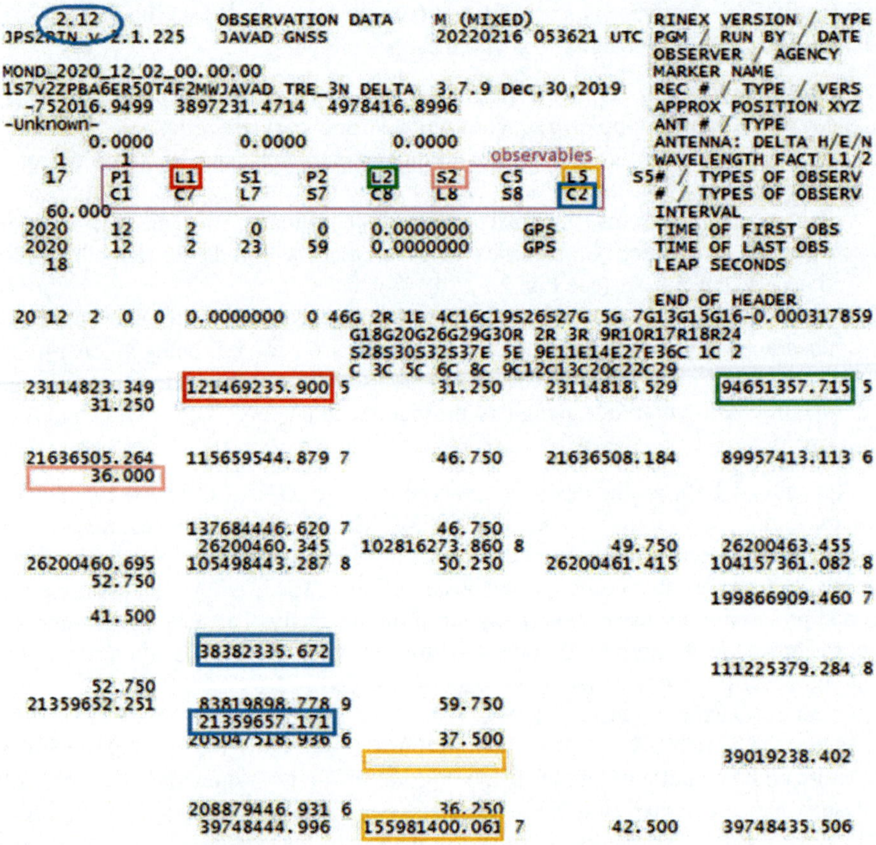

```
     2.12           OBSERVATION DATA    M (MIXED)           RINEX VERSION / TYPE
JPS2RIN v.2.1.225   JAVAD GNSS          20220216 053621 UTC PGM / RUN BY / DATE
                                                            OBSERVER / AGENCY
MOND_2020_12_02_00.00.00                                    MARKER NAME
1S7V2ZPBA6ER50T4F2MWJAVAD TRE_3N DELTA  3.7.9 Dec,30,2019   REC # / TYPE / VERS
  -752016.9499  3897231.4714  4978416.8996                  APPROX POSITION XYZ
-Unknown-                                                   ANT # / TYPE
        0.0000        0.0000        0.0000                  ANTENNA: DELTA H/E/N
     1     1                                 observables    WAVELENGTH FACT L1/2
    17    P1    L1    S1    P2    L2    S2    C5    L5    S5# / TYPES OF OBSERV
          C1    C7    L7    S7    C8    L8    S8    C2      # / TYPES OF OBSERV
    60.000                                                  INTERVAL
  2020    12     2     0     0    0.0000000     GPS         TIME OF FIRST OBS
  2020    12     2    23    59    0.0000000     GPS         TIME OF LAST OBS
    18                                                      LEAP SECONDS

                                                            END OF HEADER
 20 12  2  0  0  0.0000000  0 46G 2R 1E 4C16C19S26S27G 5G 7G13G15G16-0.000317859
                                G18G20G26G29G30R 2R 3R 9R10R17R18R24
                                S28S30S32S37E 5E 9E11E14E27E36C 1C 2
                                C 3C 5C 6C 8C 9C12C13C20C22C29
  23114823.349  121469235.900 5         31.250   23114818.529    94651357.715 5
        31.250

  21636505.264  115659544.879 7         46.750   21636508.184    89957413.113 6
        36.000


                137684446.620 7         46.750
                 26200460.345   102816273.860 8         49.750    26200463.455
  26200460.695  105498443.287 8         50.250   26200461.415   104157361.082 8
        52.750
                                                                199866909.460 7
        41.500

                38382335.672
                                                                111225379.284 8
        52.750
  21359652.251   83819898.778 9         59.750
                 21359657.171
                205047518.936 6         37.500
                                                                 39019238.402

                208879446.931 6         36.250
                 39748444.996   155981400.061 7         42.500    39748435.506
```

Fig. 4.2 Example of a RINEX 2.12 OBS-file. In the rectangular frames: red–L1W, green–L2W, pink–S2P, yellow–L5I, dark blue–C2I; the blue oval marks the format version

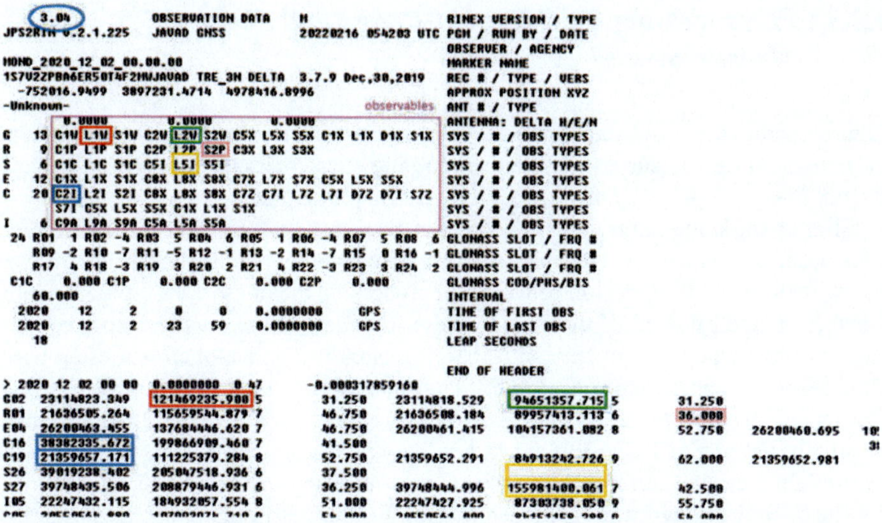

```
     3.04           OBSERVATION DATA    M                   RINEX VERSION / TYPE
JPS2RIN v.2.1.225   JAVAD GNSS          20220216 054203 UTC PGM / RUN BY / DATE
                                                            OBSERVER / AGENCY
MOND_2020_12_02_00.00.00                                    MARKER NAME
1S7U2ZPBA6ER5OT4F2MWJAVAD TRE_3N DELTA  3.7.9 Dec,30,2019   REC # / TYPE / VERS
  -752016.9499  3897231.4714  4978416.8996                  APPROX POSITION XYZ
-Unknown-                                                   ANT # / TYPE
        0.0000        0.0000        0.0000                  ANTENNA: DELTA H/E/N
G   13 C1W L1W S1W C2W L2W S2W C5X L5X S5X C1X L1X D1X S1X  SYS / # / OBS TYPES
R    9 C1P L1P S1P C2P L2P S2P C3X L3X S3X                  SYS / # / OBS TYPES
S    6 C1C L1C S1C C5I L5I S5I                              SYS / # / OBS TYPES
E   12 C1X L1X S1X C8X L8X S8X C7X L7X S7X C5X L5X S5X      SYS / # / OBS TYPES
C   20 C2I L2I S2I C8X L8X S8X C7Z C7I L7Z L7I D7Z D7I S7Z  SYS / # / OBS TYPES
       S7I C5X L5X S5X C1X L1X S1X                          SYS / # / OBS TYPES
I    6 C9A L9A S9A C5A L5A S5A                              SYS / # / OBS TYPES
 24 R01  1 R02 -4 R03  5 R04  6 R05  1 R06 -4 R07  5 R08  6 GLONASS SLOT / FRQ #
    R09 -2 R10 -7 R11 -5 R12 -1 R13 -2 R14 -7 R15  0 R16 -1 GLONASS SLOT / FRQ #
    R17  4 R18 -3 R19  3 R20  2 R21  4 R22 -3 R23  3 R24  2 GLONASS SLOT / FRQ #
 C1C    0.000 C1P    0.000 C2C    0.000 C2P    0.000        GLONASS COD/PHS/BIS
    60.000                                                  INTERVAL
  2020    12     2     0     0    0.0000000     GPS         TIME OF FIRST OBS
  2020    12     2    23    59    0.0000000     GPS         TIME OF LAST OBS
    18                                                      LEAP SECONDS

                                                            END OF HEADER
> 2020 12 02 00 00  0.0000000  0 47       -0.000317859160
G02  23114823.349  121469235.900 5     31.250  23114818.529   94651357.715 5     31.250
R01  21636505.264  115659544.879 7     46.750  21636508.184   89957413.113 6     36.000
E04  26200463.455  137684446.620 7     46.750  26200461.415  104157361.082 8     52.750  26200460.695  10
C16  38382335.672  199866909.460 7     41.500                                                          31
C19  21359657.171  111225379.284 8     52.750  21359652.291   84913242.726 9     62.000  21359652.981
S26  39019238.402  205047518.936 6     37.500
S27  39748435.506  208879446.931 6     36.250  39748444.996  155981400.061 7     42.500
I05  22247432.115  184932057.554 8     51.000  22247427.925   87303738.050 9     55.750
```

Fig. 4.3 Example of a RINEX 3.04 OBS-file. In the rectangular frames: red–L1W, green–L2W, pink–S2P, yellow–L5I, dark blue–C2I; the blue oval marks the format version

GLONASS involve the coordinates for satellites, their velocity parameters, and accelerations in the rectangular geocentric coordinate system, that are determined in reference instants in 30-min increments. A user can compute intermediate coordinates for satellites at an arbitrary instant based on those reference coordinates by using the known satellite motion equations.

Except the above data, each NAV-file includes the information on the parameters of the current correction model for the satellite onboard timescale, the ionospheric correction model parameters (GPS and BeiDou), as well as the additional information on a satellite health. The GLONASS data in NAV-files involve the information on the frequency channel number for each of satellites.

The RINEX 2 format used separate NAV-files for each of the navigation systems, whereas RINEX 3 can mix the data from different systems in one file. RINEX 4 was developed with allowance for a possibility to expand the data storage format by including various versions of ionospheric corrections, Earth rotation parameters, and the system timeshift.

A meteorological data file may contain the following parameters for the Earth atmosphere in the surface layer: Pressure, Dry temperature, Relative humidity, Wet zenith path delay, Dry component of zenith path delay, Total zenith path delay, Wind azimuth, Wind speed, Rain increment, and Hail indicator. Besides, such a file contains the information on metro sensors, whose information is used to generate a file, and their locations.

4.2.3 Preprocessing TEC Pseudo-Range and Phase Measurements

Preprocessing the TEC data involves eliminating cycle slip effects, anomalous measurement values (surges or nulls), smoothing phase measurements by code ones (smoothing code pseudo-range measurements by phase ones).

Eliminating surges and cycle slips requires, first of all, that such events be detected. This problem has been solved for years by using the known TurboEdit algorithm (Blewitt 1990). This algorithm provides preprocessing the phase measurements with a detection and removal of the carrier phase measurement anomalies separately at each carrier frequency. For the ionosphere research problems, one most often uses combinations of phase and code pseudo-delay measurements taken at two operational frequencies. Therefore, below, we describe a procedure for the measurement preprocessing based on the TurboEdit algorithm, but with allowance for this algorithm to be used for a case of difference dual-frequency measurements (4.4 and 4.5).

The TurboEdit algorithm analyzes double differences in the phase and pseudo-delay measurements for a satellite along a satellite-receiver LOS.

$$P_{\delta,i} = P_{2,i} - P_{1,i}, \tag{4.9}$$

$$L_{\delta,i} = L_{2,i}\lambda_2 - L_{1,i}\lambda_1, \tag{4.10}$$

where P_1, P_2 ($L_1\lambda_1, L_2\lambda_2$) are code pseudo-range (phase path) at the first and second operational frequencies; i is a current instant reading.

The difference between $P_{\delta,i}$ and $L_{\delta,i}$ represents the parameter $b_{\delta,i}$ including the measurement ambiguity.

$$b_{\delta,i} = L_{\delta,i} - P_{\delta,i} \tag{4.11}$$

Series of differences $b_{\delta,i}$ are subject to a characteristic estimate procedure, i.e. the current mean $\langle b_\delta \rangle_j$ and the current RMS σ_i. If a new estimate $b_{\delta,i+1}$ deviates from the mean value measured at the previous step by more than $4\sigma_i$ we establish a b_δ failure related to an anomalous measurement surge or to a cycle slip. The estimate failure has a various character depending on the cause. If the cause is an anomalous measurement surge, the failure is established at the present instant and, afterward, the difference (4.11) returns to the initial mean level. In the case of a cycle slip, the difference (4.11) transfers to another mean level after the estimate failure (above or below the previous mean level), i.e., there emerges a "step" in the measurement trend.

To distinguish between the failure related to an anomalous measurement surge and that caused by a cycle slip, we analyze two time-subsequent measurements (4.11), i.e., the $b_{\delta,i+2}$ and $b_{\delta,i+3}$ values, and estimate the following events:

- if $b_{\delta,i+2}$ and $b_{\delta,i+3}$ are $\leq 4\sigma_i$ of the mean $\langle b_\delta \rangle_i$ measured before the estimate $b_{\delta,i+1}$ failure, then we establish an anomalous measurement surge that is removed from the following processing;
- if $\left(b_{\delta,i+1} - b_{\delta,i+2}\right)$ is $> 4\sigma_i$ of the mean $\langle b_\delta \rangle_i$ and $\left(b_{\delta,i+2} - b_{\delta,i+3}\right) \leq 4\sigma_i$, then we establish a cycle slip, after which we smooth the trend between the difference (4.11) continuous measurement fragments before and after the cycle slip onset, i.e., we remove the "step" in measurements.

To eliminate an anomalous surge in the pseudo-range measurement data is possible either by casting these data, or by $\langle b_\delta \rangle_i$ linear approximation, or by using the mean:

$$P_{\delta,i} = L_{\delta,i} - \langle b_\delta \rangle_i \tag{4.12}$$

To eliminate a cycle slip is possible from the next measurements taking with allowance for the subsequent continuity of the smoothed series derivative. For all the N measurements, starting with $i + 1$, where the next cycle slip onset was recorded, one adds the constant (Δ) to the current difference (4.11). The constant smooths the step in the measurement series as follows:

$$L_{\delta,i+N}^{COR} = L_{\delta,i+N} + \Delta \tag{4.13}$$

$$\Delta = L_{\delta,i} - L_{\delta,i+1} + \frac{\left(L_{\delta,i} - L_{\delta,i-1}\right) + \left(L_{\delta,i+2} - L_{\delta,i+1}\right)}{2} \tag{4.14}$$

Then, all the subsequent values in (4.13), starting with $N + i + 1$, will look like:

$$L_{\delta,i+1}^{COR} = L_{\delta,i} + \frac{\left(L_{\delta,i} - L_{\delta,i-1}\right) + \left(L_{\delta,i+2} - L_{\delta,i+1}\right)}{2} \leq 4\sigma_i \tag{4.15}$$

Under the conditions of long-term cycle slip with no possibility of capturing the set instant in a file, there may be no phase measurement. In this case, one may use a similar procedure to eliminate a data discontinuity, but the $i + 1$ measurement corresponds to the first measurement after a phase capture. Also, the discontinuity should not be large (no more than 1–2 values are usually assumed to be missed). In this case, the values starting with the $i + 1$ are also determined, based on corrections by (4.13–4.14), while the intermediate values are linearly interpolated.

Figure 4.4 shows an example for oblique TEC series obtained from pseudo-range (4.5) and phase (4.4) measurements by the GLONASS R08 signals. The measurements were corrected by removing cycle slips (4.13–4.14). Upon correcting, the TEC phase measurements were smoothed to remove the phase measurement ambiguity (dark blue curve) by the synchronous TEC measurements obtained from code measurements (gray dots). Red curve shows the initial phase TEC measurements involving cycle slips.

At present, there are more advanced algorithms for processing the measurements of phase difference and of phase TEC series. Cai et al. (2013) proposes to use a

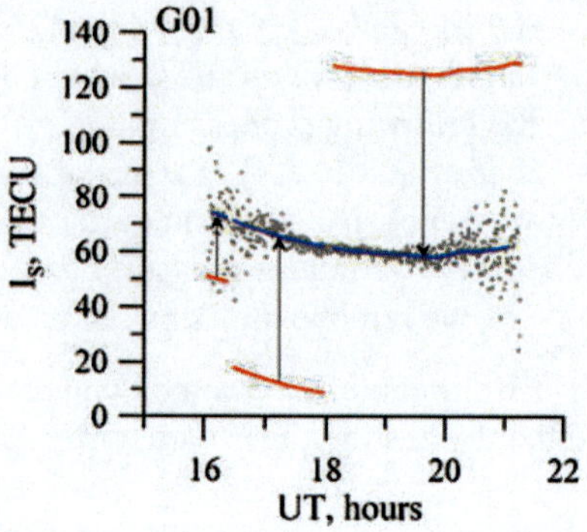

Fig. 4.4 Example of the corrected and initial TEC series obtained by pseudo-range and phase observables. Gray dots show pseudo-range TEC, red and dark blue curves represent phase TEC before and after the cycle slip elimination

forward and backward moving average algorithm and a second-order, time-difference phase ionospheric residual algorithm. Such an approach enables to directly detect anomalies in measurement series without analyzing the previous and next measurement epochs. A forward and backward moving average provides a more precise estimate for a measurement trend, by which phase measurements during the cycle slip removal are smoothed.

Liu et al. (2021) proposed to use a recurrent estimation for difference (4.11) measurements, that is regarded as a first-order Gauss-Markov stochastic process. The proposed method provides a higher sensitivity to detecting cycle slips, including those at low elevations, at a phase shift value at the cycle slip boundary $\leq \pi$.

4.2.4 *Variation Selection in the Data of Dual-Frequency Phase TEC Measurements*

Upon preprocessing the TEC series, one should detect TEC variations that correspond to the investigated classes of geophysical events in the ionosphere. Mainly, the investigation subject is the TEC variation ranges with a 2–10 min period (acoustic-gravity wave disturbances), 10–20 min, and 20–60 min (midscale, large-scale, and traveling ionospheric disturbances) (Afraimovich et al. 2013).

Setting the problem of filtering TEC variations from raw TEC series (4.4) requires that an optimal filtering procedure enabling to obtain geophysical information with a minimal effect of the filter characteristics on the result interpretation be determined.

The first stage of the filtering procedure for the TEC series obtained by GNSS MEO satellite measurements is removing the main trend of measurements. This trend is caused by a satellite motion relative to an observation point. Figure 4.5 illustrates a typical pattern of the TEC temporal dynamics during a measurement session. Apparently, the TEC mean value varies in inverse proportion to a satellite elevation, by which the measurements are conducted. There is no similar trend in

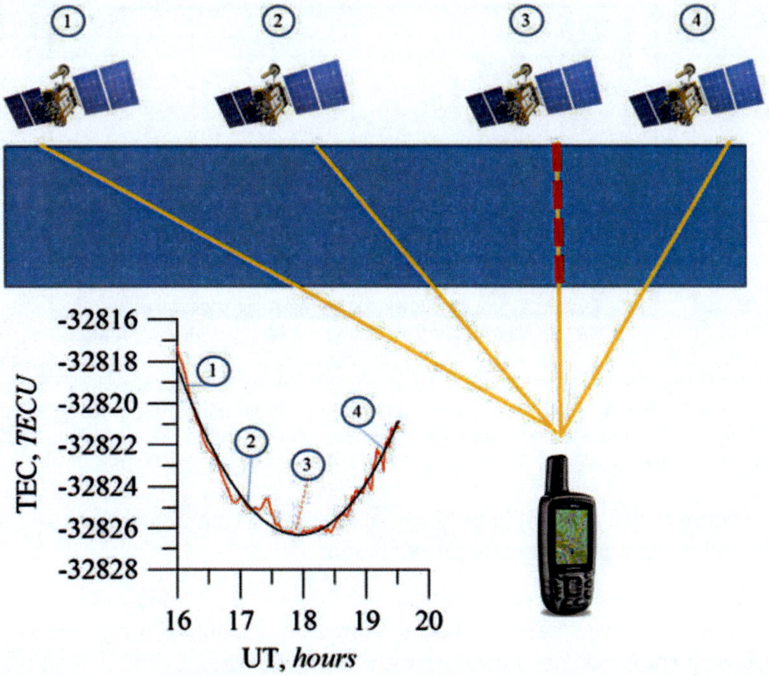

Fig. 4.5 Scheme for the trend emergence in TEC data. TEC measurements for various instants (1, 2, 3, 4) correspond to the positions of the satellites with the same numbers. Red and black line present the TEC dynamics and its trend. Yellow lines mark "satellite-receiver" beams (dotted line marks the beam, when the satellite was in zenith), and dark blue area indicates the ionosphere

the TEC measurement series obtained by GEO satellites due to the absence of their motion (Kunitsyn et al. 2016).

Selecting the type of the filter for the main trend removal determines the basic error in TEC filtering. Maletckii et al. (2020) showed that, for different filters, systematic filtering errors vary, on average, from 0.001 to 0.5 TECU. Figure 4.6 illustrates this. The vertical axis presents the mean filtering error, when removing the main trend, the horizontal axis indicates the filter type.

Analyzed were seven types of filters: Centered Moving Average Filter (CMA), Median filter (Median), 6th order polynomial filter (polyn (6)), Cubic smoothing spline (Spline), Hodrick-Prescott Filter (HP), L1 filter (l1), and double applying of CMA filter (double). Apparently, the CMA shows the greatest filtering error. The median filter shows a little lower error. However, using the double-pass smoothing CMA window already reduces the filtering error by over an order of magnitude. The other filter types show about equal efficiency and provide a minimal filtering error, because the pass function of those filters may be adapted for the form of the main measurement trend unlike the filters like the CMA and Median.

Thus, using non-linear functions to build a filter pass characteristic appears preferable for implementing the TEC measurement main trend removal. Maletckii et al.

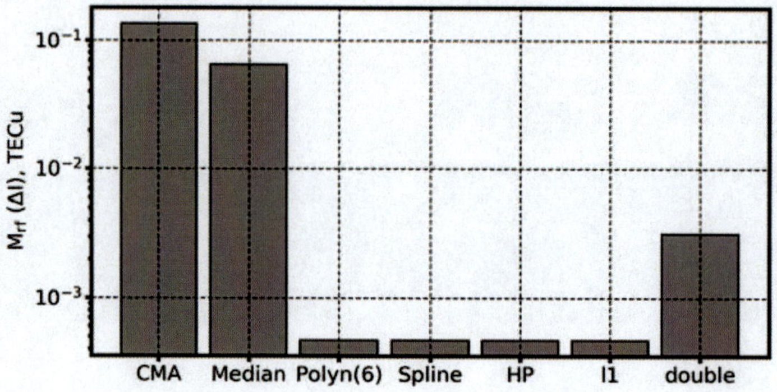

Fig. 4.6 Mean bias error of trend removal. From left to right are CMA–centered moving average filter, Median filter, polyn(6)–6th order polynomial, spline—cubic smoothing spline, HP–Hodrick-Prescott Filter, l1–L1 filter, double—is a double applying of CMA filter (Maletckii et al. 2020)

(2020) recommended a 6th order polynomial filter and a Cubic smoothing spline that have the following mathematical representation:

(1) 6th order polynomial represents an approximation of oblique TEC series (I) by a polynomial function (here, of the 6th order) over all the N measurements with a subsequent trend subtraction from the initial series.

$$\begin{gathered} \hat{I}(t_i) = \sum_{n=0}^{k} c_n \cdot t_i^n \\ \sum_{i=0}^{N} \left(\hat{I}(t_i) - I(t_i)\right)^2 \rightarrow \min; \\ \delta I(t_i) = I(t_i) - \hat{I}(t_i) \end{gathered} \tag{4.16}$$

where k is a polynomial degree (6), c_n is a ploynomial coefficient; δ is a set filtering discrepancy.

(2) Cubic smoothing spline represents an approximation of an oblique TEC trend by a continuous function that is determined in the form of local 3rd polynomials with a subsequent trend function removal from measurement series:

$$\begin{gathered} \hat{I}_i(t) = \sum_{n=0}^{3} c_{n,i} \cdot t^n \\ \sum_{i=0}^{N} \left(I(t_i) - \hat{I}_i(t_i)\right)^2 + \lambda \int \hat{I}_i''(t)dt \rightarrow \min \\ \delta I(t_i) = I(t_i) - \hat{I}_i(t_i) \end{gathered} \tag{4.17}$$

The a priori set smoothing parameter λ should be more than 0 (empirically, for the GNSS data, the value 8 is offered).

Figure 4.7 presents the systematic filtering error that is attained when using the filters of the indicated types. Apparently, of the two proposed filters, the spline-filter

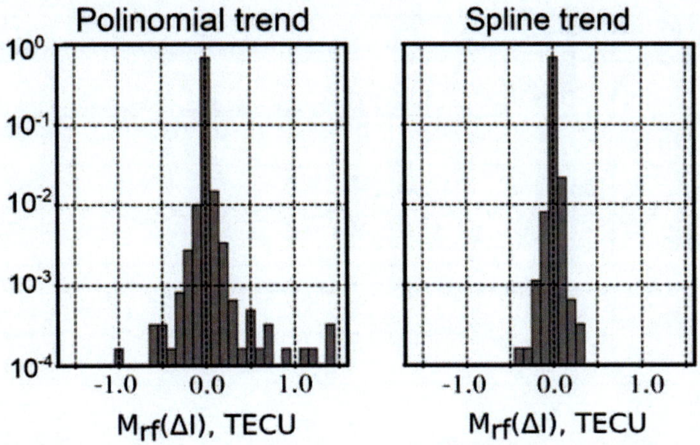

Fig. 4.7 Distribution of mean bias error of trend IRI TEC: left is for polynomial, right is for spline (Maletckii et al. 2020)

has a better filtering quality by the filtering residual error criterion. The maximal filtering error, when using that filter under the addressed test conditions, did not exceed 0.5 TECU (Fig. 4.7, right panel). It is by factor of 2–3 less than the filtering error when using the polynomial filter (Fig. 4.7, left panel).

Upon removing the trend, one should detect high-frequency TEC variations within the set range of periods. To solve this problem, Maletckii et al. (2020) proposed a centered moving average filter. That filter algorithm performs a consecutive averaging of TEC values, first, over the 2 k interval, then, over the 2 m ($m > k$) interval with the same averaging window center. Upon averaging, the results of the first and second averagings are subtracted from each other, which provides a detection of the phase TEC variations within the set frequency band (periods):

$$d\hat{I}(t_i) = \frac{1}{2k+1}\sum_{n=-k}^{k} I(t_{i+n})\cdot\omega_n - \frac{1}{2m+1}\sum_{l=-m}^{m} I(t_{i+l})\cdot\omega_l \tag{4.18}$$

Normalization for statistical measurement weights at their summation should correspond to the condition $\frac{1}{2m+1}\sum_{l=-m}^{m}\omega_l = 1$. In most cases, one neglects this condition without essential deterioration in the filtering quality and accept the statistical weight $\omega = 1$ for all the measurements relative to the smoothing window center.

A set of processing procedures, addressed above, is implemented, in particular, within the System for Ionosphere Monitoring and Researching from GNSS – SIMuRG (Yasyukevich et al. 2020b). Specifically, as a part of software-algorithmic suite, the SIMuRG uses a cubic smoothing spline to remove the trend and a centered moving average filter for three ranges of periods: 2–10 min, 10–20 min, and 20–60 min.

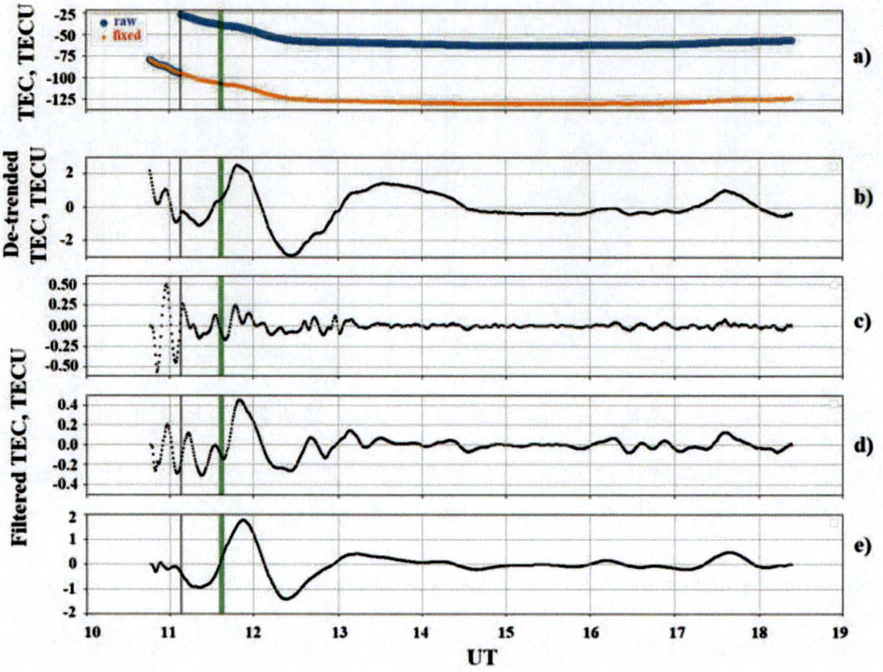

Fig. 4.8 Sequence of phase TEC data processing. Top to bottom: (1) the input data on dual-frequency phase TEC (panel a, dark blue curve) and the data after the cycle slip elimination (orange curve); (2) the data after the trend removal (panel b), (3–5) the post-filtering data in three ranges of periods 2–10 min (panel c), 10–20 min (panel d), 20–60 min (panel e). Black line marks a cycle slip instant, green line marks the magnetic storm maximal evolution

Figure 4.8 demonstrates an example for a consecutive use of processing phase TEC measurements within the SIMuRG under the 2021 Nov 4 magnetic storm conditions.

First, the phase TEC input data (top panel, dark blue curve) are corrected to eliminate cycle slips (dark blue curve presents the TEC data after the correction). The next panel demonstrates the data after the trend removal by smoothing with a cubic spline. Three bottom panels exhibit the post-filtering data in three ranges of periods: 2–10 min, 10–20 min, 20–60 min. The figure shows the correctness of processing the phase TEC series under complicated geomagnetic conditions.

4.3 Indices to Evaluate the State of the Ionosphere

4.3.1 *Ionospheric TEC-Based Indices*

GNSS TEC measurements provide, among others, enable to estimate the level of an ionospheric disturbance, as well as the status of the radiowave propagation channel for signals from separate satellites.

The ionospheric indices, whose computation is based on using absolute or relative TEC values, are split into two main classes. The first is based on initial oblique measurements of TEC. The most known index of this class is the Rate Of TEC Index (ROTI) that represents a TEC derivative RMS ($ROT = \mathrm{d}I/\mathrm{d}t$). The deviation is estimated through a recurrent mode within a running window of the set duration (Pi et al. 1997).

Over the last years, a number of indices has been developed to modify to the ROTI for solving certain research problems. These indices include: AATR (Along Arc TEC Rate) (Juan et al. 2018); DIX (Disturbance Ionosphere index) (Jakowski et al. 2012); DIXSG (Disturbance Ionosphere Index Spatial Gradient) (Wilken et al. 2018). Table 4.1 contains the basic information on the above ionospheric indices.

Except the listed, Afraimovich et al. (2008) proposed to estimate the ionospheric disturbance level based on the status of the ionosphere wave activity. For that purpose, the authors suggested one use an average RMS of the oblique TEC variations for different ranges of periods, which enables to estimate the ionospheric disturbance intensity in different parts of the spectrum. Voeykov et al. (2016) improved the proposed method: they used the TEC variation envelope amplitude instead of RMS TEC variations. This index (WTEC index) shows, sufficiently adequately, the ionospheric disturbance level and correlates well with solar and geomagnetic activities (Perevalova et al. 2016).

The second class of indices is based on vertical TEC values (re-computed from measurements or obtained when using local or global maps of vertical TEC. Yasyukevich (2021) offered the median-normalized vertical TEC RMS to estimate the ionospheric disturbance level. Nesterov et al. (2017), Gulyaeva and Stanislawska (2008) developed ionospheric indices based on vertical TEC spatial distribution. Such indices enable to estimate the TEC variability by space and by time, and may characterize a local or global ionospheric disturbance, depending on the type of a TEC data.

4.3.2 *Indices for Phase and Amplitude Ionospheric Scintillations*

One of the major manifestations of the ionosphere, in terms of signal receiving and processing in the ground hardware of different function, are scintillations of a transionospheric signal amplitude and phase. As conventional indices for a quantitative

Table 4.1 Ionospheric disturbance indices based on oblique TEC

Index	Computation principle	Formula	References
ROTI	For set line-of-sight and for station: TEC derivative RMS for five minutes	$ROTI = \sqrt{\overline{\mathrm{ROT}^2} - \overline{\mathrm{ROT}}^2}$ $\mathrm{ROT} = \mathrm{dI/dt}$	Pi et al. (1997)
AATR	ROTI analog with allowance for conversion to equivalent vertical value	$\mathrm{AATR_j(t)} = \frac{1}{\left[\mathrm{Fpp_j(\theta)}\right]^2}\mathrm{ROT_j(t)}$ $\mathrm{AATR} \equiv \mathrm{RMS_{AATR}} =$ $\sqrt{\frac{1}{N}\sum_{t=T}^{T+\Delta T}\sum_{j=1}^{Nsat}\left[\mathrm{AATR_j(t)}\right]^2}$ ΔT—averaging time, Nsat—number of satellites, N—total number of elements in the total, *F*pp—mapping function	Juan et al. (2018)
DIX	In simple version, it makes sense to use ROTI normalized RMS at various stations with some normalization. In more complicated version, it involves the ionosphere variability, and is determined by both AATR mean value and its deviations for various pairs of subionospheric points	$\mathrm{DIX^S} = \mathrm{SF^S}\sqrt{\frac{1}{\mathrm{N_P}}\sum_{k,l}\left(\frac{\mathrm{ROT_k}}{\mathrm{Fpp_k}} - \frac{\mathrm{ROT_l}}{\mathrm{Fpp_l}}\right)^2}$ *SF*—scaling factor (one usually accepts SF = 1) $DIX =$ $SF\sqrt{\frac{1}{N_P}\sum_{k,l}\left\{\begin{array}{l}\frac{1}{4}\left(\frac{ROT_k}{Fpp_k} + \frac{ROT_l}{Fpp_l}\right)^2 + \\ +\eta\left(\frac{ROT_k}{Fpp_k} - \frac{ROT_l}{Fpp_l}\right)^2\end{array}\right\}}$ η—tuning factor to balance temporal and spatial terms	Jakowski et al. (2012)
DIXSG	For line-of-sight *j* (*DIXSGj*): Analog of AATR, but, for normalization for elevation, spatial shift by IPP is used For station: $DIXSG_j$ normalized average deviation for various pairs of measurements	$cROT_i = \left\lvert\frac{ROT_i}{\Delta l_i^k}\right\rvert$, Δl_i—spatial shift of subionospheric point $DIXSG(cROT_{level})_{ij} =$ $\left(\frac{\lvert cROT_i - cROT_j\rvert}{cROT_{level}}\right)^3\left(\frac{d}{D}\right)^{-1}$ $cROT_{level}$—selectable recognition level, *d*—distance between station and IPP for *i*-th satellite, and *D* is maximal allowed distance	Wilken et al. (2018)

(continued)

Table 4.1 (continued)

Index	Computation principle	Formula	References
WTEC	TEC variation amplitude mean value (in set range of periods) with statistical weight, certain elevation	$W_{tec}(t) = \frac{\sum_{i=1}^{N} A_i(t)\omega_i(t)}{\sum_{i=1}^{N} \omega_i(t)}$ Here, A—TEC variation intensity, and ω—statistical weight	Voeykov et al. (2016)

estimate of the character and intensity of those phenomena, one uses two indices: the index of amplitude (intensity) scintillations ($S4$), and the phase scintillation index $\sigma\varphi$.

The $S4$ index represents a signal intensity variation (SI) normalized dispersion in a receiving point (van Dierendonck et al. 1993):

$$S4 = \sqrt{\frac{\langle SI\rangle^2 - \langle SI\rangle^2}{SI^2}} \tag{4.19}$$

where $\langle SI\rangle$ is a mean value for signal intensity during accumulation time T.

When averaging the signal intensity during investigations into ionospheric scintillations, van Dierendonck et al. (1993) recommended to choose the time $T =$ 60 s. However, depending on the available frequency of recording the primary measurements and on the research objective, one may choose another time of averaging.

The signal intensity (SI) received at the antenna is computed through a complex amplitude of a signal (A) field in a receiving point. The complex amplitude, in turn, is estimated based on the measurements of in-phase (I) and quadrature (Q) signal components in a navigation receiver as follows (Kintner et al. 2007):

$$SI = A^2 = I^2 + Q^2 \tag{4.20}$$

During signal digitizing, sampling, and amplifying, in a receiver, there may emerge additional nonsynchronous CNR variations in channels that generate I and Q samples, both at the stage of signal search and capture and at the stage of its tracking. Therefore, to correctly estimate signal intensity variations, one uses accumulated (over a period T) auxiliary values for a full signal strength measured in a "narrow" and "wide" frequency band. These values are computed by averaging the measurements over the $T = 20$ ms interval, like NBP (Narrow Band Power) and WBP (Wide Band Power), respectively (van Dierendonck et al. 1993):

$$\begin{aligned} WBP &= \sum_{i=1}^{20} I_i^2 + \sum_{i=1}^{20} Q_i^2 \\ NBP &= \left(\sum_{i=1}^{20} I_i\right)^2 + \left(\sum_{i=1}^{20} Q_i\right)^2 \end{aligned} \tag{4.21}$$

The *WBP* value corresponds to slow signal strength variations, while the *NBP* corresponds to fast variations. If the receiver gain during the time T does not change, the difference between the *NBP* and *WBP* will be proportional to the intensity of the signal received at a receiver antenna:

$$SI = \frac{NBP - WBP}{(NBP - WBP)_{LFT}} \tag{4.22}$$

Expression (4.21) requires that one normalize slow signal strength measurements relative to the trend, caused by a satellite motion relative to a receiver. For this purpose, the (*NBP-WBP*) difference time series is passed through a low-pass filter that detects this low-frequency trend $(NBP - WBP)_{\text{LFT}}$. van Dierendonck et al. (1993) proposed to use the cutoff frequency of this filter that equals $F_{\text{LPF}} = 1/T$, at the accumulation time $T = 20$ ms. However, in general, the filter cutoff frequency depends on the frequency of recording the signal intensity available for measurements, and on the objective of the investigation. In particular, when studying ionospheric scintillations, the following criteria should be met:

$$F_{LPF} \ll \frac{1}{T_{fr}}; T_{fr} = \frac{\sqrt{2\lambda Z}}{V_{drf}} \tag{4.23}$$

where Z is a distance from the layer of scattering ionospheric irregularities to a signal receiving point (at computations, one usually accepts $Z \approx 250$–300 km); λ is a signal wave length; V_{drf} is a mutual radial velocity of ionospheric irregularity drift and a satellite LOS ($V_{drf} \approx 300$ m/s…3 km/s).

When choosing the accumulation time T, the 20 ms interval is assumed to correspond to the nominal length of the C/A code sequence. This time can be changed. At the stages of a signal capture and of the subsequent coherent tracking, the I_i and Q_i samples are generated with a 1-ms period (Tiwari and Strangeways 2015). Then, according to (4.21), one may decrease the accumulation period within 20 ms and 1 ms. Herewith, a shorter measurement accumulation interval does not enable to implement exact signal phase tracking, and reduces the CNR. Therefore, one should post-process the accumulated measurements to obtain a robust estimate for the signal phase and for the signal code pseudo-delay. This complicates processing and interpreting the measuring information under ionospheric scintillation conditions.

Increasing the accumulation time beyond 20 ms also impedes the final estimate and interpretation of the results. This is related to that the carrier signal is additionally coded by service information with a 20-ms update period (IS-GPS-200 M 2021). Increasing the accumulation time beyond this limit assumes the presence (within the studied signal fragment) of a randomly changing phase at the instants of service message alternation. Besides, at $T > 20$ ms, a receiver clock frequency short-time instability starts to affect.

Accounting for all the above, one computes the scintillation index with allowance for the inner receiver noise as well. Also, the inner receiver noise effect on the final value of the *S*4-index is considered by the CNR measured with the quadrature components at the PLL input (van Dierendonck et al. 1993)

$$S4 = \sqrt{\frac{\langle SI\rangle^2 - \langle SI\rangle^2}{\langle SI\rangle^2} - \frac{100}{\langle CNR\rangle}\left(1 + \frac{100}{19 \cdot \langle CNR\rangle}\right)} \quad (4.24)$$

When computing here, one uses the same averaging time T to estimate $\langle CNR\rangle$ and $\langle SI\rangle$, that is selected with allowance for the above consideration.

If the measurements of the *I* and *Q* components are non-available, but there are only *CNR* measurements, one may estimate a signal intensity related to the CNR at the PLL input as follows:

$$SI = \frac{m \cdot \mu}{T \cdot n^2} 10^{0.1 \cdot CNR} \quad (4.25)$$

where m is the relation between the maximal measured value $\left(I_i^2 + Q_i^2\right)$ max and the measurement noise within the accumulation time; T is the accumulation time of the elementary measurement, ms; *CNR* is a measured signal/noise ratio, dBW; n is a number of signal samples used for accumulating the elementary measurement (e.g., at $T = 1$ ms, for a GPS C/A-code, $n = 5000$, which corresponds to the signal sampling rate of 5 MHz); m is a measurement accumulation interval duration, ms (e.g., $m = 20$, at the accumulation time of 20 ms).

Because the measurement noise, as well as the m value, depends on a satellite elevation, to more correctly compute by using Formula (4.24), one uses the model for that parameter dynamics depending on an elevation. Tiwari and Strangeways (2015) propose to use the model of the following view:

$$\mu = \frac{1}{\eta} \frac{a}{\sin(\gamma \cdot b + c)} \quad (4.26)$$

where $\eta = 4$–5 (at $T_i = 1$ ms, and $\gamma = 90°$), $a = 51.94$, $b = 0.0093$, and $c = 0.7305$ model parameters; γ is a satellite elevation, deg.

To correctly select the parameters of the model (4.26), one may use a dependence between the *S*4 index and the signal fade depth at a loop input. Conker et al. (2003) proposed the dependence (Fig. 4.9).

On the vertical axis in Fig. 4.9, we marked the signal strength fading parameter $R_{95\%}$. The parameter corresponds to the confidence interval of the instantaneous power variations measured over the set period, at the 95% confidence level.

The transionospheric signal phase scintillation index is defined as an instantaneous phase measurement series RMS with the removed measurement trend (van Dierendonck et al. 1993):

$$\sigma\varphi = \sqrt{\langle\hat{\varphi}^2\rangle - \langle\hat{\varphi}\rangle^2} \quad (4.27)$$

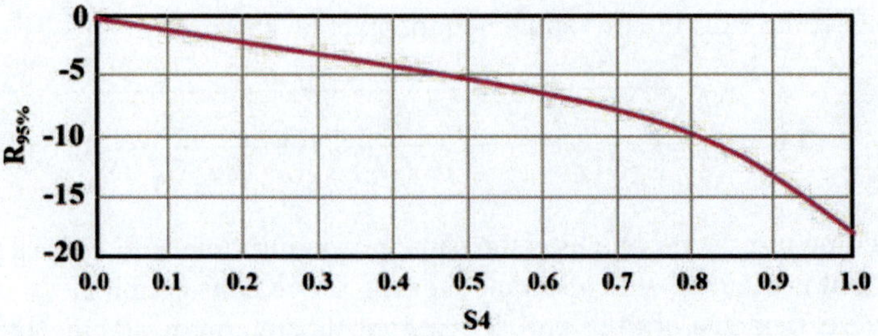

Fig. 4.9 Correlation between the loop-input signal fade depth estimated by processing the CNR measurements with a 95% confidence and the amplitude scintillation index. Re-produced and modified figure by the authors involving part of the figure which was published (Conker et al. 2003)

where $\hat{\varphi}$ is the carrier phase, from the measurements of which the trend and the phase ambiguity parameter were removed.

When estimating the phase scintillation index, the selection of the cutoff frequency during the trend removal has a great significance, as well as the phase measurement noise level does. The first parameter depends on what scale of ionospheric irregularities is to be examined. The second parameter determines the sensitivity of the phase measurements in terms of possibilities to detect the weakest ionospheric disturbances from the scintillation spectrum.

More often, the phase measurement trend removal is done by using a 6th-order Butterworth low-pass filter with the 0.1 Hz cutoff frequency (van Dierendonck et al. 1993). Depending on the objective of the study, the cutoff frequency may be specified. Phase scintillations emerge owing to a random refraction of the signal phase path when passing through small-scale ionospheric irregularities with scales of about unities to tens of km (Bhattacharrya 1992, Bhattacharrya 1999). A separate case for phase scintillation emergence is statistical focusing, when there is a scatter and an interference of diffracted waves when passing through the irregularities with the size of about the Fresnel first zone, i.e., 300–400 m at GNSS frequencies. In the phase scintillation spectrum, all those phenomena correspond to the spectral part with frequencies from 0.1 to 20–40 Hz. With allowance for the latter, we can estimate more precisely the frequency band in the phase scintillation spectrum (ΔF) as follows (Demyanov et al. 2019a, b)

$$\Delta F \approx \frac{1}{\Lambda_m \cdot V_d^{-1}} = \frac{1}{\Lambda_m \cdot (V_{db} + V_{SV} \cos \alpha)^{-1}} \tag{4.28}$$

where Λ_{m} is a scale of ionospheric irregularity, m; V_d is a relative velocity of crossing an irregularity (that moves at a V_{db} velocity) and a line-of-sight to a satellite (that moves at a V_{SV} velocity) at the height of the electron concentration maximum in the ionosphere, m/s; α is an angle between the velocity vectors V_{db} and V_{SV} (α = 0–360°).

The velocity V_{SV} at the ionosphere maximum height (250–400 km) is 70–80 m/s, and the velocity of ionospheric irregularity drift varies within an essentially wide range, from ~ 200 m/s to ~ 3000 m/s (Matsumura et al. 2011; Tsugawa et al. 2011). Therefore, the satellite motion velocity weakly affects the frequency dispersion within the phase scintillation spectrum and the range ΔF may be estimated by neglecting the satellite velocity.

Determining the cutoff frequency in the trend removal and in the phase measurement accumulation time is rather ambiguous, because the real structure of the ionospheric layer is unknown. Conker et al. (2003) proposed another procedure to compute the phase scintillation index. The procedure is based on determining the phase scintillation spectrum parameters in the set frequency band. In this case, the phase scintillation index is computed as follows:

$$\sigma\varphi^2 = 2\int_{F_{\min}}^{F_{\max}} PSD(f)df \approx 2\int_{F_{\min}}^{F_{\max}} A \cdot f^{-p}df \approx 2A\left(\frac{F_{\max}^{1-p} - F_{\min}^{1-p}}{1-p}\right) \tag{4.29}$$

where *PSD(f)* is a function of the phase scintillation spectrum; $F_{\max}$ and $F_{\min}$ are the top and bottom boundaries of a spectrum domain that is investigated; A and p are the relative amplitude and the spectrum inclination.

The parameters A and p are determined by linear approximation of the phase scintillation spectrum. During the spectrum computation, one may exclude the components that relate to the phase measurement ambiguity and to the measurement trend.

The phase measurement level may essentially restrict the measurement sensitivity when studying weak phase ionospheric scintillations. Let us address the measurement noise component contribution and estimate the minimal phase measurement error with allowance for the latter.

Above, we showed (see 2.36) that the phase measurement RMS error depends on the CNR and the PLL filter noiseband. In turn, the CNR will be, mainly, determined by the following factors: (1) the external additive noise level (electromagnetic disturbances); (2) the directional diagram and the antenna gain; (3) RF amplifier and RF front-end noise. Besides, significant is the level of the relative short-time clock frequency instability, digitation noise, and sampling in a receiver. Thus, selecting the hardware for measurements plays an important role in conducting investigations into ionospheric scintillations and weak disturbances.

When analyzing the contribution of the listed factors to the total phase measurement error, one may write a general expression to estimate the loop-input CNR level as follows (Kaplan 1996):

$$CNR = P_{rec} + G_A - N_T - L_{tr} - L_{dg} \tag{4.30}$$

where P_{rec} is a signal power received at an antenna, dBW; G_A is an antenna gain, dB; N_T is the receiver thermal noise power spectral density (noise of an RF amplifier noise, antenna, and RF front-end elements), dBW; L_{tr} is a power loss during a signal main filtering, frequency conversion, and in a cable, dB; L_{dg} is a digitizing and sampling power loss, dB.

The factors in (4.30) may be split into two groups: constant and varying. The first group depends on the hardware type and design features, and their operation does not vary during measurements. Those include the L_{tr}, L_{dg}, and G_A components with typical values $L_{tr} = -2\ldots-4$ dB, $L_{dg} = -0.55\ldots-3.0$ dB, $G_A = -2\ldots-7.5$ dB (MOPS 2017).

The second group involves the factors that have a random variation character. This refers to the hardware thermal noise N_T, whose power spectral density is estimated as follows (MOPS 2017):

$$N_T = 10 \cdot \lg\left(K \cdot T_{\sum}\right) \tag{4.31}$$

where $K = 1.38 \cdot 10^{-23}$ (W · s/K) is the Boltzmann constant; $T_{\sum}$ is a total noise temperature proportional to the total of the receiver internal thermal noise and external additive noises.

The total noise temperature $T_{\sum}$ is determined by the external noise temperature T_{EXN} that comprises the Earth noise temperature T_{EN}, galactic and cosmic radiation T_{CN}, solar radiation T_{SN}, as well as the antenna noise temperature T_A and the RF amplifier noise temperature T_{LPA}:

$$T_{\sum} = T_{EXN} + T_A + T_{LPA} \tag{4.32}$$

Under standard physical conditions, the Earth noise component regular constituent is accepted to be $T_{EN} = 300K$. However, because of the antenna directional properties, this isotropic noise passing to a PLL input has a non-uniform character. Therefore, the expression is to include the relation between the angular area of the antenna pattern, directed toward the earth surface and the full angular aperture of the directional diagram $\frac{\beta}{2\theta}$:

$$T_{EN} = 100 \cdot \left(\frac{\beta}{2\theta}\right)^2 \cdot T_{\sum,EN} \tag{4.33}$$

Typical values for the indicated parameter of navigation receivers' antennas are 0.004–0.01 (MOPS 2017).

The regular solar noise temperature equals $T_{SN} = 6000$ K at the visible angular size of the solar disk on the Earth $\beta = 0.5°$ and $\left(\frac{\beta}{2\theta}\right)^2 = 10^{-5}$.

The sky noise temperature (galactic and space radio emission) T_{CN} is accepted to be 100 K (MOPS 2017), but that noise drops isotropically on the entire angular area of an antenna pattern, i.e. $\left(\frac{\beta}{2\theta}\right)^2 = 1$.

An antenna noise temperature T_{A} depends on its physical temperature T_0 and its gain η:

Table 4.2 The receiver and external thermal noises

Noise source	Noise temperature, K	Power spectral density dBW/Hz
Pre-amplifier noise	120…300	−204…. −208
Antenna noise	30…60	−214….−211
Sky noise (all sources)	100	−208
Earth noise	1.2…3	−228…−224
Sun noise	0.06	−241
Total value	251.26…463.06	−204.6..−202.1

$$T_A = T_0 \cdot (1 - \eta) \tag{4.34}$$

The RF amplifier noise temperature depends on an amplifier noise ε and on a receiver physical temperature T_{01}.

$$T_{LNA} = T_{01} \cdot (\varepsilon - 1) \tag{4.35}$$

Table 4.2 provides estimates for different noise components of phase measurement under standard conditions (temperature being 300 K).

To compute a GPS signal phase measurement noise at different frequencies, we will choose the following loop parameters: $\Delta F_{\text{PLL}} = 18$ Hz, $T = 20$ ms, $\sigma_F(\tau) = 10^{-11}$. The minimal received signal power P_{rec} at L1, L2, and L5 may be determined from the plots in Fig. 4.10 (IS-GPS-200 2021; IS-GPS-705 2021). Computing the RMS phase measurement noise for the indicated conditions and loop settings may be done by using (2.36) and (4.30). Table 4.3 presents the computation results.

The presented computations show that an RF amplifier contributes the most to the phase measurement noise. Therefore, we may expect an essential difference in the sensitivity and the results of observations of weak ionospheric phase scintillations, depending on the chosen hardware. Besides, the results of investigations into phase scintillations and their interpretation should account for an essential difference in the measurement noise level at different GNSS operational frequencies and by using different signal components.

4.3.3 GNSS Phase Second-Order Derivative as an Ionospheric Index

The data of radionavigation parameter measurements by GNSS signals at a high temporal resolution have become everywhere more and more available (McCaffrey and Jayachandran 2017; Fanis and Stathis 2015). This opens great possibilities to

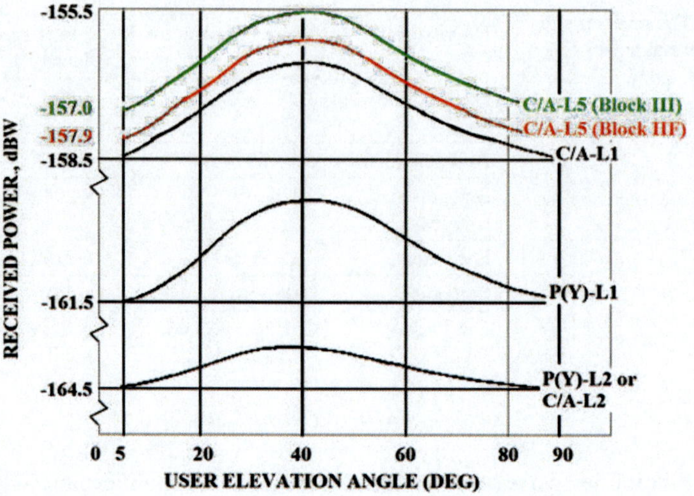

Fig. 4.10 Power of signals received by a linearly polarized antenna with 3 dBi antenna gain at the GPS L1, L2, and L5. Re-used the figure which was published in the part of the book under open licensees (https://www.intechopen.com/page/publication-agreement-chapters)

Table 4.3 Noise values of phase measurements

Frequency, MHz	Minimal value σ_φ, deg	Maximal value σ_φ, deg
L1 = 1575.42	1.59	7.22
L2 = 1227.60	3.35	14.85
L5 = 1176.45 (Block III)	1.33	6.06
L5 = 1176.45 (Block IIF)	1.42	6.43

study weak ionospheric disturbances and fine structure of the ionosphere. The ionospheric scintillation indices, $S4$ and $\sigma\varphi$, the TEC-based indices, and measurements of TEC absolute values are widely used now to diagnose ionospheric conditions of the Earth. Accuracy, sensitivity, and informativity of those indices depend on a number of parameters. The parameters involve: the accumulation time; the primary measurement recording frequency; the cutoff frequencies; the type of filtering, of data smoothing, and of measurement trend removing; the scheme for a navigation receiver; and the loop settings. Principally, TEC (and the TEC-based indices) is a product of combined synchronous measurements at two operational frequencies. This may distort the observed pattern of the ionosphere fine structure, due to smoothing of frequency-dependent effects and summing the measurements noise in separate channels.

The ionospheric condition index sensitivity is determined, first of all, by the temporal resolution of primary measurements. Herewith, one should answers the question, what recording frequency is sufficient to detect a weak ionospheric event and to detect the latter from non-informative noise. Recently, published was a number of papers based on analyzing ionospheric high-temporal resolution data (McCaffrey and Jayachandran 2017; Ghoddousi-Fard 2017; Jacobsen 2014; Demyanov et al. 2021), but there has been no complete answer to the question yet.

The manifestation of ionospheric scintillations and weak ionospheric disturbances may be different at different operational frequencies. A separate analysis of the measurements at different frequencies may yield more information on the fine structure of the ionosphere. This requires that new ionospheric indices based on using separate measurements at different frequencies be developed. Such indices should possess a higher sensitivity to weak ionospheric disturbances. It is desirable that the developed indices depend (in a simple way) on the settings and characteristics of a loop, and did not require complicated procedures of smoothing and measurement trend removing.

As one of such indices for the ionosphere fine structure status, one may use the carrier phase second derivative. This Section demonstrates an algorithm for estimating and possible applying of the above index to detect small-scale ionospheric disturbances. Note that, given a high frequency of recording phase measurements, computing the phase second derivative completely removes the phase ambiguity and the measurement trend. Another advantage of the proposed index is that it is determined during direct processing of measurements without their accumulation and smoothing. Due to this, the dependence between the index value and the loop settings, as well as the wave propagation medium parameters, is more obvious to analyze.

A key problem of using the proposed index is a possibility of detecting informative parameters from the measurement noise caused by tropospheric phase scintillations, satellite clock frequency instability anomalies, and receiver thermal noise.

To estimate a receiver thermal noise at PLL various settings, we assume that the measurements are conducted at a stationary receiver, and there are no phase variations caused by mechanical vibrations and receiver jerks. Figure 4.11 shows the dependence between the receiver thermal noise RMS and the loop-input CNR at various loop settings. When computing, we assumed that the Alan variations for the receiver clock are $\sigma_F(\tau) = 10^{-11}$.

One can see that the integration time substantially affects the thermal noise value even in the high CNR domain. Restructuring the noiseband influences on the measurement noise level considerably weakly. In general, at optimal settings and at a receiver clock frequency high stability, one can achieve the thermal noise level with RMS ~ 0.024 rad and lower. This enables (and substantially simplifies) detecting weak ionospheric phase scintillations at the thermal noise level.

Satellite clock short-time frequency instability anomalies may become another factor that restricts detecting weak ionospheric events by measurements of GNSS signal phase. Benton and Mitchel (2014) addressed such anomalies. Figure 4.12 presents an example for such an event detected when observing GPS satellite signals at a stationary observation point in Bath (UK).

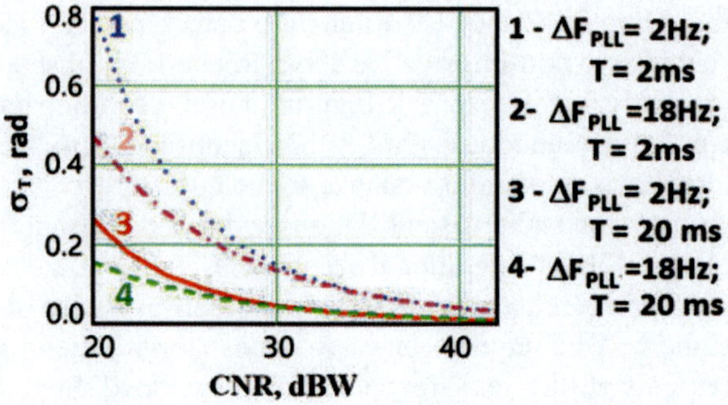

Fig. 4.11 Dependence between the phase measurement thermal noise and the loop settings in a navigation receiver

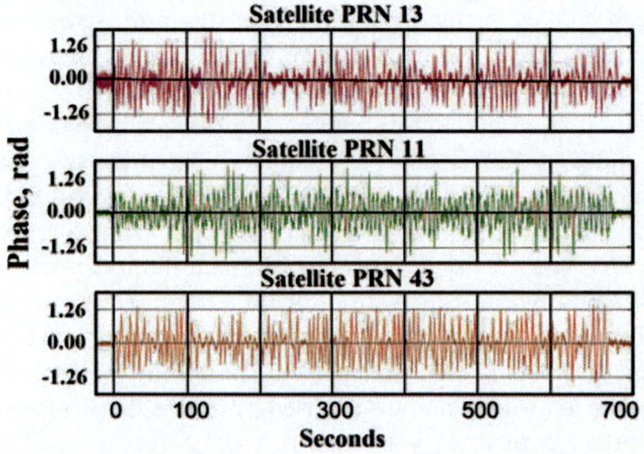

Fig. 4.12 Carrier phase short-time variations caused by satellite clock frequency instability anomalies from the 2011 May 17 (GPS PRN 13, top panel), 2012 May 18 (GPS PRN 11, middle panel), and 2012 Jun 21 (GPS PRN 43 event, bottom panel) observations. Re-produced and modified figure by the authors involving modified part of the figure which was published (Benton and Mitchel 2014)

Apparently, the phase measurement error burst caused by similar anomalies is sufficiently great, and may exceed 1.3 rad. To distinguish such an event when using a single point of monitoring from ionospheric phase scintillations is difficult. Benton and Mitchel (2014) recommend to use phase measurements from the same satellite at several spaced points. The emergence of similar (by form and duration) phase measurement error bursts at spaced points indicates that the cause is a satellite clock short-time frequency instability anomaly.

Tropospheric phase scintillations may be, in certain cases, comparable with ionospheric scintillations. The phase measurement error RMS caused by tropospheric scintillations may be estimated as follows (Gorbunov and Gurvich 1998):

$$\sigma_{TP}^2 = \frac{1}{\sin\gamma}\frac{16\pi^2}{5\lambda^2}\cdot\sigma_n^2\cdot\Lambda\cdot H_{TP} \tag{4.36}$$

where λ is a signal wavelength; σ_n is a troposphere refraction coefficient random variation RMS; Λ is characteristic size of tropospheric irregularities (typical values for this parameter are $\Lambda_0 < \Lambda \le \Lambda_m$, where $\Lambda_0 = 10$ cm, and $\Lambda_{\mathrm{m}} = 1400$ m); H_{TP} is a thickness of the tropospheric layer containing the most essential irregularities ($H_{TP} = 6$–10 km); γ is a satellite elevation.

Figure 4.13 demonstrates the tropospheric phase scintillation RMS dependence on a satellite elevation at different refraction coefficient variation RMS (left) and at different size of irregularities (right). At satellite elevations > 40° the tropospheric phase scintillation RMS does not exceed 0.1 rad. Therefore, one may recommend to use the measurements of a signal phase for the satellites with elevations ≥ 30–40° so that one could confidently distinguish ionospheric phase scintillations from the tropospheric ones.

In the geometric optics approximation, with allowance for the phenomena caused by beam refraction jitter to the entry into the layer of scattering irregularities, in the layer, and at its exit, the ionospheric phase scintillation RMS estimate σ_φ (in meters) equals (Afanasiev et al. 2001; Kolesnik et al. 2002):

$$\sigma_\varphi^2 = \frac{1}{2d_1^2}\left(\sigma_{in}^2\right)^2 + \frac{1}{d_1 d_2}\left(\sigma_{io}^2\right)^2\frac{1}{2d_2^2}\left(\sigma_{out}^2\right)^2 + \sigma_{\tilde{\varphi}}^2 \tag{4.37}$$

where $\sigma_{\tilde{\varphi}}$ is a phase scintillation RMS caused by a wave scatter and interference on small-scale irregularities of the Fresnel scale (300–400 m to 1–2 km); σ_{in}, σ_{io}, and σ_{out} are RMS of the phase scintillations caused by a beam refraction jitter to the entry into the layer of scattering irregularities, in the layer, and at its exit; d_1, d_2 are

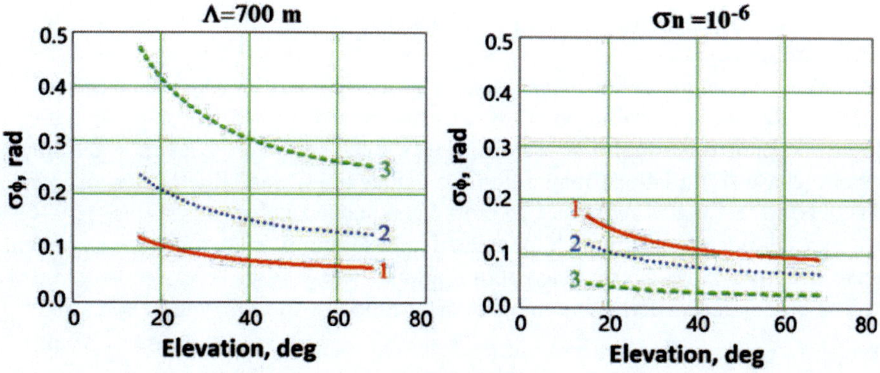

Fig. 4.13 Tropospheric phase scintillation RMS dependence on a satellite elevation at different refraction coefficient variation RMS (left) and at different size of irregularities (right). Left panel: $\Lambda = \mathrm{const} = 700$ m, $\sigma_n = 10^{-6}$ (curve 1), $\sigma_n = 2{\cdot}10^{-6}$ (curve 2), and $\sigma = 4{\cdot}10^{-6}$ (curve 3). Right panel: $\sigma_n = \mathrm{const} = 10^{-6}$, $\Lambda = 100$ m (curve 3), $\Lambda = 700$ m (curve 2), and $\Lambda = 1400$ m (curve 1)

distances between a receiver and an ionospheric layer, and between an ionospheric layer and a satellite, respectively.

The ionospheric scintillation RMS values included in (4.37) are determined with allowance for the scattering ionospheric irregularity layer parameters as follows (Kolesnik et al. 2002)

$$\sigma_{in}^2 \approx \frac{\sqrt{\pi}\cdot L\cdot d_1^2}{4(d_1+d_2+L)^2}\left(\frac{L^2}{3}+d_2\cdot(L+d_2)\right)\cdot\frac{\sigma_i^2}{l_i} \tag{4.38}$$

$$\sigma_{io}^2 = \frac{\sqrt{\pi}\cdot L\cdot d_1\cdot d_2}{4(d_1+d_2+L)^2}\left(\frac{L}{2}(d_1+d_2)+\frac{L^2}{6}+d_1\cdot d_2\right)\cdot\frac{\sigma_i^2}{l_i} \tag{4.39}$$

$$\sigma_{out}^2 = \frac{\sqrt{\pi}\cdot L\cdot d_2^2}{4(d_1+d_2+L)^2}\left(\frac{L^2}{3}+d_1\cdot(L+d_1)\right)\cdot\frac{\sigma_i^2}{l_i} \tag{4.40}$$

$$\sigma_{\varphi}^2 = \frac{\sqrt{\pi}\cdot L}{2}\cdot\sigma_i^2\cdot l_i \tag{4.41}$$

$$\sigma_i^2 = L\cdot Z\cdot\frac{0.066\cdot C_\varepsilon^2}{l_i\cdot\Gamma(p)}\left(l_i^2-\frac{1}{k_m^2}\right)^{\theta-1}\cdot\exp\left(\frac{k_0^2}{k_m^2}-k_0^2\cdot l_i^2\right) \tag{4.42}$$

where L and Z are a thickness and an extent of an ionospheric irregularity region, both are accounted for in computation ($0 < Z < L$); $\Gamma(p)$ is a Gamma function; p is an electron density disturbance spectrum degree in an ionospheric layer (typical values p for the ionosphere range from 2 to 4); C_ε is an ionospheric irregularity intensity; l_i is irregularity size; $k_0 = 2\pi/l_{max}$, $k_m = 5.92/l_{min}$ are wave and spatial numbers.

Figure 4.14 shows the ionospheric phase scintillation RMS dependences on a satellite elevation at several sets of parameters for ionospheric irregularities. The ionospheric irregularity intensity was set through an electron concentration relative deviation in an irregularity $\Delta N/N_0$ to a background value in the ionosphere $N_0 = 10^{12}$ el/m^3.

At typical parameters for Fresnel ionospheric irregularities ($\Delta N/N_0 \leq 5\%$ and $l_i \leq 1$ km), the phase scintillation RMS does not exceed 0.3 rad. These computations do not account for a multiple scattering in a thick irregularity layer, when ionospheric phase scintillations may attain the values > 1–2 rad (Bhattacharrya 1999). The obtained estimates show that, in most cases and at satellite elevations ≥ 30–40° one may confidently distinguish ionospheric phase scintillations from measurement noise and from tropospheric phase scintillations.

The phase second derivative ($d2\varphi$) is determined by the following expression:

$$d2\varphi = \frac{d^2\varphi}{dt^2} \tag{4.43}$$

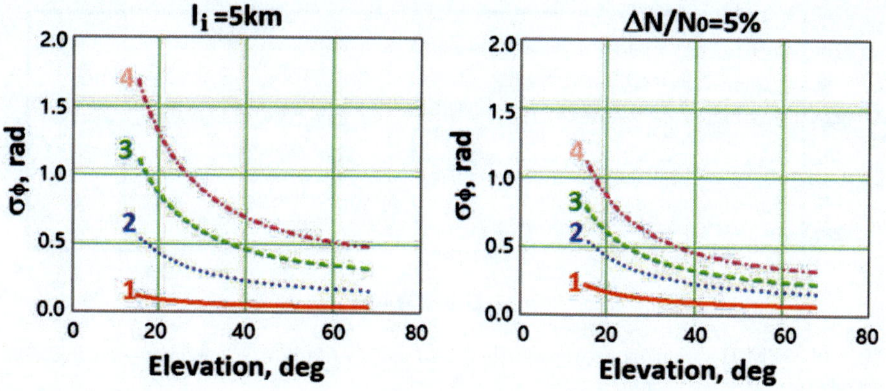

Fig. 4.14 Ionospheric phase scintillation RMS dependence on a satellite elevation. Left panel: $\Delta N/N_0 = 1\%$ (curve 1), $\Delta N/N_0 = 5\%$ (curve 2), $\Delta N/N_0 = 10\%$ (curve 3), and $\Delta N/N_0 = 15\%$ (curve 4) at irregularity invariable size $l_i = 5$ km. Right panel: $l_i = 100$ m (curve 1), $l_i = 5$ km (curve 2), $l_i = 10$ km (curve 3), and $l_i = 20$ km (curve 4) at irregularity invariable intensity $\Delta N/N_0 = 5\%$

Let us estimate the possibility of using the proposed $d2\varphi$ index to diagnose ionospheric conditions under magnetic storm conditions, and compare its behavior with that of the standard ionospheric indices *DROTI*, *S4*, and $\sigma\varphi$.

Ionospheric scintillations are most typical of the high and low latitudes. At mid-latitudes, scintillations are observed much more rarely. Therefore, detecting and studying them is a more complicated problem. For this reason, we analyzed the data under the 2015 Jun 22–25 magnetic storm conditions. The data were recorded at the mid-latitude ISTP station (Irkutsk, Russia, 52.4°N, 104.2°E). The measurements were taken with a JAVAD Delta-G3T navigation receiver (Yasyukevich et al. 2018) with a 50-Hz recording frequency at the GPS *L1* and *L2*.

The phase scintillation index $\sigma\varphi$ was computed through Expressions (4.26) after preliminary trend removal by a phase 5th order polynomial (4.16) and its filtration by a moving average with a 0–5 min window (4.18).

Selecting the parameters is related to that the phase variations caused by the function of small-scale ionospheric irregularities (1–10 km) have the period from several sec and higher (Priyadarshi et al. 2018). In particular, Pi et al. (1997) estimated this period within ~ (2–13) sec, whereas Forte and Radicella (2002) yielded ~ (0.4–5) sec.

The amplitude scintillation index was computed from the CNR data through Formulas (4.18), (4.24), and (4.25). The *DROTI* index was computed by the formula presented in Table 4.1 by using dual-frequency oblique TEC phase measurements at GPS L1 and L2. The *DROTI*, *S4,* and $\sigma\varphi$ were computed with measurement accumulation over a 1-s interval.

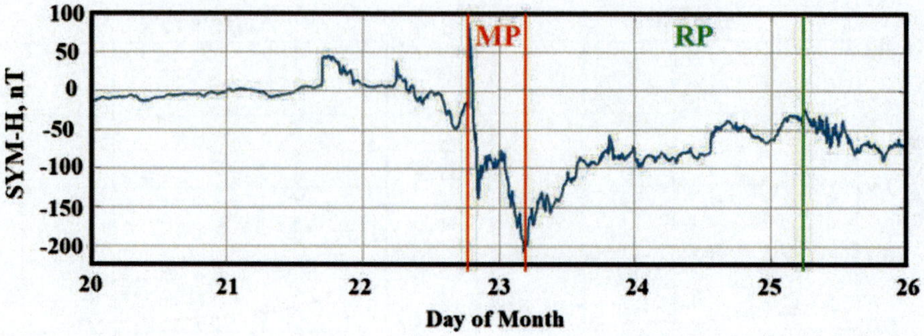

Fig. 4.15 SYM-H variations between 2015 Jun 20 and 25 (http://wdc.kugi.kyoto-u.ac.jp/aeasy/index.html)

For our analysis, we chose a period of the 2015 Jun 22–25 magnetic storm main phase. Figure 4.15 demonstrates the SYM-H index variation (http://wdc.kugi.kyoto-u.ac.jp/aeasy/index.html). The red signatures MP and RP highlight the main and the recovery phases of the magnetic storm, respectively.

Figure 4.16 presents the dynamics for the $d2\varphi$, *DROTI*, *S*4, and $\sigma\varphi$. The indices were computed from the measurements of GPS PRN 04, PRN 15, and PRN 27 signals. All the satellites were observed on Jun 22 at high (>40°) elevations. This minimizes the probability for the emergence of phase and amplitude scintillations caused by multipath and tropospheric scintillations.

The $d2\varphi$ and $\sigma\varphi$ dynamics correlates for all the noted scintillation events (including weak scintillations detected from the PRN 04 signals (Fig. 4.16b, c). In this case, the $d2\varphi$ yields a clearer peak at the scintillation maximum instant as compared with the $\sigma\varphi$. The weaker the scintillations, the less the $d2\varphi$ and *S*4 dynamics correlation (see for comparison Figs. 4.16b and d).

The $d2\varphi$ and *DROTI* variability correlation is the poorest. The *DROTI* envelope shape substantially differs from that of the $d2\varphi$. Also, the *DROTI* produces a very weak response by the PRN27 signal measurements (Fig. 4.16e, middle column).

In all the addressed cases, we can note a respond to the event in different indices at the same time interval. This evidences a possibility of using the $d2\varphi$ to investigate ionospheric scintillations.

A number of the latest papers have discussed the issue what maximal frequency for recording measurements of transionospheric signal parameters should be to extract the maximal information on the ionosphere structure from the amplitude and phase scintillation spectrum. McCaffrey and Jayachandran (2017) showed that the upper boundary between the informative part of the amplitude scintillation spectrum and the non-informative noise may attain 30–40 Hz. The same value for the phase scintillation spectrum is 40–50 Hz. Demyanov et al. (2021) estimated the most probable values of that boundary in the phase scintillation spectrum within 17–22 Hz.

The characteristic size of ionospheric irregularities that cause the emergence of scintillations is > 300–400 m at the GNSS frequencies. Such irregularities provoke both diffraction and refraction scintillations. Those scintillations are observed in the

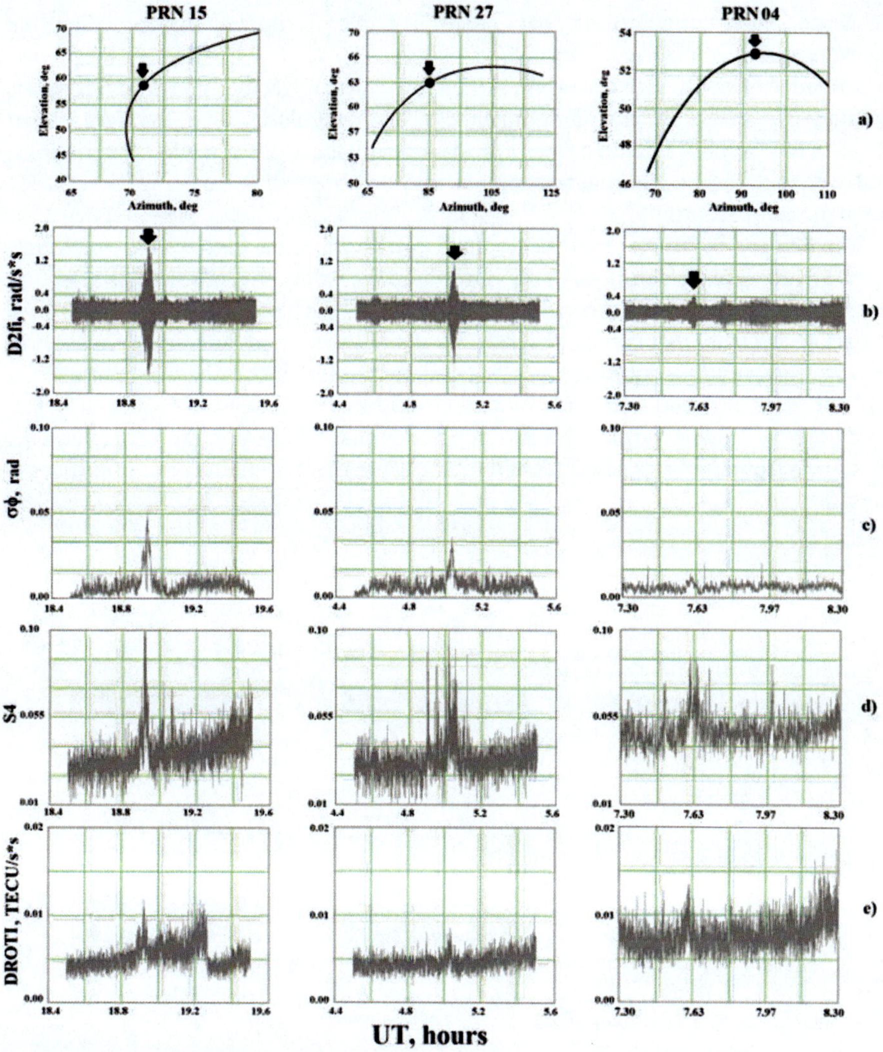

Fig. 4.16 Dynamics for the ionospheric scintillation indices at the ISTP on from GPS PRN04, PRN15, and PRN27 signals on 2015 Jun 22. The upper row shows the satellite elevation variations during the observations, where the dots mark the satellite position at the scintillation recording instant (Demyanov et al. 2019b)

series of phase direct measurements, in the TEC series, and in the TEC-derived indices (ROTI, DROTI), and also by the $S4$ and $\sigma\varphi$ that are computed based on the measurements with a > 1-Hz recording frequency.

Ionospheric smaller-scale irregularities (100–300 m), provided their sufficient intensity, cause amplitude diffraction scintillations, and, more rarely, phase ones.

To detect such irregularities, one needs a higher frequency of recording the measurements.

Irregularities with size of several km mainly cause refraction phase scintillations within 0.01–0.1 Hz. At the predominance of that irregularity type, the *S4* does not vary so considerably and has a weak correlation with the *ROTI* and *DROTI* variations, and even with the $\sigma\varphi$. Similar phenomena may be observed when using the < 10-Hz frequencies of recording.

The $d2\varphi$ is a parameter that is computed directly from the carrier phase measurements. Therefore, selecting the period of phase recording is equivalent to selecting the smoothing filter cutoff frequency and determines the frequency band in the phase variation spectrum that can be observed by the phase second derivative. To analyze this, we compared the series of the $d2\varphi$ computed based on the phase measurements with the removed trend at the measurement recording frequencies of 1, 10, and 50 Hz.

Figure 4.17 demonstrates the $d2\varphi$ series computed based on the measurements by signals from the same satellites and for the same measurement conditions that are presented in Fig. 4.16.

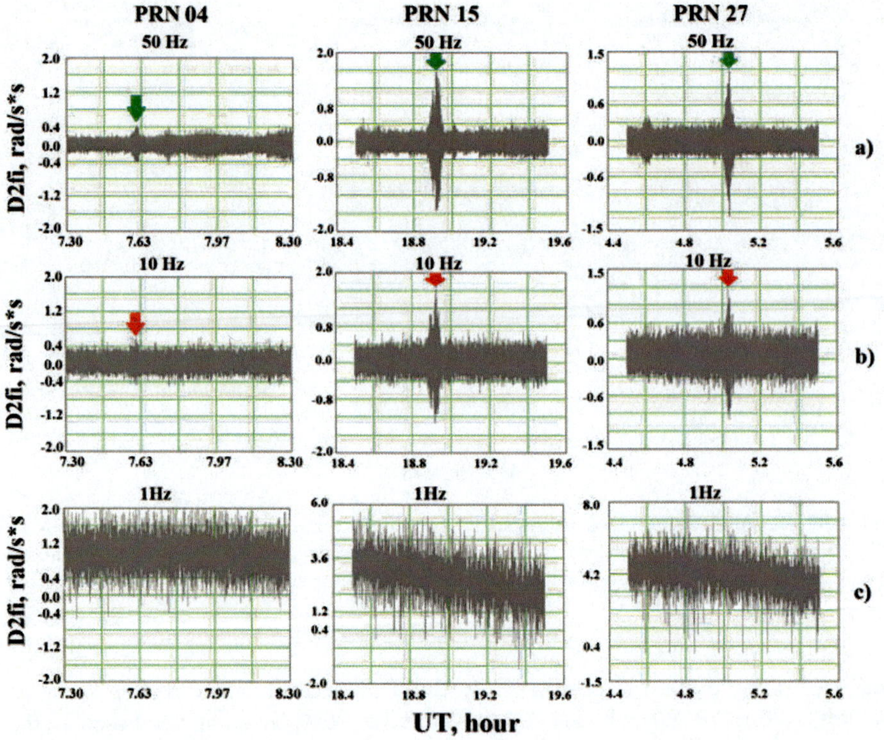

Fig. 4.17 Dynamics for the $d2\varphi$ computed from the data with 50-Hz **a**, 10 Hz **b**, and 1 Hz **c** temporal resolutions (Demyanov et al. 2019b)

The $d2\varphi$ obtained by the 1-Hz measurements does not enable to detect weak ionospheric scintillations in the addressed case (Fig. 4.17c). When using a higher (10-Hz) recording frequency, one can clearly see the event in the measurements obtained by the PRN 15 and PRN 27 signals. However, the weakest PRN 04 LOS scintillations differ poorly even in this case (Fig. 4.17b). As the recording frequency increases to 50 Hz, one can clearly see all three events, including the weakest (Fig. 4.17a).

Note, also, that when using the measurements with the lowest (1 Hz) resolution, there appears a low-frequency measurement trend in the $d2\varphi$ series. There is no such a trend when using a higher recording frequency. Besides, the recording frequency affects the $d2\varphi$ absolute value. When using the highest frequency (50 Hz), the $d2\varphi$ background values do not exceed 0.4 rad/s^2 (Fig. 4.17a). For a lower (10 Hz) frequency, the background value increases and is already 0.6 rad/s^2 (Fig. 4.17b). In the latter case, at the 1-Hz recording frequency, the $d2\varphi$ background value increases already by factor of 4–5, and does not enable to detect weak scintillations against the noise (Fig. 4.17c).

The information presented here allow us to assert that the $d2\varphi$ may be used to detect small-scale ionospheric disturbances, provided the phase recording frequency is sufficiently high to detect such events in the phase variation spectrum.

Appendix 4.1

The correspondence between radio-navigation parameters and its specification according to RINEX 3.04 format (Gurtner and Estey 2009).

GNSS system	Freq.Band/Frequency	Channel of code	Observation codes			
			Pseudo-range	Carrier phase	Doppler	Signal strength
GPS	L1/1575.42, MHz	C/A	C1C	L1C	D1C	S1C
		L1C(D)	C1S	L1S	D1S	S1S
		L1C(P)	C1L	L1L	D1L	S1L
		L1C(D + P)	C1X	L1X	D1X	S1X
		P (AS off)	C1P	L1P	D1P	S1P
		Z-tracking (AS on)	C1W	L1W	D1W	S1W
		Y	C1Y	L1Y	D1Y	S1Y
		M	C1M	L1M	D1M	S1M
		Codeless		L1N	D1N	S1N
	L2/1227.60, MHz	C/A	C2C	L2C	D2C	S2C
		L1(C/A) + (P2-P1) (semi-codeless)	C2D	L2D	D2D	S2D

(continued)

(continued)

GNSS system	Freq.Band/Frequency	Channel of code	Observation codes			
			Pseudo-range	Carrier phase	Doppler	Signal strength
		L2C (M)	C2S	L2S	D2S	S2S
		L2C (L)	C2L	L2L	D2L	S2L
		L2C (M + L)	C2X	L2X	D2X	S2X
		P (AS off)	C2P	L2P	D2P	S2P
		Z-tracking and similar (AS on)	C2W	L2W	D2W	S2W
		Y	C2Y	L2Y	D2Y	S2Y
		M	C2M	L2M	D2M	S2M
		codeless		L2N	D2N	S2N
	L5/1176.45, MHz	I	C5I	L5I	D5I	S5I
		Q	C5Q	L5Q	D5Q	S5Q
		I + Q	C5X	L5X	D5X	S5X

GNSS system	Freq.Band/Frequency	Channel of code	Observation codes			
			Pseudo-range	Carrier phase	Doppler	Signal strength
GLONASS	G1/(1602 + k*9/16), MHz k = −7.... + 12	C/A	C1C	L1C	D1C	S1C
		P	C1P	L1P	D1P	S1P
	G1a/1600.995, MHz	L1OCd	C4A	L4A	D4A	S4A
		L1OCp	C4B	L4B	D4B	S4B
		L1OCd + L1OCp	C4X	L4X	D4X	S4X
	G2/(1246 + k*7/16), MHz	C/A (GLONASS M)	C2C	L2C	D2C	S2C
		P	C2P	L2P	D2P	S2P
	G2a/1248.06, MHz	L2CSI	C6A	L6A	D6A	S6A
		L2OCp	C6B	L6B	D6B	S6B
		L2CSI + L2OCp	C6X	L6X	D6X	S6X
	G3/1202.025, MHz	I	C3I	L3I	D3I	S3I
		Q	C3Q	L3Q	D3Q	S3Q
		I + Q	C3X	L3X	D3X	S3X

GNSS system	Freq.Band /Frequency	Channel of code	Observation codes			
			Pseudo-range	Carrier phase	Doppler	Signal strength
Galileo	E1/1575.42, MHz	A PRS	C1A	L1A	D1A	S1A
		B I/NAV OS/CS/SoL	C1B	L1B	D1B	S1B
		C no data	C1C	L1C	D1C	S1C
		B + C	C1X	L1X	D1X	S1X
		A + B + C	C1Z	L1Z	D1Z	S1Z
	E5a/1176.45, MHz	I F/NAV OS	C5I	L5I	D5I	S5I
		Q no data	C5Q	L5Q	D5Q	S5Q
		I + Q	C5X	L5X	D5X	S5X
	E5b/1207.140, MHz	I I/NAV OS/CS/SoL	C7I	L7I	D7I	S7I
		Q no data	C7Q	L7Q	D7Q	S7Q
		I + Q	C7X	L7X	D7X	S7X
	E5(E5a + E5b)/1191.795, MHz	I	C8I	L8I	D8I	S8I
		Q	C8Q	L8Q	D8Q	S8Q
		I + Q	C8X	L8X	D8X	S8X
	E6/1278.75, MHz	A PRS	C6A	L6A	D6A	S6A
		B C/NAV CS	C6B	L6B	D6B	S6B
		C no data	C6C	L6C	D6C	S6C
		B + C	C6X	L6X	D6X	S6X
		A + B + C	C6Z	L6Z	D6Z	S6Z

For Galileo the band/frequency number n does not necessarily agree with the official frequency numbers: n = 7 for E5b, n = 8 for E5a + b

GNSS system	Freq.Band/Frequency	Channel of code	Observation codes			
			Pseudo-range	Carrier phase	Doppler	Signal strength
BDS (BeiDou)	B1-2/1561.098, MHz	I	C2I	L2I	D2I	S2I
		Q	C2Q	L2Q	D2Q	S2Q
		I + Q	C2X	L2X	D2X	S2X
	B1/1575.42, MHz (BDS-3 signals)	Data	C1D	L1D	D1D	S1D
		Pilot	C1P	L1P	D1P	S1P
		Data + Pilot	C1X	L1X	D1X	S1X
		B1A	C1A	L1A	D1A	S1A
		Codeless		L1N	D1N	S1N

(continued)

(continued)

GNSS system	Freq.Band/Frequency	Channel of code	Observation codes			
			Pseudo-range	Carrier phase	Doppler	Signal strength
	B2a/1176.45, MHz (BDS-3 signals)	Data	C5D	L5D	D5D	S5D
		Pilot	C5P	L5P	D5P	S5P
		Data + Pilot	C5X	L5X	D5X	S5X
	B2b/1207.140, MHz (BDS-2 signals)	I	C7I	L7I	D7I	S7I
		Q	C7Q	L7Q	D7Q	S7Q
		I + Q	C7X	L7X	D7X	S7X
	B2b/1207.140, MHz (BDS-3 signals)	Data	C7D	L7D	D7D	S7D
		Pilot	C7P	L7P	D7P	S7P
		Data + Pilot	C7Z	L7Z	D7Z	S7Z
	B2(B2a + B2b)/1191.795, MHz (BDS-3 signals)	Data	C8D	L8D	D8D	S8D
		Pilot	C8P	L8P	D8P	S8P
		Data + Pilot	C8X	L8X	D8X	S8X
	B3/1268.52, MHz	I	C6I	L6I	D6I	S6I
		Q	C6Q	L6Q	D6Q	S6Q
		I + Q	C6X	L6X	D6X	S6X
		B3A	C6A	L6A	D6A	S6A

Note When reading a RINEX 3.02 file, both C1x and C2x coding should be accepted and treated as C2x in RINEX 3.03

GNSS system	Freq.Band/Frequency	Channel of code	Observation codes			
			Pseudo-range	Carrier phase	Doppler	Signal strength
SBAS	L1/1575.42, MHz	C/A	C1C	L1C	D1C	S1C
	L5/1176.45, MHz	I	C5I	L5I	D5I	S5I
		Q	C5Q	L5Q	D5Q	S5Q
		I + Q	C5X	L5X	D5X	S5X

GNSS system	Freq.Band/Frequency	Channel of code	Observation codes			
			Pseudo-range	Carrier phase	Doppler	Signal strength
IRNSS	L5/1176.45, MHz	A SPS	C5A	L5A	D5A	S5A
		B RS (D)	C5B	L5B	D5B	S5B

(continued)

(continued)

GNSS system	Freq.Band/Frequency	Channel of code	Observation codes			
			Pseudo-range	Carrier phase	Doppler	Signal strength
		C RS (P)	C5C	L5C	D5C	S5C
		B + C	C5X	L5X	D5X	S5X
	S/2492.028, MHz	A SPS	C9A	L9A	D9A	S9A
		B RS (D)	C9B	L9B	D9B	S9B
		C RS (P)	C9C	L9C	D9C	S9C
		B + C	C9X	L9X	D9X	S9X

GNSS system	Freq.Band/Frequency	Channel of code	Observation codes			
			Pseudo-range	Carrier phase	Doppler	Signal strength
QZSS	L1/1575.42, MHz	C/A	C1C	L1C	D1C	S1C
		L1C (D)	C1S	L1S	D1S	S1S
		L1C (P)	C1L	L1L	D1L	S1L
		L1C (D + P)	C1X	L1X	D1X	S1X
		L1S/L1-SAIF	C1Z	L1Z	D1Z	S1Z
	L2/1227.60, MHz	L2C (M)	C2S	L2S	D2S	S2S
		L2C (L)	C2L	L2L	D2L	S2L
		L2C (M + L)	C2X	L2X	D2X	S2X
	L5/1176.45, MHz *(Block I Signals) **(Block II L5S Signals)	I *	C5I	L5I	D5I	S5I
		Q *	C5Q	L5Q	D5Q	S5Q
		I + Q *	C5X	L5X	D5X	S5X
		L5D **	C5D	L5D	D5D	S5D
		L5P **	C5P	L5P	D5P	S5P
		L5(D + P) **	C5Z	L5Z	D5Z	S5Z
	L6/1278.75, MHz *(Block I LEX Signals) **(Block II Signals)	L6D *,**	C6S	L6S	D6S	S6S
		L6P *	C6L	L6L	D6L	S6L
		L6(D + P) *	C6X	L6X	D6X	S6X
		L6E **	C6E	L6E	D6E	S6E
		L6(D + E) **	C6Z	L6Z	D6Z	S6Z

Note RINEX 1Z signal coding is used for both the initial Block I L1-SAIF signal and the updated L1S signal. L6D is the "code 1" of the L61 (BlockI) and L62 (Block II) signals, L6P is the "code 2" (or pilot) signal of the L61 (Block I) signal and L6E is the "code 2" of the L62 (Block II) signal as specified in IS-QZSS-L6

References

Afanasiev NT, Tinin MV, Kolesnik SN (2001) A comparison of results of a numerical simulation and an approximate calculation of statistical characteristics of radio waves in the layer with random inhomogeneities of dielectric permittivity. Waves Random Media 11:363–376. https://doi.org/10.1088/0959-7174/11/4/301

Afraimovich EL, Perevalova NP, Zhivetiev IV (2008) Relative amplitude of the total electron content variations depending on geomagnetic activity. Adv Space Res 42:1231–1237. https://doi.org/10.1016/j.asr.2007.09.003

Afraimovich EL, Astafyeva EI, Demyanov VV, Edemskiy IK, Gavrilyuk NS, Ishin AB et al (2013) Review of GPS/GLONASS studies of the ionospheric response to natural and anthropogenic processes and phenomena. J Space Weather Space Clim 3:A27. https://doi.org/10.1051/swsc/2013049

Aleshin I, Burguchev S, Kholodkov K, Alpatov V, Vasiliev A (2014) Data handling in GNSS receiver network and ionosphere monitoring service solution. In: 2014 international conference on engineering and telecommunication, pp 122–125. https://doi.org/10.1109/EnT.2014.32

Battaglia M, Zuliani D, Pascutti D, Michelini A, Marson I, Murray MH, Burgmann R (2003) Network assesses earthquake potential in Italy's Southern alps. Eos 84(28):262–264

Benton CJ, Mitchell CN (2014) Further observations of GPS satellite oscillator anomalies mimicking ionospheric phase scintillation. GPS Solutions 18:387–391. https://doi.org/10.1007/s10291-013-0338-4

Bhattacharrya A, Yen KC, Franke SJ (1992) Deducing turbulence parameters from transionospheric scintillation measurements. Space Sci Rev 61:335–386. https://doi.org/10.1007/BF00222311

Bhattacharrya A (1999) Deterministic retrieval of ionospheric phase screen from amplitude scintillations. In: Bhattacharrya A. Radio science, vol 34, pp 229–240

Blewitt G (1990) An automatic editing algorithm for GPS data. Geophys Res Lett 17(3):199–202. https://doi.org/10.1029/gl017i003p00199

BNCGG (2011) Bulgarian national committee of geodesy and geophysics. National report on geodetical and geophysical activities in Bulgaria. Sofia. June, 2011. https://www.iag-aig.org/doc/5d8dd35a05cca.pdf

Bruyninx C, Defraigne P (2018) ROB network GNSS data. Available from Royal observatory of Belgium. Observation Data. https://doi.org/10.24414/FST8-P256

Bruyninx C, Legrand J, Fabian A, Pottiaux E (2019) GNSS metadata and data validation in the EUREF permanent network. GPS Solut 23:106. https://doi.org/10.1007/s10291-019-0880-9

Cabral-Cano E, Pérez-Campos X, Márquez-Azúa B, Sergeeva MA, Salazar-Tlaczani L, DeMets C, Adams D, Galetzka J, Feaux K, Serra YL, Mattioli GS, Miller M (2018) TLALOCNet: a continuous GPS-Met backbone in Mexico for seismotectonic, and atmospheric research. Seismol Res Lett 89(2A):373–381. https://doi.org/10.1785/0220170190

Cai C, Liu Z, Xia P, Dai W (2013) Cycle slip detection and repair for undifferenced GPS observations under high ionospheric activity. GPS Solutions 17(2):247–260. https://doi.org/10.1007/s10291-012-0275-7

Chan K-K, Li BC-H (2007) The Hong Kong satellite positioning reference station network (SatRef)—system configurations, applications and services. Strategic integration of surveying services. FIG working week 2007. Hong Kong SAR, China

Chen X, Langley RB, Dragert H (1996) Investigation of annual variations in the WCDA GPS solutions. In: Proceedings of the 9th international technical meeting of the satellite division of the institute of navigation (ION GPS 1996), Kansas City, MO, pp 1809–1818

Chousianitis K, Papanikolaou X, Drakatos G, Tselentis G-A (2021) NOANET: a continuously operating GNSS network for solid-earth sciences in Greece. Seismol Res Lett 92(3):2050–2064. https://doi.org/10.1785/0220200340

Combrinck L (11–14 Nov 2008) AFREF, African geodetic reference frame; a modern and unified reference frame for Africa, presented at international symposium on global navigation satellite systems, Space-based and ground-based augmentation systems and applications, Berlin, Germany

Conker RS, El-Arini MB, Hegarty CJ, Hsiao T (2003) Modeling the effects of ionospheric scintillation on GPS/satellite-based augmentation system availability. Radio Sci 38(1):1001. https://doi.org/10.1029/2000RS002604

Davies K (1969) Ionospheric radio waves. Xerox College Publishing, p 460

Demyanov VV, Yasyukevich YV, Jin S, Sergeeva MA (2019a) The second-order derivative of GPS carrier phase as a promising means for ionospheric scintillation research. Pure Appl Geophys 176:4555–4573. https://doi.org/10.1007/s00024-019-02281-6

Demyanov VV, Sergeeva MA, Yasyukevich AS (2019b) GNSS high-rate data and the efficiency of ionospheric scintillation indices. In: Book Chapter in "Ionospheric and atmospheric threats for GNSS and satellite telecommunications", IntechOpen, ISBN 978-1-78985-996-6. https://doi.org/10.5772/intechopen.90078

Demyanov V, Danilchuk E, Yasyukevich Y, Sergeeva M (2021) Experimental estimation of deviation frequency within the spectrum of scintillations of the carrier phase of GNSS signals. Remote Sens 13:5017. https://doi.org/10.3390/rs13245017

Dow JM, Neilan RE, Rizos C (2009) The international GNSS service in a changing landscape of global navigation satellite systems. J Geod 83:191–198. https://doi.org/10.1007/s00190-008-0300-3

Duval R, Héroux P, Beck N (1997) Canadian active control system—delivering the Canadian spatial reference system. Originally presented at GIS 96, Vancouver, Canada, March 1996

Fanis M, Stathis S (2015) PLL bandwidth and noise in 100 Hz GPS measurements. GPS Solutions 19:173–185. https://doi.org/10.1007/s10291-014-0378-4

Ferraro C, Nardi A, Bianco G, Vespe F (2022) Contribution to EUREF of the ASI (Agenzia Spaziale Italiana) matera space geodesy centre "G.Colombo" (CGS). https://www.researchgate.net/publication/268359878_Contribution_to_EUREF_of_the_ASI_Agenzia_Spaziale_Italiana_Matera_Space_Geodesy_Centre_GColombo_CGS. Accessed 03 Feb 2022

Forte B, Radicella SM (2002) Problems in data treatment for ionospheric scintillation measurements. Radio Sci 37. https://doi.org/10.1029/2001rs002508

Fortes LPS, Luz RT, Pereira KD, Costa SMA, Blitzkow D (1998) The Brazilian network for continuous monitoring of GPS (RBMC): operation and products. In: Brunner FK (ed) Advances in positioning and reference frames. International association of geodesy symposia, vol 118, pp 73–78. https://doi.org/10.1007/978-3-662-03714-0_11

Gentle P, Gledhill K, Blick G (2016) The development and evolution of the GeoNet and PositioNZ GNSS continuously operating network in New Zealand. NZ J Geol Geophys 59(1):33–42. https://doi.org/10.1080/00288306.2015.1127821

Ghoddousi-Fard R (2017) Impact of receiver and constellation on high rate GNSS phase rate measurements to monitor ionospheric irregularities. Adv Space Res 60(9):1968–1977. https://doi.org/10.1016/j.asr.2017.07.039

Gorbunov ME, Gurvich AS (1998) Microlab-1 experiment. Multipath effects in the lower troposphere. J Geophys Res 103(12): 3819–13826. https://doi.org/10.1029/98JD00806

Gravelle M, Guichard M, Poirier E, Prouteau E, Testut L, Woppelmann G (2019) Status of SONEL: the GNSS at tide Gauge data assembly center for the global sea level observing system (GLOSS). American Geophysical Union, Fall Meeting 2019, abstract #G51C-0607

Gulyaeva TL, Stanislawska I (2008) Derivation of a planetary ionospheric storm index. Ann Geophys 26:2645–2648. https://doi.org/10.5194/angeo-26-2645-2008

Gurtner W (1994) RINEX: the receiver independent exchange format 1. GPS world. July 1994, pp 48–52. http://www2.unb.ca/gge/Resources/gpsworld.july94.pdf

Gurtner W, Estey L (2007) RINEX: the receiver independent exchange format version 2.11. Bern, Astronomical Institute, University of Bern, 2007. ftp://igs.org/pub/data/format/rinex211.txt

Gurtner W, Estey L (2009) RINEX: the receiver independent exchange format version 3.01. Bern, Astronomical Institute, University of Bern, 2009. https://files.igs.org/pub/data/format/rinex301.pdf

Hatanaka Y (2008) A compression format and tools for GNSS observation data. Bull Geog Surv Inst 55:21–30

Hedling G, Parker A, Wonnacott RT (2000) TrigNet: the network of active GPS base stations for South Africa. In: Proceedings of ION GPS2000, Salt Lake City, Utah, USA, pp 1865–1870

INGV (2016) INGV ring working group (2016), RETE integrata nazionale GNSS. https://doi.org/10.13127/RING

IS-GPS-200 (2021) NAVSTAR GPS space segment/navigation user interfaces: IS-GPS-200M. 21 May 2021. https://www.gps.gov/technical/icwg/

IS-GPS-705 (2021) NAVSTAR GPS space segment/user segment L5 interfaces: IS-GPS-705H. May 21, 2021. https://www.gps.gov/technical/icwg/

Jacobsen KS (2014) The impact of different sampling rates and calculation time intervals on ROTI values. J Space Weather Space Clim 4:A33. https://doi.org/10.1051/swsc/2014031

Jakowski N, Borries C, Wilken V (2012) Introducing a disturbance ionosphere index (DIX). Radio Sci 47:RS0L14. https://doi.org/10.1029/2011RS004939

Jamason P, Bock Y, Fang P, Malveaux D, Prawirodirdjo L, Scharber M (2004) SOPAC web site (http://sopac.ucsd.edu). GPS Solutions 8:272–277. https://doi.org/10.1007/s10291-004-0118-2

Jayachandran PT, Langley RB, MacDougall JW, Mushini SC, Pokhotelov D, Hamza AM, Mann IR, Milling DK, Kale ZC, Chadwick R, Kelly T, Danskin DW, Carrano CS (2009), Canadian high arctic ionospheric network (CHAIN). Radio Sci 44:RS0A03. https://doi.org/10.1029/2008RS004046

Jo JH, Cho S, Lim H, Choi B, Jo JH, Lee W-K, Baek J, Choe NJ, Park J-U (2006) The development Plan of KASI GNSS data processing software. In: Proceedings of the korean institute of navigation and port research conference, vol 1, pp 501–503

Juan JM, Sanz J, Rovira-Garcia A, González-Casado G, Ibáñez D, Perez R (2018) AATR an ionospheric activity indicator specifically based on GNSS measurements. J Space Weather Space Clim 8:A14. https://doi.org/10.1051/swsc/2017044

Kintner PM, Ledvina BM, de Paula ER (2007) GPS and ionospheric scintillations. Space Weather 5:S09003. https://doi.org/10.1029/2006SW000260

Kaplan ED (1996) Understanding GPS: principles and applications. Artech House, p 556

Kolesnik SN, Tinin MV, Afanasiev NT (2002) Statistical characteristics of a wave propagating through a layer with random irregularities. Waves Random Media 12:417–431. https://doi.org/10.1088/0959-7174/12/4/302

Kunitsyn VE, Padokhin AM, Kurbatov GA, Yasyukevich YuV, Morozov YuV (2016) Ionospheric TEC estimation with the signals of various geostationary navigational satellites. GPS Solutions 20(4):877–884. https://doi.org/10.1007/s10291-015-0500-2

Liu N, Zhang Q, Zhang S, Wu X (2021) Algorithm for real-time cycle slip detection and repair for low elevation GPS undifferenced data in different environments. Remote Sens 13:2078. https://doi.org/10.3390/rs13112078

Maletckii B, Yasyukevich Y, Vesnin A (2020) Wave signatures in total electron content variations: filtering problems. Remote Sens 12(8):1340. https://doi.org/10.3390/rs12081340

Matsumura M, Saito A, Iyemori T, Shinagawa H, Tsugawa T, Otsuka Y, Nishioka M, Chen CH (2011) Numerical simulations of atmospheric waves excited by the 2011 off the Pacific coast of Tohoku earthquake. Earth Planet Space 63:885–889. https://doi.org/10.5047/eps.2011.07.015

McCaffrey AM, Jayachandran PT (2017) Spectral characteristics of Auroral region scintillation using 100 Hz sampling. GPS Solutions 21:1883–1894. https://doi.org/10.1007/s10291-017-0664-z

McLoughlin I, Wong K, Tan S (2011) Data collection, communications and processing in the sumatran GPS array (SuGAr). Lect Notes Eng Comput Sci 2:1732–1736. https://www.lintech.org/SomeInfo/SuGAr.pdf

MOPS (2017) MOPS based procedure for minimum recommended testing of lightsquared RFI to GPS aviation receivers: appendix A, p 220

Morin P, Francisco J, Sánchez Sobrino JA, Quirós Donate R (2000) Spanish national GPS reference station network (ERGPS). Boletín Real Instituto y Observatorio de la Armada, vol 3. ISSN 1131-5040

NCEDC (2014) Northern California earthquake data center. UC Berkeley Seismological Laboratory. Dataset. https://doi.org/10.7932/NCEDC

Nesterov IA, Andreeva ES, Padokhin AM, Tumanova YuS, Nazarenko MO (2017) Ionospheric perturbation indices based on the low- and high-orbiting satellite radio tomography data. GPS Solutions 21(4):1679–1694. https://doi.org/10.1007/s10291-017-0646-1

Owen B, Zhou T, Bodor L, Fu F, Choudhury M, Mettes J et al (2018) Discovering the geoscience Australia GNSS data repository. IGS workshop 2018, Wuhan, China, PS05–03

PANGA (1996) Pacific Northwest Geodetic Array (PANGA). GPS/GNSS network and geodesy laboratory: central Washington university, other/seismic network. International federation of digital seismograph networks. https://doi.org/10.7914/SN/PW

Perevalova NP, Edemsky IK, Timofeeva OV, Katashevtseva DD, Yasyukevich AS (2016) Dynamics of disturbance level of total electron content at high and middle latitudes according to GPS data. Solar-Terr Phys 2(1):50–60. https://doi.org/10.12737/19878

Pi X, Mannucci AJ, Lindqwister UJ, Ho CM (1997) Monitoring of global ionospheric irregularities using the worldwide GPS network. Geophys Res Lett 24(18):2283–2286. https://doi.org/10.1029/97GL02273

Pinon DA, Gomez DD, Smalley R, Cimbaro SR, Lauria EA, Bevis MG (2018) The history, state, and future of the argentine continuous satellite monitoring network and its contributions to geodesy in Latin America. Seismol Res Lett 89(2A):475–482. https://doi.org/10.1785/0220170162

Priyadarshi S, Zhang QH, Thomas EG, Spogli L, Cesaroni C (2018) Polar traveling ionospheric disturbances inferred with the B-spline method and associated scintillations in the Southern Hemisphere. Adv Space Res 62(11):3249–3266. https://doi.org/10.1016/j.asr.2018.08.015

RESIF (2017) RESIF-RENAG French national geodetic network. RESIF—Réseau Sismologique et Géodésique Français. https://doi.org/10.15778/resif.rg

Romero I (ed) (1 Dec 2021) RINEX: the receiver independent exchange format version 4.00 IGS/RTCM RINEX WG Chair ESA/ESOC/Navigation Support Office Darmstadt, Germany

Sánchez L, Kehm A (2021) SIRGAS regional network associate analysis centre technical report 2020. In: Villiger A, Dach R (eds) International GNSS service: technical report 2020. IGS Central Bureau and University of Bern, Bern Open Publishing, pp 135–146. https://doi.org/10.48350/156425

Thorandt V, Wojdziak R (2011) BKG data center. In: Behrend D, Baver K (eds) International VLBI service for geodesy and astrometry 2010 annual report. NASA/TP-2011-215880, pp 157–158

Tiwari R, Strangeways HJ (2015) Regionally based alarm index to mitigate ionospheric scintillation effects for GNSS receivers. Space Weather 13(1):72–85. https://doi.org/10.1002/2014SW001115

Tsugawa T, Saito A, Otsuka Y, Nishioka M, Maruyama T, Kato H, Nagatsuma T, Murata KT (2011) Ionospheric disturbances detected by GPS total electron content observation after the 2011 off the Pacific coast of Tohoku earthquake. Earth Planet Space 63:875–879

Tsuji H, Yamaguch K (2013) Status of modernization of GEONET from GPS to GNSS. J Japan Soc Photogrammetry Remote Sens 52(3):114–120. https://doi.org/10.4287/jsprs.52.114

Smith RB, UNAVCO Community (2000) Yellowstone hotspot GPS network—NOMT-Norris Montana P.S., The GAGE facility operated by UNAVCO, Inc., GPS/GNSS observations dataset. https://doi.org/10.7283/T5B27SN9

Sorokin AA, Korolev SP, Shestakov NV, Malkovsky SI, Tsoy GI, Pupatenko VV (2017) Data management of global navigation satellite system for complex study of modern geodynamic processes in Far East of Russia. Sovremennye Problemy Distantsionnogo Zondirovaniya Zemli iz kosmosa 14(3):158–172. https://doi.org/10.21046/2070-7401-2017-14-3-158-172 (in Russian)

van Dierendonck AJ, Klobuchar J, Hua Q (1993) Ionospheric scintillation monitoring using commercial single frequency C/A code receivers. In: Proceedings of ION GPS 1993, Salt Lake City, Utah, pp 1333–1342. https://www.semanticscholar.org/paper/Ionospheric-Scintillation-Monitoring-Using-Single-C-Van-John/9ddb800085b45cfea5a579f45c095d99a9df315b

Voeykov SV, Berngardt OI, Shestakov NV (2016) Use of the index of TEC vertical variation disturbance in studying ionospheric effects of the Chelyabinsk meteorite. Geomagn Aeron 56:219–228. https://doi.org/10.1134/S0016793216020122

Wanninger L (2018) Detection of RINEX-2 files with mixed GPS L2P(Y)/L2C carrier phase observations. Sensors 18(12):4507. https://doi.org/10.3390/s18124507

Wilken V, Kriegel M, Jakowski N, Berdermann J (2018) An ionospheric index suitable for estimating the degree of ionospheric perturbations. J Space Weather Space Clim 8:A19. https://doi.org/10.1051/swsc/2018008

Yasyukevich AS (2021) Features of short-period variability of total electron content at high and middle latitudes. Solar-Terr Phys 7(4):71–78. https://doi.org/10.12737/stp-74202107

Yasyukevich YV, Vesnin AM, Perevalova NP (2018) SibNet—Siberian global navigation satellite system network: current state. Solar-Terr Phys 4(4):63–72. https://doi.org/10.12737/stp-44201809

Yasyukevich Y, Mylnikova A, Vesnin A (2020a) GNSS-based non-negative absolute ionosphere total electron content, its spatial gradients, time derivatives and differential code biases: bounded-variable least-squares and taylor series. Sensors 20:5702. https://doi.org/10.3390/s20195702

Yasyukevich YV, Kiselev AV, Zhivetiev IV, Edemskiy IK, Syrovatskii SV, Maletckii BM, Vesnin AM (2020b) SIMuRG: system for ionosphere monitoring and research from GNSS. GPS Solutions 24:69. https://doi.org/10.1007/s10291-020-00983-2

Chapter 5
Experimental Results of GNSS/SBAS Performance Under Space Weather Impacts

Vladislav Demyanov and Yury Yasyukevich

5.1 Solar Radio Burst Impact on GNSS

5.1.1 Solar Radio Flux as an Electromagnetic Jammer

The robustness of a tracking loop operation under powerful radio burst conditions depends on the characteristics and the receiver radio frequency (RF) front-end design. The RF front-end filters and, basically, amplifies GNSS satellite signals. Also, the radio frequency front-end output should provide an optimal carrier-to-noise ratio (CNR) to measure radio navigation parameters robustly and accurately. Forcing from any electromagnetic disturbances (including a solar radio burst) leads to a CNR drop at the radio frequency front-end output and to the measurement error RMS increase (2.34). Therefore, to estimate a real measurement accuracy, one should have an estimate for the equivalent CNR at the radio frequency front-end output at different values for the solar radio emission flux density.

Let us estimate the solar radio noise power at the RF front-end output, when receiving signals from GPS and GLONASS satellites, with allowance account that GPS applies the CDMA, while GLONASS—FDMA. We start computing with accounting for directional properties of the receiving antenna. Herewith, we will always have in mind the worst case, when there is solar radio emission with a right elliptical polarization.

The antenna directivity is defined as a relation of the power $P(As, \gamma)$ received by a real antenna to the power P_0 received by a reference isotropic antenna (Pistolkors 1947)

$$D(As, \gamma) = \frac{P(As, \gamma)}{P_0} \tag{5.1}$$

where As, γ are signal and noise acceptance angles in the azimuth and elevation planes.

V. Demyanov et al., *Space Weather Impact on GNSS Performance*,
https://doi.org/10.1007/978-3-031-15874-2_5

Table 5.1 Real antenna directional properties

γ, °	$G(\gamma)$, dB	$D(\gamma) = P(\gamma)/P_0$
$0 < \gamma < 5$	$-7.5 \le G \le -5.0$	$0.1775 \le D(As, \gamma) \le 0.316$
$5 < \gamma < 15$	$-4.5 \le G$	$0.354 \le D(As, \gamma)$
$\gamma > 15$	$-2.0 \le G$	$0.630 \le D(As, \gamma)$

Let us determine the directional properties of a real receiving antenna like a gain $G(\gamma)$ relative to an isotropic ideal antenna. As an example for computations, we address the directional properties of the most widespread microstrip receiving antenna located on a horizontal non-conducting surface for a 1270 ... 1625 MHz operation range. The gain of such an antenna and its directivity (D) depend only on the elevation angle, and are presented in Table 5.1 (Perov and Kharisov 2010). This antenna directivity is converted into dB as follows

$$G(\gamma), dB = 10 \lg D(\gamma). \tag{5.2}$$

Further, we account for an antenna effective area. The antenna directivity is known to relate to the antenna effective area as follows (Pistolkors 1947):

$$D(As, \gamma) = 4\pi \cdot A_e(Az, \gamma)/\lambda^2, \tag{5.3}$$

where λ is a wavelength; A_e is an antenna effective area.

Table 5.2 provides the computed values for the real antenna effective area for GPS L1 and L2 GPS, with allowance for the antenna directional properties.

Let us switch to estimating the solar radio noise power at the antenna output, accepting that:

(1) A unit of the solar radio flux power, or solar flux unit, (s.f.u) is 1 s.f.u = 10^{-22} W m^{-2} Hz^{-1} (Chen et al. 2005). Solar flare radio noise (relative to the desired signal) is addressed as white Gaussian noise with spectral density N_0. Therefore, within the GNSS signal frequency band (ΔF_{SV}), this noise has a power $P_N = \Delta F_{SV} \cdot N_0$ s.f.u.

(2) Solar radio noise has a right elliptical polarization and an L1 polarization fading in the antenna $\Delta P(L1) = 3.4$ dB at L1, and, at L2, it is $\Delta P(L2) = 4.4$ dB. The same fading in an antenna is typical of GPS and GLONASS satellite signals (ICD-GPS-200 2021; ICD-GLONASS 2008).

Table 5.2 Antenna effective area

γ, °	A_e, m^2	A_e, m^2
	λ (L1) = 0.190 m	λ (L2) = 0.244 m
$0 < \gamma < 5$	$5.099 \cdot 10^{-4} \le A_e \le 9.077 \cdot 10^{-4}$	$8.409 \cdot 10^{-4} \le A_e \le 1.497 \cdot 10^{-3}$
$5 < \gamma < 15$	$1.01 \cdot 10^{-3} \le A_e$	$1.677 \cdot 10^{-3} \le A_e$
$\gamma > 15$	$1.809 \cdot 10^{-3} \le A_e$	$2.984 \cdot 10^{-3} \le A_e$

Table 5.3 GPS signal strength at L1 and L2 (ICD-GPS-200 2021)

Frequency band	Signal	
	P(Y)	C/A
Minimal strength level for GPC received signals		
L1	−163 dBW	−160 dBW
L2	−166 dBW	–
Maximal expected strength for GPS received signals		
L1	−155 dBW	−153 dBW
L2	−158 dBW	–

Table 5.4 GLONASS signal strength at L1 and L2 (ICD-GLONASS 2008)

Frequency	Signal	
	C/A	P
Minimal strength for GLONASS received signals		
L1	−161 dBW	−161 dBW
L2	{−167 dBW}	−167 dBW
Maximal expected strength for GLONASS received signals		
L1	−155.2 dBW	−155.2 ДбВт
L2	{−161.2 dBW}	−161.2 ДбВт

{−167 dBW}—For GLONASS-M signals

(3) Solar radio noise and GNSS signals fade in the atmosphere by $P_{atm} = 2$ dB (ICD-GPS-200 2021).

We use the known standard values for the GPS and GLONASS satellite signal power within L1 and L2 at a receiving antenna input (ICD-GPS-200 2021; ICD-GLONASS 2008), provided in Tables 5.3 and 5.4.

As one can see from Table 5.3, the GPS L2 signal strength is 6 dB lower, than that of L1. Besides, the *P*-code quadrature component power is 3 dB lower at L1, than that of the C/A-code. The GLONASS minimal signal strength (Table 5.4) is lower, than that of GPS. However, a narrower radio frequency band for signals of separate satellites provides a more essential decrease in the power external radio interferences. In addition, the GLONASS quadrature components have equal strength values. This should lead to a more robust operation of GLONASS receivers under powerful solar radio noise conditions.

For convenience of the further analysis, we present the solar radio noise power at a receiving antenna output with allowance for the above losses in dBW. Also, we will account for the losses and for the fact that the noise power depends on the Sun zenith angle. Implicitly, this is reflected through the equivalent antenna area (A_e) as follows

$$L_N = 10 \cdot \lg(\Delta F_{SV} \cdot m \cdot N_0 \cdot A_e) - P_{atm} - \Delta P(L) \tag{5.4}$$

where m is a solar radio flux in s.f.u; $N_0 = 10^{-22}$ W m^{-2} Hz^{-1}.

Table 5.5 presents the computation results for the solar radio noise power at a receiving antenna output for Sun elevations >15°, and the solar emission center frequency $f = 1415$ MHz ($\lambda = 0.212$ m, $A_e = 2.253 \cdot 10^{-3}$ m^2, polarization losses being −3.4 dB). In the table second to fourth line, we present the computation for the noise power focused through the entire frequency band of the GPS signal, $\Delta F_{GPS} = 20.46$ MHz. However, the radio frequency front-end transmission band for separate satellite signals is substantially narrower. For CDMA signals, the $\Delta F_{CDMA} \leq 3$ MHz (Kaplan 1996). For FDMA signals, the $\Delta F_{FDMA} \approx 1$ MHz (Perov and Kharisov 2010). Therefore, we address further the solar radio noise power in the above frequency bands (lines 3 and 4 in Table 5.5) further.

Figure 5.1 provides a comparative plot for the powers of received GPS (Fig. 5.1a), GLONASS (Fig. 5.1b), and solar radio noise at a receiving antenna output of a user receiver (Demyanov et al. 2012a).

In Fig. 5.1a, horizontal lines show the minimal and maximal powers of GPS signal components (dotted line for the L1 C/A, dots for the L1 P(Y), solid line for the L2

Table 5.5 Solar radio noise power at a GNSS receiving antenna output

Solar radio emission flux within frequency band, s.f.u	10	10^2	10^3	10^4	10^5	10^6
$L_{N,GPS}$, dBW ($\Delta F_{GPS} = 20.46$ MHz)	−178.76	−158.76	−148.76	−138.76	−128.76	−118.76
$L_{N,GPS}$, dBW ($\Delta F_{CDMA} = 3$ MHz)	−187.1	−167.1	−157.1	−147.1	−137.1	−127.1
$L_{N,GLONASS}$, dBW ($\Delta F_{FDMA} = 1$ MHz)	−194.8	−174.8	−164.8	−154.8	−144.8	−134.8

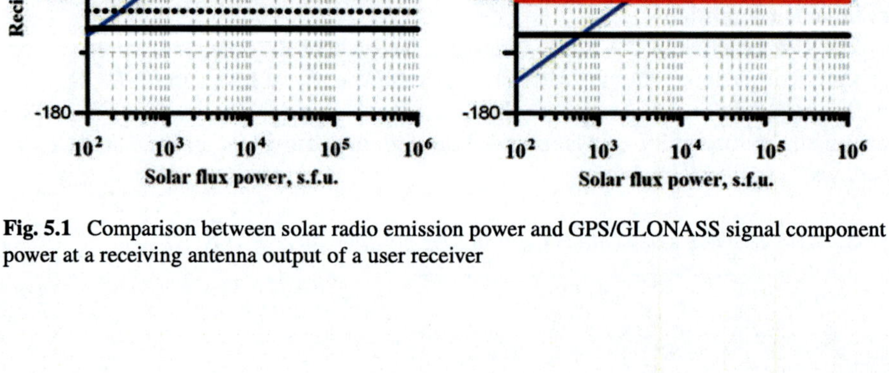

Fig. 5.1 Comparison between solar radio emission power and GPS/GLONASS signal component power at a receiving antenna output of a user receiver

P(Y). Figure 5.1b presents the same for the GLONASS L1 and L2 SP (Standard Precision) and HP (High Precision) components. In the latter case, owing to close power values of separate components, those lines on the plot partially merged.

Apparently, at a certain level of the solar radio noise (varying for different GPS/GLONASS signal components), the solar radio noise power is comparable with the satellite signal power at a receiving antenna output, and starts surpassing it. At the 10^6 s.f.u solar radio emission flux, the solar radio noise exceeds the desired signal by 26–39 dB at a GPS receiving antenna output, and by 20–32 dB at a GLONASS receiver antenna.

Let us switch to determining the solar radio noise dangerous threshold, at which the CNR at a receiving antenna output may appear inadequate for robust tracking of satellite signals.

Due to using correlative processing, a significant CNR gain is known to occur. The value for this gain (CNR_C) is determined by Expression (2.43). Because the C/A (SP) code structure is always known, the maximal noise immunity (when powerful solar radio noise impacts) will be within L1 at reception of C/A (SP) signals. At the same time, the GPS/GLONASS L2 signals are modulated by an encrypted code. To detect those codes, one uses "semi-codeless" or "codeless" processing algorithms (see Sect. 2.1.3). When processing encrypted GPS signals, there is a loss in CNR_C within 14–17 dB for "semi-codeless" algorithms, and 27–30 dB for "codeless" ones (Chen et al. 2005). A similar phenomenon is expected when processing GLONASS encrypted signals. This factor, apparently, will lead to a lower noise immunity of the channels tracking encrypted GNSS signals under the impact from solar radio emission.

After searching and capturing satellite signals for tracking, a coherent lock of the carrier phase and of the code signal delay starts. At that stage, a user receiver noise immunity is determined by the PLL noise immunity (Kaplan 1996). Therefore, the dangerous threshold for the solar radio noise that may cause a loss of lock should be determined proceeding from a minimally admissible CNR, at which a tracking loop operation is provided. A carrier phase filtering error σ_φ is determined as follows (Kaplan 1996)

$$\sigma_\varphi = \sqrt{\sigma_T^2 + \sigma_F^2} + \frac{\sigma_S}{3} \leq \sigma_{F,\mathrm{MAX}}, \tag{5.5}$$

where σ_T, σ_F, and σ_S are RMSs for a phase filtering errors caused by the effect of thermal noise, of short-time clock frequency instability, and of dynamic stress error, respectively.

The other separate components (in rad) composing expression (5.5) may be computed as per the following formulas (Kaplan 1996):

$$\sigma_T = \sqrt{\frac{\Delta F_{PLL}}{CNR} \cdot \left(1 + \frac{1}{2T \cdot CNR}\right)}, \tag{5.6}$$

where ΔF_{PLL} is a PLL noise band; *CNR* is the carrier-to-noise ratio at a PLL input; *T* is the correlation integral accumulation time,

$$\sigma_F = 160 \cdot \frac{\sigma_F(\tau) \cdot f}{\Delta F_{PLL}}, \tag{5.7}$$

where $\sigma_F(\tau)$ is a short-time clock frequency instability; *f* is a signal carrier frequency, and

$$\sigma_S = 0.4828 \cdot \frac{d^3R/dt^3}{(\Delta F_{PLL})^3}, \tag{5.8}$$

where d^3R/dt^3 is a user receiver maximal jerk along the line-of-sight to a satellite.

Formulas (5.7) and (5.8) are written for a case when a user receiver has a third-order astaticism PLL.

Based on (5.6)–(5.8), one can derive an expression that determines a minimally admissible CNR at a receiver input (CN_{thr} in dB) at the specified maximal value for a phase filtering error ($\sigma_{\varphi,\,\mathrm{MAX}}$) set in degrees (Demyanov et al. 2008)

$$CNR_{thr} = -10\lg\left(T \cdot \sqrt{1 + \frac{2}{T \cdot \Delta F_{PLL}}\left(\frac{2\pi}{360}\right)^2\left[\left(\sigma_{\varphi,\mathrm{MAX}} - \frac{\sigma_S}{3}\right)^2 - \sigma_F^2\right]} - T\right), \tag{5.9}$$

Now, let us switch to determining the equivalent CNR at a receiver input under solar radio noise conditions. For this purpose, we use the following relation in (Kaplan 1996):

$$CNR_{EQ} = -10\lg\left(10^{-0.1CNR0} + \frac{1}{\mu \cdot Q \cdot F_{PRN}} \cdot 10^{0.1JS}\right), \tag{5.10}$$

where F_{PRN} is an elementary pulse repetition rate in a code ranging sequence; *CNR0* is a carrier-to-noise ratio (dB) that corresponds to the minimal strength of the received signal in the absence of interferences (see Tables 5.3 and 5.4); *Q* is a parameter dependent on the interference spectral distribution relative to the desired signal spectrum ($Q = 1$ for a narrow-band interference, and $Q = 2$ for a broadband Gaussian interference); *JS* is the relation of the solar radio noise power to the desired signal strength (dB); μ is the coefficient accounting for CNR losses when using a "semi-codeless/codeless") method to detect the envelope of an encrypted signal (Demyanov et al. 2012a):

$$\mu = \frac{2 \cdot \Delta F_{PD}}{F_{PRN}} \cdot 10^{0.1(CNR_C - \Delta CNR_C)}, \tag{5.11}$$

Table 5.6 GNSS user receiver default settings

Setting	Value
PLL noise band	$\Delta F_{PLL} = 18$ Hz
Short-time clock frequency instability	$\sigma_F(\tau) = 10^{-11}$
Maximal jerk along the line-of-sight to satellite	$d^3R/dt^3 = 0\ m^3/s^3$
PLL correlator integration time	$T = 20\ \mu s$
GPS RF front-end frequency band	$\Delta F_{CDMA} = 3$ MHz
GLONASS RF front-end frequency band (by each satellite)	$\Delta F_{FDMA} = 1$ MHz
CNR losses when using "semi-codeless" correlative processing	$\Delta CNR_C = 17$ dB
CNR losses when using "codeless" correlative processing	$\Delta CNR_C = 27$ dB
Thermal noise spectral density power at T = 290 K°	$P_N = -204$ dBW

where CNR_C is a carrier-to-noise ratio gain due to using a correlation reception of a ranging code signal (2.43); ΔCNR_C are CNR losses when using a "semi-codeless/codeless" method for signal processing; $\Delta F_{PD} = 1/T$, Hz is a bandwidth equivalent of a signal envelope pre-detector at a DLL input. This value is determined by the lowest frequency for the satellite signal carrier phase modulation by service information bits (normally 50 Hz).

To estimate the dangerous threshold for solar radio noise that may lead to a GPS/GLONASS signal loss of lock, we will specify typical parameters for tuning a user receiver. Table 5.6 presents these settings (Perov and Kharisov 2010; Kaplan 1996).

Figure 5.2 presents the computed CNR threshold at a receiver input (CN_{thr}, horizontal red line) for the GPS/GLONASS L1 and L2 (5.9). These values in the addressed case appeared close to each other, and are 24.59 dB for L1 and 24.56 dB for L2. From the power of solar radio noise (Table 5.5), we computed the corresponding Jamming/Signal (*JS*) values and determined the corresponding values for the equivalent CNR (CNR_{EQ}, Formula 5.10).

Solid pink line in Fig. 5.2 marks the CNR_{EQ} values for the C/A (SP) code. Dark blue dots present the CN_{eq} for P(Y) (HP) code at L1. Black dotted line denotes the CN_{eq} for P(Y) (HP) code at L2. In the same Figure, horizontal dotted lines of the same colors show the CNR levels for the above signal components in the absence of solar radio emission effect. We address three cases: case one, when the ranging code at the receiving side is exactly known, and there are no losses at correlative processing (panel a); case two, when "semi-codeless" processing of an encrypted signal is used (panel b), and case three, when "codeless" processing is in operation (panel c). In all the cases, the C/A (SP) code at the receiving side was assumed to be always known, and there were no losses at correlative processing.

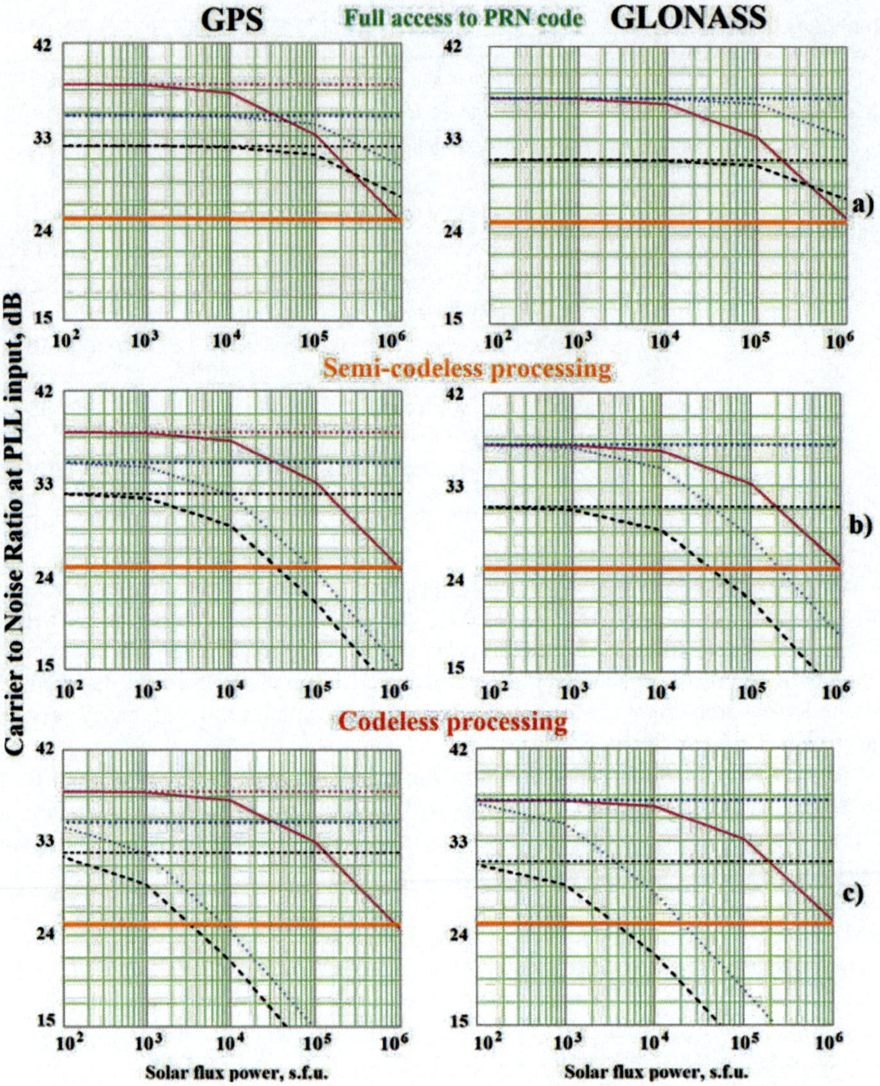

Fig. 5.2 Equivalent CNR at a GPS/GLONASS PLL input under solar radio noise conditions

Apparently, the signal component containing the L1 C/A (SP) code possesses the greatest robustness to solar radio noise. In the addressed case, the equivalent CNR for the SP signal (GLONASS) appears above the critical CN_{thr} even under the 10^6 s.f.u solar radio flux conditions. The CN_{eq}, for the C/A signal component (GPS) decreases more appreciably, and is comparable with the CN_{thr} exactly at 10^6 s.f.u. Therefore, one should expect slips of C/A tracking at $\geq 10^6$ s.f.u and above. This is caused by that the frequency band of the GPS RF front-end is wider, than that of GLONASS, and, hence, the integral power of the solar radio noise penetrating into a receiver is higher.

In the case, when the (P(Y) and HP) ranging codes at the receiving side are known, a robust reception of these signal components is provided even under the >10^6 s.f.u solar radio noise. This is due to using code sequences of large length and high CNR gain at their correlation reception (2.43).

However, when using "semi-codeless" correlative processing, one may expect a loss of lock of the above signal components already at >35.000 s.f.u solar radio noise (P(Y) component at L2, panel b). Owing to a larger power of the P(Y) and HP at L1, their tracking may be interrupted at >1,00,000 s.f.u. At the same time, a higher GLONASS noise immunity of signal detection is ensured. Probably, this is attained due to a narrower bandwidth of the tracking loop for that signal.

The problem is even more aggravated when "codeless" processing of encrypted signals is used. In this case, tracking the P(Y) at L2 may be interrupted already at >3000 s.f.u. The noise immunity of the signal reception at L1 is higher, and is at ~10,000 s.f.u. The corresponding power levels of the solar radio flux interrupting the HP (GLONASS) signal at L1 and L2 are 4000 and 20,000 s.f.u, respectively.

The obtained estimates agree with the conclusions from (Chen et al. 2005; Carrano et al. 2007). The >10^3 s.f.u solar radio noise are corroborated to affect (already negatively) the GNSS operation quality. The results here provide a serious motive to boost and deepen investigations in solar-terrestrial physics and to arrange a wide network of the modern facilities to monitor solar noise. Obvious is also the urgency of effort to increase the GNSS hardware noise immunity under solar radio noise conditions. Another relevant conclusion here may be that using FDMA increases the noise immunity of GNSS user receivers.

5.1.2 Impact on Standalone GNSS

The 2006 Dec 6 and 13 events may be vivid examples for the GNSS operation deterioration under the impact from solar radio bursts since the satellite navigation epoch start. A powerful (X6.5 class) X-ray flare occurred on 2006 Dec 6. Following that flare was a powerful radio burst. From the Owens Valley Solar Array (OVSA) spectrograph (Cerruti et al. 2006, 2008), within 19:30–19:40 UT, the 1.2–1.6 GHz solar radio emission flux exceeded 10^5 s.f.u, and reached 10^6 s.f.u at individual instants. Thus, that flare exceeded (by its power) all the radio bursts known since the GNSS epoch start by, at least, an order of magnitude. That powerful solar event led to numerous slips in operation of many broadband satellite radio systems, including GPS and GLONASS (Carrano et al. 2007; Afraimovich et al. 2009a; Demyanov et al. 2012a).

Figure 5.3 presents the 2006 Dec 6 measurement data obtained with an Ashtech Z-XII receiver (the observation point being Ancon, USA) (Carrano et al. 2007, 2009). One can see a sharp drop in the GPS L1/L2 carrier-to-noise ratio during the radio burst maximum. During the radio emission maximal power (near 19:32–19:38 UT)

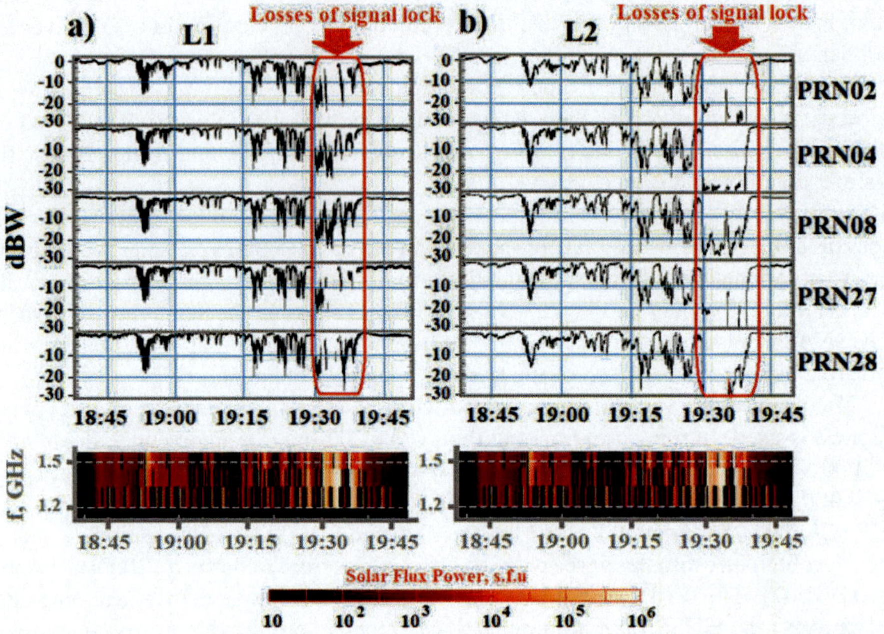

Fig. 5.3 CNR dynamics for separate GPS satellites GPS at L1 (panel **a**) and L2 (panel **b**) under the extreme 2006 Dec 6 solar flare conditions. The color scale below demonstrates the distribution for the radio flux 18:30 through 20:00 UT. Modified part of the figure which was published under open licensees (Carrano et al. 2009)

not only a sharp CNR drop was recorded, but also losses of lock for all the addressed satellites at L2, and, for some satellites (02, 04, 27, 28), at L1 as well.

On the global scale, by simultaneous measurements at 1500 IGS stations, one recorded a sharp enhancement—by 18.5% on average, regarding the total number of measurements—in the relative density of failures in the GPS signal phase measurements. This observation statistics relates to all the visible low-elevation (10°–40°) satellites when recording the failures that coincide with >100,000 s.f.u solar radio bursts.

Figure 5.4 illustrates the 2006 Dec 6 observations of failures in measurements of radio navigation parameters by signals from separate GPS satellites. Panels a, c, d in Fig. 5.4 show the data on the relative density of failures in the phase measurements, P(t), and full omissions of all the measurement samples during the 30-s data recording epoch, W(t), for different GPS satellites.

Fig. 5.4b presents the CNR variations at the GPS L1, recorded with a specialized GPS-receiver at the 50-Hz measurement recording. Figure 5.4e shows the solar noise spectrum within 1.2–2.0 GHz from the OVSA spectrograph.

One can see that, within 19:30–19:40 UT, the GPS monitoring receiver (located in a subsolar point) observed a deep CNR trough, from 45 to 10 dB-Hz (Fig. 5.4b)

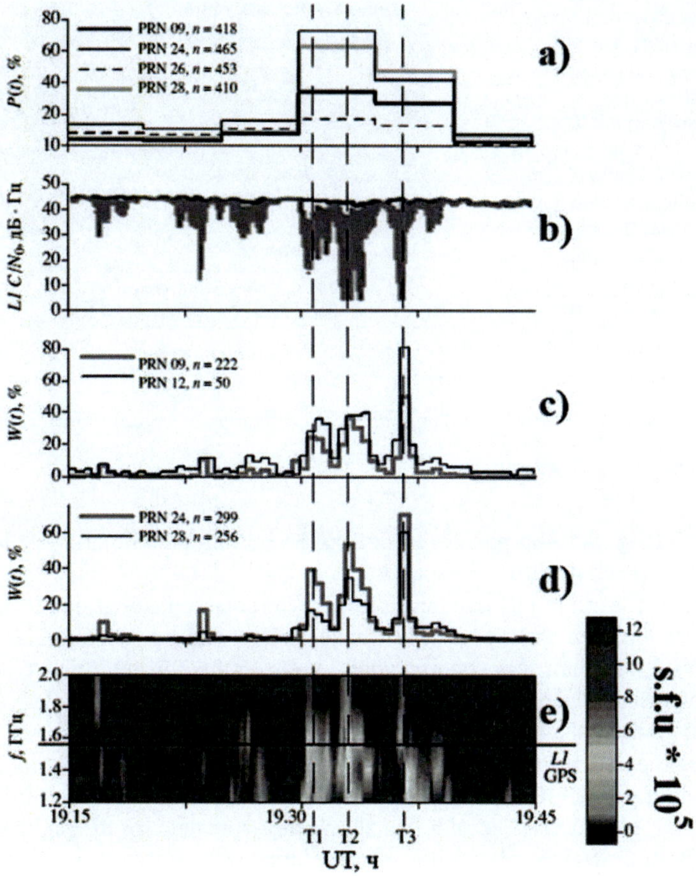

Fig. 5.4 Failures in the measurements of radio navigation parameters by GPS signals on 2006 Dec 6 (Afraimovich et al. 2009a). The solar radio bursts burst power maxima instants are marked as T1, T2, and T3 on panel **d**

(Afraimovich et al. 2008, 2009a). Nothing similar is observed for the satellites of the same group on a background day (black line in Fig. 5.4b).

Solar radio bursts (Fig. 5.4e) also correlate well with the CNR troughs (Fig. 5.4b), as well as with an increase in the measurement omission relative density $W(t)$ of all the radio navigation parameters over the measurement epoch (Fig. 5.4c and d). The measurement omission maxima $W(t)$ coincide with the solar flux maxima (T1, T2, and T3 marks in Fig. 5.4e). The radio burst effect was substantially weaker for the satellites with >40° elevations (Fig. 5.4a, PRN 26). This may be elucidated by a higher CNR in the signal receiving point from the satellites higher elevations.

Figure 5.5 shows the spatial distribution for the GPS L2 carrier phase measurement failure intensity. Apparently, during the radio emission maximum, the dual-frequency measurement mode in GPS receivers was unavailable, indeed, on the entire lit side

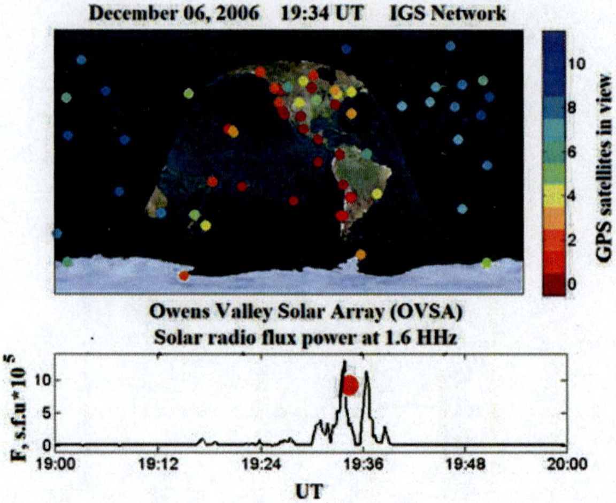

Fig. 5.5 Quantity of the observed satellites with available dual-frequency measurements of pseudo-range from the IGS network (top) and the radio emission power profile (bottom) during the 2006 Dec 6 radio burst (http://gps.ece.cornell.edu/briefs.php)

of the Earth (Fig. 5.5, top panel; red dot on the bottom panel marks 19:34 UT, to which the top panel pattern corresponds).

To estimate the minimally dangerous level of the solar radio emission for a reliable GNSS operation, one should analyze weaker solar radio bursts. The 2003 Oct 28 X17.2-class X-ray flare was accompanied by a <6500 s.f.u radio flux (Fig. 5.6b). Figure 5.6 presents the dynamics for the phase measurement failure relative density *P*(t) (panel a) under the effect of that radio burst by measurements from all the GPS satellites on the lit side of the Earth. Figure 5.6a shows (thickened black line) the *P*(t) form for the Earth dayside. The *P*(t) form was obtained for 2452 satellite-receiver beams at >10° elevations. The thin line shows the same data for the nightside.

Figure 5.6b shows the solar radio emission flux envelope from the TRST spectrograph in Trieste, Italy, at 1420 MHz (https://www.ngdc.noaa.gov/stp/space-weather/solar-data/solar-features/solar-radio/rstn-1-second/). The relative density of sample omissions $W(t)$ practically does not respond to the radio burst effect, and is not provided here.

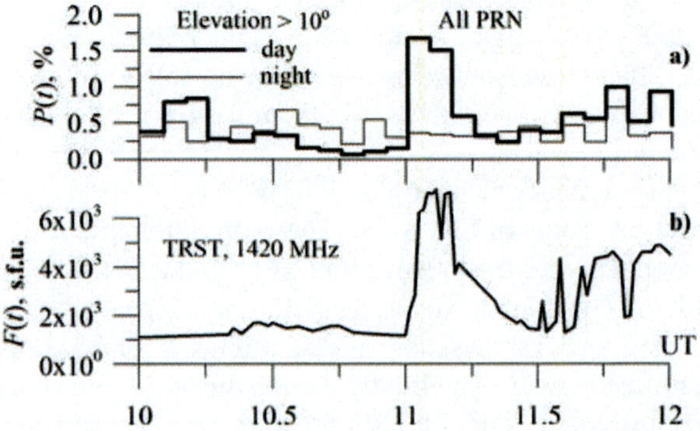

Fig. 5.6 Relative density of phase measurement P(t) failures **a** under the effect of the 2003 Oct 28 solar radio burst for all GPS satellites on the earth lit side (Afraimovich et al. 2009a)

The 2003 Oct 28 flare power appeared by two orders of magnitude lower, than that of 2006 Dec 6. However, the L2 measurement failure relative density exceeded threefold the background value and reached $P_{\max} = 1.66\%$. The form of the dayside phase measurement failure intensity curve correlates well with the form of the radio flux envelope recorded with the TRST spectrograph. The failure relative density on the Earth nightside (thin line) did not exceed the background value. Thereby, the above computations for the solar radio flux minimally dangerous level, when the GNSS operation is robust, are corroborated for the addressed event.

In terms of practice, it is interesting to analyze the dynamics for variations in the accuracy of GNSS users positioning under the effect of powerful solar radio bursts. Figure 5.7 provides the data on such dynamics under the effect of three radio bursts that occurred 2002 through 2011. We performed positioning solutions in dual-frequency modes of standard precision (C/A) (Hofmann-Wellenhof et al. 2008). The positioning error computations in the dual-frequency mode of standard precision were done by using the Navi software (Zatolokin 2020).

Figure 5.7 does not only demonstrate the features of the effect from different solar radio bursts on the positioning accuracy, but also provides an idea of the system modernization results by updating the hardware of the satellites. Thus, the 2022 observations show a sufficiently high background level of ranging error, up to 10–15 m. However, even at such significant background errors, one may note a deterioration in the positioning accuracy by order of 4–5 owing to a solar radio burst effect.

The 2006 observations show a substantially lower background level of the ranging error (2–8 m). In this case, the solar radio burst led to a deterioration in the positioning accuracy by order of 2–3 as compared with the background-day observations. The 2011 observations also demonstrate a sufficiently low background level for the ranging error (no more than 6.5 m). Under the radio burst effect, the deterioration in the positioning accuracy is by order of 8–10 lower relative to the background day. At that, the effect is very distinct and correlates well with the variation in the curve of the solar radio flux power density.

Commissioning new operational frequencies, signal components, and modulations enables to substantially improve the GNSS operation quality, including the operation under solar radio burst effects. In particular, the open civil code and L5 signals, whose strength increased, may facilitate the noise immunity growth. New GPS IIF orbital blocks are equipped with a transmitter transmitting an L5 signal. This allows one to make a comparative analysis for the signal reception noise immunity at different frequencies of separate satellites under solar radio burst conditions.

Figure 5.8 shows the dynamics for GPS signal strength at L1 (black curve), L2 (gray curve), L5 (dark blue curve) when using different signal components of the open and for the encrypted type. The left panel presents the data for the 2011 Sep 24 radio burst, the right panel—for the 2017 Sep 6 one. The red curve illustrates the variation in the solar radio emission power during solar flares: the data from the Sagamore Hill (2011 Sep 24) and from the San Vito spectrographs (2017 Sep 6) are used.

For the analysis, we used the data obtained at the ZIM2, except for the 2011 L5 data, inasmuch as the ZIM2 did not conduct the L5 measurements in 2011.

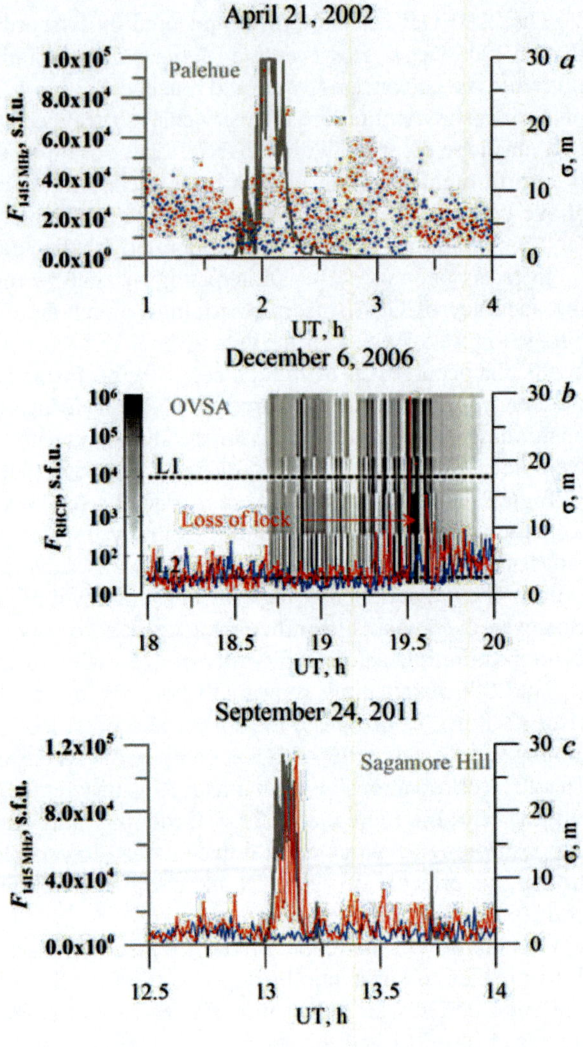

Fig. 5.7 Dynamics for positioning error by GNSS signals in the dual-frequency mode of standard precision under the effect of the 2002 Apr 21 (**a**), 2006 Dec 6 (**b**), and 2011 Sep 24 (**c**) solar radio bursts: red curve shows the data on the flare record day, dark blue curve is for a background day. Gray color shows the radiation flux at 1415 MHz (left scale) and the names of the spectrographs are provided; the right scale indicates the positioning error. GNSS stations located in a subsolar point with the Sun maximal zenith angle are: MAUI (2002), BUEN (2006), and BJCO (2011) (Demyanov and Yasyukevich 2021)

Thereupon, for the 2011, we used the L5 measurements at the WTZR (49.14° N; 12.87° E). In 2011, the WTZR was equipped with a Leica GRX1200+ with an LEIAR25.R3 LEIT antenna. Note, however, that the WTZR data are available only in the Rinex 2.11 format, therefore, we do not possess detailed information on the code, by which the strength was measured, and, accordingly, on the particular type of the measured strength (S5Q, S5I, or S5X).

In 2011, there were two IIF blocks in the GPS constellation. PRN01 was not observed during the flares in the investigated region, whereas PRN25 was at a low elevation (about 15°). We analyzed the strength measurements for L1 (S1C signal component) and L2 (S2W, S2X components) by the PRN29 signals (IIR-M). Also,

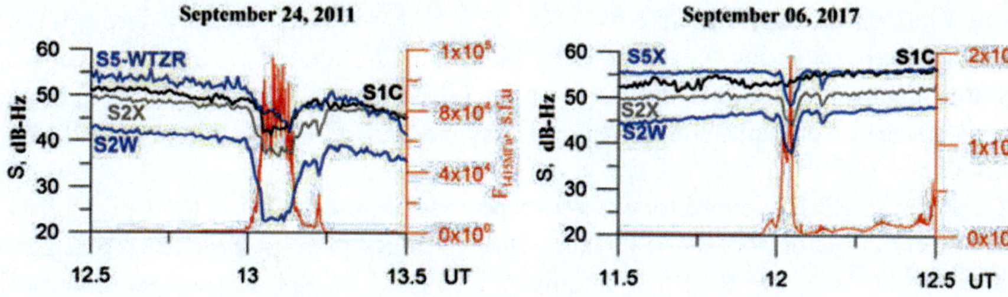

Fig. 5.8 Dynamics for GPS signal strength at L1 (black), L2 (blue (S2W) and gray (S2X)), L5 (dark blue) at the ZIM2 station, as well as for the intensity of solar radio emission (red) for the 2011 Sep 24 (left panel) and for the 2017 Sep 6 (right panel) solar flares. The left panel provides the measurements of L5 signal at the WTZR station (Yasyukevich et al. 2021)

we analyzed the L5 strength measurements (S5) from PRN25 (IIF) that was at a low elevation.

Apparently, during the 2011 and 2017 flares, the relation between the 2011 signal component strength drop and that of 2017 was in dB-Hz: 7.5/0 for L1 (S1C); 10/7 for L2C (S2X); 9/7.5 for L5 (S5/S5X), and 17/8 for L2 (S2W). Using code processing of open signals with an improved ranging code (L5 and L2C) enables to decrease the strength loss at the PLL input, as compared with semi-codeless tracking of encrypted P(Y) signal. The corresponding CNR gain was ~7 dB-Hz for the 2011 Sep 24 flare, and ~1 dB-Hz for the 2017 Sep 6 flare. At the same time, the CNR recorded at L2 when using semi-codeless tracking is the lowest.

Our comparative analysis of the GPS signal reception noise immunity showed that the solar radio flux rate difference by factor of 5.5 led a CNR deterioration by 20–50% for different frequencies and signal components. However, as the radio burst flux power increases to >1,00,000 s.f.u, one may expect an essential non-linear character of the CNR drop. Thus, despite the GNSS system improvements, the issue of the solar radio burst effect on the ultimate positioning accuracy has persisted so far.

5.1.3 Impact on Differential GNSS

Space weather effect on WAAS users has a mediated character. It manifests itself through the quality of differential corrections and reliability of alerts to the RNP availability. In Sect. 2.3.3, we showed that differential corrections to GNSS satellite ranging errors are generated by decomposition of real ranging errors measured by signals from the same satellites at a WAAS reference station network at reference instants.

Fast corrections (2.138) and corrections for ionospheric ranging error (2.140) are most susceptible to space weather effects. In particular, a correction for ionospheric ranging error ($\Delta\rho_{ION}$) is estimated based on re-processed measurements of the oblique ionospheric ranging error obtained at WAAS reference stations through the method of dual-frequency measurements (Hofmann-Wellenhof et al. 2008). Above, we showed that it is the measurements at the GNSS second operational frequency that possess a lower noise immunity and may be unavailable under powerful and extreme solar radio bursts (Figs. 5.2 and 5.5).

Alerts to the RNP availability are generated based on estimating their protection levels (PLs) in the horizontal (HPL) and in the vertical (VPL) planes (see Fig. 2.22). In the general case, the RNP PL is computed based on the current geometry of the satellite constellation (relative to the user) and variances of the ranging errors corresponding to those satellites upon applying a differential correction (Formula 2.155). We showed above that the effect of an extreme solar radio burst may lead to a loss of lock within a part of visible satellites at one or both operational frequencies (see Fig. 5.1), i.e. to a sharp DOP and to a growth in the expected positioning error.

Thus, under adverse heliogeophysical conditions, the WAAS operation deterioration in terms of generating differential corrections to ranging measurements and alerts to the RNP bound integrity may be caused by:

1. Unavailability or a sharp deterioration in the accuracy of the correction for the ionospheric error, which will lead to an increase in the variance of the residual ranging error (σ^2_{UIRE}, Formula 2.148).
2. Suppression of satellite signals from solar radio flux, which entails a CNR deterioration and the corresponding increase in the variance of pseudo-range measurements (see Sect. 2.1.3, Formulas 2.35 and 2.36).
3. Loss-of-lock of signals from a part of visible satellites under the effect of solar radio emission, which will lead to a DOP of the observed satellite constellation.

A consequence of the above phenomena may be a higher probability of a false alert or the information dangerous mismatch when generating an alert to a violation of the RNP bounds within a WAAS operation area (Demyanov et al. 2019). The first two factors also concern systems of local differential correction, such as real-time kinematic (RTK).

During extreme solar radio bursts, one may observe crucial violations in a WAAS operation in terms of RNPs within the operation area. For example, when generating alerts to violation of RNP availability as per the LPV200 requirements, one discovers an essential decrease in the area of the RNP availability (Carrano et al. 2007). Figure 5.9 illustrates this phenomenon. Here, there is an hourly-averaged map for distribution of the RNP availability over 19:00–20:00 UT. One can see that, under the 2006 Dec 6 extreme solar radio burst effect, the WAAS operation area size, within which the LPV200 RNPs were available, shrank about twofold, on average.

One also observed the cases of the WAAS EGNOS availability decrease in the European sector. There, this occurred during a weaker solar radio burst. Figure 5.10 shows an RNP availability distribution map as per APV-1 (HAL 40 m, VAL 50 m) over Europe on 2017 September 6 under the X9.3-class flare conditions. The figure panels show four hourly-averaged maps for the distribution of RNP availability, starting with 10.00 UT (panel a) to 13.00 UT (panel c). The 2017 Sep 6 radio burst power peaked over ~12.00–12.05 UT and reached ~20 000 s.f.u.

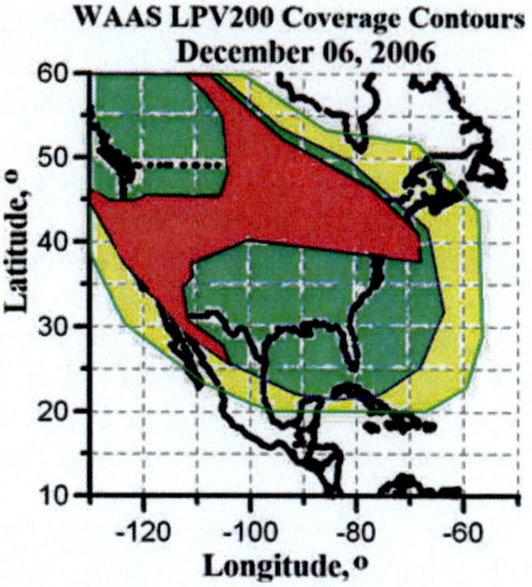

Fig. 5.9 LPV-200 RNP distribution map. Red color shows the area, within which RNPs were unavailable, green denotes the area, where RNPs were available >95% of the observational time, yellow corresponds to the area with <95% availability. Modified part of the figure which was published in open-assess sources (https://www.nstb.tc.faa.gov/DisplayArchive.htm)

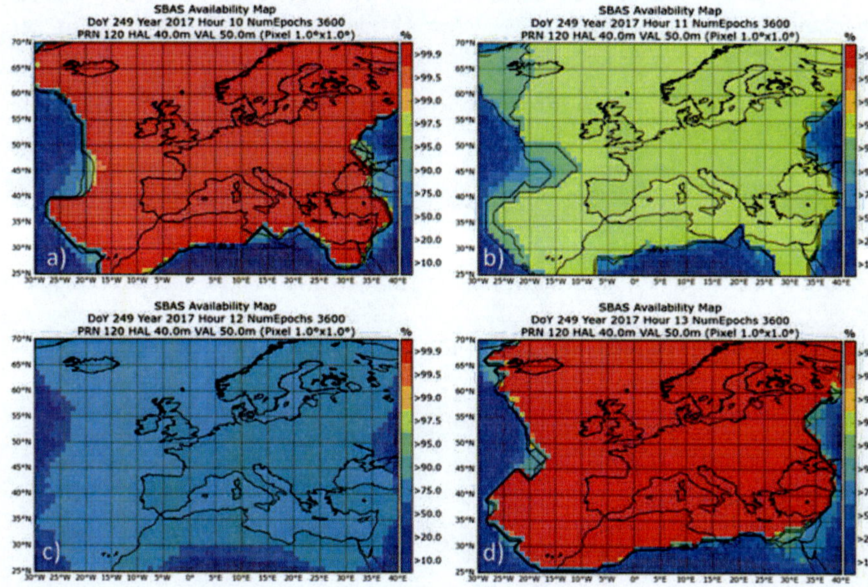

Fig. 5.10 APV-1 RNP availability for EGNOS users over Europe under the 2017 Sep 6 solar flare (Berdermann et al. 2018)

The APV-1 RNP availability drops from >99% before and after the flare/burst effect to 75% during the flare maximum (12:00 UT, panel c). The presented results explicitly demonstrate WAAS vulnerability to effects of powerful radio bursts within the GNSS operational frequency range.

One-station Precise Point Positioning (PPP) (Zumberge et al. 1997) is also susceptible to space weather negative effects, including solar radio bursts. So, during the 2017 Sep 6 solar radio burst, the mean positioning error in the kinematic PPP mode varied from 0.17 m (when measuring by GPS signals) to 0.47 (by GLONASS signals). The radio burst effect resulted in the error enhancement to ~0.5–0.6 m when measuring by signals of both systems (Yasyukevich et al. 2018). For the 2011 Sep 24 solar radio burst conditions, Sreeja et al. (2013) recorded a significant DOP deterioration that was accompanied by a considerable decrease in the accuracy in the user height determination in the kinematic PPP mode.

Figure 5.11 shows that, during the radio burst maximum power pass (after 13:00 UT), two observational stations located near the subsolar point (Recife, Sao Tome) recorded a substantial growth in the vertical DOP (VDOP). This growth was accompanied by a synchronous burst of error in determining the station height to 1.5–2 m (Recife).

Figure 5.12 demonstrates the dynamics for a GNSS positioning accuracy deterioration in the dual-frequency PPP mode and provides the results of observations under the solar radio burst conditions on 2002 Apr 21 (panel a), 2006 Dec 6 (panel b), and

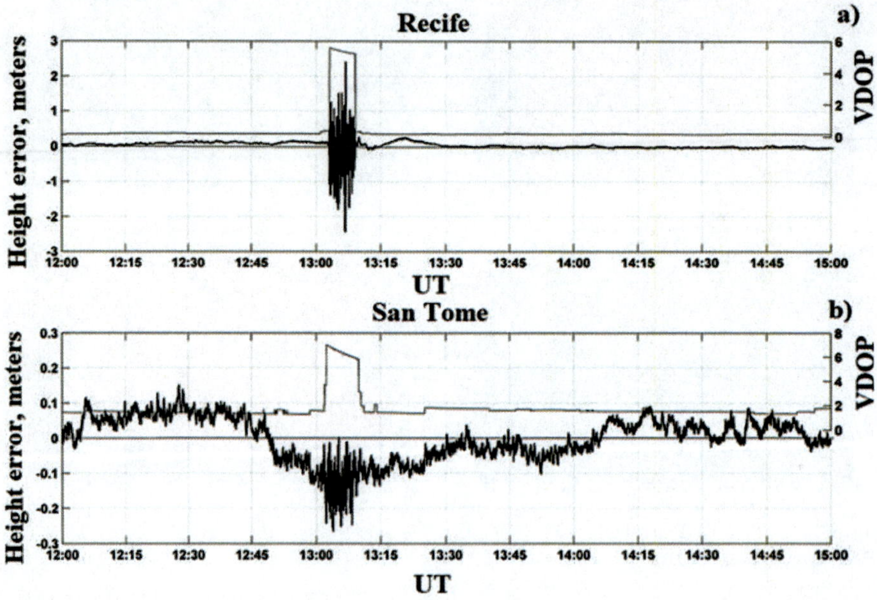

Fig. 5.11 Dynamics for the user height determination error in the kinematic PPP mode (black curve) and the VDOP (gray curve) at the Recife (top panel) and Sao Tome (bottom panel) stations (Sreeja et al. 2013)

on 2011 Sep 24 (panel c). The red curve in the figure shows the burst recording-day data, whereas the dark blue curve presents the data for the background day. Gray color marks the solar radio emission flux power envelope at 1415 MHz (left scale), and the names of radiospectrographs are presented. The right scale shows the positioning error RMS for the measurement station. The stations are located in the subsolar points with the maximal zenith angle of the Sun: MAUI (2002), BUEN (2006), and BJCO (2011) (Demyanov and Yasyukevich 2021). The station coordinates were computed based on dual-frequency measurements by using the GAMP software (Zhou et al. 2018).

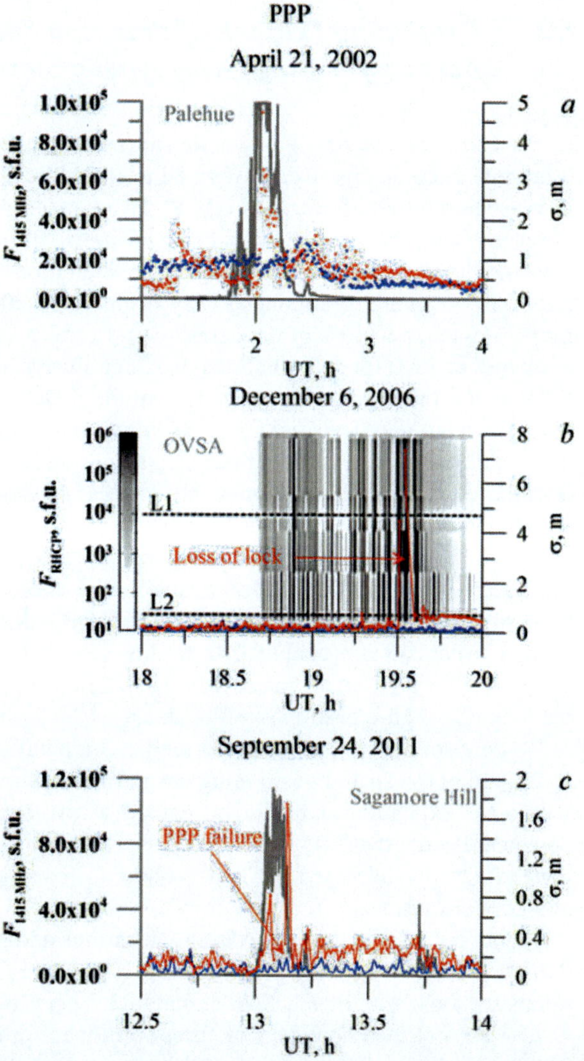

Fig. 5.12 Dynamics for the GNSS positioning error in the dual-frequency PPP mode under solar radio burst conditions on 202 Apr 21 (panel **a**), 2006 Dec 6 (panel **b**), and 2011 Sep 24 (panel **c**) (Demyanov and Yasyukevich 2021)

Apparently, an increase in the positioning error correlates with the radio burst intensity. During the radio burst maxima, the positioning accuracy in the PPP mode deteriorated to unities of meters. Under the 2011 Sep 24 burst effect, one even observes a PPP positioning failure (Fig. 5.12). One should note that the single-frequency PPP mode is more robust to negative factors of a higher radio emission, than the dual-frequency PPP mode (Berdermann et al. 2018).

5.2 Magnetic Storm Impact on GNSS

5.2.1 Positioning Dilution of Precision and Positioning Accuracy During Geomagnetic Storm

Under conditions of strong magnetic storms and storm-provoked ionospheric scintillations, there emerge losses of satellite signal lock. This may lead to a dramatic increase in dilution of precision (DOP). Even under magnetic storm conditions, one observed (with the greatest probability of up to 70%) a loss of signal lock for only one satellite within the entire observed constellation (Scone et al. 2001). Losses of lock for two satellites simultaneously occurred in 20% of the cases, for three and more – less than in 10% of the cases.

A loss of lock for 1–2 satellites does not always substantially affect the current DOP and/or the positioning error. The minimal DOP is known to correspond to the maximal volume of a spatial figure, in whose basis middle there is an observational point, and at the figure vertices there are navigation satellites (Barabanova 2010). This is possible at a simultaneous maximal angular separation of the observed satellites among themselves, as well as in the azimuth and in the elevation planes. Also, at least one of satellites should be close to the zenith relative to the observer. Therefore, if one observes ≥4 satellites, a loss of lock for the satellites with the maximal elevation most often has the strongest effect on the DOP value. For high elevated satellites, such an event is less probable, than for low elevated satellites, due to a larger signal strength. However, under intensive ionospheric scintillations, a receiver could loss the lock of a high-elevation satellite also.

To confirm the above, we performed a comparative analysis of the *PDOP* time series and of the spatial positioning error at some stations located in North America within the expanding auroral oval equatorward boundary. The observations were conducted under the 2003 Oct 29–30 powerful magnetic storm conditions (the "Halloween" storms discussed in Chap. 3), and, for comparison, under the 2003 Mar 25 quiet conditions.

Figure 5.13 designates the stations' locations and names by red triangles. There, dark blue daggers with the PRN numbers indicate the location of ionospheric pierce points for those satellites, whose signal locks were lost during the observations.

The auroral oval boundary features significant gradients of the electron density in the ionosphere. Therefore, one may expect that it is this regions, where intense

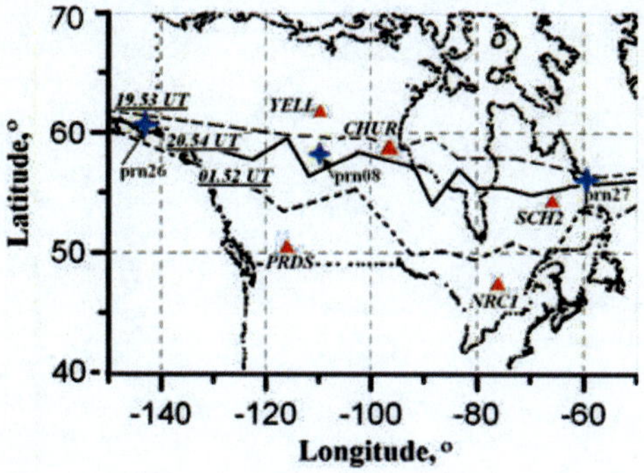

Fig. 5.13 Map showing the location of monitoring stations (triangles) and ionospheric pierce points (daggers). The dotted line indicates the auroral oval equatorward boundary position during the 2003 Oct 29–30 magnetic storm main phase and the solid broken line indicates the time from (Afraimovich et al. 2004)

multiscale ionospheric irregularities exist, including fine-scale ones. Owing to this, there is an increase in the probability of losses of signal lock for those satellites, whose observation trajectory transits through the ionosphere region within the auroral oval boundary.

Figure 5.14 provides an example for the observations at the SCH2 station, where one can clearly see the effect of a sharp PDOP increase and of the positioning error growth. The measurements were taken by using the data at a 30-s temporal resolution.

Panels (a), (b), and (c) in Fig. 5.14 present the time series for the current number of visible satellites, spherical positioning error σ(t), and the GDOP. Panels (d), (e), and (f) show the satellite positions in the sky over SCH2 before the cycle slip (20:34:30 UT, panel d), at the cycle slip instant (20:35:00 UT, panel e), and right after the tracking recovery (20:35:30 UT, panel f). Panels (g) and (h) demonstrate a simultaneous loss of signal lock for PRN 27 at L1, L2, that occurred at 20:35:00 UT.

One can see that, at 20:35:00 UT, loss of tracking of one satellite was observed (the total number of satellites changed from 8 to 7). Apparently, at the same instant, there occurred a sharp short-time GDOP increase to PDOP = 5.2 (panel c). Both these events coincide with a short-time burst of the positioning error from 11 to 23.5 m (panel b). At 20:35:00 UT, there was, indeed, a loss of PRN 27 tracking, that was near the zenith (79.2° elevation, panel e).

Due to a large period of amplitude record (30 s), one cannot analyze in detail the variations in the signal amplitude and phase before LOL. However, as one can see in the figure, PRN 27 was in the region of the auroral oval equatorward boundary during the event. Therefore, one may assume that the LOL was caused by a signal dispersion at electron density small-scale irregularities.

A different situation concerning the emergence of irregular variations in the DOP was observed under the 2003 Mar 25 undisturbed geomagnetic conditions. No one observed losses of signal lock for the high-elevation satellites at the same stations.

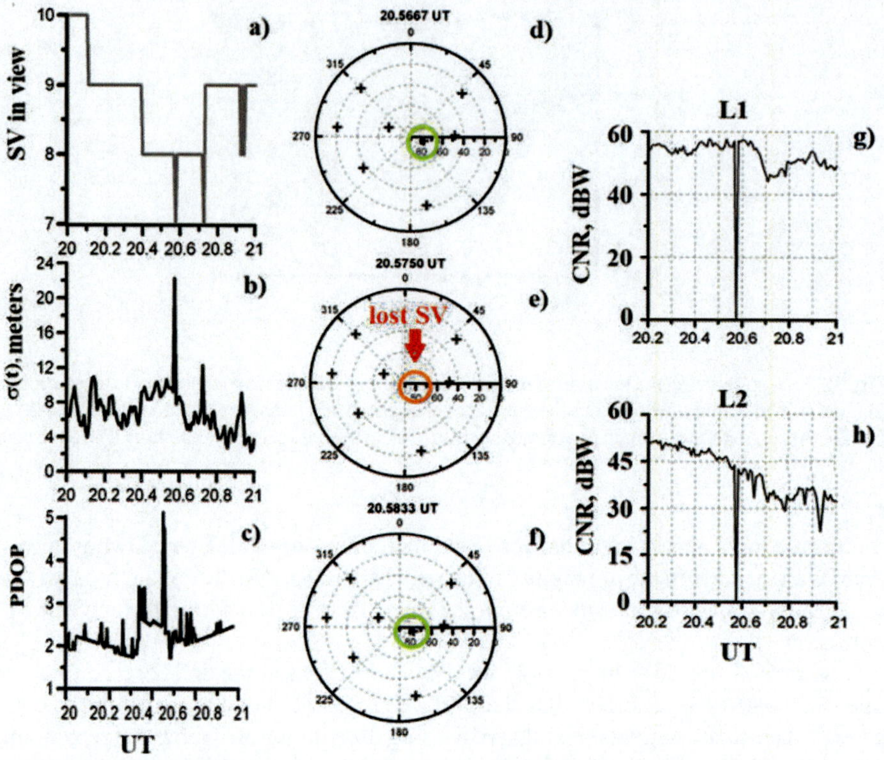

Fig. 5.14 Observations of GDOP and PDOP (SCH2, 2003 Oct 29)

The bulk of sharp PDOP cases was related to satellite rising and setting, although separate LOL cases were observed for the <20° elevation satellites.

At different latitudes, there are substantially different radiowave propagation conditions. At mid-latitudes, the conditions for transionospheric signal degradation emerge during geomagnetic disturbances in local regions (the auroral oval equatorward boundary, propagation of equatorial bubbles into the mid-latitude region). At the low and auroral latitudes, such conditions exist practically always. Thereby, one should examine the features of DOP irregular variations at different latitudes of observations under quiet and under disturbed conditions.

To perform this examination, we used the observations of GPS satellites at 350 stations located in North America. For the investigation, the data from the selected stations were sorted by latitude: high latitudes (>65°), mid-latitudes (35°…65°), and low latitudes (0°…35°). In Chapter 2, we elucidated this latitude distribution by the above ranges in terms of the structure features and the ionosphere behavior. The total scope of the DOP current values with a 30-s recording period was: 12,000 measurements at high latitudes, 1,65,000 at mid-latitudes, and 14,000 at low.

We included the cases, when the de facto observed DOP meets the conditions: $PDOP > PDOP_{max}$, $HDOP > HDOP_{max}$, and $VDOP > VDOP_{max}$. The maximally observed values for DOP were established empirically based on the analysis of the known sources (origins) (Kaplan 1996; Perov and Kharisov 2010; Wang et al. 2002), viz.: $PDOP_{max} = 5$, $HDOP_{max} = 2.1$, and $VDOP_{max} = 3.5$. At present, there is no definite standard concerning the satellite-elevation mask used when recording the measurements to the data storage RINEX files. As a rule, one uses a 5° or 10° mask. Thereby, to unify the procedures for analyzing and processing the measurements when computing DOP, we accounted for only the satellites with ≥10° elevations.

For our examination, we selected two observational periods: one during the geomagnetic disturbance main phase (2003 Oct 29, 19:00–24:00 UT) and second in the absence of strong disturbances (2003 Oct 31, 19:00–24:00 UT). Fill rectangles mark those periods in Fig. 5.15. The geomagnetic conditions are described by the dynamics for the *Dst*-index behavior (panel a). Panels b and c present the hourly-averaged TEC amplitudes with a 20–60 min disturbance period (panel b: black color shows the USA northeast and gray depicts the USA southwest), and 1–10 min (panel c: the USA northeast and southwest).

The 2003 Oct 29 19:00–24:00 UT period fell on the magnetic storm main phase. This was accompanied by the geomagnetic field sharp variations and TEC variation amplitude bursts with 20–60 min (Fig. 5.15b) and 1–10 min (Fig. 5.15c) disturbance periods. Thus, within that interval, there were favorable conditions for the emergence of intense amplitude and phase scintillations with possible losses of signal lock. At the same time, during the second observation period (2003 Oct 31, 19:00–24:00 UT), there occurred the magnetic storm recovery phase, when no such significant TEC variations with 1–60 min periods were observed. Therefore, during the second observation period, the probability for the GPS signal intense ionospheric scintillations and losses of signal lock to emerge is essentially lower.

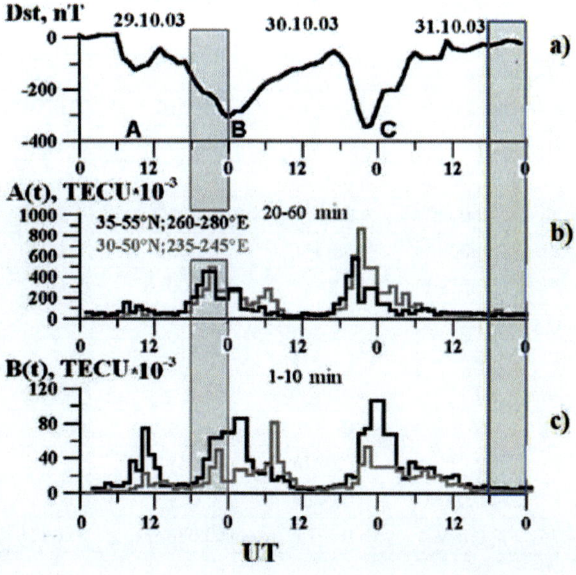

Fig. 5.15 Conditions for observations of the DOP and the GPS signal reception quality during the 2003 Oct 29–31 magnetic storm

Table 5.7 provides the generalized data on probabilities for the emergence of DOP anomalously high values at different latitudes under disturbed and undisturbed geomagnetic conditions. In the table, there are total probabilities for the PDOP > $PDOP_{max}$, HDOP > $HDOP_{max}$, and VDOP > $VDOP_{max}$ events that were determined as follows

$$P(DOP > DOP_{max}) = \sum_{i=1}^{K} P_i, \tag{5.12}$$

where P_i is a probability for the emergence of the DOP > DOP_{max} events in the ith Sturges class of the event occurrence distribution histogram; K is the total number of Sturges classes for the analyzed dataset by each group of events.

Apparently, at all the latitudes, there is an increase in the probability for the emergence of the DOP anomalous values under geomagnetic disturbance conditions. During the storm main phase, one observes an increase in the probability for the

Table 5.7 Probabilities for observing anomalously high DOP

UT	Magnetic storm main phase			Recovery phase		
	P(PDOP)	P(HDOP)	P(VDOP)	P(PDOP)	P(HDOP)	P(VDOP)
Low latitudes						
19	0.007	0.138	0.0005	0.007	0.131	0.0005
20	0	0.163	0	0	0.104	0
21	0	0.162	0	0	0.147	0
22	0	0.334	0	0	0.300	0
23	0	0.019	0	0	0.013	0
24	0.004	0.395	0	0.004	0.353	0
Mid-latitudes						
19	0.0011	0.034	0.0010	0.0001	0.0031	0
20	0.0009	0.029	0.0006	0.0009	0.016	0.0006
21	0.0020	0.019	0.0006	0.0020	0.018	0.0005
22	0.0029	0.011	0.0027	0.0027	0.012	0.0026
23	0.0032	0.056	0.0004	0.0020	0.040	0
24	0.0130	0.248	0.00016	0.0110	0.185	0
High latitudes						
19	0	0	0.014	0	0	0
20	0	0	0	0	0	0
21	0	0	0	0	0	0
22	0.01	0.0014	0.031	0	0	0.014
23	0.0046	0.0007	0.011	0	0	0
24	0.072	0	0.088	0.033	0	0.041

emergence of DOP great values. To the least extent, this is typical of the HDOP, and, to the greatest, this is characteristic of the VDOP. The latter is important, in particular, for aviation applications that use GNSS hardware.

The scope for the analyzed data at high and low latitudes is about equal. At low latitudes, the addressed phenomenon manifests itself insignificantly. The increase in the probability for the emergence of the PDOP > $PDOP_{max}$, HDOP > $HDOP_{max}$, or VDOP > $VDOP_{max}$ events is within 0.5–1%. At the same time, at high latitudes, the probability P (PDOP > $PDOP_{max}$) under magnetic storm conditions increased by 40%, and the probability P (VDOP > $VDOP_{max}$)—by 6–10%. Under quiet conditions at high latitudes, we did not note any HDOP > $HDOP_{max}$ emergences at all.

At low latitudes, the DOP great values occur, mainly, for the HDOP. In contrast, at high latitudes, this is observed for the VDOP, which does not contradict the known investigations (Perov and Kharisov 2010; Wang et al. 2002). The increase in the probability for the P (DOP > DOP_{max}) events under the geomagnetic disturbance effect fell (in the bulk of cases) on 22–24 UT. Over the same interval, we note the maximum of TEC variations with a 1–10 min disturbance period (Fig. 5.15c).

The low probability for the emergence of the PDOP > $PDOP_{max}$, HDOP > $HDOP_{max}$, and VDOP > $VDOP_{max}$ events evidences that geomagnetic disturbances at mid-latitudes do not result in numerous phenomena of increase in the DOP, but, more likely, cause local cases of the geometry deterioration for the observed constellation.

Thus, we showed that one of the possible mechanisms for of deterioration in the GNSS operation quality under geomagnetic disturbance conditions is the DOP increase due to LOLs for a part of visible satellites. This manifests itself, to a greater extent, at high latitudes as compared with the low latitudes. However, this is not always directly related to geophysical phenomena only, but may also be elucidated by the worse geometric characteristics of a navigation constellation at high latitudes (see Sect. 2.1.2, Fig. 2.8).

5.2.2 WAAS Alerts and GNSS Positioning Quality Under Geomagnetic Storm Condition

The WAAS integrity data include User Differential Range Error (UDRE) and Grid Ionospheric Vertical Error (GIVE). Considering the possibility of irregular ionospheric variations, GIVE may not be defined correctly by WAAS solutions within the expanded coverage zone. This leads to the larger confidence bounding terms and lower positioning availability expected as compared with the real availability under geomagnetic storm conditions. This raises the issue, whether WAAS algorithm of integrity monitoring provides the same efficiency for both "global" differential correction (i.e. clock, ephemeris, etc.) and "local" differential correction (i.e. ionosphere, troposphere, and multipath).

This section discusses the compliance of positioning quality in the US coverage zone (CONUS) and Canadian region during geomagnetic storms with the declared

WAAS integrity monitoring results (www.nstb.tc.faa.gov/DisplayDailyPlotArchive.htm). We compare the recorded GPS positioning quality in the single-frequency C/A ranging mode and the HPL (VPL) values corresponding to the LP, LPV, and LPV200 requirements.

The following levels of accuracy and availability for the WAAS coverage zone are addressed (GPS-WAAS-PS 2008):

1) Localizer Performance (LP) service is available given that the calculated HPL ≤ 40 m;
2) Localizer Performance with Vertical Guidance (LPV) is available given that HPL ≤ 40 m and VPL ≤ 50 m;
3) Localizer Performance with Vertical Guidance to 200-foot decision height (LPV200) is available given that HPL ≤ 40 m and VPL ≤ 35 m.

For our analysis, we selected two geomagnetic storms: the 2011 Feb 14–15 and the 2015 June 22–23. Figure 5.16 provides the 2011 Feb 14–15 (panel a) and the 2015 June 19–27 (panel b) *Dst*-index variations (from https://cdaweb.sci.gsfc.nasa.gov). The initial phase (IP, blue vertical lines), main phase (MP, red vertical lines), and recovery phase (RP, green vertical lines) of the storms are also marked on the figures. Table 5.8 provides details of the geomagnetic storms, such as storm phase durations and Tssc. The 2011 Feb 14 storm was less intense ($Dst_{\min} = -38$ nT) and shorter than the 2015 Jun 22 storm ($Dst_{\min} = -210$ nT). The first storm is weak and the second storm is intense.

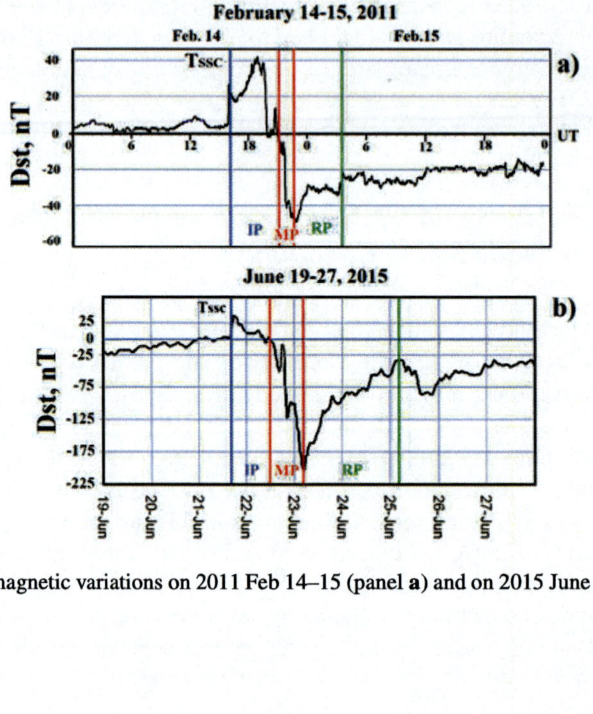

Fig. 5.16 Geomagnetic variations on 2011 Feb 14–15 (panel **a**) and on 2015 June 19–27 (panel **b**)

Table 5.8 Geomagnetic storm parameters

Date	Kp_{max}	T_{SSC}, UT	IP duration, UT	MP duration, UT	RP onset, UT
2011 Feb 14	4	16:00	16:00–20:30	20:30–23.30	23:30
2015 Jun 22	8	05:30	05:30–12:30	12:30–20:00	20:00

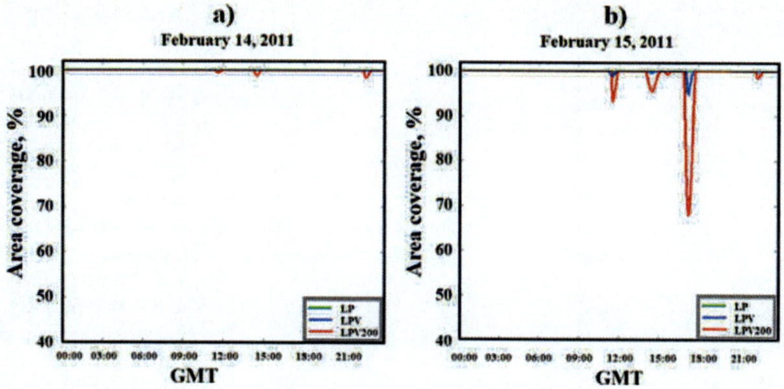

Fig. 5.17 CONUS area coverage time dependence for 2011 Feb 14 (panel **a**) and for 2011 Feb 15 (panel **b**). Modified part of the figure which was published in open-assess sources (https://www.nstb.tc.faa.gov/reports/waaspan36.pdf)

The issued alert reports for the CONUS are available at the official WAAS webpage (http://www.nstb.tc.faa.gov/DisplayDailyPlotArchive.htm). Figure 5.17 illustrates the estimations of the CONUS areas covered with LP, LPV, and LPV2000 services on 2011 Feb 14 and 15.

Although the storm main phase occurred on 2011 Feb 14, the coverage with LP, LPV and LPV200 services was fairly satisfactory throughout the day. In contrast, the WAAS alert to a sharp shrinking of the coverage zone for LPV200 service was issued approximately at 17:40 UT. This instant corresponds to the storm recovery phase.

The WAAS service degradation may be caused by not only the ionosphere impact, but also by the WAAS and GPS segment outages. Table 5.9 provides the information on the system outages on 2011 Feb 14 and 15.

GEO Uplink Subsystem (GUS) outage of geostationary satellite GEO138 was reported on 2011 Feb 14. This did not cause the LPV and LPV200 outages, because other GEOs were used for WAAS corrections that day. The next day, a long PRN21 outage (11:15–20:05 GMT) was detected, as well as a short PRN04 outage (11:32–11:43 GMT). In other words, two GPS satellites were out of service simultaneously within 11:32–11:43. The WAAS alerts for the CONUS zone in Fig. 5.17b did not exactly coincide with the reported outages. The WAAS predicted three outages of LPV and LPV200 service in California, North-Central, Arizona, and Florida on 2011 Feb 15.

Table 5.9 WAAS and GPS outages on 2011 Feb 14–15

Start date	End date	Location/satellite	Service affected	Event description
2011 Feb 14	2011 Feb 14	GEO138, Woodbine (QWE)	All	GUS switchover, Woodbine faulted. TOW 1,27,889–1,27,902
2011 Feb 15	2011 Feb 15	PRN21	All	NANU 2,011,015 There was a NANU 2,011,015 on PRN 21 from 11:15 to 20:05 GMT and SV Alert on PRN 4 (11:32–11:43), which both affected LPV and LPV200 coverage W1623D2 LPV Outages: 1. California outage due to SV alert on PRN 4 2. North-Central and Arizona outages due to NANU on PRN 21 LPV200 Outages: 1. California outage due to SV alert on PRN 4 2. North-Central, Arizona and Florida outages due to NANU on PRN 21

Figure 5.18 is the same as Fig. 5.17, but for the Canadian area on 2015 June 20 (a) and on 2015 June 22 (b).

There was a significant deterioration in the 2015 Jun 22 LP, LPV, and LPV-200 services in Canada within 19:30–21:30 GMT, which corresponds to the storm MP (Table 5.8). Table 5.10 contains the LPV (LPV200) degradation characteristics. In contrast to the previous case (see Table 5.9), the service degradation for all the North America coverage zone is caused by the ionospheric disturbances as a result of the 2015 Jun 22–23 strong geomagnetic storm. There were no GPS/WAAS segment outages on 2015 June 22, which means that the LPV and LPV200 degradation was caused by the increased storm-invoked GIVE. Probably, the observed effects were related to travelling ionospheric disturbances (TID) presence in the corresponding regions.

Let us estimate the correspondence between the WAAS alerts and the real positioning quality. In our analysis, we involved the data from 70 stations in the CONUS zone (excluding California) and 30 stations in Canada. RINEX 2.0 files with the 30-s temporal resolution are available at http://lox.ucsd.edu/. All the CONUS stations, whose data were used, are equipped with the same GNSS receivers (Trimble), which is an advantage for data interpretation. Receivers in Canada are different (TPS Net,

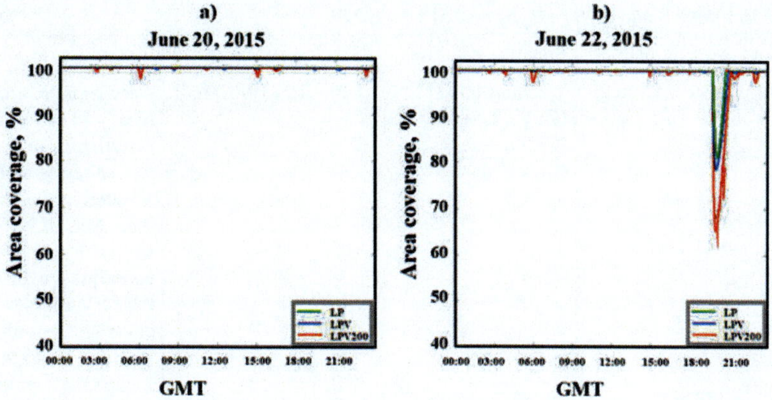

Fig. 5.18 Canadian area coverage time dependence during the 2015 Jun 22 geomagnetic storm (panel **b**) and for the quiet 2015 Jun 20 (panel **a**). Modified part of the figure which was published in open-assess sources (https://www.nstb.tc.faa.gov/reports/waaspan53.pdf)

Leica, Ashtech, Trimble, AOA and Turbo-Rogue). Figure 5.19 illustrates the station locations.

To estimate the real positioning quality in the standalone mode, we used the teqc software (Estey and Meertens 1999) to compute the positioning errors in a single-frequency standalone receiver mode. Ionospheric corrections were not applied. To calculate HPL (VAL) based on the real ionospheric ranging error data, one should perform the following data treatment (for each station separately):

- find navigation solutions and positioning error values, and compare LP, LPV, and LPV200 alarm limits (HAL and VAL) with the computed errors and estimated coverage for the CONUS and Canada as follows

$$\%\,\text{of area covered} = \frac{(N_{TOT} - N_{FLT})}{N_{TOT}} \cdot 100\% \tag{5.13}$$

  where N_{TOT} is a total number of stations within the coverage zone; N_{FLT} is a number of stations with the positioning errors exceeding VAL or HAL values according to LP, LPV, or LPV200 limits;
- compute the HPL and VPL values (see Sect. 2.3.3, Eq. 2.155) considering the real ionospheric ranging errors instead of GIVE values. To compute a ranging error, we used a "model ranging" for a real satellite constellation. The "model ranging" is determined for each satellite as follows:

$$\begin{aligned}
&\rho = R_{GEOM} + \Delta R_{ion} \\
&R_{GEOM} = \sqrt{(x_0 - x_U)^2 + (y_0 - y_U)^2 + (z_0 - z_U)^2} \\
&\Delta R_{ion} = \tfrac{40.308}{f_1^2} \cdot TEC \\
&TEC = \tfrac{1}{40.308} \tfrac{f_1^2 \cdot f_2^2}{f_1^2 - f_2^2} (C1 - P2)
\end{aligned} \tag{5.14}$$

Table 5.10 WAAS and GPS outages on 2015 June 22

Start date	End date	Location/satellite	Service affected	Event description
2015 Jun 22	2015 Jun 23	Washington D.C. (CnV), Los Angeles (CnV), Atlanta (CnV)	LPV-all, LPV200-all	Geomagnetic activity (Kp = 8) provoked disturbance in the ionosphere which resulted in the increased GIVE values. As a consequence, on June 22, LPV-200 service coverage degradation between 19:45 and 21:35 GMT and LPV service coverage degradation between 19:45 and 21:16 GMT were observed in Canada. Moreover, LPV-200 service coverage was affected in CONUS (Northeast) between 19:45 and 21:07 GMT on June 22. In addition, the increased GIVE values had a minor effect on the LPV service coverage in CONUS (Northeast) during 19:45–20:45 GMT and LPV-200 service coverage in Alaska during 20:16–20:26 GMT

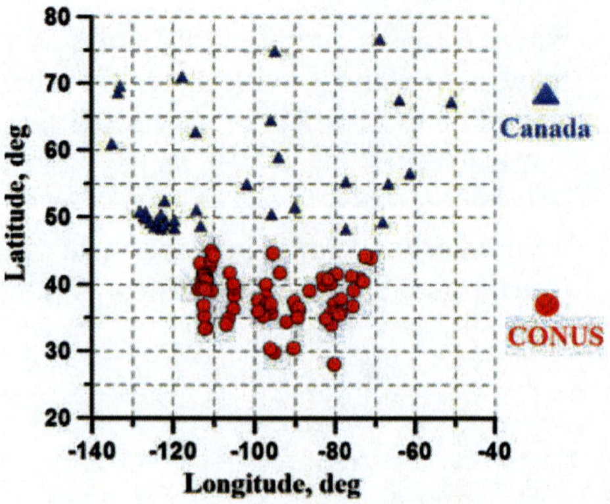

Fig. 5.19 Station locations in the Canadian area and in the CONUS (excluding California)

where x_0, y_0, z_0 and x_U, y_U, z_U are *i*-th satellite position and user position; *TEC* is TEC along the "*i*-th satellite-user" line-of-sight; $f_1 = 1575.25$ GHz and $f_2 = 1227.75$ MHz are the GPS frequencies; *C*1 is C/A-code pseudorange at L1, and *P*2 is P-code pseudorange at L2;

- compare the computed HPL and VPL values with the alarm levels of LP, LPV, and LPV200 services for each station and estimate the area covered (%) (5.13).

To model the ranging time series (5.14), we used the double-frequency P1 (or C1)–P2 "code-code" combination. This means that the current ΔR_{ion} value contains not only ionospheric errors, but also the residuals of all the frequency dependent ranging errors, such as multipath error and DCB.

We estimated the coverage for the LP, LPV, and LPV200 alert limits. Figure 5.20 shows the CONUS coverage time variation. The percent of the area covered did not decrease lower than 82% for all the protection limits both on Feb 14 (main phase) and on Feb 15 (recovery phase).

Figures 5.17 and 5.20 show that the positioning quality was poor within 15:00–23:00 UT on Feb 15 as compared with the same time on the previous day. Probably, the ionosphere variations were not responsible here. According to Table 5.9, there was a long PRN 21 outage 11:15 through 20:05 UT. The latter may be regarded as a probable source of positioning quality deterioration in the CONUS coverage zone before 20:05 UT.

There are two apparent periods of the CONUS coverage shrinking on 2011 Feb 15. The 14:30–18:00 UT interval approximately coincides with the WAAS alert (Fig. 5.17b). The positioning quality was worse 19:30 through 22.00, but the WAAS did not issue alerts for this period. According to the WAAS alerts, the poorest CONUS coverage should have been less than 70% at 17:40 UT for the LPV200 service

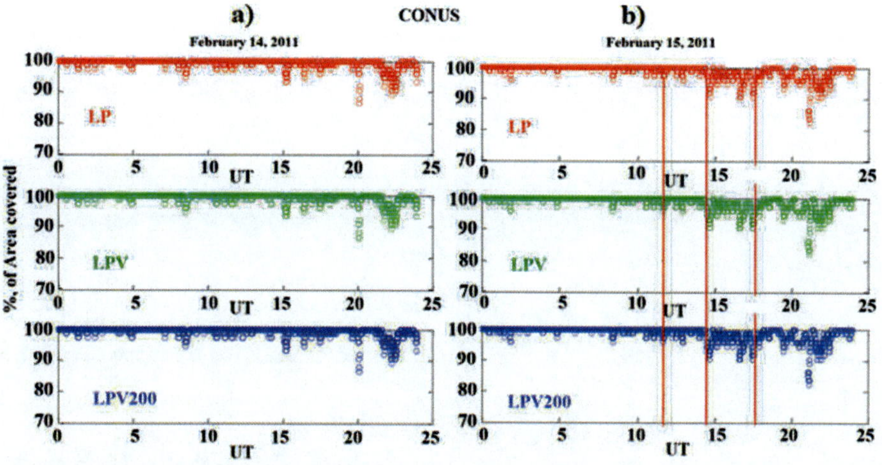

Fig. 5.20 CONUS area coverage on 2011 Feb 14 (panel **a**) and on 2011 Feb 15 (panel **b**). Vertical red lines mark the instants of WAAS alerts (see Fig. 5.17b)

(Fig. 5.15a). This value is much lower than the real percent of the CONUS coverage (90%) (Fig. 5.20b). We may regard such a mismatch as a false alarm event for WAAS services.

There are certain events of the positioning quality deterioration over 09:00–09:30 UT, at ~15:00 UT, and at ~20:00 UT. These events occurred on either day at the same time. Therefore, they probably were not related to the addressed ionospheric disturbances. Local features of satellite constellation geometry could be responsible. Such events can cause periodic regular local PDOP variations that recur in local areas (observation points).

In general, there was a significant mismatch between the observations and WAAS alerts. This is not surprising because we did not apply ionospheric corrections. Additionally, the network of GNSS station, whose data were used for the WAAS alerts, and the network that we used for our analysis differ in the number of stations and their configuration.

Figure 5.21 shows the time variations in the Canada coverage. It was not lower than 80% for LP, LPV, and LPV200 on both Jun 20 (background day) and Jun 22 (geomagnetic storm initial and main phases).

In contrast to the previous case, on 2015 Jun 22, the WAAS issued an alert to the coverage area shrinking within 19:45–21:35 UT (Table 5.10). According to Fig. 5.21, the 2015 Jun 22 positioning quality was poor, as compared with the previous quiet day, especially over 17:30–22:00 UT on Jun 22 (Fig. 5.21b). This period partially coincides with the WAAS alert (Table 5.10), but the real positioning data provide 85% as the lowest percent of the area covered. That is by 20% higher as compared with the WAAS alert figure (Fig. 5.18b).

There are certain cases of the positioning quality deterioration at 06:00 UT, 14:00 UT, 21:30 UT, and at 23.30 UT for both 2015 Jun 20 and 22. Inasmuch as these

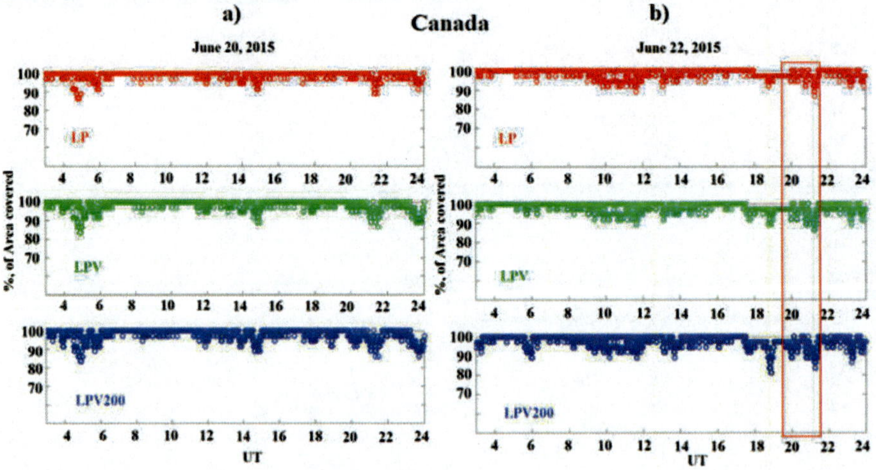

Fig. 5.21 Time variation in the Canada coverage area. The red rectangle marks the period corresponding to the WAAS alerts (see Fig. 5.18b)

events occur on either day and at the same time, they may probably be explained by regular variations in the local PDOP, or local multipath, like in the previous case.

There is a period of significant shrinking of the real percent of the area covered within 05:00–05:30 UT on 2015 Jun 20 (Fig. 5.21a). It was a quiet day, so this event was not caused by the geomagnetic storm impact. However, we should note that the most part of Canada is within the auroral oval area even on geomagnetic quiet days. Such an environment may bring ionospheric disturbances and GNSS signal quality deterioration even under geomagnetic quiet conditions.

Generally, both real positioning data and the WAAS alert figure display obvious shrinking of the percent of area covered in Canada on 2015 Jun 22, within 09:00–22.00 UT (i.e. under geomagnetic storm conditions). This period mostly corresponds to the main phase of the storm. Although we did not use any ionospheric correction, the lowest percent of the area covered from the real data is by 20% higher than that predicted by the WAAS alert figure. This may be considered as probable mismatch information of WAAS integrity service.

Figure 5.22 displays the percent of CONUS area covered versus the HPL (VPL) UT-based model computation. The time resolution of HPL and VPL computation was 5 min. This period corresponds to the WAAS GIVE computation.

The comparison between Fig. 5.22a and the Feb 14 WAAS alert figure (Fig. 5.17a) displays general accordance, especially for the LPV200 service. According to the modeling results, the lowest percent of area covered was about 93% for the LPV200 at 22:00–22:30 UT. This value was expected to be lower as compared with the correspondent WAAS LPV200 alert figure on 2011 Feb 14, because we did not use any ionospheric correction for the real data.

The HPL (VPL) modeling computation displays lower percent of the CONUS area covered on 2011 Feb 15 as compared with the previous day. The events of the lowest percent of area covered occurs within 14:00–18:00 UT (Fig. 5.22b), which

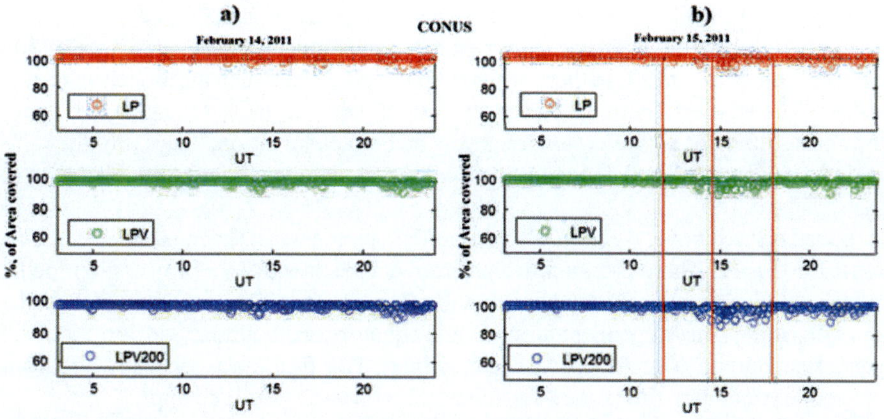

Fig. 5.22 Modeling results: percent of CONUS area covered versus time on 2011 Feb 14 (left panel) and on Feb 15 (right panel). Red lines mark approximate time points corresponding to the WAAS alerts (see Fig. 5.17a)

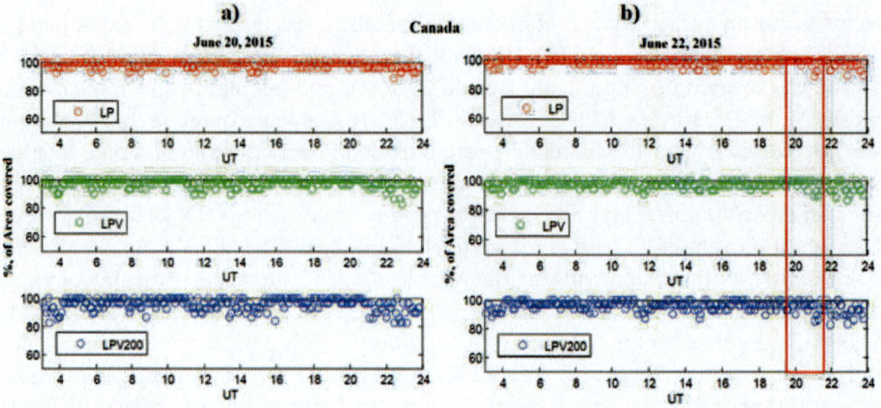

Fig. 5.23 Modeling results: percent of the Canada area covered vs UT on 2015 Jun 20 (panel **a**) and on Jun 22 (panel **b**). The red rectangle marks the period that corresponds to the WAAS alerts (see Fig. 5.18b)

generally coincides with the WAAS alert figure (Fig. 5.17b). Nevertheless, the lowest percent of the area covered was 85% at about 15:00 UT for the model computation, which is by 8–10% less as compared with the WAAS alert figure for the same time. On the other hand, the LPV200 modeling computation yields about 95% of the area covered at 17:30 UT, which is by 30% higher as compared with the WAAS alert for the same time.

Figure 5.23 displays the percent of Canada area covered versus UT based the HPL (VPL) model computation. In contrast to the previous case, the results in Fig. 5.18 do not display an obvious correspondence to either the WAAS alert figure (Fig. 5.18) or to the real positioning quality (Fig. 5.21). Generally, we can see random character of the model LP (LPV) values, but the percent of the area covered decreases sharply at 04:00 UT, 12:00 UT, and 23:00 UT on both 2015 Jun 20 and Jun 22.

There is no obvious decrease in the percent of the Canada area covered over 19:30–21:30 UT on 2015 Jun 22, so there is no correspondence between the modeling results and the WAAS alert figure. However, the lowest value of the % of coverage was not lower than 80% for all the protection levels on both observational days. For the LPV 200, the protection level is by 15–20% higher as compared with the WAAS alert figure (Fig. 5.18b).

Based on the above, we should note a random character of the modeling results on 2015 Jun 20–22. There is a significant difference between radio propagation media in the ionosphere above the CONUS and the Canada areas. Figure 5.24 a displays the position of the time-dependent auroral oval equatorward boundary on Jun 20 (solid green line) and on 2015 Jun 22 (dashed red line). This figure was reconstructed from the geophysical data (https://iswa.gsfc.nasa.gov/IswaSystemWebApp/).

According to Fig. 5.24, the bulk of Canada is within the auroral oval on both geomagnetic quiet and, especially, stormy days. That is why we think that most of the GPS satellite line-of-sights pass through high-latitude ionosphere that always

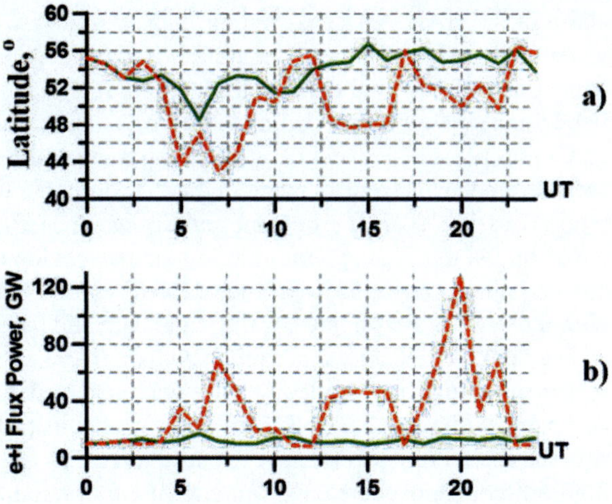

Fig. 5.24 Positioning of the auroral oval equatorward boundary on 2015 Jun 20 (solid green line) and on Jun 22 (dashed red line). Panel b) demonstrates the time-dependent OVATION Prime energy flux (electrons + ions precipitation within aurora) on 2015 Jun 20 (solid green line) and on Jun 22 (dashed red line)

contains multiscale ionospheric irregularities in contrast to the mid-latitude ionosphere in CONUS. The main conclusion is that we may not use the same solution and the same dataset of WAAS reference stations (WRS) to reconstruct the ionospheric GIVE values in the same manner for both Canada and CONUS areas.

There is no obvious correlation between the modeled LP (LPV) figure (Fig. 5.23b), dynamics of the OVATION Prime energy flux of electrons and ions (Fig. 5.24b), and the dynamics for the auroral oval equatorward boundary. On the other hand, a coincidence in time can be noticed between the WAAS alert (Fig. 5.18b) and the peak of the OVATION Prime energy flux of electrons and ions at 20:00 UT (Fig. 5.24b). It displays a contradiction between the real positioning quality and the expected protection levels based on the GIVE values.

Although the peak of the OVATION Prime energy flux occurred at 20:00 UT, the most part of the flux energy fell on the North Atlantic, Europe, and North of Russia at that time. Indeed, we can see that this energy flux did not drive the auroral oval equatorward boundary in North America (Fig. 5.24a). On the other hand, the first lower peak of the OVATION Prime energy flux at 07.00 UT (Fig. 5.24b) mostly fell on North America and caused a significant motion of the auroral oval boundary southward. But it did not bring noticeable features in both WAAS alert figure and real positioning quality results.

Our results display some cases of a significant mismatch between the WAAS alert figure and real positioning quality for both the CONUS zone and Canada. This mismatch was proved by the model computation of HPL (VPL) values based on the real ionospheric ranging errors obtained from the GPS-sites within the CONUS zone and Canada. The WAAS alerts provided the expected percent of area covered, which is much lower as compared with the real positioning data for both the GPS standalone mode and for the model HPL (VPL) computation in most addressed cases.

The WAAS integrity service provides a general response to the geomagnetic storm (Fig. 5.18b). Some particular cases, however, can be addressed as probable false alarm of the HAL (VAL) exceeding. Indeed, the WAAS alert figure provides much lower expected positioning availability under the geomagnetic storm conditions as compared with the real data (both positioning and modeling) at 17:40 UT on 2011 Feb 15 and at 19:45–21:35 UT on 2015 Jun 22. In the former case, the real positioning quality (as per the LPV and LPV200 requirements) is higher than 90% of the CONUS area covered. This value is much higher as compared with the correspondent WAAS report (less than 70%) for the same conditions. The HPL and VPL model computation based on the real ionospheric ranging errors provide the lowest percent of the area covered, that is about 93% for the CONUS zone. This is much higher as compared with the WAAS alert figure for the same date and time.

We should emphasize that a much better agreement among the WAAS alert figure, real positioning, and modeling data was achieved for the 2011 Feb 15 event as compared with that of 2015 Jun 22. This, probably, proves that the WAAS algorithm responds better to system outage events than to a geomagnetic storm, especially in the high-latitude ionosphere environment. In our opinion, such a disagreement between the WAAS alert figure and our computation results may be caused by multiscale ionospheric irregularities generated during a geomagnetic storm and can disrupt the correct GIVE values computations.

Based on the long time previous results, geomagnetic storms were accepted to cause an increase in the intensity of the ionospheric TEC all-scale disturbances within the Globe (Kinter et al. 2001; Afraimovich and Perevalova 2006; Lekshmi et al. 2011). This idea was reflected in the WAAS ionospheric standard model that underlies the GIVE value computations. Such an approach, however, may not consider the real multiscale ionospheric disturbances. These disturbances mostly affect GNSS performance in local areas associated with the travelling pattern of the ionospheric irregular structure not only at high- and low-, but also at mid-latitudes.

Although the mid-latitude ionosphere usually does not contain the sources to generate significant multiscale ionospheric disturbances, both the equatorial and the polar areas can transport ones into the mid-latitudes, especially under magnetic storm conditions (Afraimovich et al. 1999, 2009b; Ma and Maruyama 2006). The low-latitude ionosphere contains specific ionospheric disturbances referred to as "ionospheric superbubbles." These bubbles can penetrate into the mid-latitudes up to 35–37° N under geomagnetic storm conditions (Ma and Maruyama 2006). On the other hand, the polar ionosphere can be a source of strong large-scale travelling ionospheric disturbances generated within the pulsing boundary of the auroral oval area and propagate toward the mid-latitudes as well (Afraimovich et al. 2009b).

All the above, probably, evidences that it is necessary to revise the main WAAS building concept with allowance for the latitudinal features of the ionosphere. The main concept of the WAAS integrity monitoring is that the confidence bounds of the expected ranging errors should be sufficiently large to describe the error in the correction, but adequately small to allow the operation to proceed. If one cannot distinguish between nominal and disturbed conditions, then we should always assume disturbed conditions to be present. As we could see earlier, the predictions based on such a concept may generate false alerts.

In the case the main concept of the WAAS integrity monitoring maintains the same, there are two main ideas to solve the above contradiction. We can find a solution to detect ionospheric irregularities so that we can provide a high level of service during undisturbed time periods (Demyanov et al. 2019). A reduced level of the expected service would only be necessary during periods of detected ionospheric disturbances. To implement this solution in practice, the detection scheme for the ionospheric disturbances should be extremely robust to provide the necessary protection level. The integrity requirements for precision approach guidance set a probability of hazardously misleading information below 10^{-7} per approach (Rife et al. 2004). Therefore, the chance of an undetected ionospheric irregularity should be at a similarly small level. It is unlikely to achieve such a high probability of undetected multiscale ionospheric disturbances even at mid-latitudes, let alone the high- and low-latitude areas.

WAAS can also be updated to benefit from new signals of all global positioning systems (GLONASS, BeiDou, GALILEO, QZSS etc.) and GEOs. Indeed, by only upgrading the WAAS Reference Station (WRS) receivers to measure Galileo signals, we can double our sampling of the ionosphere. The increased sampling transforms into smaller broadcast confidence values on the single frequency ionospheric corrections. These lower values lead to higher availability. Similarly, the L5 signal has better noise properties and can be acquired at a lower elevation. This leads to less uncertainty in the ionospheric measurements and to smaller confidence bounds as well. In reality, however, serious unpredicted obstacles were unfolded in common multi-system signal data processing. GPS TEC measurements for the GIVE computation, although highly precise, are often rendered inaccurately due to satellite and receiver DCBs. The computed satellite DCB values are now available from a variety of sources, but receiver DCBs generally remain an undertaking of the receiver operators and processing centers (Themens et al. 2013).

Another serious problem here is a significant difference in ranging errors accuracy from all current positioning system and other ranging data sources. For example, a consistent analysis of signal-in-space ranging errors (SISREs) was conducted for all current GNSSs, considering both global average values and worst-user-location statistics. The global average SISRE values for the individual constellations equal 0.7 ± 0.02 m (GPS), 1.5 ± 0.1 m (BeiDou), 1.6 ± 0.3 m (Galileo), 1.9 ± 0.1 m (GLONASS), and 0.6 ± 0.2 m (QZSS) over a 12-month period in 2013–2014 (Montenbruck et al. 2015).

Besides, some anomalies in particular ranging are also possible. As an example of this, GPS satellite oscillator anomalies mimicking ionospheric phase scintillation can be noted (Benton and Mitchell 2012, 2014). Such ranging anomalies can be expected from an anomaly in the satellite oscillators, which is a persistent phenomenon. They have a potential to generate false alarms in the systems to forewarn of GPS outages due to the ionospheric scintillations.

In our opinion, in order to achieve a high confidence of WAAS alerts, real ionospheric measurements within the WAAS coverage zone should be involved instead of the WAAS GIVE values. A better way to implement this idea is to combine the WAAS and local area differential solutions. There are two groups of errors affecting the final UDRE and GIVE values: (1) the global ranging errors that do not depend on the user location within the WAAS coverage zone, and (2) the locally dependent ranging correction errors that are strictly dependent on the user position within the WAAS coverage zone. The first group of the errors contains the satellite clock and ephemeris correction errors, as well as DCB, and the other group includes tropospheric, ionospheric, and multipath errors. WAAS can provide high confident corrections and integrity monitoring for the first group of ranging errors within the extended coverage zone under any geophysical environment. On the other hand, locally dependent ranging errors are better to obtain from the local GNSS-networks (CORS, RTK-networks; Local Area DGPS, etc.). Combining "global" ranging correction with locally dependent ranging corrections, a user can use both WAAS and Local network ranging data statistics to compute the correspondent HPL (VPL) with higher confidence.

5.3 Effects of Ionospheric Irregularities on GNSS

5.3.1 Ionospheric Scintillation and Ranging Quality

A code pseudo-delay loop may function in the coherent or incoherent measurement mode. In the former mode, the pseudo-delay estimate RMS is substantially lower, as compared with that in the latter, but it requires simultaneous robust lock of the carrier phase. At the same time, a more robust code pseudo-delay tracking in the incoherent mode is ensured at lower CNR, because no phase lock is required. For example, from (Kaplan 1996), we can compute that, at a PLL noiseband $\Delta F_{PLL} =$ 2 Hz, at an integration time $T = 20$ ms, and a user receiver jerks $dR^3/dt^3 = 0.01$ m^3/s^3, the minimal CNR at a PLL input is 18.5 dB. At the same time, at a DLL noiseband $\Delta F_{DLL} = 0.8$ Hz, and at the same integration time $T_{COR} = 20$ ms, the signal minimal level is 16.5 dB. At the worst (by accuracy characteristics) tuning conditions and loop operation, the corresponding values are 28 dB and ~23 dB, for a PLL and DLL, respectively.

Thus, the effect of signal amplitude ionospheric scintillations may lead to a loss of carrier-phase lock a DLL will enter the incoherent tracking mode with a measurement accuracy deterioration. Let us estimate the pseudo-range measurement RMS in the code pseudo-delay incoherent tracking mode. For this purpose, we use Formula (2.35) and the plot for the dependence of the GPS minimal L1 C/A signal strength on the satellite elevation (Fig. 5.25). The plot is built with allowance for the reception by a 3-dB gain linearly polarized antenna (Kaplan 1996).

As a noise component of fast variations in the signal level, we use a standard procedure for generating random numbers with a normal distribution law in the MathCad environment. The conversion of the received signal at an antenna input into a DLL input CNR is performed as follows (Kaplan 1996):

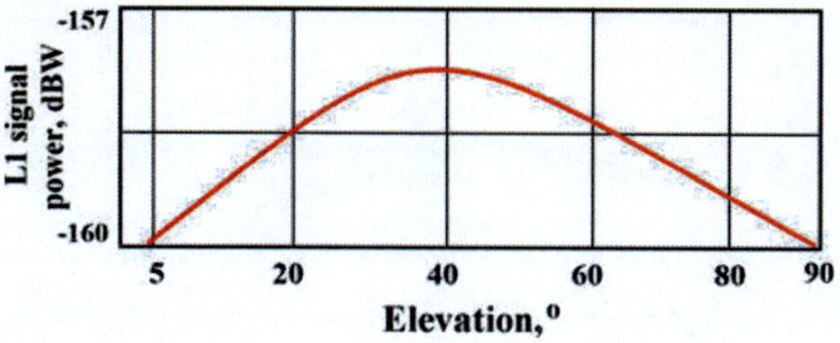

Fig. 5.25 GPS satellite L1 signal minimal received strength depending on elevation. Re-produced and modified figure based on figure from the published book (Kaplan 1996)

$$CNR = P_{rec} + G_A - N_T - N_f - L \tag{5.15}$$

where P_{rec} is an satellite signal level in a receiving point (Fig. 5.23); G_A is a receiving antenna directivity factor; $N_T = 10\ \lg(kT_0)$ is a power spectral density of receiver thermal noise; N_f is a signal strength loss in a cable; L is signal strength losses at its conversion, including analog-to-digital conversion.

At computations, we accept that the antenna does not introduce attenuation toward signal reception ($G_A = 0$); cable losses $N_f = 4$ dB; signal conversion losses $L =$ 2 dB; power spectral density of thermal noise is determined at the normal condition temperature $T_0 = 290$°K and is −204 dB (Kaplan 1996). Disposing of the calculated DLL input CNR, we can compute this parameter (2.35) measurement RMS and the corresponding $S4$-index. Figure 5.26 presents the code pseudo-delay measurement error computation results at the emergence of signal strength scintillations in a receiving antenna.

In all the cases, we simulated a situation, when satellite signal strength scintillations emerge at the 50°–60° elevations. There, the signal level is close to the maximum and the real-antenna losses are minimal. As one can see, the amplitude scintillation depth affects directly the code pseudo-range measurement accuracy. At a signal dip to the minimal threshold of 16.5 dB (Fig. 5.26, right panel), the ranging measurement error attains 30 m at the $S4 \approx 0.8$.

To corroborate the computation conclusions, we conducted joint observations of signal ionospheric scintillations and the C/A L1 code range measurement errors. The investigation was performed based on analyzing the same set of measurements and for the same conditions that were used when analyzing the DOP behavior above (Fig. 5.14).

Along with the signal levels, we also computed and recorded the C/A L1 code range measurement error absolute values (ΔR_{SV}). The total scope of each observational series *S1*, *S2,* and ΔR_{SV} with a 30-s record period for the 19:00–24:00 UT interval at high latitudes was 98,800 measurements. At mid-latitudes, we processed 1,485,500 measurements, with 126,600 at the low elevations. The *S*1, *S2,* and ΔR_{SV} observation series were grouped and analyzed separately by the observed satellite elevations. We distinguish among the groups of satellites with low ($\gamma = 0°\ldots30°$), medium ($\gamma = 30°\ldots60°$), and high ($\gamma = 60°\ldots90°$) elevations.

Tables 5.11, 5.12, 5.13, 5.14, 5.15 and 5.16 present the first Sturges classes of the histograms for distribution of *S*1, *S2,* and ΔR_{SV} ($k_{\min}$), and the corresponding

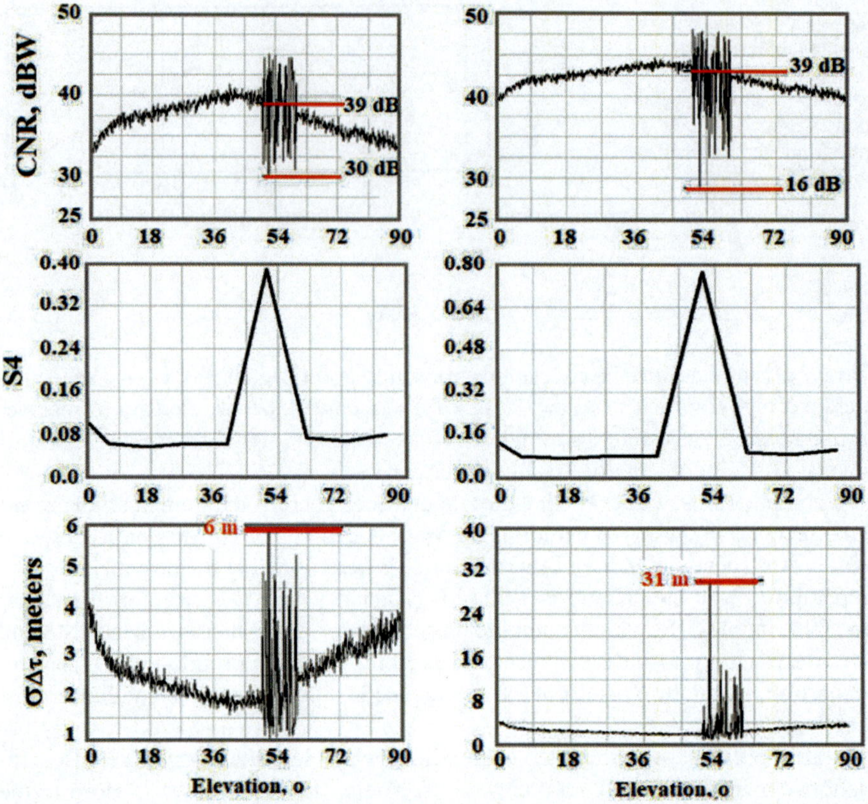

Fig. 5.26 Results for simulating the amplitude scintillations and the code pseudo-delay measurement error RMS (σΔτ) at incoherent tracking of a signal

probabilities for containment in the class ($P_{k,\,\min}$). The probability for the $S1$, $S2$, and ΔR_{SV} containment in the first class of the histograms for the distribution occurrences is dominating as compared with the remaining histogram classes. Therefore, we will analyze it in greater detail.

In the measurements relating to low elevations (Tables 5.11 and 5.12), we note an explicit growth in the probability for satellite signal containment in the minimal-value class with observational latitude increase. If, at low latitudes, the $S1$ and $S2$ values are contained in the $k_{\min}$ (from 0 to 2 to 50…65) class at the probability $P_{k,\,\min}$

Table 5.11 $k_{\min}|P_{k\ \min}$ at low elvations ($\gamma = 0\ldots30°$) under undisturbed conditions

	Low latitudes	Mid-latitudes	High latitudes
$S1_{\min}$, dB-Hz|rel.units	(2.75–65)|0.44	(0.017–53)|0.69	(0.3–63)|0.95
$S2_{\min}$, dB-Hz|rel.units	(2.00–50)|0.44	(0.013–41)|0.61	(0.2–49)|0.96
$\Delta R_{\min}$, m|rel.units	(0.1–19) |0.61	(0.15–16)|0.76	(0.1–4)|0.31

Table 5.12 k_{min}\|$P_{k\,min}$ at low elevations ($\gamma = 0\ldots30^{o}$) under magnetic storm

	Low latitudes	Mid-latitudes	High latitudes
$S1_{min}$, dB-Hz|rel.units	(3.00–69)|0.50	(0.24–53)|0.79	(0.3–63)|0.97
$S2_{min}$, dB-Hz|rel.units	(2.23–54)|0.50	(0.47–41)|0.72	(0.2–49)|0.97
ΔR_{min}, m|rel.units	(0.05–20)|0.18	(0.1–15)|0.51	(0.1–12)|0.91

Table 5.13 k_{min}\|$P_{k\,min}$ at medium elevations ($\gamma = 30\ldots60^{o}$) under undisturbed conditions

	Low latitudes	Mid-latitudes	High latitudes
$S1_{min}$, dB-Hz|rel.units	(3.31–65)|0.47	(0.170–50)|0.68	(3.00–65)|0.61
$S2_{min}$, dB-Hz|rel.units	(2.58–51)|0.47	(0.344–39)|0.59	(2.33–51)|0.63
ΔR_{min}, m|rel.units	(0.06–19)|0.67	(0.05–15)|0.70	(0.05–5)|0.45

Table 5.14 k_{min}\|$P_{k\,min}$ at medium elevations ($\gamma = 30\ldots60^{o}$) under magnetic storm

	Low latitudes	Mid-latitudes	High latitudes
$S1_{min}$, dB-Hz|rel.units	(1.31–73)|0.53	(0.30–53)|0.74	(5.0–66)|0.66
$S2_{min}$, dB-Hz|rel.units	(1.01–57)|0.53	(0.23–41)|0.65	(3.0–51)|0.69
ΔR_{min}, m|rel.units	(0.3–20)|0.18	(0.1–15)|0.49	(0.1–9.5)|0.85

Table 5.15 k_{min}\|$P_{k\,min}$ at high elevations ($\gamma = 60\ldots90^{o}$) under undisturbed conditions

	Low latitudes	Mid-latitudes	High latitudes
$S1_{min}$, dB-Hz|rel.units	(2.37–74)|0.54	(0.017–53)|0.69	(0.27–77)|0.4
$S2_{min}$, dB-Hz|rel.units	(1.84–57)|0.54	(0.013–41)|0.61	(0.21–60)|0.4
ΔR_{min}, m|rel.units	(0.11–21)|0.75	(0.02–16)|0.76	(0.1–6)|0.26

= 0.44…0.5, then, at mid-latitudes, $P_{k,\,min} = 0.61\ldots0.79$, and, at high latitudes, the $P_{k,\,min}$ attains 0.95…0.97.

With a growth in elevations, the satellite signal level, as one would expect, increases. This reflects in that, in the observational group with medium elevations, the probability for the emergence of $S1(S2) > 70$ is not less than 30–40%. In the observational group with high elevations, this probability is even more: 35 through

Table 5.16 k_{min}\|$P_{k\,min}$ at high elevations ($\gamma = 60\ldots90^{o}$) under magnetic storm

	Low latitudes	Mid-latitudes	High latitudes
$S1_{min}$, dB-Hz|rel.units	(0.92–77)|0.64	(0.16–55)|0.79	(1.4–78) |0.55
$S2_{min}$, dB-Hz|rel.units	(0.72–60)|0.64	(0.12–43)|0.65	(1.1–60)|0.55
ΔR_{min}, m|rel.units	(0.39–23)|0.30	(0.01–16)|0.51	(0.05–14)|0.90

65%. Herewith, in the observational group with ($\gamma = 30\ldots60°$) and ($\gamma = 60\ldots90°$), there is no such an obvious latitude $P_{k,\min}$ dependence on the observation latitude.

At the same time, the trend to reducing the ranging error with a growth in the elevation is also traced, albeit substantially weaker. Under the 2003 Oct 31 undisturbed conditions, the probability for the ranging error containment in the $k_{\min}$ = ~0.1…20 m class in the ($\gamma = 0\ldots30°$) observational group is 0.61 (low latitudes, Table 5.11) and 0.76 (mid-latitudes, Table 5.11). In the ($\gamma = 30\ldots60°$) observational group, the corresponding probabilities are 0.67 and 0.7 (Table 5.13), and, in the ($\gamma = 60\ldots90°$) group, already 0.75 and 0.76 (Table 5.15).

At high latitudes, as well as under magnetic storm conditions, we found no obvious ranging error dependence on an elevation. Most likely, this may be elucidated by that the final code delay measurement accuracy in the coherent tracking mode depends not only on the loop input CNR, but also on a PLL operation robustness. If, at a small signal level, the coherent tracking operation is still provided, then a sufficiently high accuracy of the code range measurement persists.

A comparative analysis for the data presented in Tables 5.11, 5.12, 5.13, 5.14, 5.15 and 5.16 for undisturbed and magnetic storm conditions shows the following. Irrespective of latitudes and elevations, one observes an explicit trend to a decrease in a signal level and to a growth in a ranging error during the magnetic storm main phase. The probability of the $S1$ and $S2$ containment in the $k_{\min}$ class increases under those conditions from 2 to 15%.

During a magnetic storm, one observes a growth in a ranging error. This manifests itself in a decrease in the probability for the minimal ranging error containment in the $k_{\min}$ class by 25–30% at mid-latitudes and by 5–10% at the high. Besides, under magnetic storm conditions, the emergence of significant (>100 m) ranging errors becomes probable even in the $\gamma = 60\ldots90°$ group of observations.

Separately, one should note the low-latitude region. Here, the probability $P_{k,\min}$ drops under magnetic storm conditions by factor of 2–3. The presence of that feature requires special investigations by engaging direct measurements of satellite signal amplitude and phase, because we may encounter different mechanisms for generating a ranging error at low and high latitudes. We can state such an assumption, in particular, based on addressing (Van Dierendonck and Hua 2001). According to the authors, the typical $S4$ in the geomagnetic equator surroundings is 0.3, whereas, at high latitudes, S4 rarely exceeds 0.2. Also, the amplitude and the carrier phase scintillation intensities at lower latitudes are about equal and are linear-correlated at $S4 < 0.5$. At overshooting that threshold, the amplitude scintillation intensity becomes more than the phase scintillation intensity (Van Dierendonck and Hua 2001). At the same time, the phase scintillation intensity in the high-latitude region is sufficiently low, and the phase scintillations linearly correlate with those of a signal amplitude.

In general, our analysis shows that the phenomenon of scatter and random refraction of satellite signals on small-scale electron density irregularities does occur during the magnetic storm main phase and may lead to significant errors in ranging measurements and decrease the positioning solution quality.

The obtained theoretical and experimental estimates for the measurement errors of code range and of the carrier phase agree with the known results from other authors and adequately elucidate the observation results. The code range measurement accuracy and the robustness of the carrier phase lock dramatically depend on the scintillation level (presented by the $S4$-index), particularly, at $S4 \geq 0.5$.

5.3.2 *Ionospheric Super-Bubble Effects on GPS Performance*

The geospace ionization plasma irregularities are known to be driven by the geomagnetic field. This may manifest itself like both in irregularity travels and in the features of their orientation in space. The latter circumstance may be essential in terms of amplifying or attenuating the effects of satellite signal amplitude and phase ionospheric scintillations.

When observing the same satellite from points 1 and 2 (Fig. 5.27), the signal, whose propagation direction features a wave vector k, crosses the geomagnetic field lines ($B_{0,1}$, $B_{0,2}$) at different angles $\xi_1 > \xi_2$. Because the electron density irregularities oriented along the magnetic field lines, then, in one case (angle ξ_1), the signal passes along all the irregularity, whereas, in the other case (angle ξ_2), it intersects the latter via the shortest track.

In terms of the signal propagation effects, the above difference under aspect conditions may manifest itself in the following:

(1) The electron density irregularities distribution region intersected by a signal increases or decreases, which entails a corresponding refraction gain in the group (phase) track and its refraction fluctuation amplitude increase.

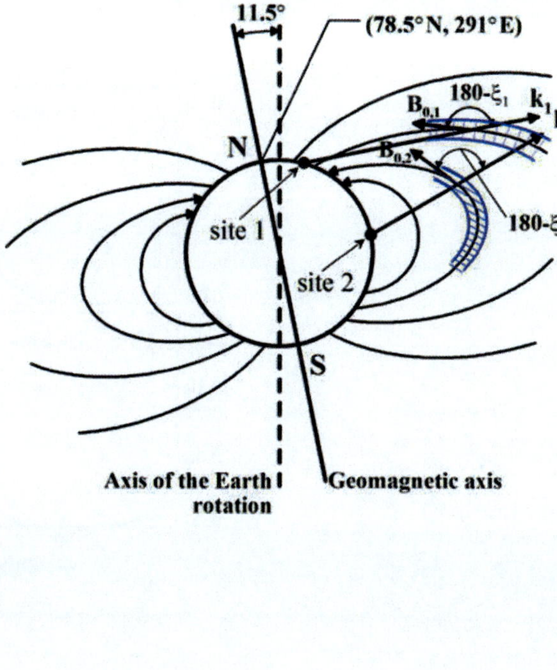

Fig. 5.27 Dependence of ionospheric scintillation effects on the orientation ξ between the wave vector (k_1, k_2) and geomagnetic field lines ($B_{0,1}$, $B_{0,2}$) (Demyanov et al. 2012b)

(2) At the boundaries of a large electron density irregularity (different-scale TIDs, as well as bubbles in the equatorial ionosphere), one observes significant electron density gradients. This is a favorable conditions for the emergence of small-scale electron density irregularities responsible for satellite signal amplitude and phase ionospheric scintillations. Accordingly, an increase in a signal passage path length through the region of small-scale irregularities within a larger irregularity also intensifies the ionospheric scintillation effect.

Anderson and Straus (2005) performed pioneering observations of the ionospheric scintillation at different signal propagation against the geomagnetic field line. They used "satellite-satellite" transionospheric paths, where GPS receivers were mounted onboard low-orbiting PICOSat, CHAMP, and SAC-C satellites with the orbit inclinations of 67°, 88°, and 98°, respectively.

All the three low-orbiting satellites showed that most of strong ionospheric scintillation events ($S4 > 0.25 \ldots 0.5$) occurred within $\pm 30 \ldots 35°$ relative to the geomagnetic equator (Fig. 5.28).

In Chap. 2, we indicated that the main source for signal ionospheric scintillations in the equatorial ionosphere are specific ionospheric irregularities—ionospheric plasma bubbles with a reduced (relative to the background) electron density values. The bubble size ranges within $0.1 \ldots 10^5$ m and they are concentrated, mainly, within $\pm 20°$ relative to the geomagnetic equator. However, there are the up-to-date data on the

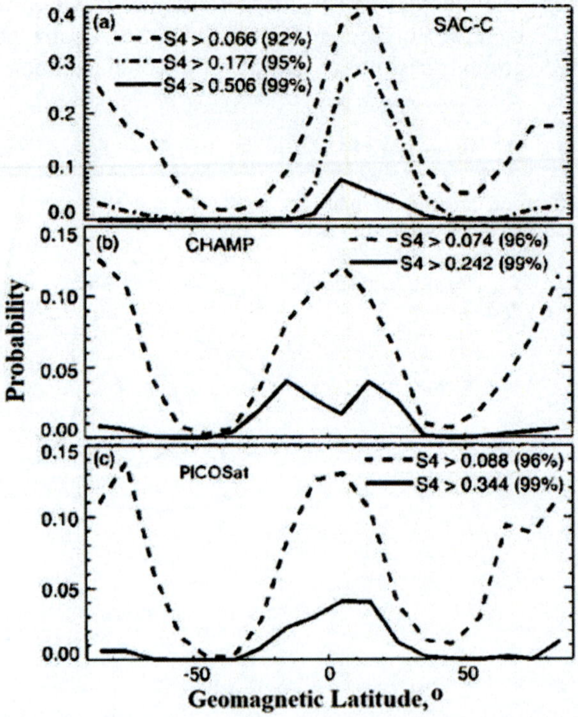

Fig. 5.28 Ionospheric scintillation intensity dependence on geomagnetic latitude (Anderson and Straus 2005)

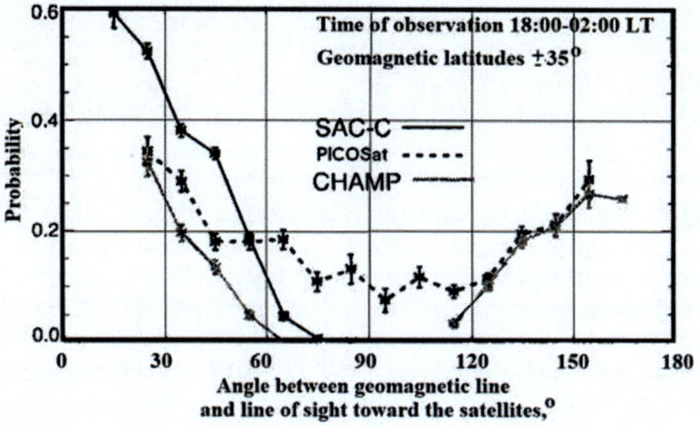

Fig. 5.29 Dependence of the GPS signal ionospheric scintillation on an angle ξ between a geomagnetic field line and a signal propagation path from observations at low-orbiting satellites (Anderson and Straus 2005)

penetration of above irregularities into the mid-latitudes to about 36°–40° N, where they also cause significant GPS signal ionospheric scintillations (Ma and Maruyama 2006).

Almost all the examined cases of the emergence of the UHF transionospheric signal parameter ionospheric scintillations (to 99%) that are caused by the equatorial bubble effect occur between 19:00 and 06:00 LT (Ma and Maruyama 2006).

In Fig. 5.29, one may note an apparent correlation between the number of $S4 > 0.1$ ionospheric scintillation observations and an angle between a magnetic field line and a satellite signal propagation path (angles ξ_1 and ξ_2, Fig. 5.27) (Anderson and Straus 2005). Apparently, the scintillation intensity considerably increases from observations at all the three low-orbiting satellites (SAC-S, PICOSat, and CHAMP) in the cases, when a GPS satellite signal propagates along a geomagnetic field line.

This is elucidated by the fact that bubbles are extended and tilted along geomagnetic field lines (Ma and Maruyama 2006). If a signal propagates perpendicular to a geomagnetic field line (angle ξ_2, Fig. 5.27), it intersects a bubble region through the shortest track, which manifests itself in a smaller level of scintillations. If a satellite signal propagates in parallel to geomagnetic field lines (angle ξ_1, Fig. 5.27), a signal path through the region saturated with small-scale electron density irregularities is longer, which manifests itself in an increase in the signal ionospheric scintillations.

Ma and Maruyama (2006) investigated the spatial–temporal dynamics for the electron density bubble evolution and travel from the equatorial into the mid-latitude ionosphere region (to ≈30°…34° N) under the conditions of the 2000 Feb 12 magnetic storm main phase. From the observations at a dense ground-based network of receivers (GEONET, Japan), the bubble was found to be extremely extended along the latitude (along a magnetic field line) and reached 36.5° N (31.5° geomagnetic latitude). During its evolution, the bubble moved eastward at about 50 m/s velocity.

Figure 5.30 illustrates this by presenting the TEC time series obtained through the measurements at a chain of the GEONET network stations. The 2000 Feb 12 measurements are designated by the line with dots. Here, for comparison, we show the averaged TEC series obtained at the same stations over Feb 9–11 (solid line). Figure 5.30 bottom panel exhibits the measurement results the plasmasphere ion concentration (at the ~860 km height) obtained by the facilities at the LEO satellite with a polar orbit—Defense Meteorological Satellite Program—DMSP-F14. For comparison, here, we provide the 2000 Feb 12 ion concentration measurements, and those for the previous undisturbed day (smooth line).

With the geomagnetic disturbance evolution, one observes a TEC growth starting with 08:00 UT and lasting to about 16:00 UT (17:00–01:00 LT) on 2000 Feb 12. The maximum is attained 10:30 through 15:00 UT (the period is marked by vertical dotted lines). Over the same period, there appear severe TEC dips relative to the general measurement trend.

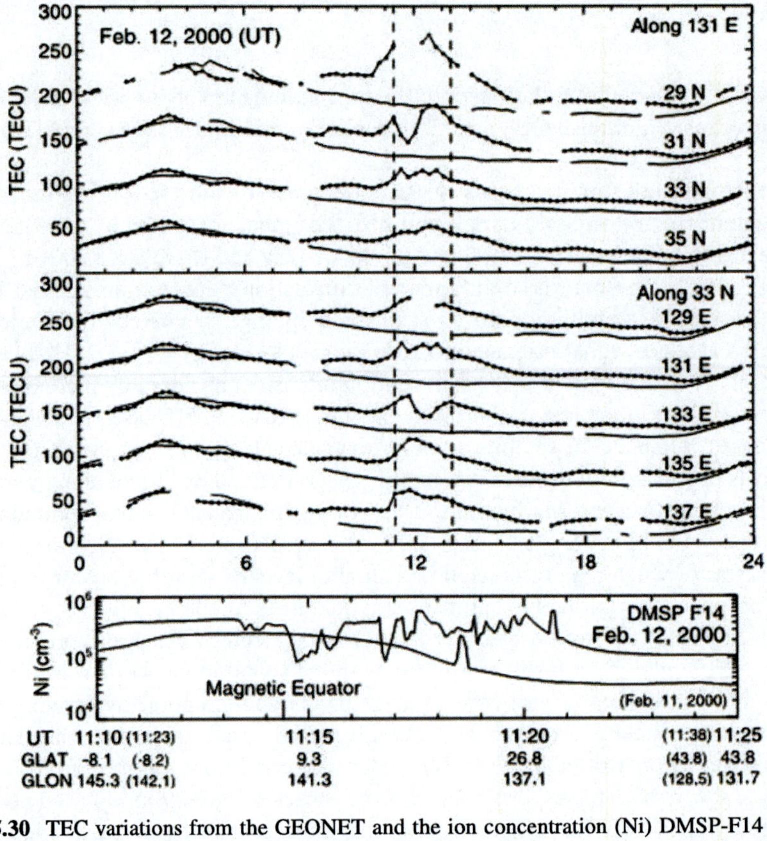

Fig. 5.30 TEC variations from the GEONET and the ion concentration (Ni) DMSP-F14 data in the magnetic equator region under the 2000 Feb 12 magnetic storm main phase conditions (Ma and Maruyama 2006)

The DMSP-F14 was over the investigated region 11:14 through 11:21 UT. By using its facilities, the ion concentration was measured within the area from the magnetic equator (8° N, 141° E) to low latitudes (29° N, 136° E). Figure 5.30 bottom panel presents the obtained measurements. The conclusion is that the TEC dips are related to the corresponding ion concentration dips that are, in turn, caused by the emergence of an equatorial bubble in the ionospheric plasma (Ma and Maruyama 2006).

In the same paper, noted was that GPS satellite signal scattering on the equatorial bubble ionospheric plasma irregularities may serve as a reason for loss of satellite signal lock in the navigation receiver. Recorded were LOLs for several satellites being within the TEC dip area over the measurement time interval (11:00 through 14:00 UT).

The above results provide a good possibility to further investigate the dependences of the radio navigation parameter measurement quality on the aspect conditions relative to the geomagnetic field lines. Let us accept them as initial, and also use the GEONET measurement data—presently, the GSI network (Tsuji and Yamaguchi, 2013)—at a 30 s temporal resolution.

We analyzed the relative density for the GPS L2 losses of phase lock. For this purpose, the separate-satellite measurements were parted split into groups, depending on the angles of geomagnetic field line intersections. Determining angles ξ was performed for each satellite in the point, where the line-of-sight intersects the 350 km height. To compute the geomagnetic field strength vector orientation in that point, we use the International Geomagnetic Reference Field (Finlay et al. 2010). The day and time were chosen the same as in (Ma and Maruyama 2006).

Right panel in Figure 5.31 presents the tracks of the GPS satellites selected for observation in the «azimuth—elevation» coordinates (α_S, γ_S). The same figure shows the location of the geomagnetic field lines, computed through the IGRF.

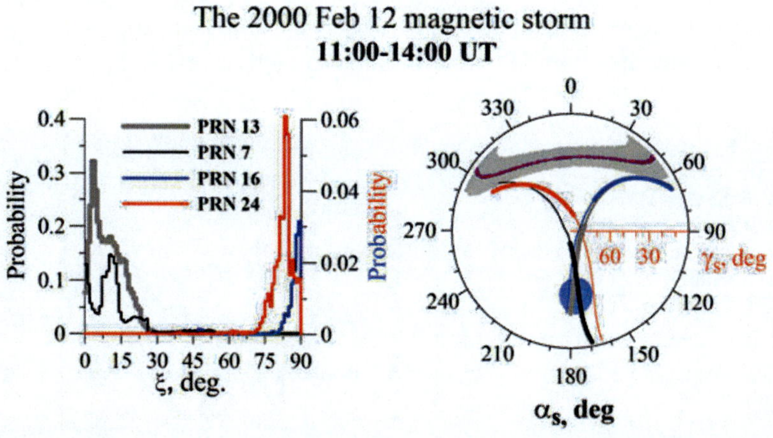

Fig. 5.31 Dependence of the density for L2 loss of phase lock on the observation aspect conditions (Demyanov et al. 2012b)

The first two satellites (PRN7, PRN13) emitted signals in the $\xi \le 30°$ angular sector (relative to the geomagnetic field line) for the most observation time. At the same time, the signal emission direction from the second two satellites (PRN16, PRN24) was, for the most observation time, within $65° \le \xi \le 90°$.

Figure 5.31 shows the dependence for the density of the carrier phase cycle slips on the angle ξ. The greatest slip number of the total measurements is observed for PRN13 (33%) and PRN7 (15%). It is those satellites, whose signal propagation direction coincided with the magnetic field line direction, which agrees well with the results in the papers mentioned above. At the same time, the slip number for the other pair of satellites did not exceed 7% (PRN24). One should note that, in the latter case, the number of cycle slips grew as the angle ξ approached 90°. There was no similar effect in (Anderson and Straus 2005).

To generalize the obtained results, we computed the total density of cycle slips for all the satellites observed within the GEONET over the 2000 Feb 12 00:00–24:00 UT time interval. Figure 5.32 presents the results. The red curve shows the occurrence probability for cycle slips depending on the angle ξ. The dark blue curve indicates the absolute full number of slips, while the thin black curve shows the full number of phase measurements for all the satellites within the measurement network. The above dependence of the slip intensity manifestation on the propagation aspect conditions relative to the geomagnetic field lines is completely corroborated.

Let us analyze the spatial dynamics for the phenomenon. For observations, we selected three instants: before the ionospheric disturbance onset (10:00 UT), during the disturbance maximum, when the electron density bubble evolution was observed (12:40 UT), and after the bubble event (13:45 UT). Figure 5.33 shows the ionospheric pierce points where cycle slips occurred. The color stands for the angle ξ between the line-of-sight and the geomagnetic field lines.

At the instants before observing the electron density bubble (10:00 UT) and after (13:45 UT), one observed separate non-numerous cycle slips not related definitely to the signal propagation aspect conditions. The slips were observed for the satellites with the $\xi \approx 35\ldots80°$ angular orientation. The situation dramatically changes during the bubble evolution (12:40 UT), when the number of the L2 cycle slips considerably

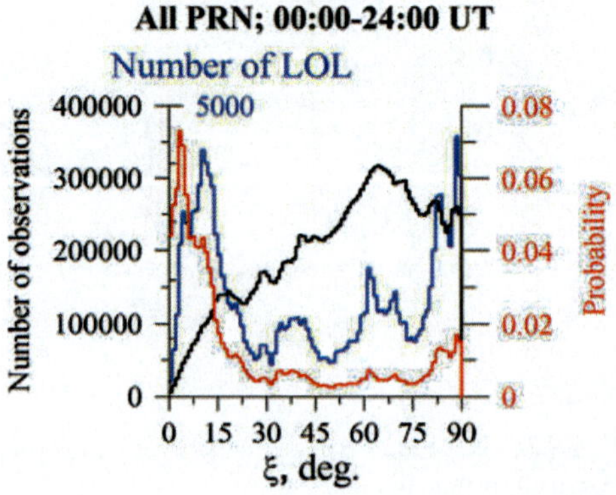

Fig. 5.32 Statistics for the density (red curve) and the number of the GPS L2 cycle slips (blue curve) depending on the observation aspect conditions within the GEONET (Demyanov et al. 2012b)

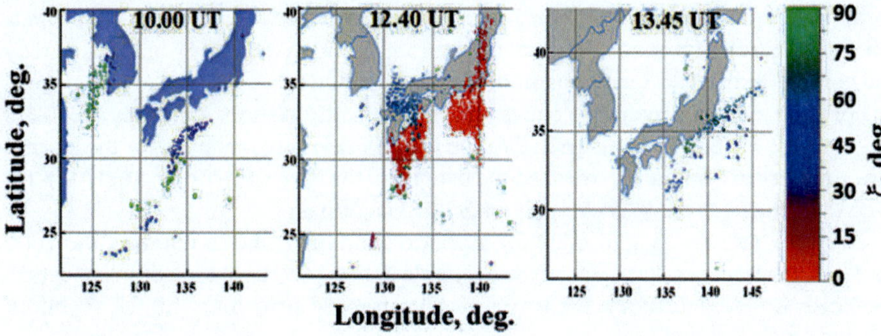

Fig. 5.33 Spatial distribution for the ionospheric pierce point positions of the GPS satellites, at which one observed the L2 cycle slips owing to the angle between the signal propagation path and the geomagnetic field line (the colors corresponding to the intersection angle are shown on the color scale) (Demyanov et al. 2012b)

increases for the satellites, the direction to which is oriented along a geomagnetic field line within angles $\xi \approx 0\ldots20°$. One can clearly see two localized regions, in which the main portion of slips is concentrated. Those fields correspond to the site, where the bubble addressed in (Ma and Maruyama 2006) evolved and traveled (Fig. 5.30).

Thus, using a dense network of ground stations enables to investigate in detail the spatial–temporal dynamics for the emergence of radionavigation parameter slips related to evolution of the characteristic electron density irregularities in the equatorial ionosphere. The obtained observations corroborate and supplement the results in (Ma and Maruyama 2006; Anderson and Straus 2005). The GPS signal propagation aspect conditions relative to the magnetic-oriented electron density irregularities in the ionosphere are shown to affect essentially the final quality of radionavigation parameter measurements.

The electron density emergence and evolution are mainly characteristic of only the equatorial ionosphere (within ±20° relative to the geomagnetic equator). However, it is shown that they are possible to penetrate considerably northward into the region of lower mid latitudes (to about 38–40° N). The latter circumstance should also be considered when designing the up-to-date WAAS.

5.3.3 Auroral Oval and GNSS Positioning Errors

At mid latitudes, the cases of the GNSS operation quality degradation have, preferentially, a local character. Normally, the regions of lower positioning accuracy at mid latitudes coincide with the ionosphere regions, within which there are favorable conditions for production of intense small-scale electron density irregularities. This relates, in particular, to the auroral oval equatorward boundary, where one observes high TEC gradients (Themens et al. 2013). There is a necessity for a detailed experimental research of a spatial–temporal pattern presenting variations in the accuracy and continuity of GNSS user positioning at mid latitudes.

For such an experiment, we chose two magnetically disturbed periods (2000 Jul 15–16 and 2001 Sep 25–26). To determine the position and dynamics for the auroral oval equatorward boundary, we used the data from the Space Environmental Monitor (SEM) mounted at the NOAA and POES LEO satellites.

To independently diagnose the ionospheric conditions and to interpret the 2000 Jul 15 experiment results, we used the available data of the HF radar measurements in the North America territory: the Haystack Observatory, Millstone Hill, USA (http://haystack.mit.edu/madrigal).

Figure 5.34 shows the configuration for the station locations and the dynamic for the auroral oval equatorward boundary in the North America territory. The rhomb marks the HF radar at Millstone Hill. The shaded ellipse denotes the HF radar signal back-scatter volume horizontal projection. Black circles show the GPS-station locations, where carrier-to-noise ratios (S1 and S2) were measured (lower-case letters indicate their names). Rectangles mark the GPS-stations, for which the positioning error (capital letters indicate the station names). Triangles present the position of ionospheric pierce points at instants t_{max}, when one observed the maximal density for the reference $P_{max}(t)$ omission. Near, provided are the station names, the satellite numbers, and the t_{max} values. Solid and dotted polygonal lines mark the auroral oval equatorward boundary position for the instants next to the lines. The measurements obtained at the network of those stations have a 30 s temporal resolution and are at the SOPAC resource (Sect. 4.2.1).

Figure 5.35 a shows the dynamics for the 2000 Jul 15–16 Dst index that features the variation maximal amplitude of up to −295 nT. A dotted vertical line marks that magnetic storm SSC instant according to (http://ftp.ngdc.noaa.gov/stp/solardata/suddencommencements/).

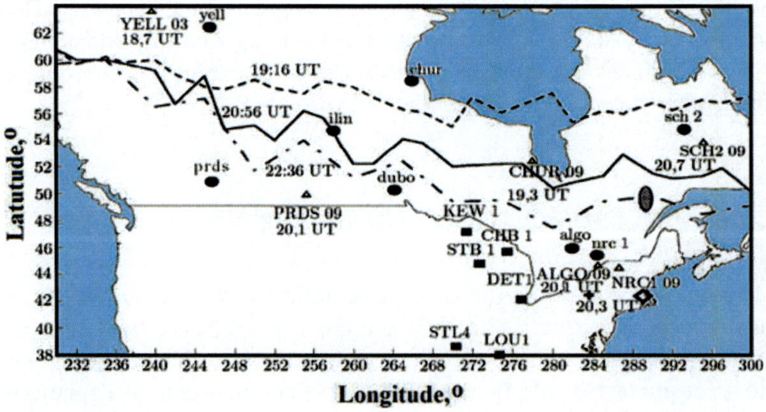

Fig. 5.34 Experiment geometry in the North America during the 2000 Jul 15 magnetic storm (Afraimovich et al. 2004)

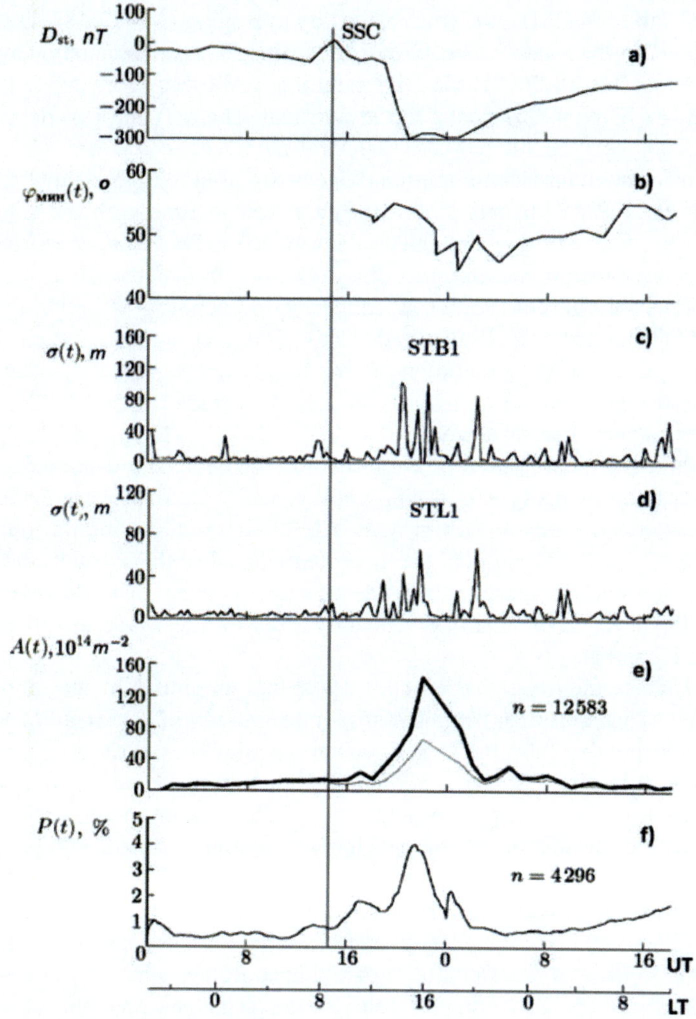

Fig. 5.35 **a** Dynamics for the *Dst*-index during the 2000 Jul 15 magnetic storm; **b** variations in the auroral oval equatorward boundary position $\varphi_{min}(t)$; **c, d** observation series of spatial positioning error RMS at the STB1 and STL1 GPS stations; **e** TEC variations ($A(t)$): thin line for the 2–10 min variation periods, thick line—for the 20–60 min ones; **f** time dependence of the L2 signal cycle slip relative density for the visible satellites (Afraimovich et al. 2004)

The oval boundary $\varphi_{min}(t)$ for 288° E was determined by the 0.1 erg/sm^{2}s particle flux power level accurate within 1°. After 2000 Jul 15 20:00 UT, the auroral oval extends rapidly southward and reaches ~43° N at 0:00 UT on Jul 16. The oval equatorward boundary persists at 44–46° N to about Jul 16 04:00 UT, and then, owing to the auroral activity decrease, recedes northward. A recurring extension of the oval equatorward boundary is noted on late Jul 16. We obtained the series of data on the TEC variation RMS $A(t)$ (Fig. 5.35e) by averaging over 12,583 TEC series. The estimates for the relative density of cycles slips P(t) are obtained by averaging the observations for all the visible GPS satellites and types of receivers (the total of observations being 4296) during the experiment. The $A(t)$ and $P(t)$ dependences were built for 30–50° N and 200–300° E (Afraimovich et al. 2009b).

Apparently, the dynamics for the relative density of cycle slips (panel f) and the positioning error RMS (panels c and d) agree well in time with the $\varphi_{min}(t)$ and $A(t)$ variations. During the geomagnetic disturbance main phase, one observes an essential decrease in the positioning accuracy. This occurs over the same time interval as the maximal shift of the auroral oval equatorward boundary, as well as during the maximal TEC fluctuations (19:00–00:00 UT). This fact evidences again in favor of the assumption that the generation of small-scale electron density irregularities at the auroral region boundary may cause signal degradation and GPS operation deterioration as a navigation system.

The high positive correlation of the growth in the cycle slip density $P(t)$ and in the TEC variation intensity $A(t)$ during geomagnetic disturbances indicates that a possible reason for the cycle slip growth is GPS signal scattering on ionospheric irregularities. Also, the duration of the time intervals, when the signal level becomes below the noise level, increases, which causes a loss of signal lock. However, for the L1 signal, the duration of such intervals is by order of magnitude less, than that of the weaker L2 signal.

Let us address the dependence of a GPS signal amplitude at the fundamental frequency $S1(t)$ and the integral number of sample omissions $D(t)$ for radionavigation parameters during the 2000 Jul 15 geomagnetic disturbance at the stations marked by bullet points in Fig. 5.34. By a radionavigation parameter sample omission, we understand a full absence of code delay and phase measurements at both operational frequencies during the current epoch of observations. Figure 5.36 presents the dynamics of $S1(t)$ and $D(t)$ at the indicated stations. The left scale shows the signal level in relative units, and the right scale exhibits the relative density for sample omissions. The dots mark the $S1(t)$ dynamics, while the vertical lines indicate the $D(t)$ dynamics. Panels a–f present the geographic coordinates for the stations and the GPS satellite numbers. The temporal trend for the $S1(t)$ measurements corresponds to the elevation.

19:00 through 21:00 UT, one observed a higher number of omissions at the above stations. Note the instants $t_{max,}$ when the maxima of the omission density $D_{max}(t)$ were recorded at separate GPS stations. For the same instants, we mark the positions of the GPS satellite ionospheric pierce points and their numbers in Fig. 5.34 (triangles).

As per the figure, the ionospheric pierce points of the satellites, for which sample omissions were recorded, are located near the auroral oval equatorward boundary and are close to instants t_{max}. For the latitude-separated stations, the t_{max} instants are different. At the YELL (PRN03) high-latitude station, the omissions started between 19:10 and 19:30 UT. At the ALGO, NRC1, PRDS, CHUR, SCH2 (PRN 09) mid-latitude stations, the sample omission maximum was observed within 20:00 and 21:00 UT. Such a time difference may be elucidated by the motion of the auroral oval equatorward boundary north-to-south over the 19:00–21:00 UT interval.

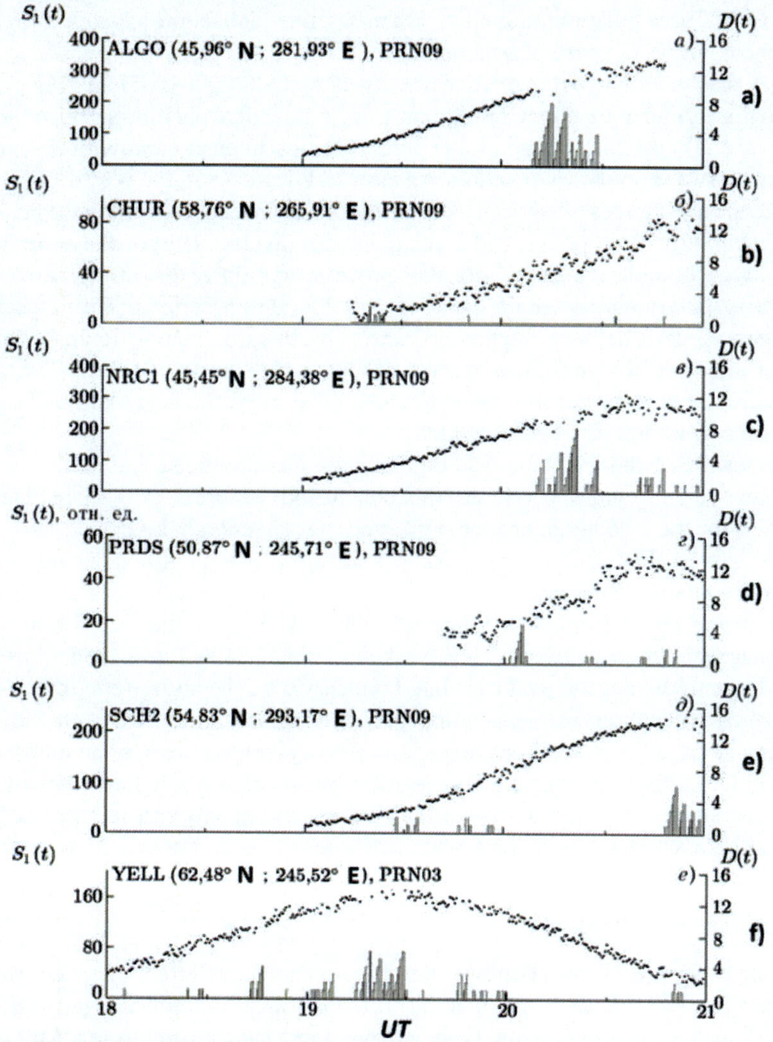

Fig. 5.36 Dependence between a GPS signal amplitude at the first frequency ($S_1(t)$) and the distribution of the omissions ($D(t)$, vertical lines) during the 2000 Jul 15 magnetic storm. Panels indicate the names and geographic coordinates of the stations and the numbers of satellites, for which those observations were recorded (Afraimovich et al. 2004)

Note that the omissions were observed, in most cases, at a maximal or sufficiently high satellite signal level, when the elevations were large. Thus measurement omissions might be caused by an essential wave propagation medium change that led to a sharp deterioration of the satellite signal reception conditions. This evidences in favor of the implication that ionospheric scintillations may lead to a decrease in the GNSS user positioning quality, when satellite signals propagate through the ionosphere within the auroral oval boundary.

Thus, the dynamics for the relative density of the GPS L2 loss of lock (Fig. 5.36f), the integral number of count omissions (Fig. 5.36), and positioning error RMSs (Fig. 5.35c, d) correlate with both the TEC variation intensity and with the auroral oval equatorward boundary position. An essential decrease in the positioning accuracy and the most losses of signal lock at mid-latitudes is recorded during the geomagnetic disturbance main phase, and is related to the displacement of the auroral oval equatorward boundary. Along with that process, one observes an increase in the TEC variation amplitude within the 1–10 and 20–60 min periods. With a decrease in the geomagnetic activity and the oval retreat northward, the positioning error, the number of losses of signal lock, and the TEC variation amplitude drop. Thus, they are the processes at the auroral region boundary that cause the degradation of signals and the deterioration of GPS navigation.

The results of the 2000 Jul 15 simultaneous measurements the Millstone Hill incoherent scatter radar corroborate this conclusion. Figure 5.37 demonstrates the dynamics for the incoherent scatter signal and the reflection height.

One can see that, within 20:00–00:00 UT, the Millstone Hill radar observed a powerful signal of back-scattering (coherent echo) from electron density irregularities in the upper F-region of the ionosphere. The horizontal projection of the radar-diagnosed ionosphere region is marked in Fig. 5.34 by the shaded ellipse and coincides with the auroral oval traveling boundary over the above time interval.

A coherent echo originates at anomalously powerful radiowave scatter in the meter and decameter range at small-scale electron density irregularities in the ionosphere. On 2000 Jul 15, the scattering areas were near the auroral oval equatorward boundary. The observed HF signal variations occur almost synchronously with an increase in the TEC variation intensity, with the growth in the density of radionavigation parameter measurement failures, and with an augmentation in the positioning error.

Summary of this Chapter

1. Our experimental observations corroborate a negative effect of space weather factors on GNSS operation in the stand-alone and differential navigation modes. The main causes of that are: GNSS signal direct suppression from solar L-band

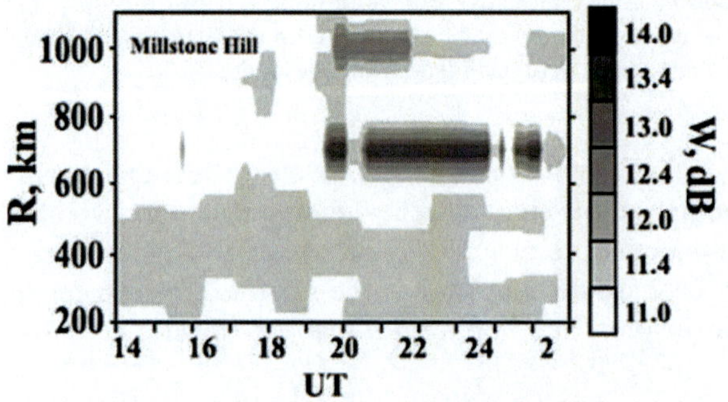

Fig. 5.37 Observations of the 2000 Jul 15–16 incoherent echo at the Millstone Hill back scatter radar (part of the figure from Afraimovich et al. 2004)

radio bursts with a >10^3 s.f.u flux density, and signal ionospheric scintillations in the ionosphere regions containing higher values for the electron density gradients. Both may cause an increase in ranging errors or losses of lock for a part of visible satellites (including those at large elevations). The latter serves as a cause for a sharp GDOP and for a proportional increase in the positioning error.
2. The effect of solar radio bursts leads to an increase in the relative density of radionavigation parameter measurement failures and to an increase in the positioning error in the navigation receivers located on the sunlit surface of the Earth. The effect extent on each navigation receiver depends on the Sun zenith angle and on the elevation of each satellite. When the zenith angle and the satellite elevation increase, the negative effect grows both for GNSS users in a standalone navigation mode and for WAAS users.
3. The robustness of GPS satellite signal tracking under the solar radio bursts essentially depends on the operational frequency, modulation, ranging code characteristics, as well as on signal processing algorithms of a receiver. Semi-codeless tracking of the GPS L2 signals may already be interrupted at a solar radio noise power ~4000 s.f.u. A receiver may lose the lock of a civilian access code at the GPS L1 at the >10,000 s.f.u solar radio-flux. For GLONASS signals, the corresponding values are 10,000 and 13,500 s.f.u, respectively, when processing L1 and L2 signals. The GLONASS higher noise immunity is due to a narrower band for the receiver RF front-end.
4. Increasing a GNSS satellite transmitter power at separate operational frequencies and commissioning longer open-access ranging codes considerably improves a user receiver noise immunity to the effect of solar radio bursts.
5. A difference in the solar flux power by a factor of 5.5 led to a carrier-to-noise ration drop by 20–50% for different frequencies and signal components. With an increase in the radio burst flux power to >1,00,000 s.f.u, one may expect an essential non-linear character of the carrier-to-noise ratio drop. Thus, despite the increase in the transmitter power of the GNSS modern orbital blocks, the problem of solar radio burst effect on the final positioning accuracy has persisted so far.
6. The magnetic storm effect on the GNSS operation quality and navigation differential systems has a global character, but the pattern for the GNSS operation quality degradation depends principally on the observation point latitude. At mid latitudes, an increase positioning errors and radionavigation parameter measurement failures do not have a universal character, but are related to local ionosphere regions, in which there are conditions for generating intensive small-scale electron density irregularities. Such regions are traveling ionospheric disturbances and the auroral oval boundary. Besides, the equatorial bubbles may penetrate to 35–40° N of the mid-latitudes.
7. At low latitudes, cycle slip density depends on the angle between line of sight and the major axis of the field-aligned irregularities (equatorial bubbles). Most cycle slips (15 through 33%) are observed for the satellites, whose LOS is parallel to the geomagnetic field line (along the field-aligned irregularity).

8. When the "receiver-satellite" LOS is orthogonal to the field-aligned electron density irregularity axis, we note that cycle slips locally increase up to 5%. No other authors corroborated the effect.
9. Under geomagnetic disturbance conditions, the probability increases for a short-time sharp GDOP due to a loss of lock for satellites with >30° elevations. At low latitudes, such an effect is insignificant: there is no more than 0.5–1% growth in the occurrence probability that a DOP exceeds a DOP_{max} (for PDOP, HDOP, VDOP). At the same time, at high latitudes, the probability *P* ($PDOP > PDOP_{max}$) under magnetic storm conditions increased by 40%, and the probability P ($VDOP > VDOP_{max}$)—by 6–10%. Under quiet conditions, at high latitudes, no one noted occurrences of the $HDOP > HDOP_{max}$ events at all. At low latitudes, anomalous DOP values are manifest themselves, primarily, for the HDOP, whereas, at high latitudes, for the vertical, in contrast.
10. When we track the code range and carrier phase, their accuracy and tracking robustness depend on the ionospheric scintillations, especially under strong amplitude scintillations, when *S*4 exceeds 0.5. Regardless of a receiver latitude and satellite elevations, a signal strength decreases, and the ranging error increases at the magnetic storm main phase. Under geomagnetic storms, the probability increases from 2 to 15% so that L1 and L2 signal strength is at a minimal level and high ranging errors occur for the satellites with high elevations ($\gamma = 60\ldots90°$). The later events are not characteristic of quiet conditions.

References

Afraimovich EL, Perevalova NP (2006) GPS monitoring of the earth's upper atmosphere. SC RRS SB RAMS, Irkutsk, Russia, p 480, ISBN: 5-98277-033-7, in Russian

Afraimovich EL, Boitman ON, Zhovty EI (1999) Dynamics and anisotropy of traveling ionospheres disturbances as deduced from tran ionospheres sounding data. Radio Sci 34(2):447–487

Afraimovich EL, Astafieva EI, Berngardt OI, Demyanov VV, Kondakova TN, Lesyuta OS, Shpynev BG (2004) Mid-latitude amplitude scintillations of GPS signals and GPS failures at the auroral oval boundary. Radiophys Quantum Electron 47:453–468. https://doi.org/10.1023/B:RAQE.0000047237.67771.bc

Afraimovich EL, Demyanov VV, Ishin AB, Smolkov GYa (2008) Powerful solar radio bursts as a global and free tool for testing satellite broadband radio systems, including GPS–GLONASS–GALILEO. J Atmos Solar-Terr Phys 70(15):1985–1994. https://doi.org/10.1016/j.jastp.2008.09.008

Afraimovich EL, Demyanov VV, Gavrilyuk NS, Ishin AB, Smolkov GYa (2009a) Malfunction of satellite navigation systems GPS and GLONASS caused by powerful radio emission of the Sun during solar flares on December 6 and 13, 2006, and October 28, 2003. Cosmic Res 47:126–137. https://doi.org/10.1134/S001095250902004X

Afraimovich EL, Astafieva EI, Demyanov VV, Gamayunov IF (2009b) Mid-latitude amplitude scintillation of GPS signals and GPS performance slips. Adv Space Res 43(6):964–972. https://doi.org/10.1016/j.asr.2008.09.015

Anderson PC, Straus PR (2005) Magnetic field orientation control of GPS occultation observations of equatorial scintillation. Geophys Res Lett 32(21):L21107. https://doi.org/10.1029/2005GL023781

Barabanova LP (2010) Minimization of GNSS geometric factors. J Comput Syst Sci Int 49(2):310–317

Benton CJ, Mitchell CN (2012) GPS satellite oscillator faults mimicking ionospheric phase scintillation. GPS Sol 16:477–482. https://doi.org/10.1007/s10291-011-0247-3

Benton CJ, Mitchell CN (2014) Further observations of GPS satellite oscillator anomalies mimicking ionospheric phase scintillation. GPS Sol 18:387–391. https://doi.org/10.1007/s10291-013-0338-4

Berdermann J, Kriegel M, Banys D, Heymann F, Hoque MN, Wilken V, Borries C, Heßelbarth A, Jakowski N (2018) Ionospheric response to the X9.3 Flare on 6 September 2017 and its implication for navigation services over Europe. Space Weather 16:1604–1615. https://doi.org/10.1029/2018SW001933

Carrano CS, Groves KM, Bridgwood CT (2007) Effects of the December 2006 solar radio bursts on the GPS receivers of the AFRL-SCINDA network. In: Doherty PH (ed) Proceedings of the international beacon satellite symposium, Boston College, June 11–15, 2007

Carrano CS, Bridgwood CT, Groves KM (2009) Impacts of the December 2006 solar radio bursts on the performance of GPS. Radio Sci 44:RS0A25. https://doi.org/10.1029/2008RS004071

Cerruti AP, Kintner PM, Gary DE, Lanzerotti LJ, de Paula ER, Vo HB (2006) Observed solar radio burst effects on GPS/WAAS carrier-to-noise ration. Space Weather 4:S10006. https://doi.org/10.1029/2006SW000254

Cerruti AP, Kintner PM, Gary DE, Mannucci AJ, Meyer RF, Doherty P, Coster AJ (2008) Effect of intense December 2006 solar radio bursts on GPS receivers. Space Weather 6(10):S10D07. https://doi.org/10.1029/2007SW000375

Chen Z, Gao Y, Liu Z (2005) Evaluation of solar radio bursts' effect on GPS receiver signal tracking within international gps service network. Radio Sci 40(3):RS3012. https://doi.org/10.1029/2004RS003066

Demyanov VV, Afraimovich EL, Jin S (2012a) An evaluation of potential solar radio emission power threat on GPS and GLONASS performance. GPS Sol 16:411–424. https://doi.org/10.1007/s10291-011-0241-9

Demyanov VV, Yasyukevich YV, Ishin AB, Astafyeva EI (2012b) Effects of ionosphere super-bubble on the GPS positioning performance depending on the orientation relative to geomagnetic field. GPS Sol 16:181–189. https://doi.org/10.1007/s10291-011-0217-9

Demyanov VV, Zhang X, Lu X (2019) Moderate geomagnetic storm condition, WAAS alerts and real GPS positioning quality. J Atmosp Sci Res 2(1):10–23. https://doi.org/10.30564/jasr.v2i1.343

Demyanov VV, Yasyukevich YuV (2021) Space weather: risk factors for global navigation satellite systems. Solar-Terr Phys 7(2):28–47. https://doi.org/10.12737/stp-72202104

Estey L, Meertens C (1999) TEQC: the multi-purpose toolkit for GPS/GLONASS data. GPS Sol 3(1):42–49. https://doi.org/10.1007/PL00012778

Finlay CC, Maus S, Beggan CD, Bondar TN, Chambodut A, Chernova TA et al (2010) International geomagnetic reference field: the eleventh generation. Geophys J Int 183(3):1216–1230. https://doi.org/10.1111/j.1365-246X.2010.04804.x

GPS-WAAS-PS (2008) Global positioning system wide area augmentation system (WAAS) performance standard, 1st edn. https://www.gps.gov/technical/ps/2008-WAAS-performance-standard.pdf

Hiromichi, TSUJI Kazunori, YAMAGUCHI (2013) Status of Modernization of GEONET from GPS to GNSS. Journal of the Japan society of photogrammetry and remote sensing 52(3)114–120. https://doi.org/10.4287/jsprs.52.114

Hofmann-Wellenhof B, Lichtenegger H, Wasle E (2008) GNSS—global navigation satellite systems: GPS, GLONASS, Galileo, and more. Springer-Verlag, Wien, p518. https://doi.org/10.1007/978-3-211-73017-1

IS-GPS-200 (2021) NAVSTAR GPS space segment/navigation user interfaces. Revision M, 21 May 2021

ICD-GLONASS (2008) Global navigation satellite system GLONASS. Interface control document. Navigational radiosignal In bands L1, L2. Edition 5.1

Kaplan ED (ed) (1996) Understanding GPS: principles and applications. Artech House, Boston, p 556

Kintner PM, Kil H, de Paula E (2001) Fading time scales associated with GPS signals and potential consequences. Radio Sci 36(4):731–743. https://doi.org/10.1029/1999RS002310

Lekshmi DV, Balan N, Ram ST, Liu JY (2011) Statistics of geomagnetic storms and ionospheric storms at low and mid latitudes in two solar cycles. J Geophys Res Space Phys 116:A11328. https://doi.org/10.1029/2011JA017042

Ma G, Maruyama T (2006) A super bubble detected by dense GPS network at East Asian longitudes. Geophys Res Lett 33(21):L21103. https://doi.org/10.1029/2006GL027512

Montenbruck O, Steigenberger P, Hauschild A (2015) Broadcast versus precise ephemerides: a multi-GNSS perspective. GPS Sol 19:321–333. https://doi.org/10.1007/s10291-014-0390-8

Perov AI, Kharisov VN (2010) GLONASS. The principles of construction and operation. Radiotekhnika Publ., Moscow, Russian, p 800

Pistolkors AA (1947) Antennas—Moscow: Svyazizdat. Communications Publishers, Russian, p 259

Rife J, Walter T, Blanch J (2004) An overbounding SBAS and GBAS error distributions with excess-mass functions. In: Proceedings of the international symposium on GNSS/GPS, December 2004, Sydney, Australia. http://web.stanford.edu/group/scpnt/gpslab/pubs/papers/Rife_IGNSS_2004.pdf

Scone S, Knudsen K, de Jong M (2001) Limitations in GPS receiver tracking performance under ionosphere scintillation. Phys Chem Earth Part A 26(6–8):613–621. https://doi.org/10.1016/S1464-1895(01)00110-7

Sreeja V, Aquino M, de Jong K (2013) Impact of the 24 September 2011 solar radio burst on the performance of GNSS receivers. Space Weather 11(5):306–312. https://doi.org/10.1002/swe.20057

Themens DR, Jayachandran PT, Langley RB, MacDougall JW, Nicolls MJ (2013) Determining receiver biases in GPS-derived total electron content in the auroral oval and polar cap region using ionosonde measurements. GPS Sol 17:357–369. https://doi.org/10.1007/s10291-012-0284-6

Van Dierendonck AJ, Hua Q (2001) Measuring Ionosphere scintillation effects from GPS signals. In: ION 59-th annual meeting, Albuquerque, New Mexico, 11–13 June 2001, pp 391–395

Wang J, Iz BH, Lu C (2002) Dependency of GPS positioning precision on station location. GPS Sol 6:91–95. https://doi.org/10.1007/s10291-002-0021-7

Yasyukevich Y, Astafyeva E, Padokhin A, Ivanova V, Syrovatskii S, Podlesnyi A (2018) The 6 September 2017 X-class solar flares and their impacts on the ionosphere, GNSS, and HF radio wave propagation. Space Weather 16(8):1013–1027. https://doi.org/10.1029/2018SW001932

Yasyukevich YV, Yasyukevich AS, Astafyeva EI (2021) How modernized and strengthened GPS signals enhance the system performance during solar radio bursts. GPS Sol 25:46. https://doi.org/10.1007/s10291-021-01091-5

Zatolokin DA (2020). Program to solve the GNSS navigation problem “Navi”: certificate of state registration of the software no. 2020612010. Russian

Zhou F, Dong D, Li W, Jiang X, Wickert J, Schuh H (2018) GAMP: an open-source software of multi-GNSS precise point positioning using undifferenced and uncombined observations. GPS Sol 22:33. https://doi.org/10.1007/s10291-018-0699-9s

Zumberge JF, Heflin MB, Jefferson DC (1997) Precise point positioning for the efficient and robust analysis of GPS data from large networks. J Geophys Res Solid Earth 102(B3):5005–5017. https://doi.org/10.1029/96JB03860

Chapter 6
Real-Time Data Assimilation for Space Weather Effects Mitigation on GNSS/SBAS

Artem Vesnin

6.1 Introduction Remarks

This chapter discusses different approaches to global navigation satellite systems (GNSS) data processing. The quasi-real time data processing and the data processing on a global scale are needed to estimate the impact of propagation medium parameter variations on GNSS performance. First, let us introduce two concepts: "systematic bias" and dispersion of the coordinate determination. The "systematic bias" is a result of the inaccurate estimate of the electron content on the satellite-receiver line-of-sight. Some applications, for example GNSS receivers on board of spacecraft, ignore ionosphere contribution (Kovář 2017). However, ground-based applications must account for the ionosphere. There is a different bias at every satellite-receiver line-of-site. "Dispersion" is the phenomenon caused by inhomogeneities of the electron density distribution in the ionosphere. Both effects are related to errors of electron density determination along the line-of-sight. Both can occur/increase under disturbed Space Weather conditions. It is important to understand the difference between these two concepts because the approaches for their detection and mitigating their effects on GNSS performance are different.

The major part of the systematic bias may be taken into account with the help of the empirical models, for example the International Reference Ionosphere (IRI) (Bilitza et al. 2017). An output of empirical models is referred as "space climate" or "climate representation" and introduces well known trends of electron density distribution. The residual between climate representation and real state of the ionosphere could be estimated by the Space Weather models. "Weather representation" should capture trends and variation that are smoothed by climate representation. Space Weather models could be based on the same principles as climate ones but fed with observations (data assimilation). One can also improve climate representation by the correction of the model driving parameters, which is much easier than data assimilation. Both Space Weather and space climate models have global scale. That means that the model basis could fit observation on a global scale. Climate representation might introduce "systematic bias" of coordinate determination. Weather

V. Demyanov et al., *Space Weather Impact on GNSS Performance*,
https://doi.org/10.1007/978-3-031-15874-2_6

representation should reduce it on the global scale. Hence model results could be interpreted as global representation of electron density, on top of which we can implement other ionospheric effects to mitigate the "dispersion".

In this chapter, we consider that the irregular, short-living ionospheric structures such as traveling ionospheric disturbances (TID) are responsible for dispersion. It is possible to fit observation that exhibit some sort of irregularities at the particular observation point. However, this is more the exception than the rule and the global models should not account for such effects. For example, IRI uses harmonics with periods from 4 to 24 h to represent temporal behavior of peak density. Large-scale TID can have a period up to 2 h. This means that the majority of TID is smoothed out from IRI representation of electron density. TID modeling is needed to be performed separately from global electron density modeling. TID and other irregularity models should use electron density from climate or weather representation as a background.

As it was mentioned in the previous chapters, TEC is one of the most popular parameters for ionosphere state estimation. Both the ionospheric slab and TID superimposed on it could be estimated using TEC. It is not always possible to make conclusions on GNSS operation nuisance based on TEC and data products derived from TEC. We can calculate TEC from GNSS phase measurements assuming refractive distortions only. In the presence of a diffractive effect this condition is violated. In the presence of amplitude scintillations TEC estimates are not applicable. However, TEC can be calculated at the epochs prior to amplitude scintillations. Also, it could be calculated near the region that exhibit amplitude scintillations given dense network of GNSS receivers. Using TEC data products (where they are available) we can build irregularity propagation model to get this data product at latter time or in different region. Understanding the connection between data product and irregularities that cause scintillations allows to make conclusions about scintillation occurrence. Yasyukevich et al. (2020a, b) showed the presence of small-scale irregularities in the ionosphere during the passage of a large-scale TID was demonstrated. Those small-scale irregularities cases phase scintillations of GNSS signal at observation sites. Large-scale TID was shown using TEC variation data product. We can assume phase scintillations for locations with no observation sites based on the TID propagation. Hence, we can use TEC data products as a proxy to the conditions that cause GNSS signal distortions where/when those conditions could not be estimated directly.

There is a set of parameters/inputs that are needed for the estimation of the positioning accuracy (see Fig. 6.1): TEC along line-of-sight, TIDs/irregularity presence based on data, and scintillation probability. The positioning model is fed by coordinate timelines obtained on a receiver. The coordinate timelines allow us to see coordinate definition accuracy for any time.

We can involve additional measurements to increase awareness of the state of the environment. Instrument measurements should allow us to correct the media in order that, for example, TEC calculations are more accurate. One such instrument is the ionosonde. An ionosonde measures electron density profile in the ionosphere up to a height above the ionosonde. GNSS and ionosonde data can be compared, although with some assumptions. Ionosonde data could be used directly to bring

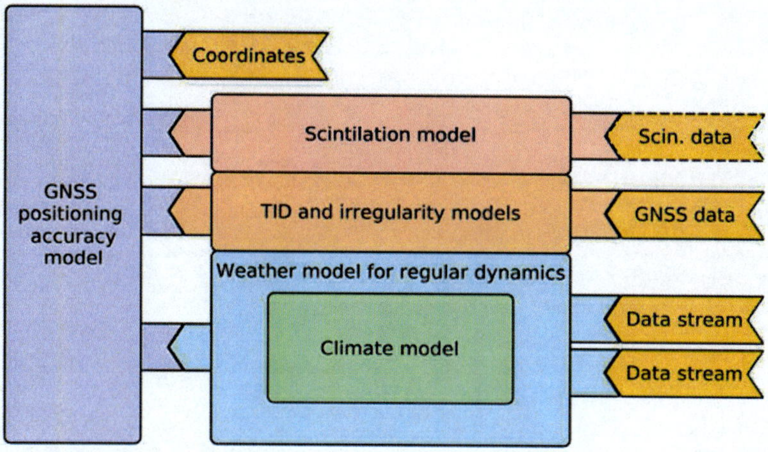

Fig. 6.1 Block diagram of GNSS positioning accuracy model components

climate representation to weather representation. In the latter case we use the result of ionosondes measurement assimilation into the model. We calculate TEC using 3D electron density distribution and no assumptions are needed.

6.2 GIRO—Global Ionospheric Radio Observatory

The Global Ionospheric Radio Observatory is the largest open-source database for ionosonde measurements of electron density and drift velocities in the ionosphere (Reinisch and Galkin 2011). You can give it a try: google "ionogram" and most probably you'll see data from GIRO. The number of locations where ionosondes are installed varies with time. Each location is unsigned with a unique ID. One can find data from 120 ionosondes on the web page that grants access to the physical characteristic derived from measurements https://giro.uml.edu/didbase/scaled.php. Digisondes constitute the major part of the network. There are ionosondes of different models that join the GIRO ecosystem and provide data in required data format. GIRO provides a data access interface as well as a database that can be used as a protection from data loss on observation sites. The majority of ionosondes (more than 60 in 2018) provide access to real-time data. The current development of GIRO is shown in Fig. 6.2.

Unlike GNSS, there is no single exchange format for ionosonde data such as RINEX or NTRIP. Raw data, ionograms, are stored in manufacturers' proprietary formats, and integration of systems from different manufacturers can occur at the level of electron density profile parameters: NmF2(foF2), hmF2, and so on. Ionograms are also called height-frequency characteristics, given vertical sounding. Vertical sounding ionograms show how the amplitude of a reflected signal depends on height.

Fig. 6.2 GIRO location with real-time data streams (Galkin et al. 2018)

Signals have a particular frequency. Hence the amplitude of the reflected signal displayed using frequency-height axes. Frequencies are unambiguously converted into electron density. Height of reflection does not directly correspond to the real altitude of this electron density. One should solve the inverse problem in order to refer the density to the correct height. The heights on the ionograms are so-called effective or virtual heights, since the radio signal is slowed down in the plasma. Inverse problem solution accounts for entire plasma bulk in the signal propagation path to obtain real height. Solution of the inverse problem is called ionogram processing or ionogram scaling (Galkin et al. 2008; Huang and Reinisch 1996). The term scaling may be read as "scaling raw measurements down to several meaningful characteristics".

Parameters of the profile obtained from ionograms are of interest for real-time applications. Raw ionograms could be used for some special processing. GIRO automatically processes ionograms and provides all standard profile parameters defined by URSI (Piggott and Rawer 1978). GIRO data are used for the IRI Real-Time Mapping model to provide real-time Space Weather information (now-casting). The update rate of weather representation corresponds to the ionograms measurements rate. The majority of the ionosondes provide measurements every 15 min.

In addition to the absolute parameters of the electron density ionograms show variability of the ionosphere on a certain scale. Ionospheric disturbances can manifest themselves in different ways on ionograms. Disturbance with structure that depends on height can be observed at a single ionogram sounding session. In this case we see a distortion of the reflection trace compared to the expected one. Figure 6.3 shows ionograms before (left) and during (right) the passage of the disturbance through the ionosonde directional diagram. An additional layer is visible, indicated by a dotted arrow, which was not observed before. Data on a single ionogram are not obtained instantly, a sounding session can take up to 10 min. However, it is conventional to

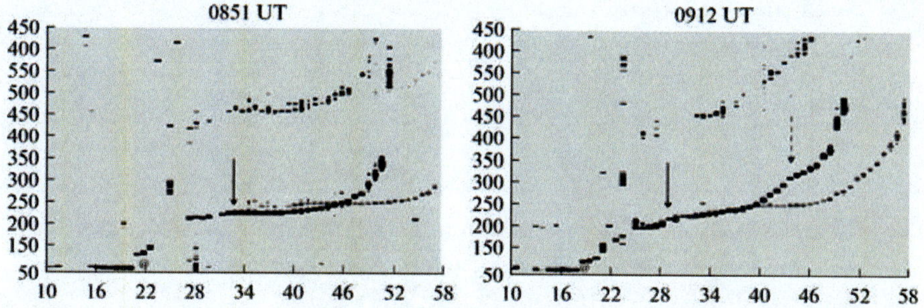

Fig. 6.3 Traveling ionospheric disturbance according to the ionosonde data. Ionograms obtained on September 25, 2009, at the Irkutsk DPS-4 ionosonde during the radar-progress experiment. TID is observed as additional ionogram trace inflection marked by a dotted arrow. Modified and translated from Russian based on figure from the published book (Vesnin and Ratovsky 2011)

attribute them with a single timestamp. For fine processing the acquisition time of one ionogram might be accounted for in order to compare GNSS TEC variations and electron density variations.

When the perturbation is large-scale one and with negligible vertical structure, we will not see a significant distortion of the reflection trace. Entire sounding ionospheric bulk falls into approximately the same TID phase. We can consider the time course of the reflection heights of signals with different frequencies to analyze such disturbances. Figure 6.4 shows the reflection heights for sounding signals with frequencies from 2 to 6 MHz. A wave pattern is seen, which can be interpreted as a vertical motion of the entire ionosphere. This is possible when longitudinal atmospheric waves act on the ionospheric plasma, entraining the plasma up/down along magnetic field lines, into zones of less/greater recombination. Such large-scale disturbance can be observed as TEC variations with appropriate filtration window.

Ionosonde signals of a given frequency are reflected from the surface with equal electron density. Vertical sounding forms a signal that propagates in a cone corresponding to the directional diagram. The reflection happening in this cone may come to the receiving antennas of the ionosonde. This cone can be considered as a volume where joint measurements with GNSS can be carried out. The vertical sounding arises from the fact that signal to reflection point and from reflection point travels in vertical (radial in spherical coordinate system) direction, given ionospheric layers are parallel to the earth surface. Vertical propagation could break if the ionospheric layers are not horizontal as on Fig. 6.5.

We can even obtain multiple reflections given that layers are tilted. Direction of arrival can be measured if an ionosonde has multiple receiving antennas. Such a slope, deviations of the electron density gradient vector from the vertical (radial), is of particular interest and may indicate the presence of TIDs in the ionosphere. To measure the tilt of the ionosphere, it is more convenient to use the ionosonde in the drift measurement mode. This mode is different from the ionogram measurement mode. When an ionogram is obtained, sounding is performed on a linearly distributed set of frequencies with a small (<0.1 MHz) step. The size of the set and the signal

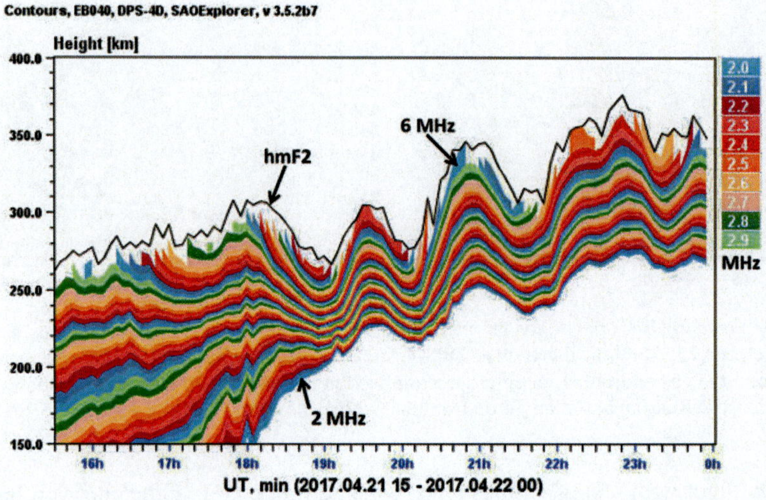

Fig. 6.4 Large-scale TID affects the electron density at all altitudes that the ionosonde signal can reach. No ionogram visible trace distortion would be observed, however the entire trace could move on an ionogram due to TID. Black line shows the position of hmF2 below which no sounding is possible. Data are for Ebro Observatory, Roquetes, Spain (Reinisch et al. 2018)

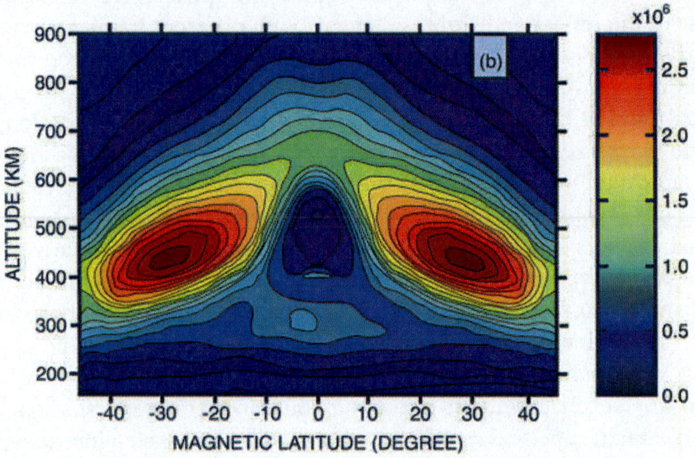

Fig. 6.5 Electron density altitude-latitude slice near the magnetic equator. Due to horizontal gradients ionospheric layers appear to be tilted. The tilt appears due to morphological features of the ionosphere. TID could also cause a tilt

accumulation time at each frequency determine the recording time of each ionogram. The drift mode employs a reduced number of frequencies, while increases probing time for each of them. For each received signal, the angles of arrival are measured. The data can be displayed on a polar coordinate system, where the distance from the center is related to the zenith angle of arrival, see Fig. 6.6. Such a view also shows azimuth of the reflected signal. Each dot on Fig. 6.6 corresponds to a reflection from the ionosphere. Velocity plasma drift along the beam is estimated based on the Doppler shift. The drift velocity is shown with the color of dots. In the case of a horizontal ionosphere, reflections can only come from the vertical direction and the points will fall into the center. A TID with a propagation vector from east to west will cause measurement distribution to be also stretched from east to west. Reflections come from a larger range of zenith angles. Estimation of velocities allows one to obtain not only the structure of the perturbation, but also the propagation direction. Using GNSS measurements, we will see these perturbations by means of TEC as variations of different periods.

We can avoid thin single layer approximation when using a dense network of stations used to map. Hence it is possible to make estimates of the ionospheric height variation in horizontal direction. Paznukhov et al. (2020) used ionosonde and GNSS data to compare ionospheric tilt. The horizontal TEC gradients were related to the ionospheric tilt that occurs due to wavelike disturbances (Fig. 6.7).

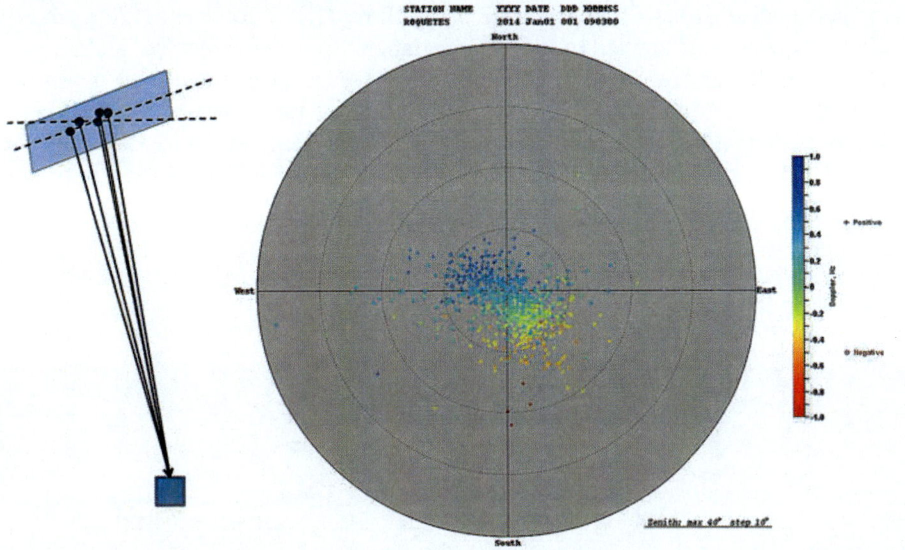

Fig. 6.6 Illustration of ionosphere tilt measurements using the ionosonde. Measurements are made at several frequencies. The distance from the centre is related to the zenith angle of arrival of reflected signal. Azimuth of arrival could also be estimated from a figure (Paznukhov et al. 2020)

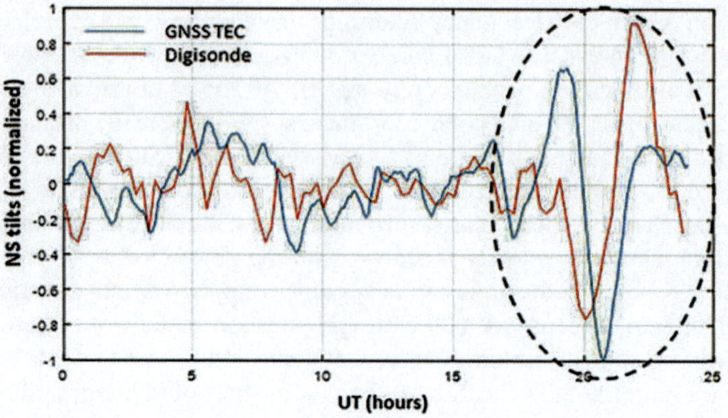

Fig. 6.7 Ionospheric tilt by means of ionosonde and GNSS measurements (Paznukhov et al. 2020)

The pilot project to determine TID presence uses oblique sounding on two ionosondes (Reinisch et al. 2018). Figure 6.8 shows the implementation of the experiment. In this mode, ionosondes use signals transmitted from neighboring stations. Sounding sessions are synchronized. Such systems make it possible to determine the presence of ionospheric disturbances on the path between two ionosondes. The method that is designed to catch TID exists for GNSS measurements. The method built a virtual interferometer using three GNSS receivers (D1 method). Cross-correlation is applied to the TEC variations from three receiver and a satellite (Maletckii and Astafyeva 2021). In the presence of distinguished wave-like disturbance correlation is higher that in presence of random noise. When correlation exceeds a threshold, it is assumed a TID is present. TEC variation could be calculated for different time scales, thus TID of different scales could be detected, given we can use interferometers with the base corresponding to TID scale.

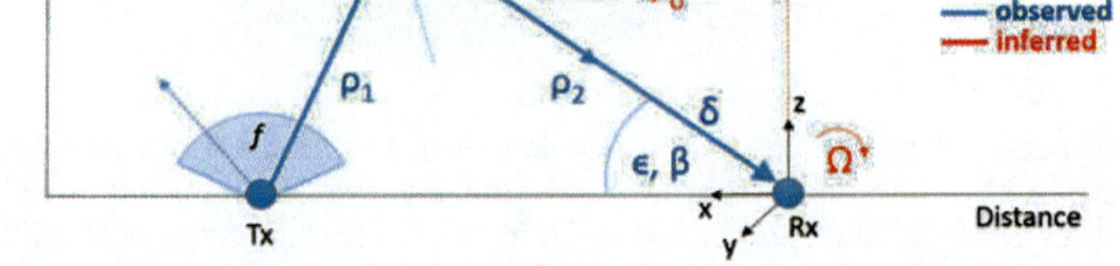

Fig. 6.8 Experiment setup for TID detection using two ionosondes. The transmitter and receiver are synchronized (Reinisch et al. 2018)

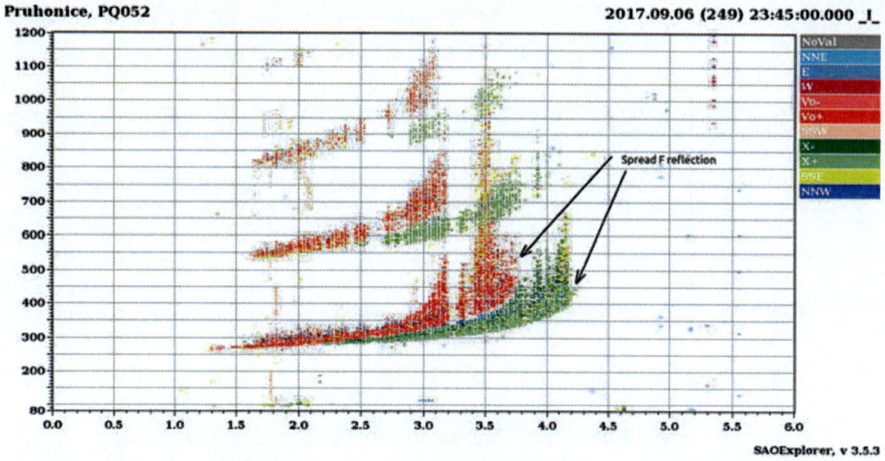

Fig. 6.9 Spread F on the ionogram. Signals of a given frequency have been emitted several times. Also, signals could return to the receiving antennas from different parts of the directional diagram. These factors result in noticeable different travel time in the ionosphere for different probes and rays within a directional diagram (https://www.ufa.cas.cz/irregularities-in-the-ionosphere/)

We consider ionospheric distribution that consists of layers. The layers could be parallel to the Earth surface or tilted. Sometimes irregularity could be nested in plasma distribution. The irregularity scale could be different. The signal could interact with irregularities of a certain order. Each frequency is probed several times. Irregularities cause different probes to have different travel time. Hence, we see ionogram trace broadening, see Fig. 6.9.

We observe a phenomenon known as F-spread. This is often associated with small-scale irregularities in the ionosphere, which lead to diffraction phenomena. In contrast to the case of a layered ionosphere, when a signal propagation is refractive. Irregularities could have different scales. There could be conditions when both ionosonde and GNSS signals propagation is diffractive. When GNSS signals propagate through refractive media we observe scintillations. At the same time, due to the different scale of inhomogeneities that affect the F-spread signal and scintillation, they do not have to coincide in time. Irregularities, known as equatorial plasma bubbles, could be observed in the equatorial ionosphere. Plasma bubbles are quite a regular event. This allows climatology studies of bubbles (Gentile et al. 2006a). Climatology of plasma bubbles in the topside ionosphere (Gentile et al. 2006b) will contribute to the understanding of GNSS signal perturbations. The equatorial plasma bubbles are not a single source of the small-scale irregularities. Large and middle scale TIDs do not lead to signal diffraction by themselves but can be accompanied by small-scale irregularities. In (Yasyukevich et al. 2020a), a connection was found between the passage of a large-scale disturbance observed globally and the occurrence of both GNSS phase scintillations and F-spread (Fig. 6.10). Large-scale disturbance was observed by means of TEC variations. Although TEC could not be estimated on satellite-receiver line-of-sights subject to scintillations, most data are not affected by

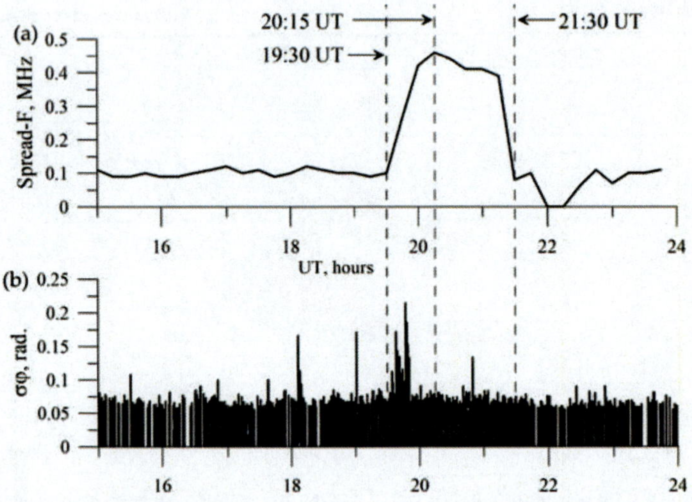

Fig. 6.10 Effects of ionospheric irregularities on different-band radio signals. Panel **a** presents the spread-F at the Irkutsk ionosonde; panel **b** displays the σφ for the L1 GPS signal during the 22 June 2015 magnetic storm. The dashed lines mark 19:30 UT, 20:15 UT, 21:30 UT (Yasyukevich et al. 2020a)

scintillation and are used for TEC variations mapping. Thus, we can conclude that large-scale TID and small-scale irregularities are observed simultaneously.

Real-time data streams from GIRO are assimilated into the IRI model, which makes it possible to switch from the climatic representation of the ionosphere to the weather representation. The results of ionosonde data assimilation and limits of weather representation and model basis are described in Sect. 6.4. The data from observation sites received at GIRO servers in an almost real-time model (7-min delay). Timestep for data acquisition is 15 min, which corresponds to operation mode for the majority of ionosondes. Assimilation is performed by calculating corrections to the climatology coefficients of the IRI model. Climatology representation gives monthly median parameter values for any location and time. Assimilation is performed for four main parameters of the ionosphere electron density profile: peak density, NmF2, (critical frequency, foF2), peak height, hmF2, half-thickness, B0, and profile shape parameter, B1. Weather representation of these parameters can significantly improve ionospheric plasma bulk modeling. However, the global model is still insufficient to reproduce strong gradients, TID or irregularities. The latter should be accounted for as a separate model. In global simulations, the order of the expansion is always a compromise between accuracy and stability of the simulation. In the machine learning field the same issue arises when a model is over trained when the model fits the training dataset well but its performance over the test dataset is rather poor. Back to mathematical expansion, when higher orders are turned on, the accuracy can increase over data used for assimilation. At the same time we bring noise to the representation. This causes the model performing worse outside observation sites

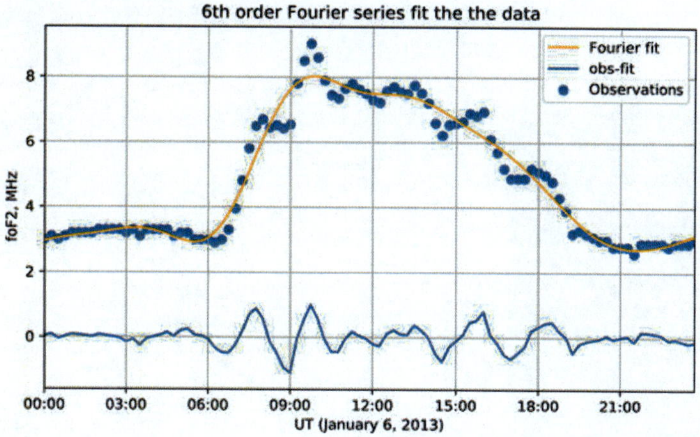

Fig. 6.11 Illustration of the inability of 6th order harmonics to resolve fast variations. The dots are the measurements at El Arenosillo (37.1° N, 6.7° W) site for the time interval indicated on the axis. The bold line is 6th order Fourier series fitted to the data, the thin line is residues left after fitting. Residues originate from a limitation of the Fourier series order and show periodic variations

and makes it difficult to analyze the phenomena in a physical way. Figure 6.11 shows the residual data after modeling the critical frequency series with a 6th order Fourier series. Same series is used in the IRI model. Wave disturbance with significant, up to 1 MHz, amplitude stays outside model representation. The wave-like residual is caused by TID passage.

A similar effect is observed in the spatial domain. Data from two nearby stations may differ significantly. Model basis could not catch up to the significant spatial gradients when data from both sites are fed into the model. Thus we may lose accuracy in ionosphere representation. Figure 6.12 presents data from two stations JEJU (JJ433) and ICHEON (IC437). JEJU was inside the crest equatorial anomaly and ICHEON is outside it. The model basis could not catch a difference of 3 MHz at 03:00 UT at a distance of 500 km. The difference is left as global model error.

The GIRO data assimilation into IRI is the perfect tool for ionospheric weather representation on a global scale. Median nature of climate representation could be greatly improved by the observations. Weather representation provides us with good background distribution of electron density. The background is not a quiet time reference, since the data are in there. The background means "all what model basis could follow". As shown above a weather representation could be limited, yet there is no concern it better fits the real state of the ionosphere. Weather representations coefficients have the same order for spatial and temporal expansion as IRI does. It grants compatibility with IRI. The mathematical expansion in analytical form allows us to compute electron density profiles at any given location at any time rather fast. TID, irregularity or special gradient model could be superimposed given this background electron density. These phenomena should not be modeled on the global scale, since those are effects limited in space and time. Additional modules must

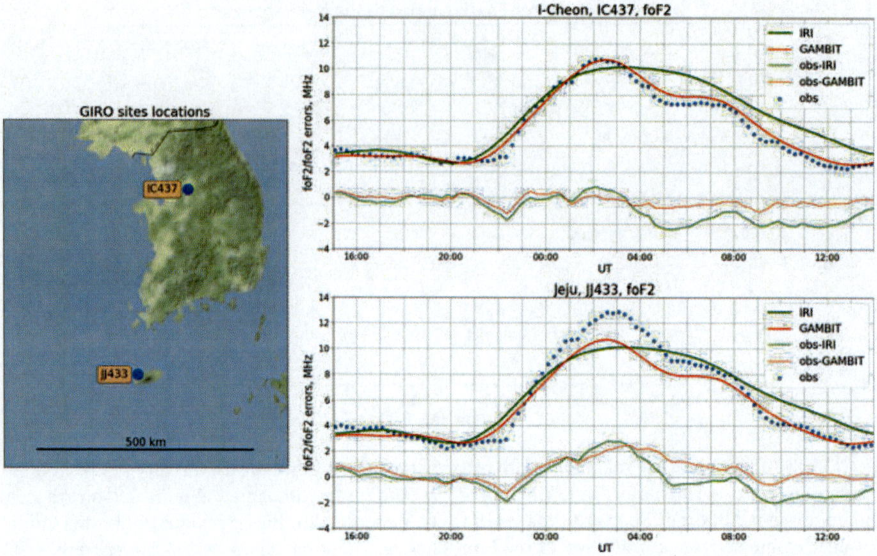

Fig. 6.12 Severe spatial gradients of foF2 near equatorial anomaly. Left panel shows two spatially close stations in South Korea: ICHEON (IC437) and JEJU (JJ433). The panels on the right show observations (blue dots), weather representation (red curve) and climate representation (green curve) models. Red curve follows ICHEON data while underestimates JEJU data. The data model mismatch is due to the global nature of the model basis and severe difference in spatially close data

be included in the simulation as shown on Fig. 6.1. The inclusion of a perturbation model can improve the quality of the determination of the electron density and TEC. Perturbations model could also use the corresponding climate model as a median proxy and then be fed by data to improve irregularity and disturbance presentation over particular areas. GNSS scintillation as a major operation nuisance could be proxied by the different data: F spread and TEC variation. Both of them could be accompanied by the small-scale irregularities that cause GNSS scintillation. This could be of great help since GNSS scintillations observations are less common than TEC observations.

TEC distributions over the globe are called global ionospheric maps (GIM), but the presence of other plasma shells of the Earth should be kept in mind. The plasmasphere can extend from the upper boundary of the ionosphere to GNSS satellite heights. The electron density is significantly lower in the plasmasphere than in the ionosphere, but due to its large size, it can make a significant contribution to the TEC. To refer of the plasmasphere size L-shells are used. L-shell is determined in approximation of the dipole magnetic field. L-value describes the set of magnetic field lines which cross the Earth's magnetic equator at L Earth-radii from the Earth. Under quiet conditions, the boundary of the plasmasphere, the plasmapause, is located at 4–5 L. Figure 6.13 shows the plasmaspheric electron content (PEC) and its contribution to the TEC. PEC is obtained by subtracting the ionospheric electron content from TEC. The ionospheric electron content is obtained by integrating the electron density profile obtained from ionosonde measurements. The ionosonde makes it possible to obtain the electron density profile up to the ionospheric peak height. The top-side profile is obtained in the way to correspond to the bottom-side profile; hence it represents the ionosphere quite good but is still a kind of assumption.

The plasmasphere is a "smoother" medium than the ionosphere, since the neutral atmosphere is orders of magnitude thinner at plasmaspheric heights. On the other side the plasmasphere is the last plasmaspheric shell that corotates with the Earth, outer shells are subjected to the solar wind change more. During a geomagnetic storm, a significant drop in the concentration in the plasmasphere can be observed. Figure 6.13 shows a comparison of the plasma concentration for L-shells 2.08 and 2.41 during quiet time and during a geomagnetic storm. Data are derived from plasmagrams of RPI on board IMAGE (Reinisch et al. 2000). One can see a threefold

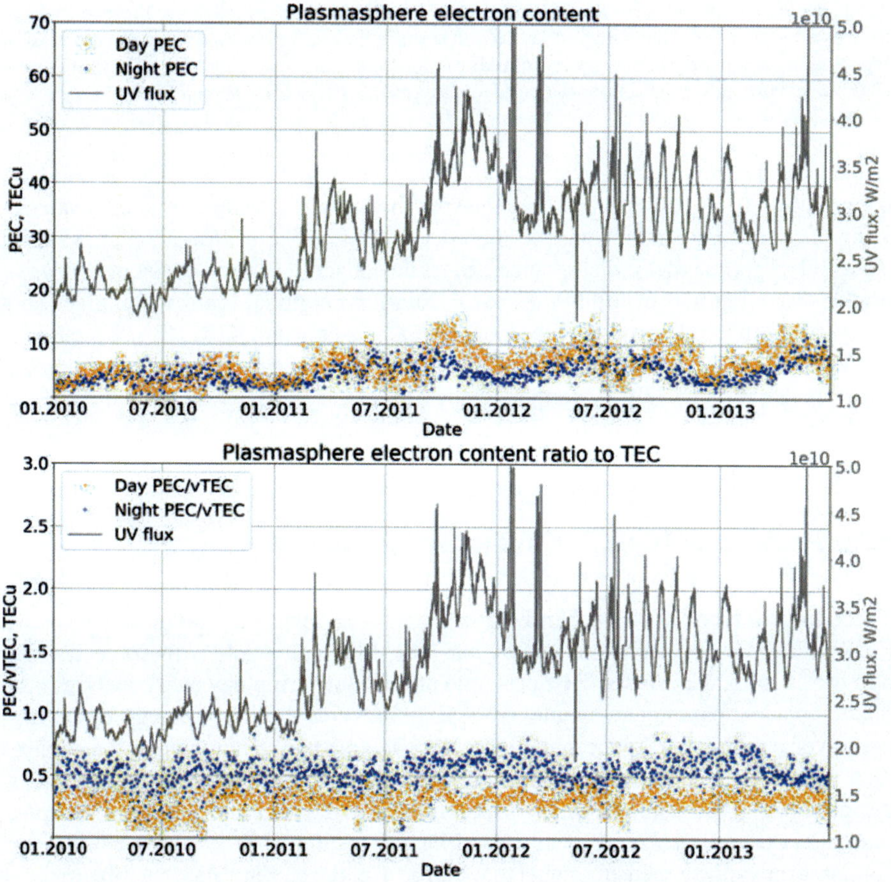

Fig. 6.13 Changes in daytime (black) and nighttime (blue) PEC values (top panel) and PEC/TEC ratio (bottom panel) in Irkutsk from 2010 to 2013. Red lines show dynamics in the solar UV central flux (redraw Yasyukevich et al. 2019)

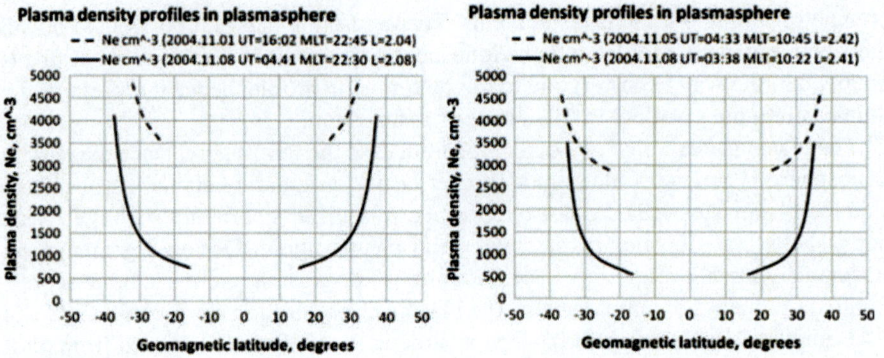

Fig. 6.14 Plasma density profile along magnetic field lines derived from plasmagrams of RPI on board IMAGE. Plasma density is plotted versus magnetic latitude of the particular position at the magnetic field line. Dashed lines are profiles derived at quiet times (see figure for time), which serve as a baseline. Solid lines are profiles derived at storm time on 11.08.2004 when plasma density depletion in the ionosphere was observed. Left panel shows profile measured for ~22:30 MLT and ~2.06 L, and the right panel shows the profile measured at ~10:30 MLT and ~2.42 L

drop in the electron concentration. Also, during a storm, the plasmapause, boundary between co-rotating plasma and outer plasma shells, shifts towards smaller L-values. A significant change in the concentration in the plasmasphere, after a long time after the onset of geomagnetic disturbances, must be considered when consider propagation environment in transition time spans. Data-driven global ionospheric maps and the measurements themselves incorporate PEC change into TEC. Climate models of the ionosphere should consider the dynamics of the plasmasphere to improve the representation of TEC (Cherniak and Zakharenkova 2016; Jakowski and Hoque 2018) (Fig. 6.14).

6.3 Real-Time Data Assimilation Approaches

There are two types of ionospheric models: empirical ones and first principles ones. Empirical helps to catch data trends and keep them in track with some limited number of so-called drivers. First principle models start from physical equations and obtain physical parameters evolution keeping the system self-consistent. Equations are solved numerically, hence first principle models are more computational complex, then empirical ones. Data assimilation is a technique that allows to enhance models performance given data streams. Assimilation is different from machine learning, interpolation and statistical analysis. While the results of those three could look similar, bring sparse measurements to a uniform grid, see Fig. 6.15.

Assimilation requires two ingredients: data and a model. Machine learning converts data to a model. Interpolation/extrapolation tends to bring observations to a place (or time) where there are no observations, using mathematics rather than

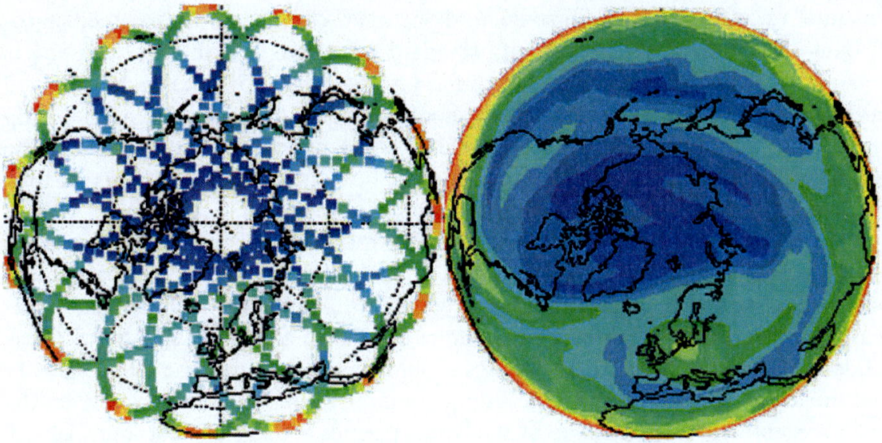

Fig. 6.15 Sparse satellite measurements are brought to a uniform grid. While the result could look like interpolation, the assimilation make us confident data gaps are filled based on physics

physics. In contrast, assimilation uses a physical based model and the data. The assimilation has a great advantage since observations may or may not be of the same observation type used for model development. There were no GNSS TEC at the time the IRI model was created, however we could incorporate TEC data since we understand it is integral to electron density distribution. We can employ other sources of observation, given that we understand how the model and observations are related. Data assimilation case of the ionosphere could seek for several goals:

- to make a model representation as close to real state of the media, from space climate to space weather;
- to determine initial or boundary condition for first principle modeling in order to perform forecasting;
- to perform physical based interpolation/extrapolation of observations, for example TID parameters;
- to find effective parameters-drivers, for example solar flux or sunspot number.

As we described in the previous session there are effects on GNSS by plasma bulk that could be obtained from weather representation of the ionosphere. Weather representation gives background distribution of the electron density. There are also perturbations introduced by the irregularities and traveling disturbances. The effects of the latter could be more severe on GNSS than strong but smooth change of electron density distribution. There should be different approaches for data assimilation for global weather representation and for local irregularity/disturbances. Global weather representation can employ the basis of climatology models. Empirical models could perform rather well for space climate representation: median to median errors are kept low. Damboldt and Suessmann (2011) used ionosondes data for 7 solar cycles and studied IRI performance. The results were overwhelming: foF2 median-to-median separation was 0.01 MHz with standard deviation 0.78 MHz and hmF2 median-to-median separation was 1.51 km with standard deviations 25 km. Hence the basis obtained during the 4-year interval from 1954 to 1958 from 150 ionosondes survived far beyond the epoch it was created. It should be noted that IRI uses magnetic dip angle, by means of the IGRF, which is updated independently. So it keeps track of the geomagnetic field. The weather starts when we move from median data to real-time data, while the basis is kept the same as for climate representation.

Data assimilation is especially useful under real-time conditions. New data are coming, and we should make calculations immediately. There is no option to build new models, so an existing set of physical equations or previously defined mathematical expansion are used. Retrospective studies are different, we have time to build new models on the data collected. Two things should be distinguished: model complexity and the data fitting procedure. The fitting could be iterative and non-iterative. The model complexity is an ability to fit small peculiarities in data. While it sounds good, this is undesirable behavior of the model. It should represent physical phenomena and keep as much noise outside the model representation as possible. Figure 6.16 refers to both issues when the model is simple and over complex. Let us consider IRI temporal representation for foF2, which is Fourier series of 6th order. Simple models (smaller order of Fourier harmonics) would show high bias on the data, since some physical behavior is missed from model representation. The complex model (greater order of Fourier harmonics) would show high variance. A complex model follows perturbations in data and brings them beyond the place and time they are observed. Thus, a model shows good results for data assimilated but rather poor results for other data used. Figure 6.16 summarizes model complexity concerns.

Fitting procedures could also cause the same issue. When overfitting and underfitting (see Fig. 6.17) are both issues. The models with fixed basis would not rather encounter the fitting issues. However, we should be aware of it since it could arise at some step of assimilation. Overfitting or underfitting should not arise in global weather representation by means of IRI. They could arise when we build TID and

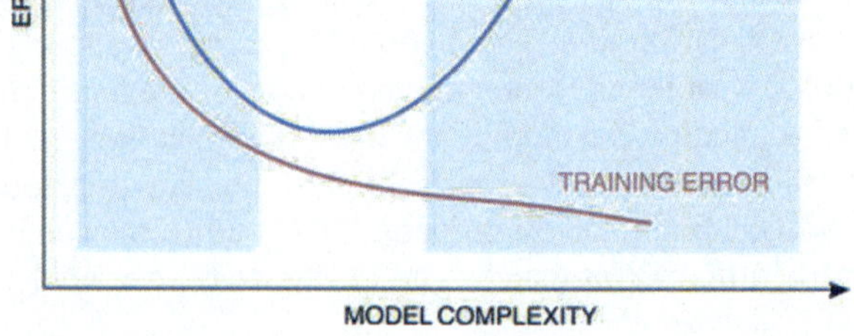

Fig. 6.16 Representation errors dependence on model complexity; for assimilated data (blue curve) and data used for validation (green curve)

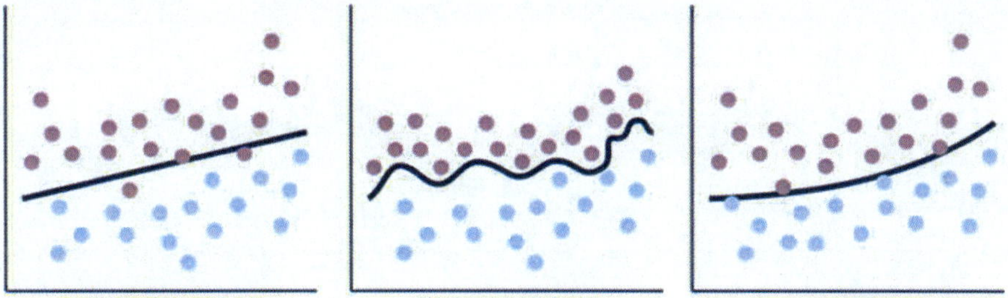

Fig. 6.17 Fitting issue. It is common for iterative procedures

irregularities models on top of it. Since the spectra of irregularities is continuous it is up to us to develop a bound. The bound will limit us to irregularities of scale we model. Narrowing used scales will lead to underfitting. Vice versa overfitting will arise if set broad limits.

The proxy for data assimilation is model drivers update. While drivers are defined independently, the update rate could be rather small. Average number of sunspots for 12 months, Rz12, is used as IRI driver. There are other options currently, however it is still used and it is illustrative to consider Rz12 and IRI model. By the definition the lag of such a driver value is 12 months. Two models runs at two distinct times, say 01/01/2021 and 01/01/2022, use different values for Rz12 to represent the ionosphere at 01/01/2021. Such an approach is applicable for climate application and retrospective analysis. Another issue is that some timescales are lost in an analysis. For example, seasonal variations of the ionospheric plasma concentration (see top panel of Fig. 6.18) are built in coefficient sets. Coefficients are given for every month. Diurnal variations are built on a basis. However, 27-day variations (see bottom panel of Fig. 6.18) could not be reproduced by such an approach: Rz12 is a 12-month average so 27-day cycle is whipped out. We can define effective Rz12 to better fit the data at particular epochs or use another driver such as radiation flux from the Sun. The effective Rz12 could be found by simple enumeration, given model runs are computation easy. It is not quite assimilation since we use observations only for testing. As we mentioned above assimilation itself could be used for model drivers definition. We could fit data in the model and then use its parameters (expansion coefficients) to find Rz12, given knowledge of coefficients-driver relation.

Manipulation with a driver could make the model to be zero shifted compared to the data on the global scale. However, variance could stay high even with zero average or median error (Fig. 6.19). This is due to driving parameters change affects entire harmonics basis. Data assimilation could perform fine tune of the model mechanisms in order to provide small variance of the model representation error.

Global weather representation could be achieved by using ionosonde data. The results are in the next section. However, GNSS data are of interest for ionospheric disturbances and irregularities monitoring and accounting. 4D representation of the

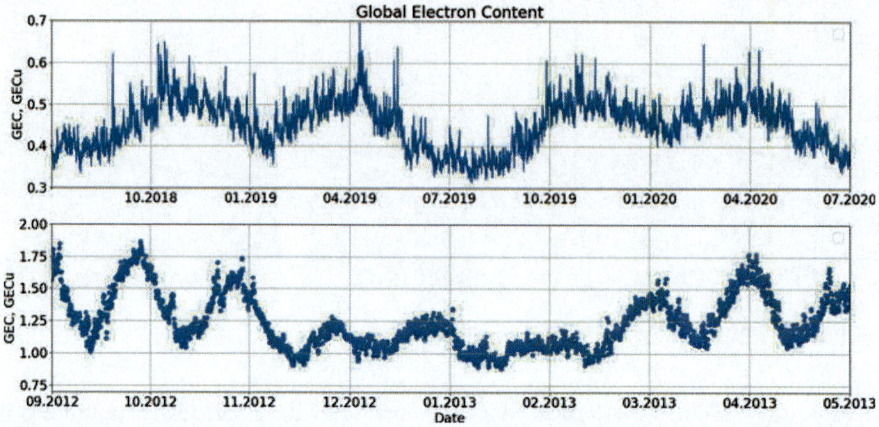

Fig. 6.18 Global electron content showing annual and seasonal variation (top panel) and 27-days variations due to the sun rotation (bottom panel) (taken from http://simurg.space)

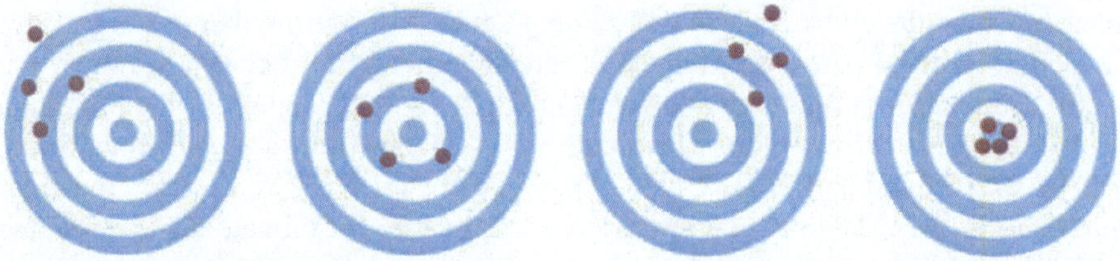

Fig. 6.19 Model representation improvement using driving parameter manipulation (on the left) and using data assimilation (on the right)

ionosphere is more useful than 2D representation by means of global ionospheric maps, since it allows to implement irregularity and disturbance models on top of it.

GNSS receivers provide a wide variety of data to consider. Given the phase and pseudorange measurements are available there are several ways we can use them. The most known ionospheric community is a geometry free combination that allows to obtain TEC. The PPP community also widely uses wide-lane and narrow-lane combination, which could also be useful for ionospheric studies via Melbourne-Wübbena combination, that allows cycle slip detection. Ionospheric free combination to exclude the majority of the error introduced by plasma on the way of signal propagation. Cycle slips are negative factors for both navigation and ionospheric studies. It is an instrumental issue rather than the influence of propagation media.

GNSS TEC based data products can fall into two main groups: absolute parameters and parameters of the ionospheric variability. Absolute parameters are harder to obtain, since measured slant TEC on individual line-of-sights is affected by different propagation and the instrumental effects. Absolute TEC is the solution of the equation system that includes the observations. The equations are designed in such a way to separate instrumental effects from the propagation media influence. The parameters of variability could be obtained on the individual line-of-sights and are easier to calculate. Among those parameters are rate-of-TEC index (ROTI) and TEC variations in different filtration windows. Table 6.1 summarizes data products that could be obtained from raw GNSS measurements and mostly from TEC. Scintillations should be considered separately since the TEC is undefined during scintillations times. The real-time capability accounts for time lag needed to obtain data product. It does not account for computation time, since the time could be different at different hardware.

We need to consider the type of data since different assimilation procedures might be more suitable in different cases. To assimilate the absolute measurements, it is possible to use the approach implemented in the global mapping of the parameters of the electron density profile obtained during probing with an ionosonde (GAMBIT). The results of the assimilation are described in the next session (Brunini et al. 2013) applied a formula relating the two parameters foF2 and M3000F2 and the global representation of these two parameters in order to calculate the global representation of hmF2. The model for hmF2 was different from foF2, however the migration to another basis was successful. Same basis makes it possible to obtain the expansion coefficients of hmF2 in the same form as foF2. This gave a serious advantage, since now hmF2 is represented by the same basis as foF2. The total electron content, which contains most of the ionospheric delay, can be represented in the same way. Further we consider assimilation approaches that are also used in geophysical studies.

Kalman filter. The data assimilation uses Kalman filter (Gelb 1974; Welch and Bishop 1995) which is a way to use sparse and incomplete observation in a model. Theoretical models such that Global Assimilation of Ionospheric Measurements (GAIM) extensively employ Kalman filter (Scherliess et al. 2004). The problem to be solved with this approach should be written in a form:

$$x_{k+1} = A_k x_k + B u_k + w_k \tag{6.1}$$

where x is the n-dimensional state vector of the system (read ionosphere) at different epochs; A_k is $n \times n$ matrix that translates system state at time k to state at time $k+1$; u is called control input in a form of l-dimension vector and $n \times l$ matrix B relates control vector at time k to state vector at time $k+1$; w_k is n-dimensional vector noise of the system state. Looks good, we have the system state as a vector. Have a rule to transit between different time epochs in a form of matrix–vector multiplication. Also, we have control input to drive system state by processes that are not represented in the current model. Finally, we have some uncertainty in the form of a noise vector that adds to system state. The system state could be 3D electron density distribution or 2D TEC maps. The control vector could be neutral atmosphere composition or value of solar ionizing radiation. If B reflects physical laws there should be no subscript, since the form of laws is unchanged from epoch to epoch. Matrix A is time dependent and is evaluated from epoch to epoch. We defined system state and its evolution in time, however no measurements are introduced. The measurements are related to as system state through another matrix:

$$y_k = H_k x_k + v_k \tag{6.2}$$

Table 6.1 Data product based on GNSS TEC

Data product	Order (compared to raw observables which considered as 0-order)	Usage	Real-time capability
Slant TEC	1—geometry free combination	Ionosphere space gradients estimation	Real-time
TEC slips	1—sTEC derivative or		
Absolute TEC	2—derived from linear system build from sTEC	Solution of the linear system of sTEC	Up to several hours to converge than near real-time
DCB	2—accompany data product for absolute TEC	Correction of sTEC series to provide non-negative TEC for every observation time	same as absolute TEC
ROTI	2—STD of TEC change rate	Local variability along line-of-sight	5 min for standard definition
TEC variations	2—sTEC deterending and filtration in a give window	TID manifestation. While ROTI show ionospheric inhomogeneity, variations show wave like structures	5–30 min (half of window used for filtration)
wTEC	2—combination of TEC on different line-of-sights	Global variability above the station	
TID velocities	3—TEC variations	Detection of TID	Same as TEC variations
GIM	3—combination of absolute TEC for a single station or a group of stations	Global regular representation of the absolute TEC	Several days up to real-time
Adjusted TEC	3—correction of slant TEC series with appropriate shift	Global non-regular representation of the sTEC series	Up to real-time
Effective layer height	3—space correlation of the TEC variations on the receiver network		Same as TEC variations

Real-time capability refers to the time delay data could be obtained, given real-time GNSS TEC streams where x_k is still state of the system; H_k is $m \times n$ matrix that connects m-dimensional measurement vector y_k to n-dimensional state vector x_k; v_k measurement noise as m-dimensional vector. We can see that from a single state vector we can obtain different measurements. Consider x_k to be 3D electron density we could retrieve both foF2 and TEC measurements from it by appropriate H_k matrix. GNSS sounding geometry could be incorporated inside the H_k by the navigation information that defines satellites location.

Control input vector could be omitted in case no data for governing processes are available. Noise estimates v_k and w_k should be normally distributed. The Kalman filter is an iterative technique. We make prediction of the system state at time $k + 1$ using system state and noise estimates. Further we find errors between predicted state and measurements when the latter are available. Using errors we correct system state vectors and the noise estimate. The cycle is repeated. The detailed description could be found in (Welch and Bishop 1995) and in the references therein. The same set of equation could be found in (Scherliess et al. 2004), which is not surprising since the Kalman filter provides a framework and a particular form of the matrix depends on the problem. Transition model A_k is where we define underlying principles of how system states are simulated. Scherliess et al. (2004) stated: "replacement of this statistical transition model with a physics-based model of the ionosphere/plasmasphere system should lead to significant improvements in the 3-D plasma density reconstruction and will, in particular, be important in data sparse regions". GAIM used a physical-based way to make state transitions between epochs. However, any other approach to make data transition, e.g. statistical transition, could be used in the Kalman filter.

The matrix approach for the Kalman filter implementation assumes some sort of discretization. Thus some sort of grid is introduced. The dimensions of the grid are trade off between accuracy and computational cost. Figure 6.20 presents an example of such a grid. The grid is not required to be rectangular and uniform. The region of interest, for example equatorial ionosphere, could be denser populated with grid points than others. One can see the capability to build regional and local models using different grids (Aa et al. 2015).

Codrescu et al. (2018) used the definition for Ionosphere-Thermosphere-Plasmasphere system as "strongly forced system" versus "chaotic system" and stated that a strongly forced system sensitive to external force, e.g. solar wind. They used an Ensemble Kalman Filter to approximate certain steps in Kalman filter to make calculations easier. Takasu et al. (2010) used the Kalman filter to enhance the range of Real-Time Kinematic. GNSS TEC data assimilation with use of Kalman filter is common since this approach allows to simulate 3D media and use TEC, which is line integral of 3D media (Hajj et al. 2004; Anghel et al. 2008, 2009). Time dependent state-measurement transform H_k the Kalman filter is friendly to measurement of different types. We already mentioned GNSS slant TEC assimilation in order to improve the electron density distribution, satellite in situ measurements could also be incorporated (Lee et al. 2012; Codrescu et al. 2018; He et al. 2020). The Kalman filter use is not limited to 3D media specification, it can be used for single line-of-sight measurement to monitor quality of a signal propagating through irregular media (Zhang et al. 2010; Susi et al. 2017). The positioning as result of calculations

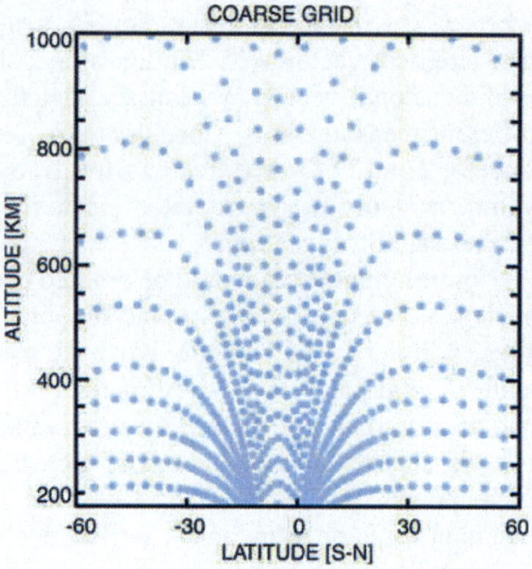

Fig. 6.20 Grid used for ionospheric electron distribution modelling. Grid size is chosen under computational efficiency requirements (inspired by Scherliess et al. 2004)

based on GNSS observables is also assessed by the Kalman filter procedure in order to improve the accuracy (Psychas et al. 2018).

Spherical-harmonics-like decomposition. Spherical harmonics are natural for media representation in a spherical coordinate system (Blais 2008). Fourier series is natural for temporal representation of periodic process. Spherical harmonics and temporal harmonics constitute the basis of the model if they are chosen to be representation of some data:

$$I(\theta, \lambda) = \sum_{n=0}^{N} \sum_{m=0}^{n} U_{nm} Y_n^m(\theta, \lambda) \tag{6.3}$$

$$I(t) = a_0 + \sum_{j=1}^{J} (a_j \cos(2\pi jt/T) + b_j \sin(2\pi jt/T) \tag{6.4}$$

Y_n^m are spherical harmonics as a function of colatitude θ and longitude λ; U_{nm} is the coefficients of a spherical harmonic; a_j and b_j are coefficients of Fourie series; I is any ionospheric parameter. We see that $I(\theta, \lambda)$ and $I(t)$ are the separate models: we can obtain spatial snapshot or time series of a parameter. We need $I(t, \theta, \lambda)$ for the model to be global. Coefficients U_{nm}, a_j and b_j are obtained as data convolution with complex conjugate of corresponding function:

$$U_{nm} = \int I(\theta, \lambda) \underline{Y}_n^m(\theta, \lambda) \sin\theta \, d\theta \, d\lambda \tag{6.5}$$

$\sin\theta d\theta d\lambda$ is an element of a spherical surface. While the entire 3D ionosphere could be modeled using the spherical harmonics, it is conventional to reduce the dimension. It allows us to consider not a single 3D problem but several 2D problems. That is why we use only (θ, λ) variables. Ionosphere could be reduced to a vertical profile model, given by several parameters: foF2, hmF2, B0 and B1. The parameters distribution are 2D spherical surfaces given by some set of spherical harmonics. 3D representation is calculated by calculating a profile for grid points, since the exact values of parameters for any point are known. The spatial model is then an appropriate set of harmonics and coefficients for these harmonics. The time dependence can be introduced by having 2D parameters distribution for some time grid. The more efficient way is to give time dependence to the harmonic's coefficients. The latter approach allows us not to be limited with the size of the time grid. With the time dependence of spatial harmonics coefficients, we could obtain continuous 4D specification of the ionospheric electron density: for any point in space and for any timestamp. The model then will look like:

$$I(t, \theta, \lambda) = \sum_{n=0}^{N} \sum_{m=0}^{n} U_{nm}(t) Y_n^m(\theta, \lambda) \tag{6.6}$$

$$U_{nm}(t) = a_0^{nm} + \sum_{j=1}^{J} (a_j^{nm} \cos(2\pi jt/T) + b_j^{nm} \sin(2\pi jt/T) \tag{6.7}$$

Alternatively, we can make Fourier series coefficients depend on space coordinates through spherical harmonics expansion:

$$I(t) = a_0 + \sum_{j=1}^{J} (a_j(\theta, \lambda) \cos(2\pi jt/T) + b_j(\theta, \lambda) \sin(2\pi jt/T) \tag{6.8}$$

The plasma in the near-Earth environment is strongly governed by the geomagnetic field. That is why a different set of functions (Jones and Gallet 1962) is developed to make a model basis following geomagnetic field geometry and evolution. The model basis moves from pure spherical harmonics to so-called geographical functions:

$$I(\theta, \lambda) = \sum_{k=1}^{K} W_{jk} G_k(\theta, \lambda) \tag{6.9}$$

$G_k(\lambda, \theta)$ includes modified magnetic dip angle, χ, that help keep in track with the geomagnetic field:

$$G_k(\lambda, \theta) = (\cos m\theta + \sin m\theta) \sin^j \chi \cos^m \lambda \tag{6.10}$$

$$\chi(\lambda, \theta) = \arctan \frac{I(\lambda, \theta)}{\sqrt{\cos \lambda}} \tag{6.11}$$

where $I(\lambda, \theta)$ is magnetic inclination. While additional parameter breaks basis orthogonality, it gives great improvement to the ionosphere representation (Brunini et al. 2013). The orthogonality is important when we perform data decomposition in the basis functions. With non-orthogonal functions, we can "count" data behavior several times, which gives incorrect distribution. In order to solve this problem authors introduce another set of functions $F_k(\lambda, \theta)$ and transition between $F_k(\lambda, \theta)$ and $G_k(\lambda, \theta)$. Thus, expansion uses orthogonal set of functions:

$$V_k = \int I(\theta, \lambda) F_k(\theta, \lambda) \sin\theta \, d\theta \, d\lambda \tag{6.12}$$

Then V_k are used to find W_k. Let us repeat it, $W_k G_k(\theta, \lambda)$ are preferred over $V_k F_k(\theta, \lambda)$ since the $G_k(\theta, \lambda)$ depend on the geomagnetic field. Data assimilation in case of spherical-harmonics-like representation seems straightforward; however, it requires observation and the model output to be of the same type. There is no such thing as state-to-measurements transform as in the Kalman filter. We should use foF2 observations or somehow transform other observations into foF2 ones if we want to update the foF2 model. The procedure of data assimilation is not iterative, given uniform data coverage (Galkin et al. 2012), hence the result is uniquely defined. Since the data are still sparse we can expand them on the global scale using some assumptions Galkin et al. (2012) uses "spectra interpolation" rather than "data interpolation" to achieve uniform coverage. It could be achieved by simply setting (model-observations) error drop off to zero as we move away from the observational site. This approach is referred to as spot correction. We can do something different: given good data availability on each site we could obtain Fourier expansion for all sites. The coefficients of Fourier series are then put on mat at site locations and used for interpolation using Non-linear Error Compensation Technique with Associative Restoration (NECTAR) technique. This approach is referred as morphing (Fig. 6.21).

Since the harmonics are global it seems reasonable to expand the amplitude of these harmonics on the global scale. The difference between results of assimilation by simply including the data (Fig. 6.22).

Sparse data could also be used then we need to apply iterative optimization methods to fit model results into the data (Galkin et al. 2012; Bilitza et al. 2017).

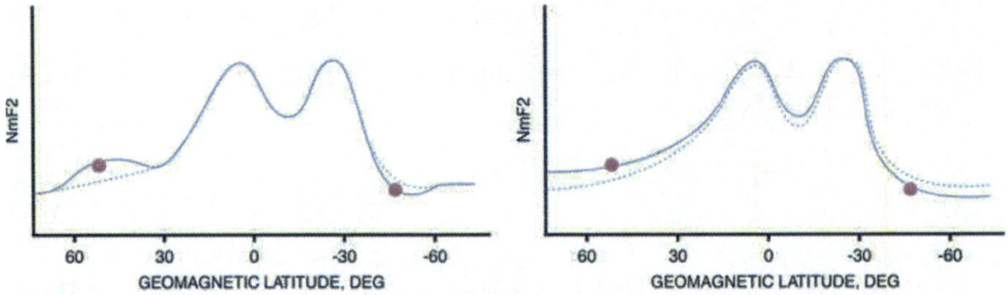

Fig. 6.21 Illustration of the different approaches for data assimilation. The left panel illustrates assimilation in the data domain, "spot correction". Only a tiny vicinity around the site could be affected. The right panel illustrates assimilation by morphing (inspired by Galkin et al. 2012)

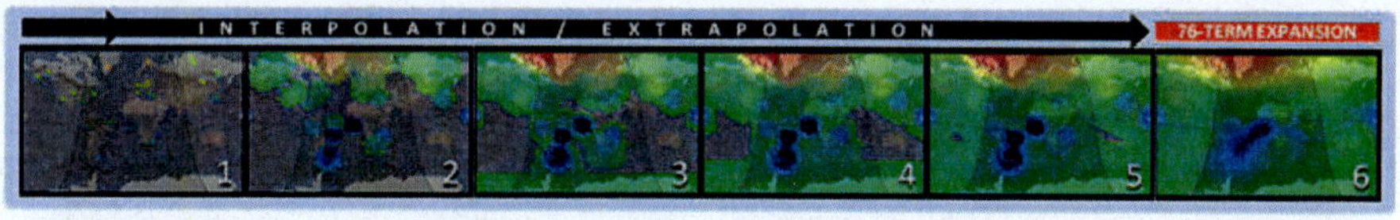

Fig. 6.22 Expansion of sparse data on global scale (Galkin et al. 2020)

Empirical orthogonal functions analysis. Spherical harmonics as well as geographical functions have analytical form. Geographical functions have shown better performance of ionospheric parameters modeling. There were several candidates for the geomagnetic field term in the geographical functions. Tus they were assumed to be better model for the ionosphere, and then were tested and the winners was chosen, and we use through IRI. There was a human made decision of how the term should look like. What if we have the technique that allows us to find interrelation in the data without making hypothesis about the closed form of this relation? If we could, the result can be treated as a basis for data decomposition. The technique that does it is called empirical mode decomposition (EMD) or empirical orthogonal function (EOF) analysis (Monahan et al. 2009).

EOF analysis operates over some temporal-spatial distributed parameter $I(t, x, y)$. With some assumption we can plot $I(t, \theta, \lambda)$ on a rectangular grid, Mercator projection. Hence, EOF is applicable for ionospheric parameters analysis. The regional model requires even less assumptions to use a rectangular grid. Grid points flatten to the m-dimensional position vector $\boldsymbol{X}_i$. Given the data are available for n timestamps, the model is constructed as follows:

$$I(X_i, t) = \sum_{q=1}^{Q} \alpha_q(t)\phi_q(X_i) \tag{6.13}$$

where α_q is time dependent amplitude of spatial pattern ϕ_q. It looks very much like the case when we make coefficients of spherical harmonics time dependent. EOF distinguishes time dependence and spatial pattern by the design:

- Data matrix $n \times m$ matrix is formed. This matrix contains all spatial and temporal variations of the data.
- Then $m \times m$ covariance matrix is calculated from the data matrix.
- Components of EOF analysis (EOF modes) ϕ_q arises as an eigenvector of the covariance matrix.
- EOF modes are chosen such that they cover some percentile of the data variability.

Orthonormality arises naturally, since the eigenvectors are orthonormal. Since we made no assumptions of the closed form of EOF modes, they are given in empirical form in the same dimensions as input data. Say we have a global ionospheric map with dimension 71×73 cells. Then we have a position vector with length equal to 5183. Then the orthogonal functions would be of the same 5183 dimensions.

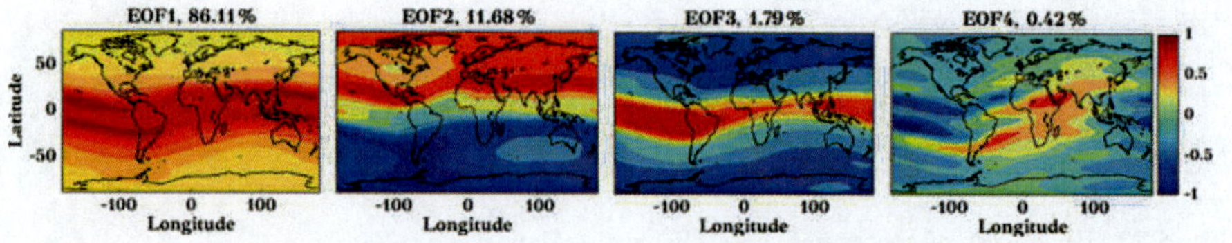

Fig. 6.23 First two EOF eigen patterns (**b**) and (**c**) of TEC map decomposition (Vaishnav et al. 2019)

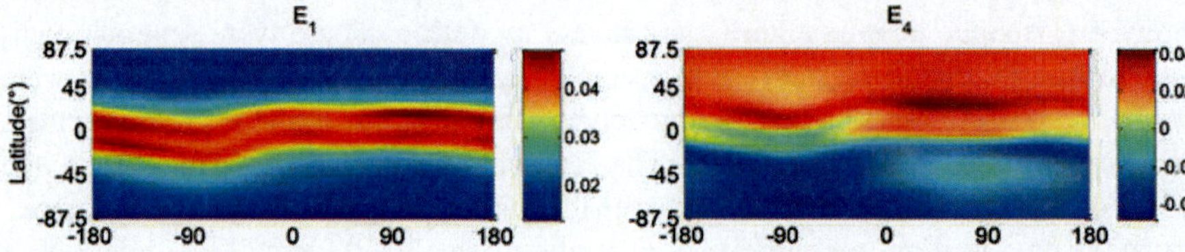

Fig. 6.24 The distribution of the first two EOF base functions through EOF decomposition of vertical TEC (VTEC) as a function of modified dip latitude and local time coordinate (Li et al. 2020)

Given 20 years of the data history with 15-min resolution, we result in 7,00,800 observation timestamps. That gives us 5183 × 7,00,800 data points available for a given ionospheric map. We can reduce data dimension by two orders if we limit the analysis to the first 76 functions as in IRI spatial representation of foF2 (or even less This analysis is also called principal components analysis (PCA) due to the reason described above. Since the ionospheric data exhibit similar patterns. The EOF modes could be reused in different studies. For example, Wan et al. (2012) and Zhang et al. (2012) expand TEC map using EOF. Wan et al. (2012) results are consistent with the Zhang et al. (2012) that report a similar form of EOFs.Vaishnav et al. (2019) calculated EOFs for TEC maps using IGSG data from 1999 to 2017. The contribution of the first 4 harmonics covers 98% of TEC variance (Fig. 6.23).Li et al. (2020) used different approach and obtained same form of EOFs, however their numeration is different: E1 and E4 on the Fig. 6.24 correspond to E1 and E2 on the Fig. 6.23. Wan et al. (2012) and Zhang et al. (2012) used geomagnetic coordinates and Li et al. (2020) and Vaishnav et al. (2019) used geographical. For the geomagnetic coordinates the EOF pattern are more similar than for geographical.

The first EOF looks like a TEC map. However, the important role of EOFs is that they are orthogonal to each other. The latter allows us to decompose the sparse data into EOF modes obtained using the continuous (in both time and space) dataset. Sparse measurements prevent us from performing EOF analysis. However, they could be used on a prepared set of EOF to find time dependent amplitude.

Jamjareegulgarn et al. (2020) build a regional ionospheric model using EMD. Ansari et. al. (2019) have used a regional GNSS network to perform EOF decomposition of TEC dynamics over South Korea. The regional EOF models have capability to perform better than global models due to the denser grid. The regional models are also capable to for TEC variations and another TEC-based product mapping, given the latter exhibit some climatology. Climatology means there are noticeable patterns that could be caught by EOF analysis, for example existence of wave-like disturbances with persistent propagation time, scale and time they arise. In cases of random processes, for example TEC variation can appear any time with arbitrary propagation direction and scale, it is more useful to use detection and then try to fit a disturbance model into the data.

6.4 GAMBIT—Global Assimilative Model of Bottomside Ionosphere Timeline

6.4.1 GAMBIT Description

GAMBIT (Global Assimilative Model of Bottomside Ionosphere Timeline) is a public repository for data obtained using IRI mathematical basis and technology for mapping measurements in real time by means of IRTAM (IRI Real-Time Assimilative Mapping). By mapping technology we mean the "extrapolation" of data from a limited number of points to a global spatial scale. There are several ways to do this, for example, minimizing the model error on the data using the least squares method (Reinisch et al. 2012). GAMBIT data were obtained with use of Non-linear Error Compensation Technique with Associative Restoration (NECTAR) (Galkin et al. 2020). NECTAR spreads the data on a global scale and then data is convoluted with a known function of mathematical basis. Thus we obtain corrections to the coefficients of the IRI model to make a representation of the space weather based on data. The IRI model is one of the best performers of the Coupling, Energetics, and Dynamics Of Atmospheric Regions (CEDAR) Electrodynamics Thermosphere Ionosphere (ETI) Challenge http://ccmc.gsfc.nasa.gov/support/CETI2010/. This program was designed to evaluate the current state of the thermosphere/ionosphere models on a number of test cases. The CEDAR Challenge shows a surprising result: with the current state of our understanding of space weather, the performance of the best assimilation theoretical model is not much different from the empirical model that describes the mean climatology of the ionosphere. Theoretical models are expected to be more comprehensive and reliable due to their design and use of basic physical laws. Earth's ionosphere is a complex physical system. It is subjected to various driving forces both from above through interaction with solar radiation and the magnetosphere, and from below through interaction with the neutral atmosphere. This allows empirical models, being data-driven ones, to capture the features of the ionosphere, at the level of theoretical models or even more accurately. GAMBIT uses IRI coefficients as climate reference and ionosonde network measurements to derive corrections to climate coefficients. The corrections are determined in such a way that when they are added to the climate coefficients, deviations of model representation for observations are minimized. Models for foF2, hmF2, B0, B1 are available. Those

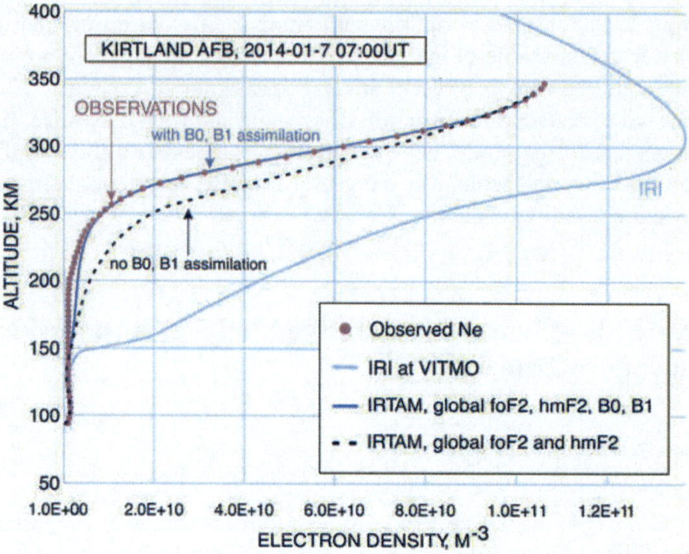

Fig. 6.25 B0 and B1 role when modelling the bottom-side electron density profile. Inspired by Bilitza et al. (2017)

are critical frequency, concentration maximum height, half-thickness of the ionospheric layer and the profile shape parameter, respectively. The mathematical basis used for models is originally foF2 basis. Later it was extended to modeling other quantities. Climatology models of hmF2, B0, B1 were turned to foF2 basis in order to perform assimilation in the same framework. Assimilation of B0 and B1 together with foF2 and hmF2, leads to a more accurate description of the electron density profile, and hence more accurate values of the ionospheric TEC obtained from the profile. The role of B0 and B1 is presented on Fig. 6.25. Use of foF2 and hmF2 only correction may lead to the overestimation or underestimation of TEC. The data are given for the ionosonde at Kirkland AFB for January 17, 2014, 07:00 UT.

The mathematical framework used to model the above parameters of the electron density profile includes dependence on the asymmetric magnetic field of the Earth. The dependence is based on modified magnetic inclination $\chi(\lambda, \theta)$. Compact form of representation of spatial distribution:

$$a_j(\lambda, \theta) = \sum_{k=0}^{75} U_{jk} G_k(\lambda, \theta) \tag{6.14}$$

where geographical functions $G_k(\lambda, \theta)$ are defined by (6.10) so-called. The choice of j and k for a certain function is made in accordance with the table from (Jones et al. 1966). $G_k(\lambda, \theta)$ do not form an orthogonal set of functions, so the least squares minimization is not applicable. Therefore, another set of functions $F_k(\lambda, \theta)$ is formed, the latter are orthogonal. The transition between $F_k(\lambda, \theta)$ and $G_k(\lambda, \theta)$ is known.

Error minimization occurs on $F_k(\lambda, \theta)$ basis and then we return to the original basis. Magnetic inclination helps keep on track with the Earth's magnetic field through a separate model. The advantage is that the climate coefficients of the profile parameters remain unchanged from the moment they are obtained. For example, for foF2, there are two modifications of the URSI coefficients (Rush 1989) and CCIR (CCIR 1967, 1991). The change due to the migration of the Earth's magnetic field occurs due to the inclusion of magnetic inclination in the basis functions. Climate representation behavior associated with solar radiation is tied to the 12 month average number of sunspots, Rz12. Rz12 is used to calculate the weighted average between two sets of coefficients for high (Rz12 = 100) and low (Rz12 = 0) solar activity. Seasonal changes are given as a different coefficient set for each month. Thus, to describe for example *foF2* at any point at any time, 988 * 12 * 2 = 23,712 numbers are needed.

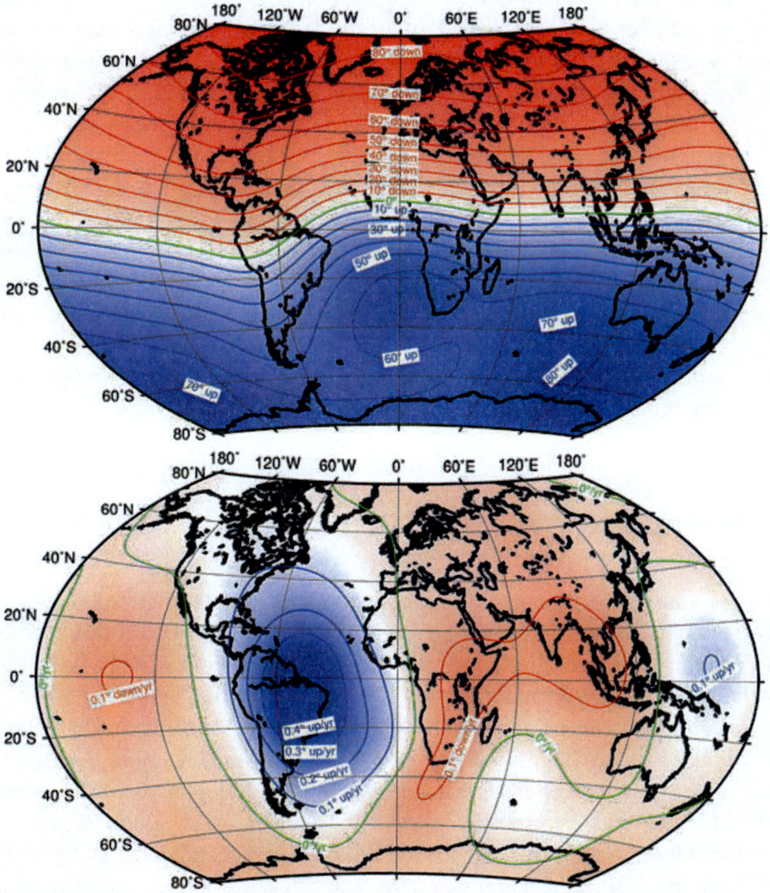

Fig. 6.26 Map of magnetic inclination and predicted changes in magnetic inclination from 2020 to 2025 according to the British Geological Survey (taken from http://www.geomag.bgs.ac.uk/education/earthmag.html)

Accounting for the magnetic field greatly simplifies the use of the model, since the magnetic inclination can change by several tenths of a degree in one year. See Fig. 6.26 for distribution and predicted dynamics for the inclination of the geomagnetic field.

Thus, we have three key ingredients for building a climatology model, which is the basis for building a global space weather model:

- Climate model with the ability to update parameters-drivers or coefficients.
- Magnetic field model that is updated on regular basis.
- Real-time data streams.

One can see that irregularities and disturbances are not included in the list above. When analyzing disturbances in the ionosphere, it is customary to choose or calculate a quiet-time reference. It represents such a state of the ionosphere relative to which we can estimate the amplitude of disturbances. In terms of observations quiet-time reference can be the distribution of the electron concentration a day ago or the distribution averaged over a month. In the case of modeling, the distributions obtained in IRI using climate coefficients can also be a quiet-time reference. The latter would work if there is no systematic bias in climatology representation. Otherwise, the resulting perturbations cannot be interpreted as physical disturbances, but rather will be processing artifacts. Hence if we want to estimate TID or ionospheric irregularities we would rather use weather representation than climate one. Also, TID or irregularities should not go to the global space weather model since they are limited in space and should not interfere on a global mathematical basis. The further of this section is devoted to the assimilation results of ionosonde data into the IRI foF2 basis. In other words we want to estimate how the basis designed for climate representation of the ionosphere can perform for weather representation. While the assimilation to different model bases could be different, the same processing framework could be applied to any model validation.

6.4.2 *From Climate to Weather Representation*

Here we evaluate the performance of ionosonde data assimilation into IRI. While Bo and B1 are important for accurate density profile representation, in this section we cover foF2 and hmF2 assimilation results. One could question why we use foF2 and hmF2 while GNSS is of primary interest. The reason for that is that ionosonde data is more determined in terms of sounding geometry. Also, ionospheric characteristics foF2 and hmF2 are much better determined, foF2 is almost absolute measurements and hmF2 determination depends on inverse problem solution but is well determined. While GNSS measurements are subjected to numerous assumptions before we can locate them. Given Sect. 6.4.1 ionosonde data are best candidates to be analyzed together with the GNSS measurements. We want to develop processing framework below, hence we are using ionosonde data and than can spread developed principles on GNSS measurements.

Table 6.2 Model errors for $foF2$ and $hmF2$ for the 11 years time interval for different datasets (the description for datasets is in the text below)

	$foF2$		$hmF2$	
	ε_C (MHz)	ε_W (MHz)	ε_C (km)	ε_W (km)
Manual	0.69	0.31	27.0	15.0
Full	0.76	0.38	33.0	18.0
Disturbed	1.05	0.49	40.0	23.0

We denote IRI results as **climate representation** and model results after an assimilation is done as **weather representation**. The statistics are a necessary piece of information for GAMBIT users, since they are interested in the range of applicability and trustworthiness of the models. The information below is on Non-linear Error Compensation Technique Associative Restoration (NECTAR) performance since there are other techniques to achieve assimilation. First, we calculated average value of absolute error ε_M, using the expression:

$$\varepsilon_M = \frac{\sum_{n=1}^{N} |O_n - M_n|}{N} \tag{6.15}$$

where O_n is an observation, M_n is a model value, and N is the number of measurements. The model can be calculated anywhere at any given time; hence N corresponds to the number of observations. Since there are two models under investigation, we used both ε_W and ε_C to refer to weather and climate representations. We use representation instead of model since mathematically they are the same. Assimilation uses the one-day-history of the observation. We calculate ε_W for the last point in one-day-history to perform analysis below, where no other algorithms are mentioned.

Statistics. We examined assimilation on three different datasets: (1) using only manually scaled measurements (by human) which is ~1 million points dataset; (2) using all measurements including auto-scaled ones (by software); which is ~13 million points dataset, and; (3) using only measurements made during geomagnetic disturbed time (with DST <-50 nT). Terms "auto-scaled" and "manually scaled" refer to the processing of an ionogram in order to obtain electron density profile. Datasets cover 11 years of observation history. Important point here is that we use the same observation for assimilation and validation. Machine learning people could argue that approach, however we use it to understand model limits. We perform some other test with the division of "training" and validation dataset later in this section. The results are given in Table 6.2.

We see that data assimilation gives twice model representation improvement calculated as ratio between ε_W and ε_C. The improvement is the highest for manually scaled data and $foF2$. The description of the datasets is below.

Manual: Assimilation technique is robust to auto-scaling errors, but obviously auto-scaling errors contribute to error. If auto-scaling errors are considered ε_W

increases. However, it is due to data retrieval being inaccurate, rather than assimilation being faulty, which does not reflect real assimilation performance. The best way to estimate pure assimilation accuracy is to exclude all auto-scaled measurements and leave only manually scaled measurements. This allows us to minimize the influence of autoscaling errors. Unfortunately, manually scaled data are not continuous and do not have global coverage to run both assimilation and validation using purely manually scaled data. Hence auto-scaled data are still assimilated, but we use only manually scaled data to calculate ε_W and ε_C. For *foF2* and *hmF2*, we have obtained ε_W equal to 0.31 MHz and 15 km respectively. The improvement factor is 2.25 for foF2 and 1.82 for hmF2. Only auto-scaled errors were excluded and only from ε_W and ε_C calculations, other types of errors were not inspected.

Full: All available measurements represent a more significant statistical set, which includes 13 million data points. Manually scaled data are available only for some locations, and timelines of such data, except few locations, are not continuous. Users will obtain GAMBIT data based on dominantly auto-scaled measurements. To estimate the assimilation performance users get, we use all data available for 11 years. For *foF2* ε_W is 0.38 MHz and ε_C is 0.76 MHz. The data assimilation improves climatology representation twofold. For hmF2 ε_W is 33.0 km and ε_C is 18.0 km, which gives the improvement factor equals to 1.83. The performance of both models are worse for this dataset than for the regime when only manually scaled data were used. As noted above, auto-scaling errors contribute to the performance. One can see contribution due to auto-scaling error by comparing the first and the second rows for each characteristic.

Disturbed: The GAMBIT is designed to incorporate ionospheric weather into the IRI. This is the reason for testing the assimilation during geomagnetic disturbed time. The third row in Table 6.2 was obtained using only measurements made during disturbed time. Both manually and auto-scaled measurements were used. Geomagnetic activity of moderate level or more severe was understood by disturbed time. This is equivalent to DST index less than −50 nT. The results show that absolute values of ε_W and ε_C are larger, but their ratio, i.e. improvement factor, stays almost the same when compared to the values from the second row. For the foF2 improvement factor is even larger than the improvement factor during quiet time (shown in the second row). This is reasonable since the GAMBIT is designed to capture perturbations in the ionosphere by data assimilation, while the IRI does not have this capability.

Error distribution. While average absolute error gives some sense of assimilation performance. We can see one major feature when looking at the error histograms for foF2 and hmF2. Error distributions are Gaussian-shaped however width and center could vary. Both ε_W and ε_C are zero-centered for foF2 on Fig. 6.27, which means climate and weather representation have zero mean error. Weather representation has narrower distribution as expected since model components are tuned to follow the data as close as possible.

Histogram for hmF2 on Fig. 6.28 is more interesting since the climate representation is shifted: most probable error is not zero. Bad for climate representation, but assimilation could almost remove model-observations mismatch. The hmF2 model

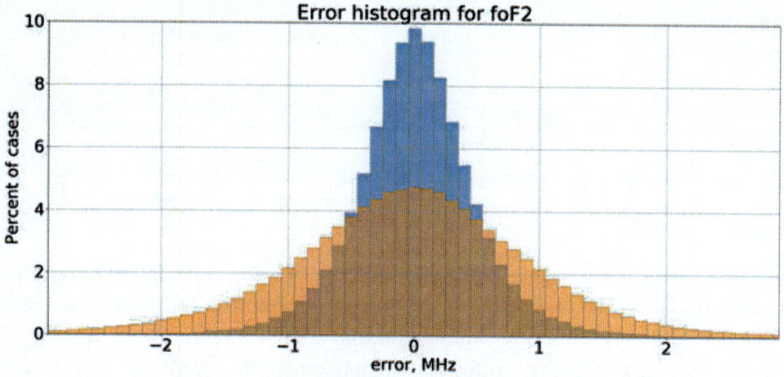

Fig. 6.27 Histogram of (weather-climate) deviation for foF2 for full dataset. Gray bars—errors of climate representation, black bars—weather representation

used for assimilation validation is non-standard and proposed by (Brunini et al. 2013). This model allows to work with foF2 and hmF2 in the same mathematical framework. Leaving the source of non zero mean aside, we would like to pay attention to assimilation results. Both standard deviation and mean error are fixed by assimilation.

One can ask why weather representation is not ideal (gaussian width is zero). This happens always when consider model representation of noisy real-world data. We could make the model more complex and have good model-observation agreement on "training" dataset, i.e. datasets used in assimilation procedure. It doesn't allow physical analysis of the model representation. Also, it makes model representation worse farther and further from assimilated data. Figures 6.27 and 6.28 show performance of model basis functions for foF2 and hmF2 resentation. Parameters of distributions from these figures are listed in Table 6.3.

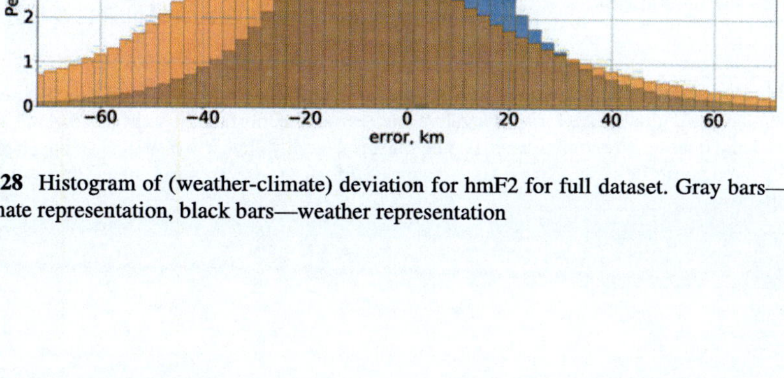

Fig. 6.28 Histogram of (weather-climate) deviation for hmF2 for full dataset. Gray bars—errors of climate representation, black bars—weather representation

Table 6.3 Model error for $foF2$ and $hmF2$

	foF2 (MHz)	hmF2 (km)
GAMBIT mean	0.01	-1.7
GAMBIT standard deviation	0.56	-14.3
IRI mean	0.01	20.3
IRI standard deviation	1.05	41.6

The mean value of ε_W and ε_C distributions for foF2 are both 0.01 MHz. They can be counted equal to zero since 0.01 MHz is several times smaller than the ionosonde's resolution for critical frequency measurements, which is usually in the range from 0.025 to 0.05 MHz. The mean values of ε_W distribution for hmF2 is -1.7 km, which means that IRTAM corrected the IRI mapping to minimize (observation-model) bias. These results are consistent with the results of Damboldt and Suessmann (2011). The standard deviation is higher for the present study, since Damboldt and Suessmann (2011) performed median-to-median comparison, and we compared individual data points.

Geomagnetic latitude dependence. We expect model representation to depend on latitude due geomagnetic field configuration and other reasons. On Fig. 6.29 we see dependence of the ε_W and ε_C on geomagnetic latitude. Since ionosonde locations are not uniformly distributed. The labels of the horizontal axis correspond to the geomagnetic latitude of an ionosonde. The errors were calculated for every ionosonde separately. There is asymmetry in ionosondes distribution between hemispheres: the northern hemisphere is more populated with the observation sites.

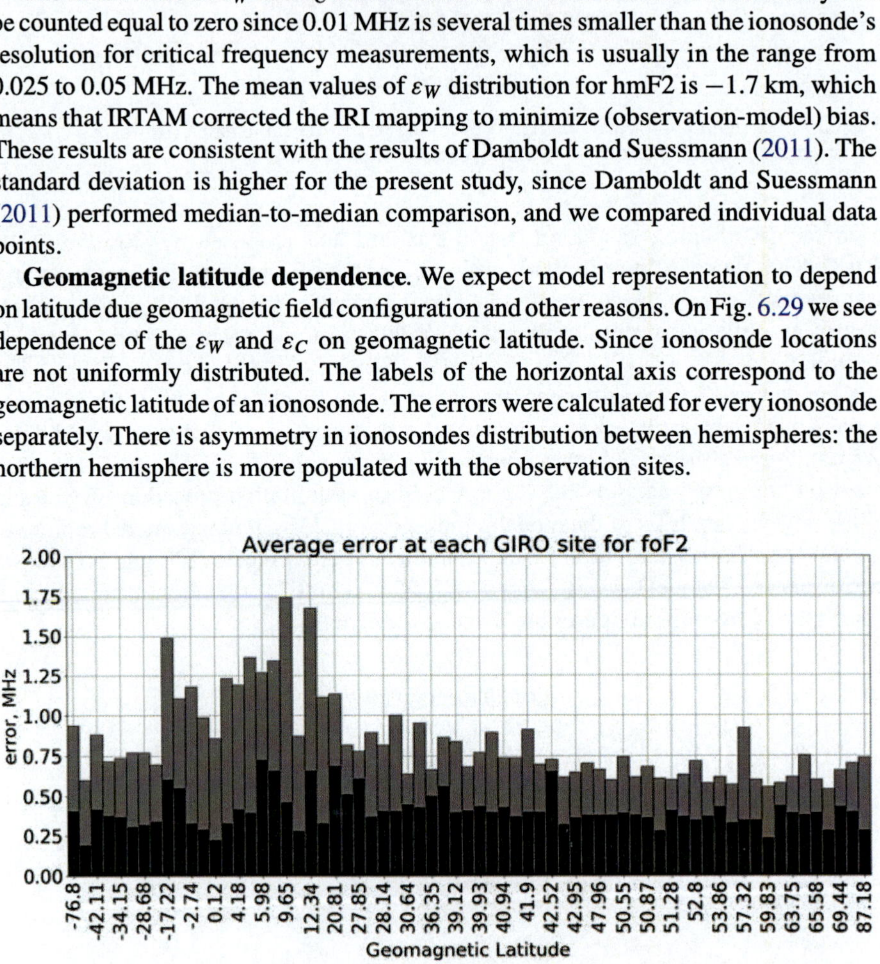

Fig. 6.29 Dependence of average ε_W and ε_C on geomagnetic latitude for foF2. Each bin corresponds to an individual GIRO site. Negative values on the horizontal axis correspond to the southern hemisphere. Horizontal axis is not uniform, and values correspond to observational sites of geomagnetic latitude. Gray bars—errors of climate representation, black bars—weather representation

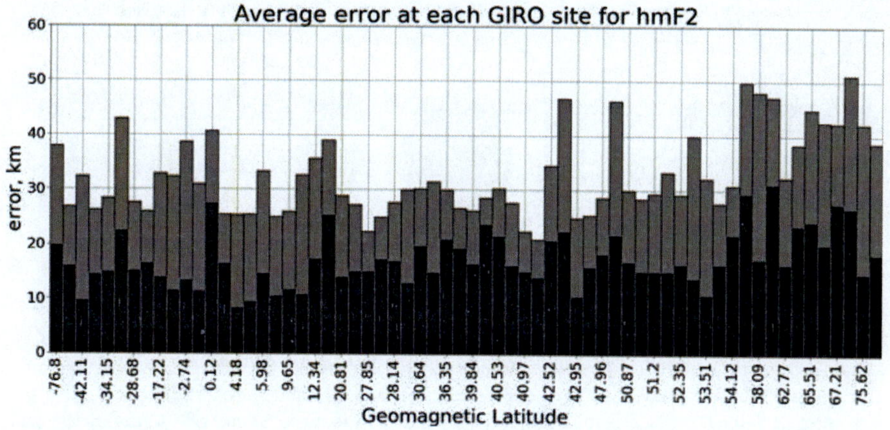

Fig. 6.30 Dependence of average ε_W and ε_C on geomagnetic latitude for hmF2. Each bin corresponds to an individual GIRO site. Negative values on the horizontal axis correspond to the southern hemisphere. Horizontal axis is not uniform, and values correspond to observational sites of geomagnetic latitude. Gray bars—errors of climate representation, black bars—weather representation

The strongest error corresponds to the equatorial ionosphere for geomagnetic latitude in the range −20 to 20 degrees. This is due to high ionospheric variability in this region. Assimilation could fix model representation above the geomagnetic equator. Some mismatch stays for locations that are off the equator to the North and South. For middle latitude and high latitude model representation is nearly the same in terms of absolute errors. Relative error may change the picture since electron density is lower on high latitudes than on lower latitudes. The analysis of relative error is not provided here since GNSS accuracy depends on TEC. Positioning error will suffer the same way from X TECu difference on low and high latitudes. Also, climate representation shows on average higher model-observation deviations in middle-latitudes in the southern hemisphere (~0.8 MHz) than in the northern (~0.6 MHz). The coefficients for climate representation were created at the time when there were not that many observations from the southern hemisphere (Jones et al.1969).

We can also inspect hmF2 dependence on the geomagnetic latitude on Fig. 6.30. Here high latitudes contribute to the error of climate representation the most. The reason for this arises from the geomagnetic field geometry. At high latitudes, the field lines are nearly perpendicular to the Earth's surface. This makes plasma vertical motion more probable than at low latitudes, where the field lines are nearly parallel to Earth's surface. The weather representation could achieve better results. However for some locations the error is twofold lower than for most of the locations. The height of the ionospheric maximum, hmF2 is critical for TID mapping by means of GNSS observations. It is conventional that most.

Local time dependence. Further we studied models accuracy dependence on local time. The ionosphere was divided into three regions: low, middle and high latitude ionosphere. Geomagnetic latitude was used to attribute a particular site to

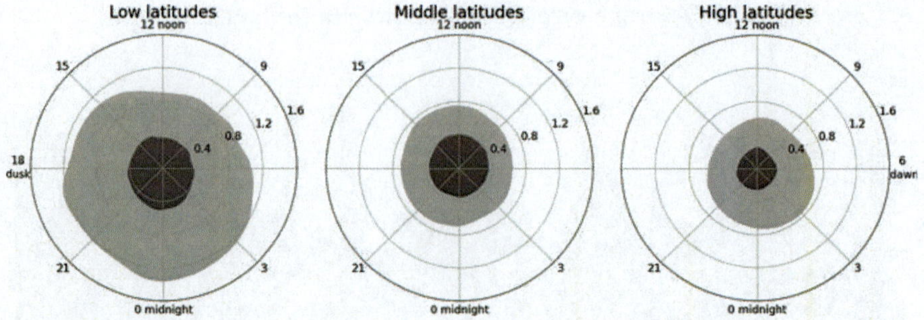

Fig. 6.31 Dependence of ε_W (dark gray) and ε_C (light gray) for foF2 on local time in three latitudinal regions. Low latitudes: geomagnetic latitude < 20°; middle latitudes: 20° < geomagnetic latitude < 60°; high latitudes: 60° < geomagnetic latitude. Numbers around the circle show the local time: 12 is local noon, 0 is midnight, 6 is dawn and 18 is dusk. Values of ε_W (dark gray) and ε_C (light gray) are shown as a distance from the origin

a particular region. Low latitude region contains all sites with geomagnetic latitude less than 20°. Sites with geomagnetic latitudes greater than 60° are prescribed to high latitude regions. All other sites belong to the middle latitude region. Each individual data point was attributed to a particular local time. Local time for each data point was calculated using expression

$$LT = UT + 1hour\frac{longitude}{15^{\circ}} \tag{6.16}$$

where LT is local time, UT is universal time, longitude is the geographical longitude of an ionosonde location. Full dataset is used. The results for *foF2* are on Fig. 6.31.

We see Fig. 6.31 and Fig. 6.29 are consistent. The left plot of Fig. 6.31 shows that the equatorial ionosphere is a very dynamic region. Ionospheric conditions there can be far from the climatology representation, even when no extreme space weather events occur. The assimilation greatly improves climate representation. For the equatorial ionosphere, the improvement is the largest. The ionospheric characteristics are further from the climatology average around dusk and midnight time. The weather representation is relatively inaccurate during the nighttime, but still it is ~4 times better than the climatology representation. For the other two regions, there is no significant IRTAM accuracy dependence on local time, and the improvement is about a factor of two.

Same analysis was performed for *hmF2*. Results are on Fig. 6.32.

The left panel shows that ε_C is not as uniform on local time at low latitudes as in the other two regions. The highest deviations appear at nighttime, and weather representation did not show significant dependence on local time. Both weather and climate representations show weak local time dependence in the mid-latitude ionosphere. Although ε_W are bigger than that for the low latitude ionosphere. As expected, the accuracy of climate representation is the lowest in the higher latitudes

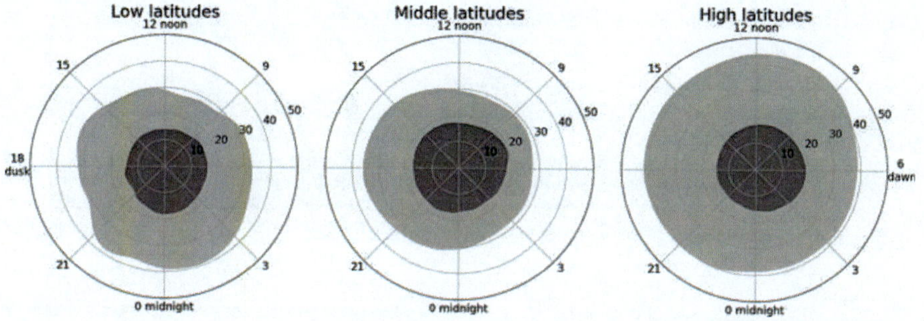

Fig. 6.32 Dependence of ε_W (dark gray) and ε_C (light gray) for hmF2 on local time in three latitudinal regions. Low latitudes: geomagnetic latitude < 20°; middle latitudes: 20° < geomagnetic latitude < 60°; high latitudes: 60° < geomagnetic latitude. Numbers around the circle show the local time: 12 is local noon, 0 is midnight, 6 is dawn and 18 is dusk. Values of ε_W (dark gray) and ε_C (light gray) are shown as a distance from the origin

due to the topology of the magnetic field. There is no significant dependence of LT for weather representation. The improvement is the highest for high latitudes.

Seasonal dependence. The seasonal variations of the ionospheric characteristics, *foF2* and *hmF2*, for a given location appear due to different levels of solar ionizing radiation over the course of the year. To study seasonal dependence of model representations, it is illustrative to divide each year into four parts centered at: summer solstice (June 21), winter solstice (December 21), spring equinox (March 21) and fall equinox (September 21). We divide the full dataset to collect data separetly for the seasons. The ε_C did not show significant discrepancy between these four periods. The deviation for foF2 for all seasons fall into the range 0.38 ± 0.025 MHz. The deviation for hmF2 falls into the range 18.0 ± 1 km. The uncertainties (0.05 MHz and 2 km) are of foF2 and hmF2 measurement accuracy. Thus, the dependence of the IRTAM modeling on season is negligible.

Solar cycle dependence. Variations due to solar activity appear due to different levels of solar ionizing radiation over the course of the solar cycle. The sunspot number varies even more significantly than the radiation level, from near zero at solar minimum to more than hundred at solar maximum. Sunspots are the sources of the fast solar wind, and the fast solar wind causes perturbation of the ionospheric characteristics. Even when no extreme events occur (flares and coronal mass ejections), the ionosphere is modulated by the Sun rotation (~27-day period). At solar cycle maximum, larger sunspot numbers enhance the amplitude of the 27-day cycle. Hence, during solar maximum the deviations from climatology are larger than during solar minimum. Climatology representation could follow the 27-day cycle given that real-time drivers are available. To study solar cycle variations of the ε_W and ε_C were collected for each epoch (15-min time step) over 11 years from 2003 to 2014. Then we apply a running window median filter with window size equals to one 30 days. The results are on Fig. 6.33.

The IRI performance is more sensitive to space weather events, and the weather representation performs rather well during disturbed time, which is seen from the

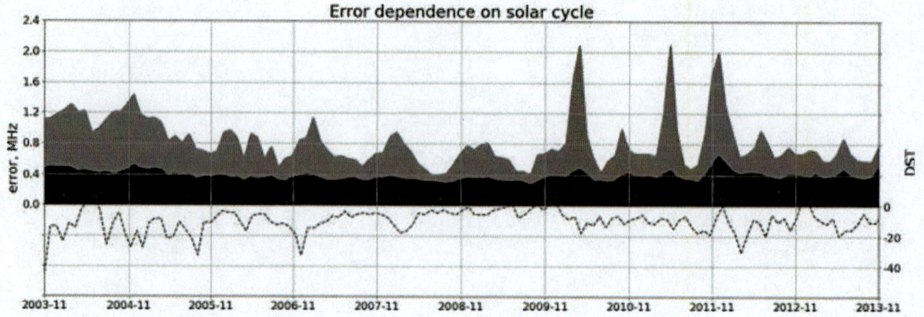

Fig. 6.33 Timelines of median-filtered ε_W (black) and ε_C (grey) from November 1 2003 to April 1 2014. Each vertical grid line corresponds to year timespan. DST index (dashed line) is given for reference to see geomagnetic activity. DST is given as 30-day median

2003–2004 data. During high geomagnetic activity the climatology representation is less accurate, which is obvious from the comparison of DST index and IRI. This is reasonable and was explained above. There are three major spikes shown by ε_W for 2010–2011 years that do not correspond to high geomagnetic activity, and average DST stays below -10 nT. This means that the source of the high deviations from climate is not in geomagnetic activity.

Post Sunset Uplift. We can consider the performance of the weather representation at the single station. We have looked for the event which repeats every day to obtain meaningful statistics. This allows us to accumulate enough data to exclude noise and show how assimilation can improve even regular ionosphere behavior. Though it is regular in time, it can be smoothed in space and hence underestimated by a model. Hence, we also require that the event is limited in space domain.

It is a well-known problem that the IRI underestimates the evening uplift of the ionosphere at magnetic equator in post-sunset time (Adeniyi et al. 2003). To understand this phenomenon, consider solar ionizing radiation and plasma recombination to neutral particles. These two governing processes drive the ionospheric plasma production rate. After the solar terminator passes at a sunset there is no more ionizing radiation from the Sun. The ionization is cut off. The recombination rate depends on density and drop off slower compared to ionization rate. Sunset occurs at smaller altitudes first and only then at larger altitudes. The density drops with altitude, so the recombination rate is larger at smaller altitudes. Thus, around sunset, plasma production rate maximum moves towards larger altitudes. We are interested in the production rate peak height rather than the magnitude of the peak. It causes ionospheric peak density to move to higher altitudes, which is equivalent to the increase of *hmF2*. To avoid a confusion, *foF2* is continuously decreasing, but moves to larger altitudes. This phenomenon is called the post sunset uplift of the ionosphere at the geomagnetic equator. The time scale of this process is about one hour, and the IRI basis includes as high as 6th order time harmonic, which allows to describe as fast as 4-h period variations. Adeniyi et al. (2003) assumed that higher order harmonics are needed to be able to track post-sunset uplift. However using the set of harmonics,

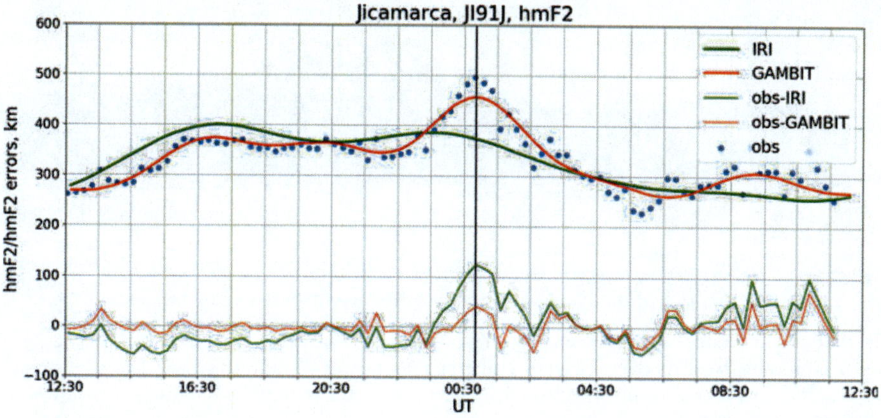

Fig. 6.34 Post-sunset uplift of the equatorial ionosphere. Data are for site Jicamarca, JI91J, (12.0° S, 76.8° W). Dots represent observations, red and green bold curves are IRTAM and IRI representations respectively. Red and green thin curves are differences between observations and model's representations for IRTAM and IRI respectively. Data are plotted versus UT. The vertical line corresponds to sunset time at 300 km altitude

we can reproduce sharper (than 6th order time harmonic) gradients at a given point. As a reminder *hmF2* originally has a different from *foF2* model in the IRI, with lower resolution. Although hmF2 was moved to use the same basis as foF2, it was built using original representation which is of lower resolution. Hence, fast variation might be missed due to the limited basis of hmF2 and can be reproduced by the *foF2* basis. Figure 6.34 shows a single post-sunset uplift as it is drawn by observations as well as climate and weather representations of this event. Observations are for March 13–14 2011. Geomagnetic latitude of the Jicamarca station is 1.7°.

Weather representation can capture the uplift. We examine weather representation on ability to catch the uplift routinely. Jicamarca data were retrieved for 2003–2014. We calculate sunset time at 300 km altitude daily since sunset time depends on the day of year. Each ε_W and ε_C was attributed with time mark observations that are calculated on. Time mark was calculated as a time shift relative to the sunset time. Observations that appear 12 h prior to sunset have negative shifts and the observations that appear after sunset have positive shifts. Data were averaged for every time shift. We collect ε_W and ε_C account for their sign, which gives us four sets negative ε_W and ε_C, positive ε_W and ε_C. We find the average for every time shift for every positive/negative set. This allows us to see how many observations correspond to the case when the model underestimates observations (uplift event). We expect to see positive maximum of ε_W just after sunset given ε_W and ε_C to be (observation - model). The result of this analysis is shown by Fig. 6.35.

Indeed, there is a maximum shown by ε_C just after sunset time. It is not clear from Fig. 6.35, but the count of data points for which ε_C are positive is twice as high as the count of data points for which ε_C are negative. This means that the IRI tends to underestimate observations two times more frequently than overestimate

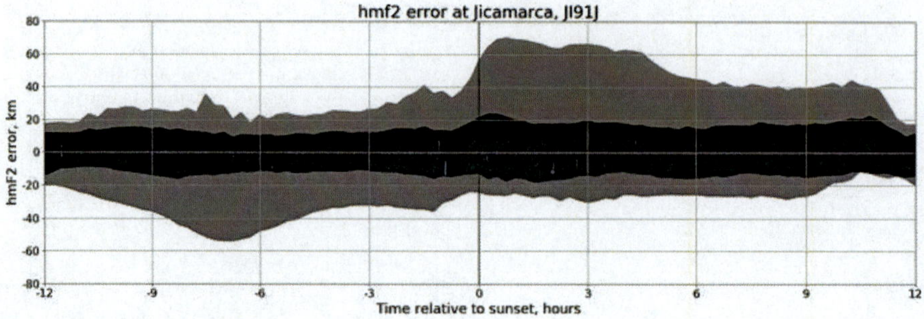

Fig. 6.35 Dependence of ε_W (black) and ε_C (gray) on time relative to sunset time at JI91J site. The time 1:00 means that data correspond to 1 h after sunset, and −2:00 means that data correspond to 2 h before sunset. Zero corresponds to the time of local sunset at 300 km altitude. Upper/lower envelopes correspond to positive/negative ε_W and ε_C

observations. When we consider positive ε_C we see it maximum just after sunset is greater than values of negative ε_C (in absolute values) for the same time. Hence the climate representation underestimates observations more frequently and with greater error. There is no strong positive peak in ε_W data at near sunset time, while some residual error remains. Basis used in the model is enough to reproduce a major fraction of the uplift. The residual error is assumed to be due to lack of the higher harmonics in the basis. Weather representation successfully reproduces post-sunset ionospheric uplift on the equator. The only concern is that the ratio between occurrences of positive and negative errors stays the same as for climate representation. At this point, the assumption of Adeniyi et al. (2003) about the inability of the IRI formalism to reproduce short time scale uplift is only partially true. As we see in Fig. 6.35, the uplift can be almost resolved.

Forecast capability. GAMBIT is designed for ionospheric weather nowcasting. However, it assimilates data in frequency domain thus we can assume weather could work on the short range forecast. Figure 6.36, shows how we estimate forecast capability. One day observation history is used for assimilation. Calculation for any time in one-day-history denoted is backcast—we perform calculations that are “surrounded” with data. We expect uniform accuracy of model representation in backcast mode. Now we take some time beyond this one-day-history, all observations are to the left of the time of interest. We expect gradual degradation of weather representation while moving further from observations. We expect no climate representation dependence on time. How far can we move from observations to have any improvement? We stated ε is calculated for the last point in the one-day-history window in the beginning of the current section. Here we use a different approach. We collect ε_W and ε_C for every observation in the window and attributes them with time shift with respect to assimilation time. Positive shift means forecast mode, negative—backcast mode.

The backcast performance is horizontal, which means that, on average, all observations in the past are presented equally accurate. Now, consider the part of the plot corresponding to the positive hours. The labels on the horizontal axis show the

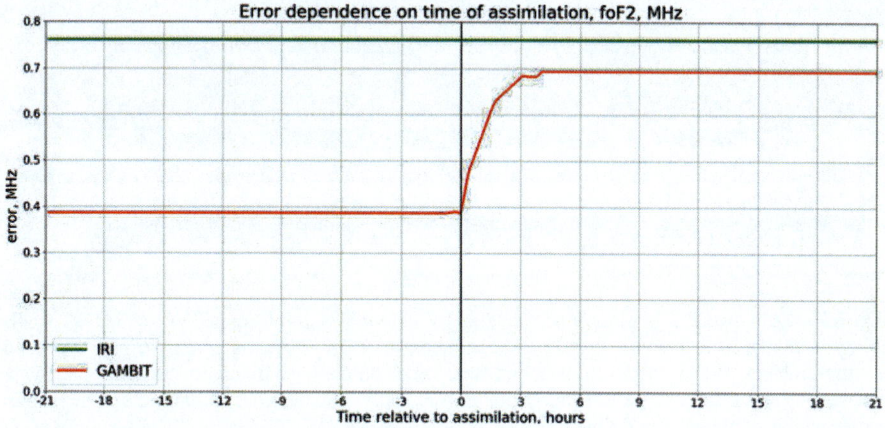

Fig. 6.36 Weather representation forecast capability and "backcast" accuracy by means of ε_W (red). Labels on the horizontal axis show hours after/before assimilation. Plot is based on observations collected for 2010–2013 years. Value of ε_C (green line) is given for reference and does not depend on time. Dashed line shows time of assimilation with respect to which backcast and forecast are denoted

number of hours passed since the assimilation time. For example, a value at +3 h means that coefficients were calculated using observation obtained three hours ago. Weather representation accuracy is higher as time of interest is closer to assimilation time. It is more interesting what we can observe 4 h after assimilation time. Although ε_W takes a constant value around 0.69 MHz, it is not equal to ε_C, which is 0.76 MHz. The reason for that is the model driver mismatched. For this analysis we used 12-month average sunspot numbers, Rz12, to drive the IRI. Measured sunspot number was available only for 2010, and for 2011–2013 sunspot numbers were extrapolated, thus were not precise. While it could be considered an analysis flaw, it shows climate representation dependence on driving parameters update. Rz12 is available with a 12-month delay, so some extrapolation is always performed. We can substitute this driver with another one whose update rate is faster. However, the point here is that data assimilation is robust to such things as model driver mismatch. Since sunspots numbers are not available immediately in real-time, weather representation avoids this problem using real- time observations that have solar activity "built in".

Spatial limits. We have considered the performance of the weather representation in different cases. We used the same data for assimilation and validation as we noted in the beginning of the section. All analysis completed above can be considered as a test on the mathematical basis ability to reproduce observations in various conditions. However, the study wouldn't be full without having a different validation set. To fill this gap we now "remove" some stations from the assimilation dataset and then bring them back when we perform validation. This allows us to see how far in space assimilation is still valid. We perform multiple assimilation runs to achieve weather representation with different site sets. The sites are excluded due to the reason that

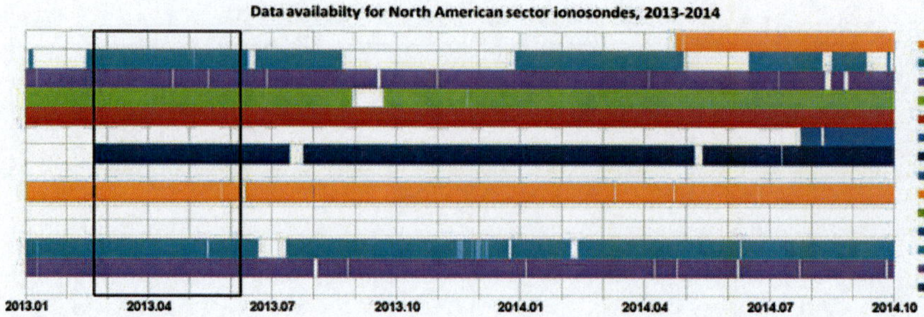

Fig. 6.37 Data availability plot for sector from 15° N to 45° N and from 65° W to 130° W. Availability is shown from January 1, 2013 to September 30, 2014. Different colour strips correspond to different sites, the legend is on the right. Sites order on the legend and plot is the same. Black rectangle denotes the time period chosen for further study. Sites HAJ43, DS932, CS839 were not operating during the analysed time span

we still need ionosonde observations for validation; any other model or observation type would show assimilation performance as we want it.

The ionosphere is a medium which displays high variability, but it does not change randomly. ionospheric characteristic, *foF2* or *hmF2*, at two spatially close locations should not be much different from each other. It could change during extreme space weather events. To understand how far from observational sites locations the weather representation is still reliable, we assigned some sites to be control sites. Control sites do not participate in the assimilation, they are only used to calculate ε_W and ε_C. We used only foF2 observations in this study, since hmF2 and foF2 mathematical basis is the same. The foF2 characteristic was preferred over hmF2, because foF2 measurements are more precise, which is illustrated by histogram in Fig. 6.27.

We examine different regions and choose the North American sector, since it allows us to draw conclusions from data analysis. Ionosondes within geographical region from 15° N to 45° N by latitude and from 65° W to 130° W by longitude. We are interested in uniform data coverage over the time interval under investigation. We can see data availability for the region and 2013–2014 years in Fig. 6.37. All sites except PRJ18 belong to the middle latitude region by means of geomagnetic coordinates. We assume data are available for a day if there are at least 24 data points and data points are distributed uniformly in time. Uniform data distribution allows us to avoid processing artifacts.

We chose a time interval from 2013.02.19 to 2013.06.11, when data from eight sites were continuously available. The sites are PA836 (Pt. Arguello 34.8° N, 120.5° W), IF843 (Idaho National Lab 43.8° N, 112.7° W), BC840 (Boulder 40.0° N, 105.3° W), AU930 (Austin 30.4° N, 97.7° W), EG931 (Eglin 30.5° N, 86.5° W), PRJ18 (Ramey 18.5° N, 67.1° W), WP937 (Wallops Island 37.9° N, 75.5° W), MHJ45 (Millstone Hill 42.6° N, 71.6° W). The geographical locations of the sites are shown in Fig. 6.38.

We performed several tests using following configurations:

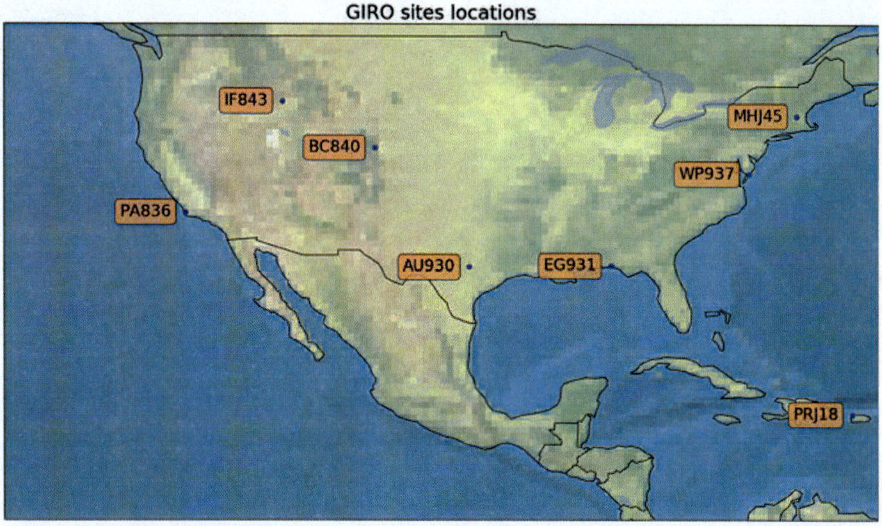

Fig. 6.38 Geographical locations of the sites used for spatial limits study in the North American sector is shown. Each site location is shown by star and labelled appropriately

1. Internal test: observations from PA836, IF843, PRJ18 and MHJ45 sites were assimilated, BC840, AU930, EG931 and WP937 sites were assigned control sites.
2. External test: opposite of the internal test.
3. Advanced external: only observations from AU930 and EG931 sites took part in the assimilation, BC840, PA836, IF843, PRJ18, WP937 and MHJ45 were assigned control sites.
4. Higher latitude test: observations from PA836, AU930, EG931 and PRJ18 sites took part in the assimilation, IF843, MHJ45, BC840 and WP937 were assigned control sites.
5. Lower latitudes test: opposite of the high latitude test.

To estimate weather representation on control sites (ε_W, control), we also provide errors when all data were assimilated (ε_W, control) as well as climate representation errors (ε_C). The results for all tests are combined in Table 6.4.

As it is expected, ε_W for control sites is lower than ε_W when this site participates in assimilation. For the internal test improvement factors for the sites BC840 and WP937 are larger than for other two sites AU930 and EG931. This resulted from the fact that BC840 and WP937 are geographically closer to the nearest site participating in the assimilation. The accuracy is worse for the external test than for the internal test. We consider a pair of sites that are closest to each other: BC840-IF843. The improvement for BC840, when IF843 was excluded, was 1.75 and the improvement factor for the IF843, when BC840 is excluded, is 1.39. We assume it is due to the geometry. In the internal test control sites are wrapped into polygons formed by PA836, IF843, PRJ18 and MHJ45. In the external test control sites are outside the polygon formed

Table 6.4 Results of five spatial limits tests

Control site	ε_W, control	ε_W, assim	ε_C	Improvement, control	Improvement, assim
Internal test					
BC840	0.44	0.35	0.77	1.75	2.20
WP937	0.54	0.37	0.79	1.46	2.14
AU930	0.64	0.41	0.86	1.34	2.10
EG931	0.64	0.41	0.82	1.28	2.00
External test					
IF843	0.51	0.37	0.71	1.39	1.92
MHJ45	0.63	0.37	0.78	1.24	2.11
PA836	0.63	0.31	0.81	1.29	2.61
PRJ18	1.17	0.43	1.23	1.05	2.86
Advanced external test					
BC840	0.62	0.35	0.77	1.24	2.20
WP937	0.72	0.37	0.79	1.10	2.14
IF843	0.60	0.37	0.71	1.18	1.92
MHJ45	0.75	0.37	0.78	1.04	2.11
PA836	0.69	0.31	0.81	1.17	2.61
PRJ18	1.17	0.43	1.23	1.05	2.86
Higher latitude test					
BC840	0.49	0.35	0.77	1.57	2.20
WP937	0.60	0.37	0.79	1.32	2.14
IF843	0.51	0.37	0.71	1.39	1.92
MHJ45	0.63	0.37	0.78	1.24	2.11
Lower latitude test					
PA836	0.67	0.31	0.81	1.21	2.61
AU930	0.73	0.41	0.86	1.17	2.10
EG931	0.65	0.41	0.82	1.26	2.00
PRJ18	1.19	0.43	1.23	1.03	2.86

We calculate average of ε_W and ε_C over each site, units are MHz. Time interval for averaging is highlighted on Fig. 6.37. Two left columns contain unit less improvement factors

by BC840, AU930, EG931 and WP937. Relative position of the sites also matters. For the external test, the geometry was such that the second close sites, which are AU930 for IF843 and EG931 for MHJ45, are on the one line with the closest sites BC840 and WP937 correspondingly. We assume that the effect of assimilation of AU930 (for IF843) and EG931 (for MHJ45) observations is suppressed by BC840 and WP937 correspondingly. For the internal test, the geometry is different, and both IF843 and PA836 contribute the value at BC840, which turned out to give better results.

From the advanced external test we see that even two sites, AU930 and EG931, can give a positive effect for weather representation at spatially far sites, although this effect is small.

The results of the higher latitude test and lower latitude test show almost no improvement over the climatology. Assimilation of the data from the lower latitudes gives a larger positive effect on the simulation result at higher latitudes than vice versa. The results of the two last tests show that ionospheric characteristics vary more significantly with latitude than with longitude. In other words, the assimilation results can be trusted further in zonal direction than in meridional direction.

We performed qualitative analysis of spatial limits of assimilation. Now we could estimate it qualitatively. It is illustrative to see site-to-site correlation of observations. Since ionospheric characteristics are local time dependent, we use a local time reference frame. We calculate local time offset between any two sites, using their geographical longitudes. We assume that data for the same local time on similar geomagnetic latitude should have higher correlation coefficient. To remove the main trend (climate), we subtract IRI model values from the observations and consider ε_C as an input for correlation analysis. Since space weather is contained in ε_C, correlation of ε_C for two sites should lead to weather representation correlation for the same sites. The correlation factor was calculated using Pearson product-moment correlation coefficient:

$$r = \frac{\sum_{i=1}^{n}(x_i - \overline{x})(y_i - \overline{y})}{\sqrt{\sum_{i=1}^{n}(x_i - \overline{x})^2 \sum_{i=1}^{n}(y_i - \overline{y})^2}} \tag{6.17}$$

where x_i and y_i are ε_C for different sites. Correlation coefficient is in the range from -1 to 1. Using the same station as for analysis above we can obtain following results listed in Table 6.5.

Data from spatially close sites correlate at higher extent, for example, sites AU930 with EG931 and BC840 with IF843. It is interesting that PRJ18 station almost does not correlate with others. This site is close to low latitude region, while the others are pure middle-latitude stations. In general, data in Table 6.5 agree with data in Table 6.4. Consider two sites with high correlation and assume that one of these sites takes part in the assimilation and another does not. Then the weather representation at the control sites will be better than the climatology representation. However, this behavior is expected only if two sites are isolated from others. We can make this simplified conclusion for model representation at any given location when data from multiple sites are assimilated.

We can summarize spatial limits tests by introducing the parameter that shows how the weather representation works at control sites data with respect to the same data when they assimilated. We construct following parameter τ:

$$\tau = \frac{\varepsilon_C - \varepsilon_{w,control}}{\varepsilon_C - \varepsilon_{w,assim}} \tag{6.18}$$

Table 6.5 ε_C correlation for the sites in the North American sector

	AU930	BC840	EG931	IF843	MHJ45	PA836	PRJ18	WP937
AU930		0.560	0.671	0.355	0.150	0.479	−0.024	0.363
		1270	1070	1990	2690	2190	3350	2200
BC840	0.560		0.480	0.641	0.217	0.534	−0.062	0.347
	1270		2000	740	2820	1460	4360	2575
EG931	0.671	0.480		0.242	0.376	0.325	0.064	0.500
	1070	2000		2730	1890	3200	2370	1300
IF843	0.355	0.641	0.242		0.200	0.552	0.012	0.194
	1990	740	2730		3305	1205	5080	3170
MHJ45	0.150	0.217	0.376	0.200		0.147	0.187	0.517
	2690	2820	1890	3305		4280	2710	620
PA836	0.479	0.534	0.325	0.552	0.147		−0.013	0.245
	2190	1460	3200	1205	4280		5540	4005
PRJ18	-0.024	−0.062	0.064	0.012	0.187	−0.013		0.072
	3350	4360	2370	5080	2710	5540		2305
WP937	0.363	0.347	0.500	0.194	0.517	0.245	0.072	
	2200	2575	1300	3170	620	4005	2305	

Each cell is divided in two: the top part contains the correlation factor; the bottom part contains distance between sites on the sphere (great circle distance) in kilometers. Table is diagonal symmetric

τ equals 1 when weather representation at the site is equally accurate for cases when it is a control site and when the data from the site are assimiated. Factor τ is equal 0 if weather representation is the same as climatology representation. The letter means assimilation of observation from other sites did not contribute to weather representation at the control site. For each row in Table 6.4 we calculate τ. The plot of τ versus average distance to two closest neighbors is in Fig. 6.39.

We see that τ drops to 0.5 at about 1300 km distance. Thus, within 1300 km ionospheric weather representation is noticeably improved by the assimilation. Weather representation converges to the climatology representation beyond 1300 km.

6.5 Factors Affecting GNSS Positioning Accuracy

6.5.1 Instrumental Approaches to Improve Positioning

The global model of GNSS positioning accuracy should not be purely global. We have seen that global modeling is rather smooth even when in weather representation. The smooth ionospheric distribution can be removed by ionospheric models or combinations of phase measurements known as ionospheric-free combination. TEC is

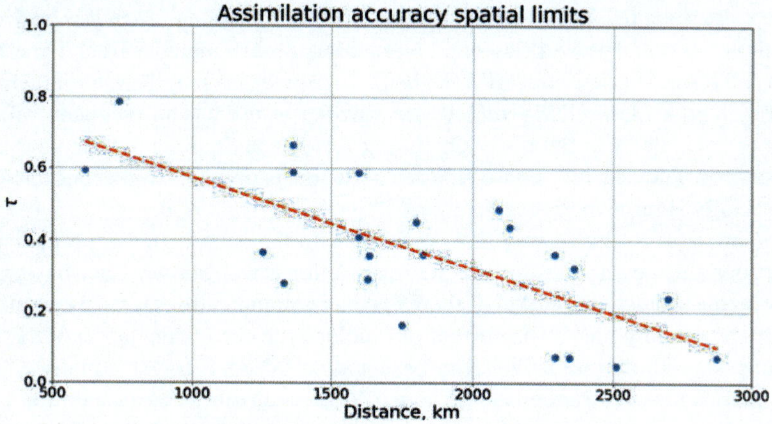

Fig. 6.39 Spatial limits of data assimilation. Points correspond to the rows of Table 6.4

derived from geometry free combination excluding geometry of sounding. Contrary ionospheric free combinations exclude the influence of plasma on signals. Both ionosphere-free and geometry-free combination uses dual frequency measurements. Single frequency receivers still need ionospheric models for correction. Anyway, the most "GNSS-effective" events appear on the regional and local scales. Traveling disturbances and irregularities should not be modeled on a global basis as we consider in the previous sections, while global weather representation can be used as a background. The local models should be implemented in order to bring them into the game. The models might be of the same concepts as global models, e.g. Kalman filter or EOFs analysis, given that the concepts are suitable for regional and local use. The models also could arise as explicit forms of the TID or ionospheric plasma bubbles, given theoretical understanding of their original shape and evolution in time. Any type of irregularity or TID models are more sensitive to data streams than global models, since we want them to track processes that are tinier. TID may have an amplitude of 0.1–0.5 TECU superimposed on 50 TECU background plasma, by means of integral GNSS measurements.

Given expensive massive GNSS receivers, generating 30+ GB data in a day, we can do all this fancy processing and obtain good stable solutions. The problem of GNSS accuracy monitoring is to give users with affordable GNSS receivers and with limited computational resources the ability to perform correct coordinate definition and to understand that accurate definition is not possible under current conditions a receiver is subjected to. The latter might be even more important, since it may cause equipment damage, if GNSS receiver is for moving platform positioning. The number of users whose positioning we should improve could reach millions of devices. Not all of them are critically dependent on GNSS positioning. However, the users that do critically depend on GNSS may operate the equipment that is several times more expensive that a receiver itself. For such users it is a matter of survival to have robust positioning services and understanding when they should not be trusted.

There are several approaches to improve GNSS positioning. They can be divided into autonomous and networking ones. Network approach includes Real-Time Kinematics (RTK) and Differential GPS (DGPS). To provide each of them, ground infrastructure of high class GNSS receivers networks is used. The receivers of these networks are called Continuously Operating Reference Stations (CORS) or simply base stations. The density of base stations networks is defined from economical and other socially significant demands.

RTK operates in such a way that corrections to the coordinates are broadcasted. Corrections are known since the base station location is well known. Each base station calculates the corrections and sends them via the communications networks including cellular and satellite links. The corrections include the current deviation of the XYZ coordinates from the true value. Non-base-station GNSS receiver, usually called a rover, applies those corrections to its own XYZ coordinates. The assumption is that media influence is the same at the rover station and the base stations. Closer base stations should have better correlation with the rover station. Basic RTK requires only one directional data stream from the base stations to the rover. There are modifications to basic RTK. Chen et al. (2011) describes the approach called Virtual Reference Station (VRS). There is an additional server between base stations and rover stations. The server receives data from the base stations and using appropriate models expand real data to a rover location. Each rover will receive its own data stream in standard format. Basically, we create the virtual base station at the rover location using physical or empirical modeling to estimate the state of propagation media. Chen et al. (2011) also introduce RTX service that should break the limits of VRS approach, mainly to ensure network work with 100 and more base stations. Base stations could greatly improve the positioning accuracy, however there are errors sources that are almost uniform over RTK covered area. Bogusz et al. (2015) used the 5-year data form base stations located in Poland to estimate common mode errors for this region.

With DGPS, satellite phase and group measurements are broadcast using the RTCM protocol, which are combined at the target receiver to eliminate the influence of instrumental errors and inaccuracies introduced by the propagation medium. DGPS allows for finer calculations than RTK and can be used at large distances from base stations, which reduces the requirements for base station infrastructure (Puente et al. 2013). Since GNSS base station networks are currently operating it seems like a solved problem and there are no needs in GNSS accuracy improvement models. However, it is only a cursory look. Such models or parts of the models might be implemented in such systems. Also, continuous improvement and development of tools that enhance positioning is needed. New autonomous application challenges the existing state-of-art algorithms. The food delivery rovers might be a good example. While self-driving cars are already part of our lives, they operate in well controlled environment, roads, among similar objects, cars. The drones should be more flexible hence should rely more on GNSS positioning to supply other “sense organs”.

The Precise Point Positioning (PPP) approach uses data from a single receiver. Additional information about the state of the GNSS satellite constellation and the state of the environment is used to improve the quality of the coordinates. To refine the positions of the satellites and the time scale, corrections to the ephemeris and time of the onboard clock are incorporated into the solution. The state of the ionosphere includes the influence in PPP calculations through models or other obtained distributions of TEC, such as global ionospheric maps. A distinction should be made between real-time and post-processing applications. During the construction of infrastructure facilities, we can allow long-term measuring campaigns, which increases the accuracy of coordinate referencing. A rover makes subsequent measurements in new coordinates. Having a pair of receivers eliminates most of the instrumental errors and some propagation environment ambiguity for moving receivers. For stationary receivers, post-processing is still the most accurate method. Accurate 3D electron density distribution obtained by observations assimilation enhanced by TID models also driven by observations could greatly improve awareness of positioning accuracy. However smaller irregularities are of random nature and are harder to deal with. Let's us consider first what we can deal with using combination of the signals and then return to the propagation environment issues. Figures 6.40 and 6.41 represent different measurements combinations that are used to improve positioning by removing some instrumental and propagation media errors (Fig. 6.41).

Single Difference (between-receivers)—two receivers and one satellite are used, p_r^a and p_q^a beams. This combination allows us to level the receiver's clock error, as well as the inaccuracy of the satellite's location determination. To some extent, it is possible to exclude the influence of the propagation media. Since the baseline formed by two receivers is much smaller than the distance to the satellites (~20,000 km), the distance between satellite-receiver beams in the ionosphere can be considered equal to the baseline. The condition under which the rays pass through approximately

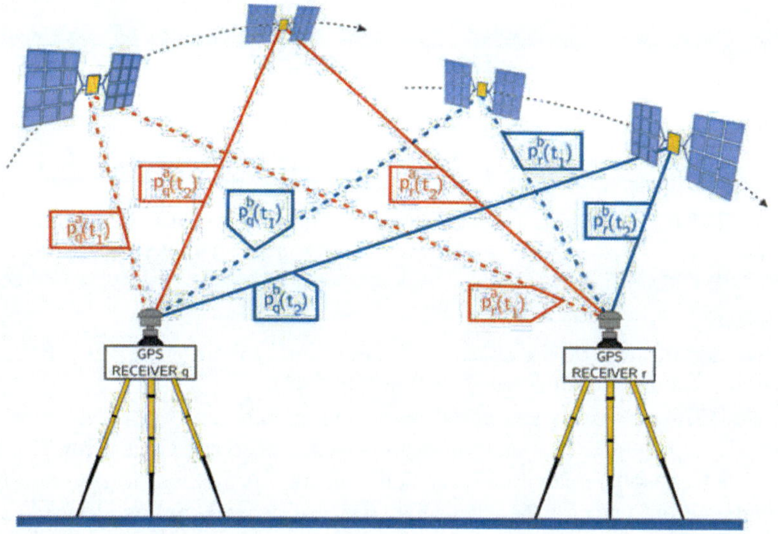

Fig. 6.40 Receiver satellite beams for various DGPS combinations. Shown are two receivers (q) and (r), and two satellites (**a**) and (**b**). Satellites are shown for two different epochs

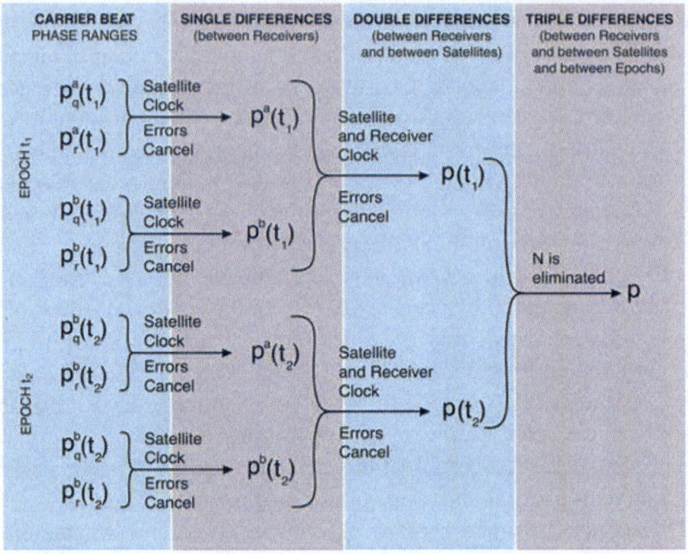

Fig. 6.41 Single, double и triple difference relations (https://www.e-education.psu.edu/geog862/node/1727)

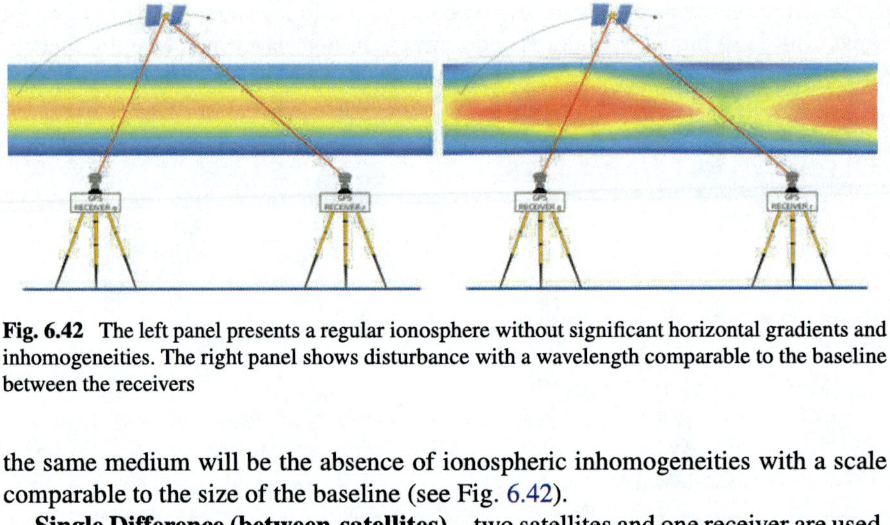

Fig. 6.42 The left panel presents a regular ionosphere without significant horizontal gradients and inhomogeneities. The right panel shows disturbance with a wavelength comparable to the baseline between the receivers

the same medium will be the absence of ionospheric inhomogeneities with a scale comparable to the size of the baseline (see Fig. 6.42).

Single Difference (between-satellites)—two satellites and one receiver are used, p_r^a and p_r^b line-of-sight. This method eliminates the receiver's clock error. The influence of the troposphere can also be excluded using this method. It is more difficult to exclude the influence of the ionosphere, since it depends on the elevation angle. Subionospheric points for various satellites can be up to several thousand kilometers apart. The ionosphere at such distances can exhibit different behavior, even when there are no space weather effects. For example, one line-of-sight contained in the mainly dayside ionosphere and another in the nightside ionosphere, during the passage of the solar terminator. The receiver's clock error is also not fixed. This method does not significantly improve the positioning accuracy.

Double Difference—two receivers and two satellites are used, beams p_r^a, p_r^b, p_q^a and p_q^b. Allows you to eliminate the satellite clock error that remains after the between-receivers single difference.

Triple Difference—two receivers and two satellites are used, beams p_r^a, p_r^b, p_q^a and p_q^b, for two epochs. Thus, difference allows us to eliminate integer phase ambiguity.

The electron density changes relatively smoothly during TID propagation, although the amplitude can reach several TECu Fig. 4.8. The relative velocity of the sub ionospheric point and the direction of TID propagation are such that the GNSS measurements even with 30 s sampling rate makes it possible to obtain a smooth series. The existence of irregularities, such as equatorial plasma bubbles, change observations significantly. Two successive samples may travel through the electron density regions, such that the line-of-sight integral differs significantly (Fig. 6.43).

According to Kintner et al. (2007) inhomogeneities with a size equal to the first Fresnel zone are the source of phase scintillations. However, Yeh and Liu (1982) consider the inhomogeneities with a size 10–100 times larger than the first Fresnel zone have the strongest effect on the phase. In the first case, we are dealing with diffraction effects and cannot calculate TEC during phase scintillations. In the case when distortions can be considered refractive, TEC can be calculated. McCaffrey and Jayachandran (2019) considered refractive variations of the signal due to

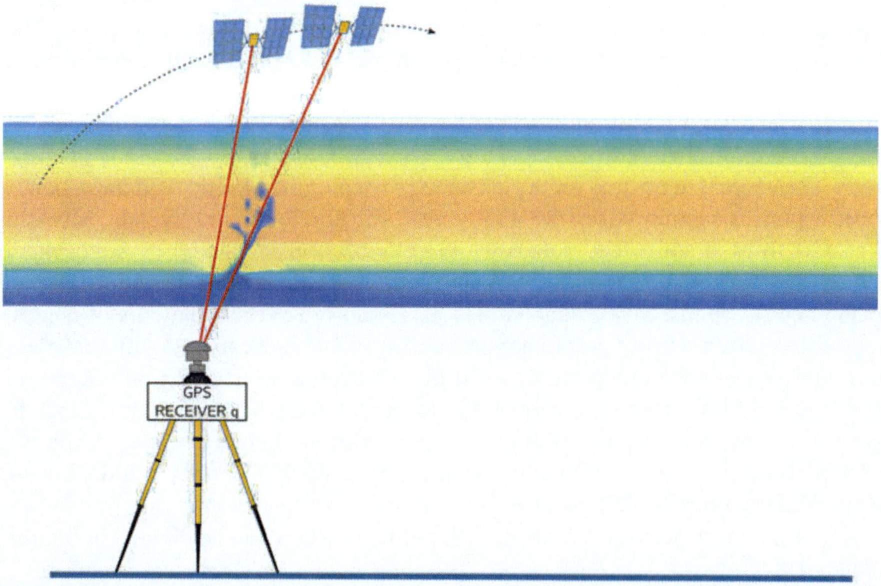

Fig. 6.43 Two line-of-sight in a media with strong irregularities

enhanced plasma motion could appear in phase scintillations. Caused by different physical processes these variations appear in scintillations index due to calculations. GNSS signals amplitude scintillations are considered as a result of signal diffraction: "changes in the refractive index along the ray path may cause small decreases in the amplitude, or power, of the signal due to reflection and absorption, but will not cause variations in the amplitude of the signal". This means that the entire ionospheric slab could change the amplitude compared to the propagation in free space, however variations that contribute to phase variations give negligible contribution to amplitude variations. In the absence of a mechanism for separating phase scintillations into diffractive and refractive ones, one should be careful when calculating these derivatives of TEC, since the TEC itself may be undetermined. For example, in (Carrano et al. 2019), the ratio of ROTI to the amplitude scintillation index is considered. Although it is technically possible to calculate TEC and then ROTI, it will no longer have its original physical meaning of rate-of-TEC. Scale of the scintillations-effective irregularities is 365 m on 350 km height (McCaffrey and Jayachandran 2019); scales (wavelength) of the TID casing TEC variations with 2–60-min periods are 6–250 km; ROTI effective irregularities are of the 1 km in size (Jacobsen 2014). Jacobsen (2014) used statistical analysis of sampling rate influence on the ROTI value. ROTI is more than 5 times smaller for 30 s sample rate than for 1 s, which is purely calculation issues. ROTI can be used as a proxy to scintillations, while it is not directly related to irregularity causing scintillations, given irregularities spectrum is continuous. ROTI and smaller scintillation-effective irregularities can cause random TEC and phase variations respectively if presented on a line-of-sight.

It is necessary to consider not only the influence of the ionosphere (space weather) but also the influence of the troposphere (weather), when studying the quality of positioning. Troposphere is accounted for through the so-called dry and wet delays. The latter considers the amount of water vapor in the atmosphere, and it draws more dynamic than dry dela. Appropriate models can be used to estimate the magnitude of this delay (Tunalı and Özlüdemir 2019). Another important influence of the neutral atmosphere are winds. Neutral winds could move plasma due to collisions. The geomagnetic field force plasma to move along magnetic field lines, hence vertical plasma motion arises eventually even for horizontal only winds. Miyoshi et al. (2018) used numerical models with Ground-to-topside model of Atmosphere and Ionosphere for Aeronomy (GAIA) to reproduce TEC and Ne variations driven by the neutral winds. Sivakandan et al. (2021) showed vertical patterns of neutral wind and electron density perturbations (see Fig. 6.44). We can note different electron density perturbations patterns for different neutral wind direction relative to magnetic field line. One case correspond to downward plasma motion another to upward plasma motion given plasma is captured in the magnetic field. While the models may not be able to predict phase and amplitude of TID, they could allow us to build climatology of TID occurrence.

As for the influence of the plasma shells of the Earth it can be divided into three main ones:

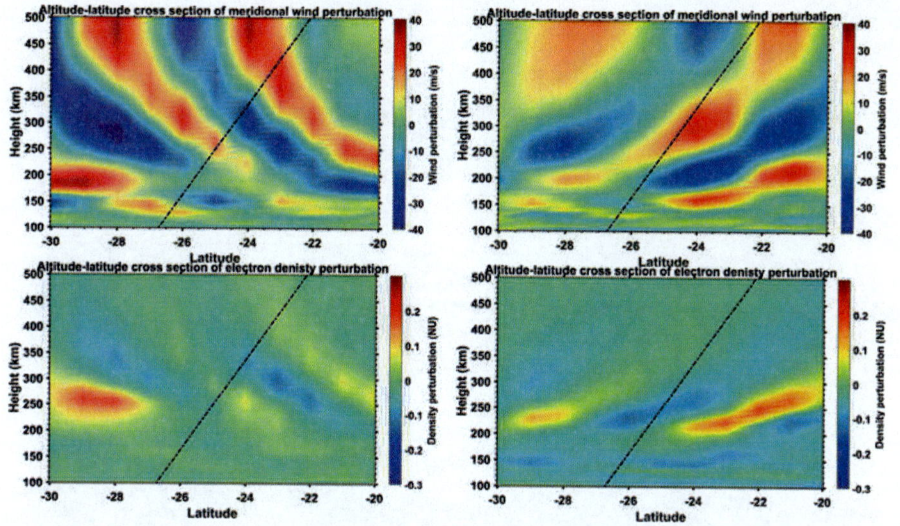

Fig. 6.44 Results of neutral wind and electron densisty simulations at South Africa at 13:42 LT on 8 January, 2011. Altitude–latitude cross sections of perturbations are shown. Magnetic field line with constant dip angle (~ 45°) at 250 km is shown as black dotted lines in all panels.Left and rigth panels show different wind direction relative to magnetic field line.(Sivakandan et al. 2021)

- signal phase delays in a smoothly changing medium—the medium changes more slowly than the movement of a subionospheric points. Global climatology and weather models improve the definition of the environment for the rays passing through different paths in space. For example, 10 TECu induces a delay of 5.4 ns (1.6 m) on the travel time of GPS L1 signals, and a delay of 8.9 ns (2.7 m) on the travel time of GPS L2 signals (Bergeot et al. 2014),
- the influence of moving ionospheric disturbances—the movement of disturbances is comparable to or greater than the movement of a subionospheric points. TID models are required to account for this kind of disturbances. A model might be of generic view and then observations are used to fit parameters,
- the influence of ionospheric inhomogeneities and drifts, that can significantly change phase observable in a short time (phase scintillations) or even cause diffraction effects and as the result signal interference at the location of the receiver (amplitude scintillations).

6.5.2 *Spatial and Temporal of Factors that Affect GNSS Positioning*

The geometry of GNSS sounding for ground-based receivers varies from location to location. The strongest dependence is on the latitude. Figure 6.45 shows the diurnal distributions if the ionospheric pierce points mapped on the earth surface.

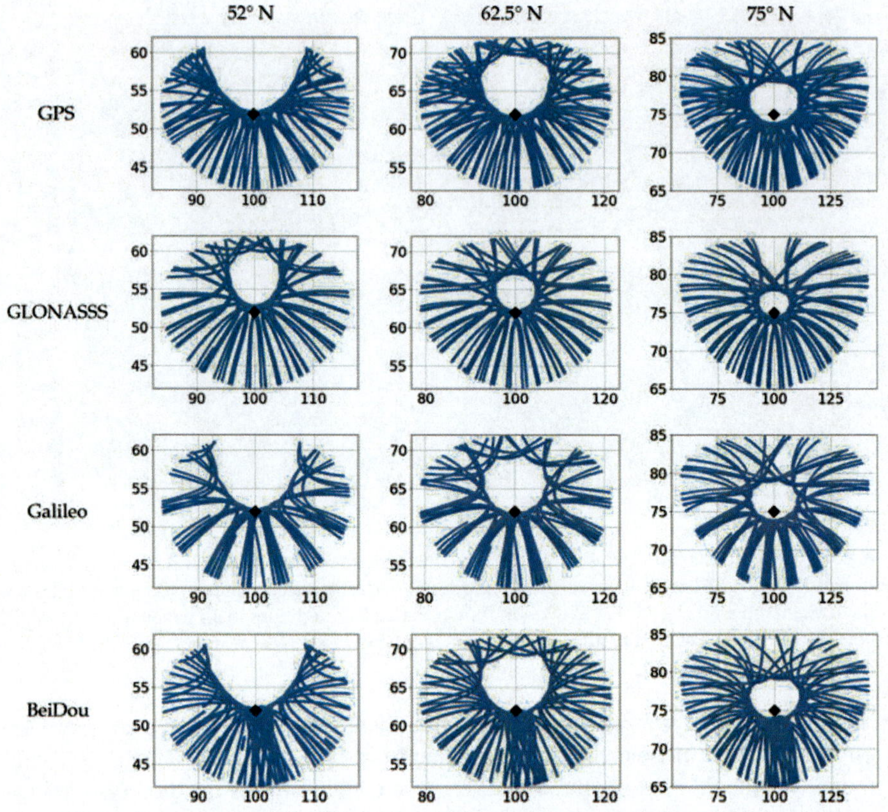

Fig. 6.45 Diurnal ionospheric points for GPS, GLONASS, Galileo, BeiDou (top to bottom). The rhomb marks the station position. The station longitude is 100° E. Left column: at 52.2° N, mid column: at 62.5° N, right column: at 75° N (Yasuykevich et al. 2022)

It is clearly seen that the local ionosphere, right above the receiver, does not contribute to the observations for high latitude receivers. For the middle and high latitude ionosphere there are always blind spots, where observations do not come from. GNSS signal monitoring for specific receivers requires accurate disturbance and irregularities models supported by the real-time observations from this receiver and from the neighbor receivers. Neighbor receivers allow tracking of the ionosphere in blind spots and in the area not currently covered by the observations (see Fig. 6.46 for 1-h observation coverage).

The local and regional nature of the factors that may cause disruption of GNSS signal accompanied by changing sounding geometry lead researchers for local and regional models for GNSS quality monitoring. Abdelazeem et al. (2017) describes a regional model for TEC obtained using GPS and BeiDou measurements. TEC models are proposed for single-frequency users that could construct an ionosphere free combination. The input for the models is the vertical TEC calculated using

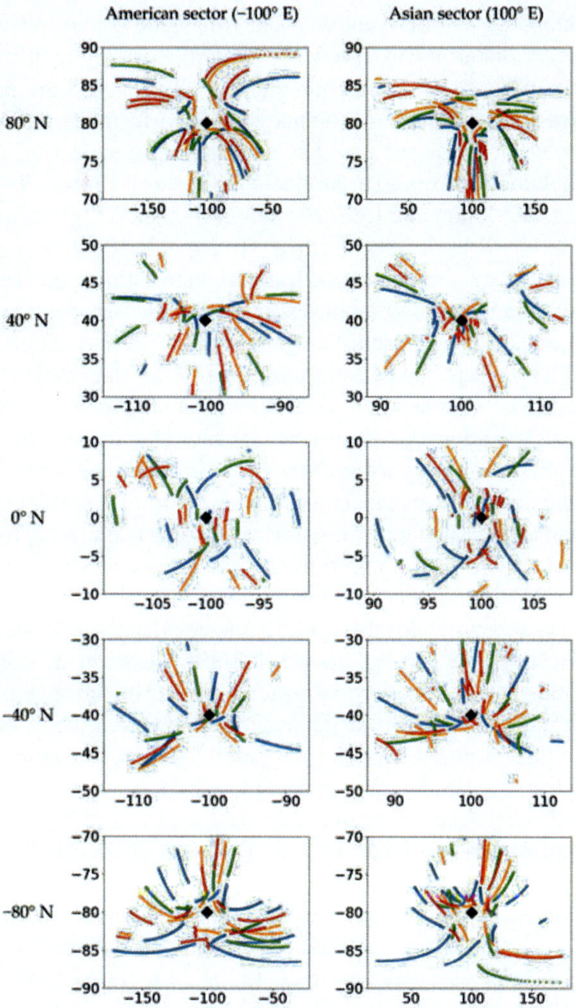

Fig. 6.46 Ionospheric points within 12–13 UT 10 October 2020 at different latitudes in the American (mid column, −100° E) and in the Asian (right column, 100° E) sectors. Dark blue points correspond to GPS, orange—GLONASS, green—Galileo, red—BeiDou. The black rhomb marks the station position. The station latitude is indicated on the left (Yasuykevich et al. 2022)

slant TEC from receivers, 1° × 1° resolution is obtained using weighted sum of the vertical TECs. Compared to 2.5° × 5° resolution of global ionospheric maps, this model may contain more details. Cheng et al. (2021) used tomography to recover 3D distribution with 2° × 2° × 20 km spatial resolution every 1 h. IRI electron density and TEC are the inputs for the calculations. The resulting data driven distribution is used to monitor large-scale long-living ionospheric dynamics up to 1000 km during the geomagnetic storm. Bergeot et al. (2014) introduce 0.5° × 0.5° near real-time data products including absolute vertical TEC maps for current time and difference of latter with 15-days median. TEC map is converted to L1 range error for convenience. TEC variability reaches 7 TECU for the time span of 15-min shown during the 2003 Halloween storm. This rapid localized change caught by regional mapping can be smoothed out from global ionospheric maps. Medium and small-scale disturbances should be caught using regional mapping with high spatial and temporal resolution. Yoon and Lee (2014) used the Korean GNSS network to map medium-scale TIDs. The mapping is performed based on slant TEC series converted to vertical ones using standard mapping function. High resolution regular maps are obtained and could be subjected to spectral analysis to retrieve modes and refer them as TID. Yoon and Lee (2014) found that TID passages were also accompanied with the ionospheric irregularities. The irregularities intensity was derived based on ROTI, however no scintillations indexed were estimated. The latter study used a technique for ionospheric gradient estimation in the automatic regime proposed by Jung and Lee (2012) proposed. The gradients of different nature including gradients caused by TID passage can be estimated. The GNSS data as well as space weather indexes, Kp and DST, serve as an input for the technique. Space weather only indexes limit the study to geomagnetic effective events, leaving such intensive events as sudden stratospheric warming aside. Nevertheless, gradients are good candidates for the proxy to the irregularities that can cause severe impact on GNSS signals quality. Some gradients such as solar terminator caused ones can be easily located using sunset/sunrise times. Gradients caused by the ionosphere morphology could be estimated from the global ionospheric models for both weather and climate representation. Gradients that appear randomly should be detected and tracked using observations. Hernández-Pajares et al. (2006) showed the link between 20-min-period TID occurrence rate and error of ionospheric correction. TID occurrence is higher during the daytime in winter. The local-time dependence during summer is not so noticeable and is slightly higher at nighttime. There is small TID intensification on sunrise over the span of a year. We can obtain the same picture for any given GNSS receivers' network, which allows us to estimate local time dependence of TID occurrence probability and thus ionospheric correction accuracy. Lee et al. (2021) showed longitude dependence of the middle scale TID occurrence rate. They also showed that TID occurrence rate increases with the decrease of solar activity (see Fig. 6.47). The results hold for both hemispheres and seasonal patterns are pronounced for any solar activity.

Hernández-Pajares et al. (2006) also introduced the method to estimate TID parameters from slant TEC series using small networks. After all necessary preprocessing of TEC series is done, they use linear detrending in a window. The window size depends on the TID time period under investigation. The wave-like residuals were subjected to the FFT. To ensure low influence of geometry factors at low elevation angles elevation mask of 50° was applied. Assuming the same TID is observed by the receivers of a small network we can estimate the wave vector and velocity of the TID based on time it needs to travel from one receiver to another. The latter research used slowness (inverse of phase velocity) introduced by Jacobson et al. (1995) studied TID by means of TEC derived from geosynchronous satellite signals with 136/137 MHz. The signals were received at the interferometer located in Mexico. While the signal frequencies were different from GNSS ones, geometry-free combination takes care of it. They found the same local time vs. season TID confirmed by Hernández-Pajares et al. (2006). Also, they studied TID period vs. local time distribution for four seasons centered on solstices equinoxes. They found that TID intensity

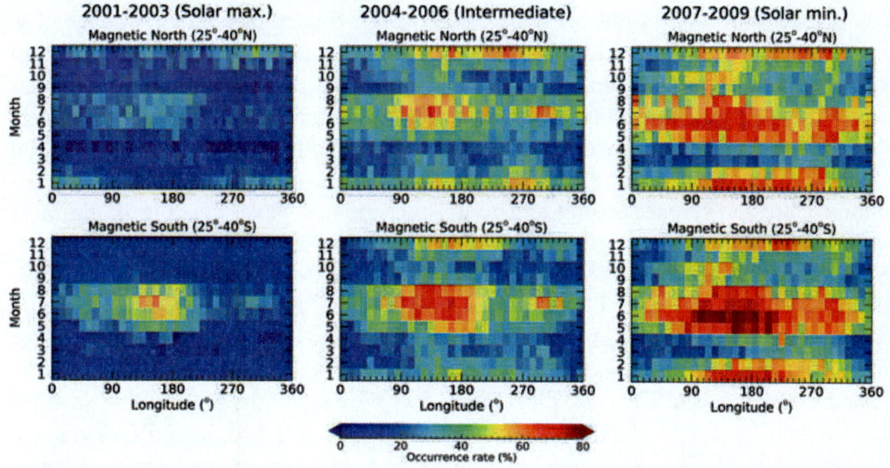

Fig. 6.47 Dependence of the MSTID occurance rate on solar activity in the (top three plots) magnetic North (25°–40°N) and (bottom three plots)magnetic South (25°–40°S). The latitude range is limited to exclude equatorial plasma bubbles influence on the data. (Taken from (Lee et al. 2021)).

gradually decays from 50-min (0.34 milliHz) TID to 10-min ones (1.68 milliHz). Also, they reveal TID occurrence dependence on local time and amplitude (in terms of root mean square of the series). The occurrence rate is the highest for daytime. Maximum amplitude depends on season and can appear at sunset for autumn equinox and at noon during spring equinox. Propagation direction of the TIDs we also studied. The propagation is almost southward around noon and west/northwest after local sunset. Latter results were obtained using an interferometer for four receivers aimed to serve as interferometer. Modern GNSS systems such as Beidoi include geostationary and geosynchronous satellites. This allows us to build virtual interferometers composed of different receivers using both moving and stationary ionospheric pierce points (Tereshin et al. 2018; Maletckii and Astafyeva 2021). The choice of the receivers is defined based on TID scale and find medium scale TID occurrence along with large scale TID one. The data from a single receiver could be used in several interferometers improving data coverage (Figs. 6.48 and 6.49).

Equatorial plasma bubbles are large-scale irregularities that occur in the equatorial ionosphere mostly during the sunset to evening period. While being a large-scale structure that should not cause diffraction of GNSS signals, the turbulence that accompanies the plasma bubble can generate smaller structures down to Fresnel scale. Plasma bubbles being manifestations of instability in the ionosphere are harder to track as discussed above. Climate and data-driven weather distribution of plasma in the ionosphere could be modeled using global representation. We can predict where TID will move using local models and real-time data streams. Plasma bubbles and other irregularities that could appear sporadically should be monitored "online". Extension of plasma bubbles outside the place where it is observed could be easily performed. We can also try to estimate occurrence probability through observing events that could trigger plasma bubbles. Rayleigh–Taylor instability is believed to be a source of plasma bubbles (Ott 1978). Plasma bubbles refers to the event when Irregularities of various scales appear. Both spread F and GNSS scintillations appear due to the bubbles. Instability creates the conditions for the plasma bubbles generation. However unstable media could exist long without external forcing or initial irregularities seeding. Huang and Kelley (1996) have studied plasma bubbles development under different seeding and found that gravity waves seeding is the most effective one. The wavelength of the gravity waves was 200 km during simulation. We remember that gravity waves also cause TIDs that are observed in GNSS measurements. The medium scale TIDs (those correspond to 200 km) are unambiguously observed in GNSS measurements. Less than 10 min are enough to generate bubbles in the ionosphere according to isodensity contours in Huang and Kelley (1996). Same behavior is demonstrated on 2D plot (Fig. 6.50). We can use the time bubble need for development to observe TID by means of TEC series and high-pass filter (to leave 10-min and faster TEC variations). TEC variations are then can be used to estimate the probability of bubbles generation.

Barros et al. (2018) used GNSS data from South America for more than 3 years to study seasonal and latitudinal frequency of plasma bubbles occurrence. The definition they define plasma bubble from observation viewpoint as combination of four factors:

- a TEC depletion with an amplitude exceeding 10 TECU;
- a TEC perturbation with at least one well-developed valley;

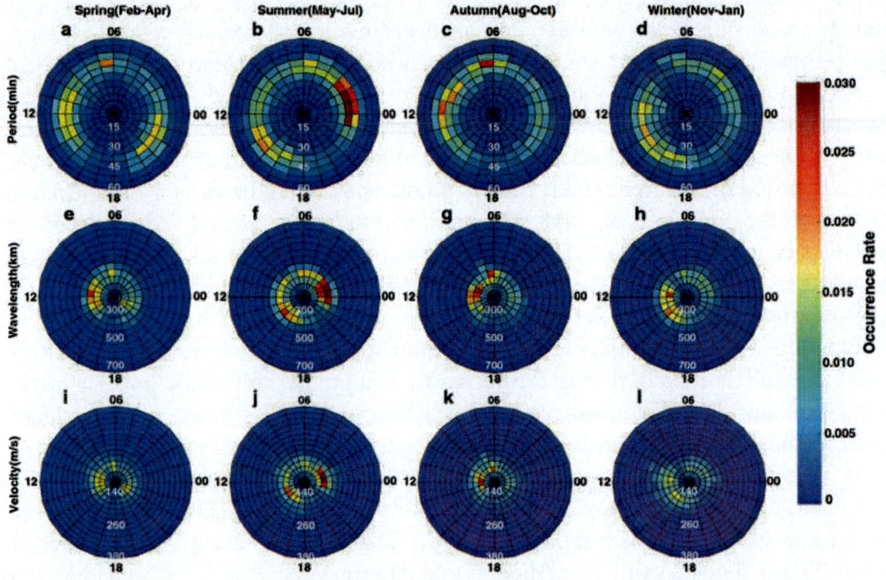

Fig. 6.48 TID occurrence dependence on local time and amplitude. Panels show four quarter-year periods centred on **a** winter solstice, **b** spring equinox, **c** summer solstice, and **d** autumn equinox. The value 15 on the horizontal axis corresponds to 0.1 TECu (Chen et al. 2019)

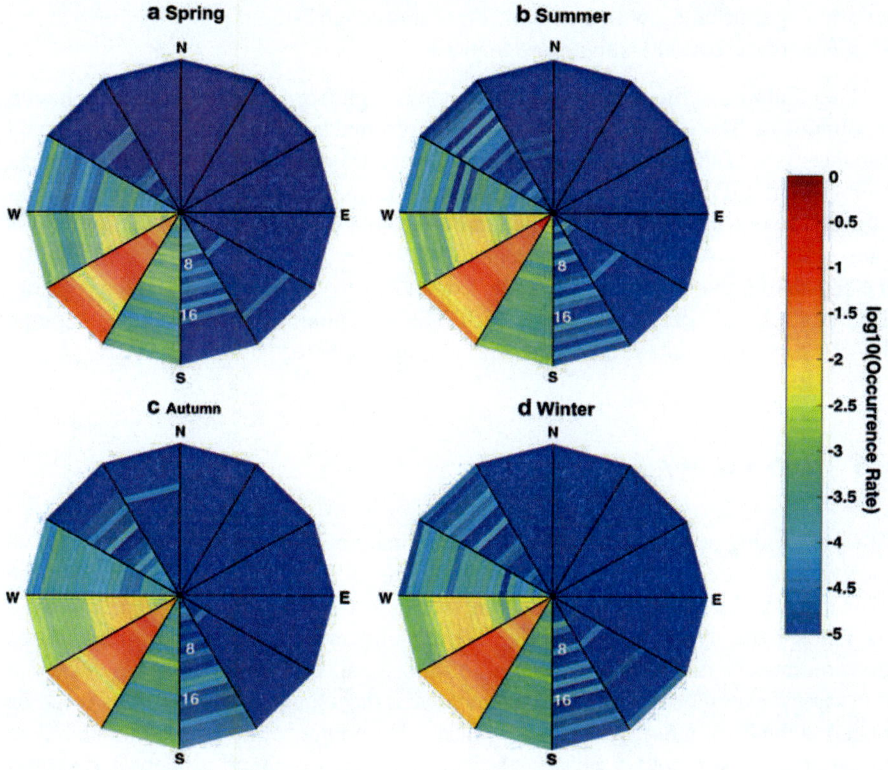

Fig. 6.49 TID occurrence dependence on local time and azimuth. Panels show four quarter-year periods centred on **a** winter solstice, **b** spring equinox, **c** summer solstice, and **d** autumn equinox. The value 15 on the horizontal axis corresponds to 0.1 TECu (Chen et al. 2019)

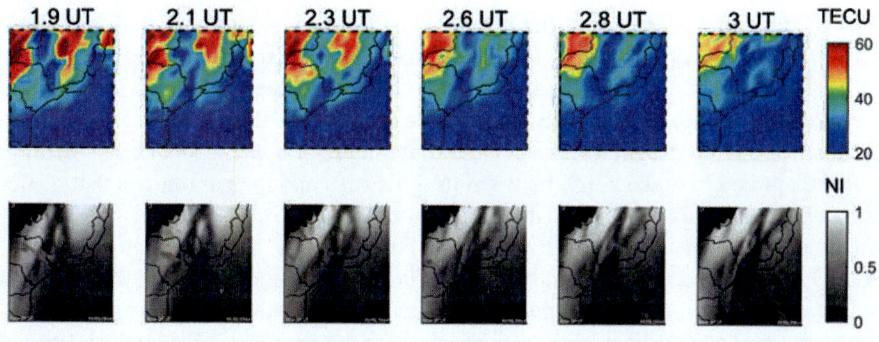

Fig. 6.50 Time development of equatorial plasma bubbles over south America at Junary 3, 2014 according to tomographic reconstruction of ionosphere by means of vertical TEC (top panels) and optical instrumnets (bottom panels) (Taken from (dos Santos Prol et al. 2018)).

- a TEC perturbation with an extension of more than 500 km;
- a TEC perturbation lasting more than 1 h.

They found that the occurrence frequency is high from September to March with maximum on December and January. Authors make a link between pre-reversal enhancement of the electric field and bubble occurrence. The correlation of plasma bubbles and TID occurrence is a more interesting relationship. It allows us to get a proxy to plasma bubbles through more reliable observation of TEC variations caused by TIDs. As for latitudinal dependence of the plasma bubble occurrence rate, the plasma bubble region is in range from 1200 to 2500 km from the magnetic. The range slightly depends on the month. The maximum occurrence rate is in range 15–20° geomagnetic latitude in the southern hemisphere.

6.6 Conclusions

The positioning accuracy depends on the accuracy of the TEC estimation along the receiver satellite line of sight and hence on (a) the accuracy of the distribution of the main electron density fraction in both the ionosphere and the plasmasphere (climate); (b) ionospheric dynamics not included in the climate assessment; (c) irregularities that could cause diffractive destruction of the signal. The (b) includes fast transient processes (several hours long) as well as several days long processes. TIDs could be linked to the former and the recovery phase of a geomagnetic storm could be linked to the latter. The recovery phase, when the ionosphere-magnetosphere system has not yet reached a steady state, could last several days. Long leaving dynamics can be accounted for the same basis as climatology. Transient processes should be accounted for using different models. Transient processes models should have local or regional scale. Both (b) and (c) can be considered manifestations of space weather. However, TID could be tracked in time using physical justified model and GNSS observation. Conversely, irregularities triggering is the matter of statistics, although they have well defined generation mechanisms. GNSS only measurements could provide the information about the irregularity appearance on the line-of-sight, however, do not say much about the irregularities aside the sounding region. Hence machine learning could be applied to make a link between irregularities and the parameters that could be reliably tracked in real time using observation.

Climate assessment makes the main contribution to TEC, but at the same time can be easily eliminated instrumentally using DGPS. The time spectrum of the climatic estimate of electron density in the ionosphere often has several modes from 24 to 4 h. If the model is driven with appropriate data sources the daily cycle could be modulated by various slower processes such as the rotation of the sun (with a period of 27 days), the rotation of the Earth (with periods of 12 and 6 months) and the 11-year solar cycle. Data assimilation could bring all ionosphere effective processes into modelling. GNSS observation could be assimilated to provide 2D maps of vertical TEC. Other data, such as ionosonde measurements, could be used to improve 3D electron density distribution and then the 3D ionosphere can be converted to both vertical and slant TEC.

Transient processes of space weather can occur regularly in space and time. Waves that are generated in the ionosphere during the passage of the solar terminator regularly appear from day to day. TIDs of various scales can arise in arbitrary locations and then propagate, drawing periodic structure in time and space. Plasma bubbles in the nighttime equatorial ionosphere and disturbances, such as inhomogeneities in the polar regions can arise and decay spontaneously, form GNSS observations viewpoint.

Acknowledgements We thank Dr. Ivan Galkin from Space Science Laboratory, University of Massachusetts Lowell for supervising work described in Sect. 6.4. We also acknowledge Alexandra Kustavinova for artist's view and implementation of some figures.

References

Aa E, Huang W, Yu S, Liu S, Shi L, Gong J, Shen H, et al (2015) A regional ionospheric TEC mapping technique over China and adjacent areas on the basis of data assimilation. J Geophys Res Space Phys 120(6):5049–5061

Abdelazeem M, Çelik RN, El-Rabbany A (2017) An efficient regional ionospheric model using combined GPS/BeiDou observations. J Spat Sci 62(2):323–335

Adeniyi JO, Bilitza D, Radicella SM, Willoughby AA (2003) Equatorial F2-peak parameters in the IRI model. Adv Space Res 31(3):507–512

Afraimovich EL, Edemskiy IK, Voeykov SV, Yasyukevich YV, Zhivetiev IV (2009) The first GPS-TEC imaging of the space structure of MS wave packets excited by the solar terminator. In: Annales geophysicae, Vol 27, No 4. Copernicus GmbH, pp 1521–1525

Akmaev RA (2011) Whole atmosphere modeling: connecting terrestrial and space weather. Rev Geophys 49(4)

Anghel A, Astilean A, Letia T, Komjathy A (2008) Near real-time monitoring of the ionosphere using dual frequency GPS data in a Kalman filter approach. In: 2008 IEEE international conference on automation, quality and testing, robotics, Vol 2. IEEE, pp 54–58

Anghel A, Carrano C, Komjathy A, Astilean A, Letia T (2009) Kalman filter-based algorithms for monitoring the ionosphere and plasmasphere with GPS in near-real time. J Atmos Solar Terr Phys 71(1):158–174

Ansari K, Park KD, Panda SK (2019) Empirical orthogonal function analysis and modeling of ionospheric TEC over South Korean region. Acta Astronaut 161:313–324

Barros D, Takahashi H, Wrasse CM, Figueiredo CAO (2018) Characteristics of equatorial plasma bubbles observed by TEC map based on ground-based GNSS receivers over South America. In: Annales geophysicae, Vol 36, No 1. Copernicus GmbH, pp 91–100

Belehaki A, Jakowski N, Reinisch BW (2004) Plasmaspheric electron content derived from GPS TEC and digisonde ionograms. Adv Space Res 33(6):833–837

Bergeot N, Chevalier JM, Bruyninx C, Pottiaux E, Aerts W, Baire Q, Huang, W, et al (2014) Near real-time ionospheric monitoring over Europe at the Royal

Bilitza D, Sheikh M, Eyfrig R (1979) A global model for the height of the F2-peak using M3000 values from the CCIR numerical map. Telecommun J 46(9):549–553

Bilitza D, Altadill D, Truhlik V, Shubin V, Galkin I, Reinisch B, Huang X (2017) International reference ionosphere 2016: from ionospheric climate to real-time weather predictions. Space Weather 15(2):418–429

Blais JA (2008) Discrete spherical harmonic transforms: numerical preconditioning and optimization. In: International conference on computational science. Springer, Berlin, Heidelberg, pp 638–645

Bogusz J, Gruszczynski M, Figurski M, Klos A (2015) Spatio-temporal filtering for determination of common mode error in regional GNSS networks. Open Geosci 7(1)

Brunini C, Conte JF, Azpilicueta F, Bilitza D (2013) A different method to update monthly median hmF2 values. Adv Space Res 51(12):2322–2332

CCIR (1967) Atlas of Ionospheric Characteristics, Comité Consultatif International des Radiocommunications, Report 340-4. Int Telecommun Union Geneva

CCIR (1991) Atlas of ionospheric characteristics, Comité Consultatif International des Radiocommunications, Report 340-6. Int Telecommun Union Geneva

Carrano CS, Groves KM, Rino CL (2019) On the relationship between the rate of change of total electron content index (ROTI), irregularity strength (CkL), and the scintillation index (S4). J Geophys Res Space Phys 124(3):2099–2112

Chen G, Zhou C, Liu Y, Zhao J, Tang Q, Wang X, Zhao Z, (2019) A statistical analysis of medium-scale traveling ionospheric disturbances during 2014–2017 using the Hong Kong CORS network. Earth Planets Space 71(1):52 https://doi.org/10.1186/s40623-019-1031-9

Chen X, Allison T, Cao W, Ferguson K, Grunig S, Gomez V, Talbot N, et al (2011) Trimble RTX, an innovative new approach for network RTK. In: Proceedings of the 24th international technical meeting of the satellite division of the institute of navigation (ION GNSS 2011), pp 2214–2219

Cheng N, Song S, Li W (2021) Multi-scale ionospheric anomalies monitoring and spatio-temporal analysis during intense storm. Atmosphere 12(2):215

Cherniak I, Zakharenkova I (2016) NeQuick and IRI-Plas model performance on topside electron content representation: spaceborne GPS measurements. Radio Sci 51(6):752–766

Codrescu SM, Codrescu MV, Fedrizzi M (2018) An ensemble Kalman filter for the thermosphere-ionosphere. Space Weather 16(1):57–68

Damboldt T, Suessmann P (2011) Information document on the analysis and validity of present ITU foF2 and m (3000) f2 maps. Int Telecommun Union 212-1

dos Santos Prol F, Hernández-Pajares M, Tadeu de Assis Honorato Muella M, de Oliveira Camargo P (2018) Tomographic Imaging of Ionospheric Plasma Bubbles Based on GNSS and Radio Occultation Measurements. Remote Sens 10(10):1529. https://doi.org/10.3390/rs10101529

Galkin IA, Khmyrov GM, Kozlov AV, Reinisch BW, Huang X, Paznukhov VV (2008) The artist 5. In: AIP conference proceedings, Vol 974, No 1. American Institute of Physics, pp 150–159

Galkin IA, Reinisch BW, Huang X, Bilitza D (2012) Assimilation of GIRO data into a real-time IRI. Radio Sci 47(04):1–10

Galkin IA, Reinisch BW, Bilitza D (2018) Realistic ionosphere: real-time ionosonde service for ISWI. SunGe 13:173–178

Galkin IA, Reinisch BW, Vesnin AM, Bilitza D, Fridman S, Habarulema JB, Veliz O (2020) Assimilation of sparse continuous near-Earth weather measurements by NECTAR model morphing. Space Weather 18(11):e2020SW002463

Gelb A (1974) Applied optimal estimation. MIT press

Gentile LC, Burke WJ, Rich FJ (2006a) A climatology of equatorial plasma bubbles from DMSP 1989–2004. Radio Sci 41(5)

Gentile LC, Burke WJ, Rich FJ (2006b) A global climatology for equatorial plasma bubbles in the topside ionosphere. In: Annales geophysicae Vol 24, No 1. Copernicus GmbH, pp 163–172

Hajj GA, Wilson BD, Wang C, Pi X, Rosen IG (2004) Data assimilation of ground GPS total electron content into a physics-based ionospheric model by use of the Kalman filter. Radio Sci 39(1):1–17

He J, Yue X, Le H, Ren Z, Wan W (2020) Evaluation on the quasi-realistic ionospheric prediction using an ensemble Kalman filter data assimilation algorithm. Space Weather 18(3):e2019SW002410

Hernández-Pajares M, Juan JM, Sanz J (2006) Medium-scale traveling ionospheric disturbances affecting GPS measurements: spatial and temporal analysis. J Geophys Res Space Phys 111(A7)

Huang CS, Kelley MC (1996) Nonlinear evolution of equatorial spread F:2. Gravity wave seeding of Rayleigh-Taylor instability. J Geophys Res Space Phys 101(A1):293–302

Huang X, Reinisch BW (1996) Vertical electron density profiles from the digisonde network. Adv Space Res 18(6):121–129

Jacobson AR, Carlos RC, Massey RS, Wu G (1995) Observations of traveling ionospheric disturbances with a satellite-beacon radio interferometer: seasonal and local time behavior. J Geophys Res Space Phys 100(A2):1653–1665

Jacobsen KS (2014) The impact of different sampling rates and calculation time intervals on ROTI values. J Space Weather Space Clim 4:A33

Jakowski N, Hoque MM (2018) A new electron density model of the plasmasphere for operational applications and services. J Space Weather Space Clim 8:A16

Jamjareegulgarn P, Ansari K, Ameer A (2020) Empirical orthogonal function modelling of total electron content over Nepal and comparison with global ionospheric models. Acta Astronaut 177:497–507

Jones WB, Graham RP, Leftin M (1966) Advances in ionospheric mapping by numerical methods. Inst Telecommun Sci

Jones WB, Graham RP, Leftin M (1969) Advances in Ionospheric Mapping by Numerical Methods, National Bureau of Standards US Dept. of Commerce NBS Tech. Note 337, May (Reissued as ESSA Tech. Note ERL 107 ITS 75 May 1969)

Jones WB, Gallet RM (1962) Representation of diurnal and geographic variations of ionospheric data by numerical methods. Telecommun J 29(5):129–147

Jung S, Lee J (2012) Long-term ionospheric anomaly monitoring for ground based augmentation systems. Radio Sci 47(04):1–12

Kintner PM, Ledvina BM, De Paula ER (2007) GPS and ionospheric scintillations. Space Weather 5(9)

Kovář P (2017) piNAV L1—GPS receiver for small satellites. Gyroscopy Navig 8(2):159–164

Lee IT, Matsuo T, Richmond AD, Liu JY, Wang W, Lin CH, Chen MQ, et al (2012) Assimilation of FORMOSAT-3/COSMIC electron density profiles into a coupled thermosphere/ionosphere model using ensemble Kalman filtering. J Geophys Res Space Phys 117(A10)

Lee WK, Kil H, Paxton LJ (2021) Global distribution of nighttime MSTIDs and its association with E region irregularities seen by CHAMP satellite. J Geophys Res: Space Phys 126(5):e2020JA028836

Maletckii B, Astafyeva E (2021) Determining spatio-temporal characteristics of coseismic travelling ionospheric disturbances (CTID) in near real-time. Sci Rep 11(1):1–16

Miyoshi Y, Jin H, Fujiwara H, Shinagawa H (2018) Numerical Study of Traveling Ionospheric Disturbances Generated by an Upward Propagating Gravity Wave. J Geophys Res: Space Phys https://doi.org/10.1002/2017JA025110

Li S, Xu J, Zhou H, Zhang J, Xu Z, Xie M (2020) Global total electron content prediction performance assessment of the IRI-2016 model based on empirical orthogonal function decomposition. Ann Geophys 38(2):331–345. https://doi.org/10.5194/angeo-38-331-2020

McCaffrey AM, Jayachandran PT (2019) Determination of the Refractive Contribution to GPS Phase "Scintillation". J Geophys Res: Space Phys 124(2):1454–1469. https://doi.org/10.1029/2018JA025759

Monahan AH, Fyfe JC, Ambaum MH, Stephenson DB, North GR (2009) Empirical orthogonal functions: the medium is the message. J Clim 22(24):6501–6514

Ott E (1978) Theory of Rayleigh-Taylor bubbles in the equatorial ionosphere. J Geophys Res 83(A5):2066. https://doi.org/10.1029/JA083iA05p02066

Paznukhov V, Altadill D, Juan JM, Blanch E (2020) Ionospheric tilt measurements: application to traveling ionospheric disturbances climatology study. Radio Sci 55(2):e2019RS007012

Piggott WR, Rawer K (ed) (1978) U.R.S.I. Handbook of ionogram interpretation and reduction, WDC-A for STP Rep. UAG-23A, Boulder, CO

Psychas D, Verhagen S, Liu X, Memarzadeh Y, Visser H (2018) Assessment of ionospheric corrections for PPP-RTK using regional ionosphere modelling. Meas Sci Technol 30(1):014001

Puente I, González-Jorge H, Martínez-Sánchez J, Arias P (2013) Review of mobile mapping and surveying technologies. Meas 46(7):2127–2145. https://S0263224113000730.10.1016/j.measurement.2013.03.006

Reinisch BW, Haines DM, Bibl K, Cheney G, Galkin IA, Huang X, Reiff P (2000) The Radio Plasma Imager investigation on the IMAGE spacecraft. In: the IMAGE Mission (pp 319–359). Springer, Dordrecht

Reinisch BW, Galkin IA (2011) Global ionospheric radio observatory (GIRO). Earth Plan Space 63(4):377–381

Reinisch BW, Huang X, Galkin I, Bilitza D (2012) A real-time assimilative model for IRI. In: EGU general assembly conference abstracts, p 12828

Reinisch B, Galkin I, Belehaki A, Paznukhov V, Huang X, Altadill D, Ishii M, et al (2018) Pilot ionosonde network for identification of traveling ionospheric disturbances. Radio Sci 53(3):365–378

Rovira-Garcia A, Juan JM, Sanz J, González-Casado G, Bertran E (2016) Fast precise point positioning: a system to provide corrections for single and multi-frequency navigation. NAVIGATION J Inst Navig 63:231–247

Rush C (1989) Ionospheric mapping-an update of foF2 coefficients. Telecomm J 56:179–182

Scherliess L, Schunk RW, Sojka JJ, Thompson DC (2004) Development of a physics-based reduced state Kalman filter for the ionosphere. Radio Sci 39(1):1–12

Sivakandan M, Otsuka Y, Ghosh P, Shinagawa H, Shinbori A, Miyoshi Y (2021) Comparison of seasonal and longitudinal variation of daytime MSTID activity using GPS observation and GAIA simulations. Earth Planets and Space 73(1):35 https://doi.org/10.1186/s40623-021-01369-5

Susi M, Andreotti M, Aquino M, Dodson A (2017) Tuning a Kalman filter carrier tracking algorithm in the presence of ionospheric scintillation. GPS Sol 21(3):1149–1160

Takasu T, Yasuda A (2010) Kalman-filter-based integer ambiguity resolution strategy for long-baseline RTK with ionosphere and troposphere estimation. In: Proceedings of the 23rd international technical meeting of the satellite division of the institute of navigation (ION GNSS 2010), pp 161–171

Tereshin NA, Kozlovtseva EA, Padokhin AM and Kurbatov GA (2018) Monitoring Traveling Ionospheric Disturbances Using GNSS Interferometry with Geostationary Satellites. In: 2018 Progress in electromagnetics research symposium (PIERS-Toyama), pp 2414–2416. https://doi.org/10.23919/PIERS.2018.8597898

Tunalı E, Özlüdemir MT (2019) GNSS PPP with different troposphere models during severe weather conditions. GPS Solutions 23(3) 82 https://doi.org/10.1007/s10291-019-0874-7

Vaishnav R, Jacobi C, Berdermann J (2019) Long-term trends in the ionospheric response to solar extreme-ultraviolet variations. Ann Geophys 37(6):1141–1159. https://doi.org/10.5194/angeo-37-1141-2019

Vesnin AM, Ratovsky KG (2011) Modeling of the effect of the "apparent F1 layer" observed in the radar-progress experiment with the help of the software for automated processing of vertical sounding ionograms. Geomag Aeron 51(7):923–929

Wan W, Ding F, Ren Z, Zhang M, Liu L, Ning B (2012) Modeling the global ionospheric total electron content with empirical orthogonal function analysis. Sci China Technol Sci 55(5):1161–1168

Welch G, Bishop G (1995) An introduction to the Kalman filter

Yasyukevich AS, Vesnin AM, Yasyukevich YV, Padokhin AM (2019) Correlation between Total and Plasmasphere Electron Content and Indexes of Solar and Geomagnetic Activity. In 2019 Russian Open Conference on Radio Wave Propagation (RWP) (Vol. 1, pp. 87-90). IEEE.

Yasyukevich Y, Vasilyev R, Ratovsky K, Setov A, Globa M, Syrovatskii S, Vesnin A, et al (2020a) Small-scale ionospheric irregularities of auroral origin at mid-latitudes during the 22 June 2015 magnetic storm and their effect on GPS positioning. Remote Sens 12(10):1579

Yasyukevich Y, Mylnikova A, Vesnin A (2020b) GNSS-based non-negative absolute ionosphere total electron content, its spatial gradients, time derivatives and differential code biases: bounded-variable least-squares and taylor series. Sensors 20(19):5702

Yasyukevich YV, Vesnin AM, Kiselev AV, Mylnikova AA, Oinats AV, Ivanova VA, Demyanov VV (2022) MITIGATOR: GNSS-Based System for Remote Sensing of Ionospheric Absolute Total Electron Content. Universe 8(2):98–https://doi.org/10.3390/universe8020098

Yeh KC, Liu CH (1982) Radio wave scintillations in the ionosphere. Proceedings of the IEEE 70(4):324–360.https://doi.org/10.1109/PROC.1982.12313

Yoon M, Lee J (2014) Medium-scale traveling ionospheric disturbances in the Korean region on 10 November 2004: Potential impact on GPS-based navigation systems. Space Weather 12(4):173–186. https://doi.org/10.1002/2013SW001002

Zeiler A, Faltermeier R, Keck IR, Tomé AM, Puntonet CG, Lang EW (2010) Empirical mode decomposition-an introduction. In: The 2010 international joint conference on neural networks (IJCNN). IEEE, pp 1–8

Zhang D, Ridley AJ, Xiao Z, Hao Y (2012) A global model: empirical orthogonal function analysis of total electron content 1999–2009 data

Zhang L, Morton YT (2010) A variable gain adaptive Kalman filter-based GPS carrier tracking algorithm for ionosphere scintillation signals. In: Proceedings of the 23rd international technical meeting of the satellite division of the Institute of Navigation (ION GNSS 2010), pp 3107–3114

Chapter 7
How to Improve GNSS/SBAS Reliability Under Space Weather Impacts: Analysis

Vladislav Demyanov and Yury Yasyukevich

Based on the above, one can see that space weather effects on GNSS hardware operation have a complicated character. The consequences of those effects are hard to predict. Nevertheless, we demonstrated a number of features and basic mechanisms for deterioration of GNSS performance characteristics under the conditions of geomagnetic disturbances, of ionospheric scintillations, and of solar radio bursts. From the results obtained, one may propose a system of technical-organizing measures that will enable to improve the GNSS performance quality with allowance for current helio-geophysical conditions. The system includes:

1. Fore- and nowcasting positioning quality under geomagnetic disturbances.
2. Real-time alert of users to a geomagnetic disturbance in the signal propagation medium.
3. Real-time alert to powerful solar radio bursts at GNSS frequencies.
4. Using a complex technical system to provide the GNSS RNPs under space weather unfavorable conditions.

Let us switch to addressing the above measures in greater details.

7.1 Fore- and Nowcasting GNSS Performance Quality Under Geomagnetic Storm Conditions

During geomagnetic disturbances, one observes an unambiguous trend to a decrease in the positioning quality for GNSS users. This manifests itself like a sharp increase in a user positioning error and a decrease in the availability of required navigation parameters (RNP) at the specified requirements for positioning accuracy. At the same time, there is no exact quantitative relation among the geomagnetic disturbance level, positioning accuracy, and continuity. The problem of the positioning quality prediction under geomagnetic disturbance conditions comes to generating long-term

V. Demyanov et al., *Space Weather Impact on GNSS Performance*,
https://doi.org/10.1007/978-3-031-15874-2_7

and real-time alerts to an expected deterioration in RNP availability for different GNSS customers.

Using *long-term alerts (several hours through 1–2 days)* enables to define an optimal configuration to provide navigation solutions at the preparation stage for solving applied problems. *Real-time alerts or nowcast (at present or a few minutes prior to the forcing)* may be used for optimal reconfiguration and integration of navigation data sensors. They may also be used as a part of a navigation suite (navigation receiver, inertial navigation facilities, radio-beacon systems, etc.). Besides, a nowcast may be regarded as a supplement to algorithms for the Receiver Autonomous Integrity Monitoring (RAIM), and an integrity check of differential systems.

Let us address possibilities to provide a forecast for positioning quality under magnetic storm conditions. This forecast should be based on reliable prediction of geomagnetic disturbances proper. For a geomagnetic disturbance to emerge, a geo-effective coronal mass ejection (CME) on the Sun is required. The CME geo-effectiveness is determined by a flare region location on the Sun relative to the earthward direction and by the interplanetary magnetic field (IMF) characteristics (Bazarzhapov et al. 1979). For long-term (1–2 days) forecast of geomagnetic disturbances, observations of the Sun surface in extreme ultraviolet ($250 < \lambda < 1350$ Å) are especially relevant. These observations enable to detect a CME and to identify geoeffective events on the Sun. Presently, one uses the Extreme Ultraviolet Imaging Telescope (EIT) onboard the Solar Heliospheric Observatory, SOHO (https://sohowww.nascom.nasa.gov/). This instrument enables to observe solar activity at the stage of CME (including geoeffective) emergence and manifestation.

Upon detecting a geoeffective solar event, one can compute the disturbance arrival time in the Earth surroundings by the known shockwave velocity in the interplanetary space. This shockwave (hereinafter—shock) is initiated by a plasma cloud ejected from the Sun surface (Akasofu and Chapman 1972). The shock arrival time at the Earth magnetosphere boundaries is directly dependent on the shock energy and on the time of coronal mass flare injection. Figure 7.1 presents the dependence of the shock propagation time at 1 astronomical unit (AU, 1AU = 150,000,000 km) as a function of CME duration and energy in the interplanetary space. One can see that this time ranges within significant limits: about 20 through 85 h. However, one cannot forecast the exact time of a disturbance arrival and a magnetic storm onset with a satisfactory accuracy for the time being.

To estimate the expected geomagnetic disturbance power, one needs (except for the disturbance arrival time) a forecast for the IMF *Bz*-component. Now, to reliably forecast the IMF disturbances has been an open problem. The IMF is known to emerge at an ejection of the solar magnetic field by the coronal mass expanding into the interplanetary space (Akasofu and Chapman 1972). Therefore, IMF disturbances related to flare CMEs are determined by a configuration of magnetic field lines within a flare region. Thus, forecasting IMF disturbances is closely related to predicting solar flares and related CMEs.

Since the early twenty-first century, forecasting CME events has been developed based on observations of S-shaped regions on the solar surface (X-ray sigmoids) (Berghmans et al. 2005). X-ray sigmoid emergence is supposed to indicate a gain in

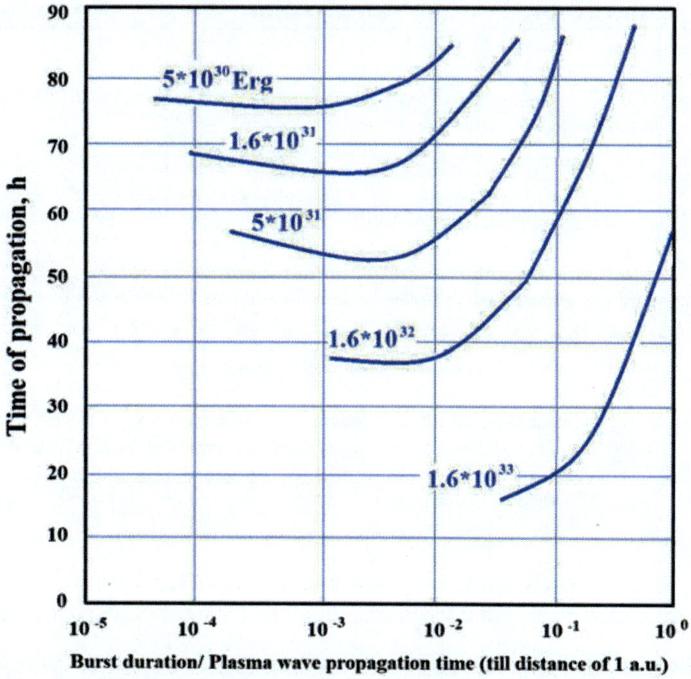

Fig. 7.1 Propagation time to 1 AE as a function of the parameter Δ (initial disturbance duration divided by the propagation time to 1 AU) for shocks of constant energy. Re-produced and modified based on the figure published in the book (Akasofu and Chapman 1972)

an emerging flow of superheated plasma from the lower regions of the Sun interior. Such an event precedes (albeit optionally) a flare with a coronal mass ejection into the interplanetary space. X-ray sigmoids are confidently observed in the soft X-ray ($\lambda > 1$ Å), in particular, by the geostationary GOES facilities, as well as from observations of the Sun in the Hα line. However, nowadays, these observations do not provide high-reliability CME forecasts. Figure 7.2 illustrates this, where there are the estimation results for reliability of forecasting different-class X-ray flares associated with CMEs. The observations were conducted in the X-ray band $\lambda =$ 1–8 Å within 22 months 2003 through 2005 (Michalek et al. 2009).

As one can see, the greatest forecast accuracy (with a <20% error) is provided for the weakest C-class flares. More powerful flares are less reliable to forecast: the maximal error (<40%) is for M-class flares, and a <55–60% error is for the most powerful X-class flares.

From the observational statistics for magnetic storms (presented in Sect. 4.1), one can see that:

(1) the probability of a magnetic storm emergence drops with a flare-class increase (i.e., with an increase in the flare power);

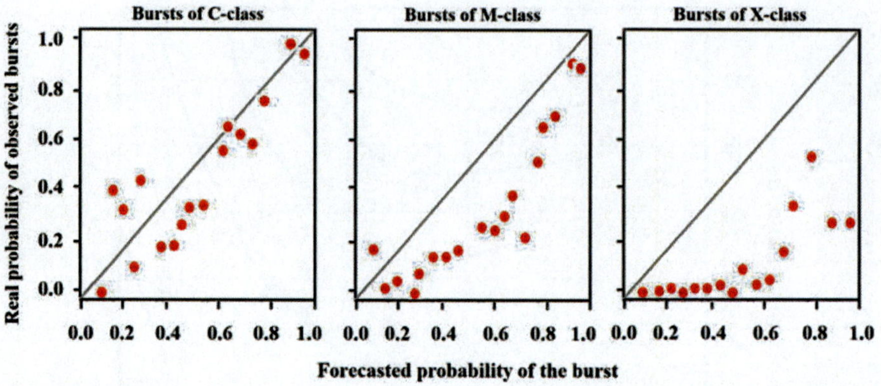

Fig. 7.2 Estimating the forecast reliability for different-class flares from the 2003 Apr—2005 Feb observations. Dotted line shows the complete agreement between the forecast and observation, while rectangles indicate the real observation results. Re-produced and modified figure based on the part of the figure published in (Michalek et al. 2009)

(2) for all the magnetic storms, there is an explicit periodicity in a storm probability increase and decrease, that has a statistical dependence on the sunspot number.

The latter circumstance simplifies forecasting powerful and extreme magnetic storms, based on analyzing long-period observations of solar activity. However, as one can see from the observational statistics, strong magnetic storms are sufficiently rare.

Some new studies provide additional possibilities in augmenting the accuracy of the magnetic storm forecast. Thus, in (Lekshmi et al. 2011), based on the data analysis over the last two solar activity cycles, noted is the presence of a "UT-effect" for strong magnetic storms: the magnetic storm onset most often corresponds to 00:00 UT, although nobody has provided an exhaustive elucidation for this effect yet.

In general, one may state that the technique to forecast magnetic storms of a given intensity has not met practical requirements yet, and needs further improving. Developing this research trend is possible now, based on using a suite of ground- and space-based facilities to monitor the status of the Sun and the "Space weather" (Williamson et al. 2000). The latter combination of monitoring facilities is supposed to provide a reliable forecast of geomagnetic disturbances 48 h prior to their manifestations on the Earth.

If there is a forecast of a magnetic storm onset and power, the strategy to alert a GNSS user may be based on estimating geomagnetic activity indices. In (Datta-Barua et al. 2005), the authors analyze a possibility to use the *Kp*- and *Dst*-indices to detect the deterioration of the WAAS availability under magnetic storm conditions.

A deterioration of RNP availability for WAAS users may occur for different reasons. Therefore, one should identify the cases related to only ionospheric geomagnetic disturbances. To identify such events, the authors in (Datta-Barua et al. 2005) use a long-term statistics covering the records for reference values of "vertical" ionospheric ranging errors in IGP points (see (2.140), (2.141)) within a WAAS operation area. Based on these reference values (τ_{PP}), one estimates "vertical" ionospheric errors ($I_{\text{V, IPP}}$) within an 800–2000 km radius. These estimates are computed through the linear interpolation method relative to each IGP point (2.141) for every satellite visible within this area, for the same observational period.

Thus, the total statistics for the analysis involves all the "vertical" ionospheric errors available within the WAAS operation area and their estimate RMS (Walter et al. 2001)

$$I_{V,IPP} = \begin{vmatrix} \tau_{U,1} \\ \tau_{U,2} \\ \vdots \\ \tau_{U,k} \end{vmatrix}, \quad \sigma^2_{V,IPP} = \begin{vmatrix} \sigma^2_{U,1} \\ \sigma^2_{U,2} \\ \vdots \\ \sigma^2_{U,k} \end{vmatrix} \tag{7.1}$$

where $\tau_{U,i}$ is a "vertical" ionospheric error for an i_{th} satellite in a sub-ionosperic point computed for each IGP at a fixed ionospheric height (the ionosphere is presented as a thin layer model at 350 km high); $\sigma^2_{U,1}$ is a $\tau_{U,i}$ error RMS; k is the number of sub-ionosperic points used for the analysis.

The reference values of the "vertical" ionospheric ranging error in IGP points are obtained through the method of linear interpolation for measurements of the ionospheric ranging errors obtained at WAAS reference stations. On the "North" (y)—East (x)" plane, this interpolation may be presented like a plane inclined northward or eastward in the following general view (Walter et al. 2001)

$$I_{V,IGP}(x, y) = a_0 + a_1 x + a_2 y \tag{7.2}$$

The plane inclination (7.2) determines the distribution of "vertical" ionospheric errors ($\tau_{U,i}$) on the "North—East" plane in the form of an observational matrix as follows

$$G = \begin{vmatrix} 1 & d_1^E & d_1^N \\ 1 & d_2^E & d_2^N \\ \vdots & \vdots & \vdots \\ 1 & d_k^E & d_k^N \end{vmatrix} \tag{7.3}$$

where d_1^E and d_1^N are spatial gradients for a "vertical" ionospheric error.

Describing the field for ionospheric ranging error distribution in the form of (7.2) and (7.3) presents the state of that error under the undisturbed ionosphere conditions. As a geomagnetic disturbance intensifies, the spatial gradients substantially vary (see Table 2.7). Thereby, the model for the spherically stratified ionosphere and the linear presentation of the ionospheric layer within the WAAS operation area do not describe real values of an ionospheric ranging error anymore. Eventually, this leads to generating alerts to deterioration of the RNP availability due to ionospheric disturbances.

To detect such events, the authors in (Walter et al. 2001) propose to use a likelihood estimate for an ionospheric error presentation described above as follows

$$\chi^2 = I^T_{V,IPP} \cdot W \cdot \left[I - G^T \cdot \left(G \cdot W \cdot G^T\right)^{-1} \cdot G \cdot W\right] \cdot I_{V,IPP} \tag{7.4}$$

where W is a measurement weight matrix determined like

$$W^{-1} = \begin{vmatrix} \sigma^2_{U,1} + \sigma^2_{DR} & r_{1,2} & \ldots & r_{1,k} \\ r_{2,1} & \sigma^2_{U,2} + \sigma^2_{DR} & \ldots & r_{2,k} \\ \vdots & \vdots & \vdots & \vdots \\ r_{k,1} & r_{k,2} & \ldots & \sigma^2_{U,k} + \sigma^2_{DR} \end{vmatrix} \tag{7.5}$$

where σ^2_{DR} is a priori variance for an unaccounted estimation error for a $\tau_{U,i}$ (for the continental WAAS (CONUS) operation area, this error RMS nominal level is σ_{DR} = 0.35 cm); $r_{i,k}$ are coefficients for cross correlation of ionospheric ranging errors, that are determined like (Walter et al. 2001)

$$r_{i,k} = \begin{cases} 0, & if\ i\ and\ k\ share\ no\ common\ satellite\ and\ receiver \\ r^2_{SV}/(F_{PP,i} \cdot F_{PP,k}), & if\ i\ and\ k\ share\ common\ satellite \\ r^2_{REC}/(F_{PP,i} \cdot F_{PP,k}), & if\ i\ and\ k\ share\ common\ receiver \end{cases} \tag{7.6}$$

where $F_{PP,i}$ is an inclination factor accounting for the ith satellite elevation angle in the observation point (2.81).

To detect a HAL/VAL cylinder deterioration (see Fig. 2.22) owing to an ionospheric disturbance, one uses the condition

$$\chi^2 \geq \chi^2_{RATIO} \cdot \chi^2_{THR} \tag{7.7}$$

where the χ^2_{RATIO} is a set ratio determined by a probability of missed detection (P_{MD}) and a false alert probability (P_{FA}), when the availability deterioration alert is generated while the RNP is, indeed, available; χ^2_{THR} is a detection criterion threshold value that is selected depending on admissible probability of detection missing.

Figure 7.3 provides the elucidations for the values in Condition (7.7) (Walter et al. 2001). Comparing the criterion χ^2 time variation, HPL/VPL estimates (see Fig. 2.22), and the geomagnetic activity index behavior, one may estimate the adequacy of using this or that index as an indicator for the expected RNP availability deterioration under geomagnetic disturbance conditions.

In (Datta-Barua et al. 2005), the authors performed a comparative analysis for using geomagnetic *Kp*- and *Dst*-indices to predict the RNP availability for the

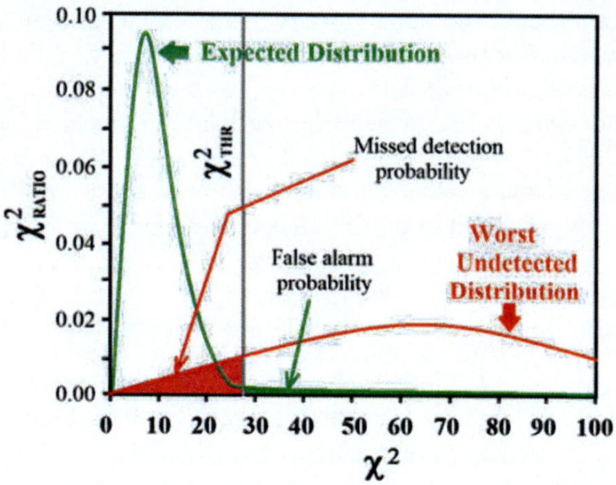

Fig. 7.3 Ratio between the probability of the detection missing (P_{MD}) and the probability of false alert (P_{FA}), that determines a threshold for making a decision on RNP availability (unavailability) for WAAS users (elucidation to the event detection condition (7.7)). Re-produced and modified figure based on the part of the figure published in (Walter et al. 2001)

LPV/LPV200 CAT aviation users. The analysis was done from available data on measurements at the WAAS network of monitoring stations, and from the information on actual navigation service availability (HPL/VPL estimates with a 95% confidence level) within 2003 Jan 7 and 2005 Jan 3. When using Criterion (7.7) as a measure of positioning availability, the authors found a 0.69 correlation between the *Dst* and $\lg(\chi^2)$ hourly values. Herewith, using the *Dst*-index to forecast the RNP availability deterioration provides a lower value for false alert, than by using the *Kp* at the set probability of detection missing.

The false alert probability when detecting the positioning availability deterioration depends on the selected minimal threshold value for the *Dst*-index, at which there emerges an RNP unavailability due to an ionospheric disturbance. So, at the $Dst = -250$ nT threshold, the false alert probability is $P_{FA} = 0.17\%$, and at the minimal $Dst = -77$ nT, this probability considerably increases and reaches $P_{FA} = 6.1\%$ (Datta-Barua et al. 2005).

One may partially elucidate the advantages of using the *Dst*-index as an indicator for the expected RNP availability at the set level so that the *Kp* is obtained by averaging over the 3-h interval. At the same time, the *Dst* is a result of averaging over only 1-h interval. Another possible elucidation may be that the *Kp* is obtained based on the measurements within the Northern hemisphere, and, consequently, may be subject to auroral substorm effects to a greater extent (Akasofu and Chapman 1972). The *Dst*-index is obtained from equatorial measurements of the geomagnetic field, that are more sensitive to ring currents (Akasofu and Chapman 1972).

A sufficiently high probability of a false alert, when using the *Dst*-index, does not enable to apply it for real-time alert to the current level of positioning availability. However, this index may be recommended for a forecast ($\geq$1 h) of the expected positioning availability at the required level.

In general, we may say about a possibility to create a system of long-term alert to a deterioration in the GNSS/WAAS integrity and the RNP availability under geomagnetic disturbance conditions, based on the analysis of geomagnetic activity indices.

One should additionally study this to reveal more effective indices of geomagnetic activity for predicting and real-time alerting of GNSS users. In particular, it may appear more expedient to involve regional *a* and *K* indices for this purpose, as well as, probably, individual geomagnetic field components (specifically, the *H*-component). The *Dst* variation rate is another identifiable factor that may appear better, than the *Dst* absolute value. The intention to use any geomagnetic activity index for long-term prediction of the GNSS operation quality also requires addressing the issue of a disposable time to predict these indices and the issue of the reliability of such a prediction.

7.2 Alerts to a Geomagnetic Disturbance in the Radio Wave Propagation Medium

The purpose for a real-time alert of GNSS users is a possibility to adapt a navigation receiver to dramatically changing signal propagation conditions under geomagnetic and sporadic disturbances in the ionosphere. Adapting the user receiver to such effects may be attained due to:

(1) adjusting the tracking loops of radio navigation parameters;
(2) excluding the ranging measurements containing anomalously large errors from positioning solutions (as a part of RAIM algorithms).

Under geomagnetic variation and storm conditions in the ionosphere, the intensity of electron concentration turbulences at all spatial–temporal scales increases (Afraimovich and Perevalova 2006). For real-time monitoring of the GPS signal propagation channel state, one should determine an index or a system of indices. This index (indices) should be simultaneously related to the radio wave propagation ionospheric channel status and to the quality of radionavigation parameter measurements. Let us address a possibility to use the index for amplitude scintillations, *S*4, for the present purposes. Using the *S*4-index here is stipulated by technical simplicity of its estimate directly based on processing the amplitude measurements in the GNSS user receiver.

The *S*4-index enables to estimate the values for ionospheric scintillations of transionospheric signal intensity in a receiving point. This estimate may be translated into an appropriate CNR value at the tracking loop input (see Sect. 5.3.2). In its turn, the CNR determines the RMS for the carrier phase measurement errors and for the code pseudo-delay (2.35, 2.36). At the same time, the current *S*4-index depends on the key parameters of the radio wave propagation trans-ionospheric channel (Bhattacharrya 1999). Therefore, one may regard this index as an adequate indicator to alert to a deterioration in the current propagation conditions of GPS satellite signals under disturbed or essentially inhomogeneous ionospheric conditions.

Let us test the adequacy of using the *S*4-index to estimate the radio wave propagation ionospheric channel status under geomagnetic disturbance conditions in real time (Demyanov and Yasyukevich 2014). For that purpose, we use the data on measuring radionavigation parameters and coordinates of a stationary point, the Institute of Solar-Terrestrial Physics (ISTP), Irkutsk, Russian Federation. The measurements were taken with a single-frequency navigation receiver GPS/GLONASS MRK-19L (Demyanov and Yasyukevich 2014). The survey control for the MRK-19L offset antenna was done by an ASHTECH Z-XII3T navigation dual-frequency receiver. Simultaneous TEC measurements on the satellites' line of sight (LOS), that were used for the positioning solution in the MRK-19L, were provided with the above dual-frequency receiver. The antennas of the receivers were mounted 10 m from each other on the same building. The reconstruction of absolute TEC values and its variations was done by measuring the GPS satellite signal phase at L1 and L2 frequencies. In this case, we used the methods and processing facilities described in Chap. 4 (Sect. 4.2.3). To mitigate the multipath effect on the measurement results, the satellites with a <15° elevation were excluded from the analysis.

Figure 7.4 presents the structure for the testing stand. The stand comprises: an MRK-19L navigation receiver (MRK19L 1999) with an antenna; a computer to control MRK-19L (Control Computer); an ASHTECH Z-XII3T dual-frequency receiver with its authorized antenna; a computer to record and process the measured information (Data proc. computer). The Control Computer has specialized software to interact with the MRK-19L hardware. This software selects the MRK-19L operation mode, as well as the data communications mode and protocol between the MRK-19L and the Data proc. computer. The binary-format measuring information is transmitted from the user receiver output to the Data proc. computer. We have developed recording-and-processing software (RPS) installed on the Data proc. computer. The RPS primarily process the measurements obtained from the MRK-19L output and compute absolute TEC values and their variations.

The RPS converts the data entering the Control Computer COM-port in a data exchange standard unified format (NMEA-0183) with a record to a primary data file at a 4-s record period. Further, the RPS detects record time samples, current coordinates of the receiver, and the PDOP. The coordinates are computed into rectangular geocentric coordinates of the receiver, the total spherical positioning error and absolute errors in determining those coordinates are estimated as

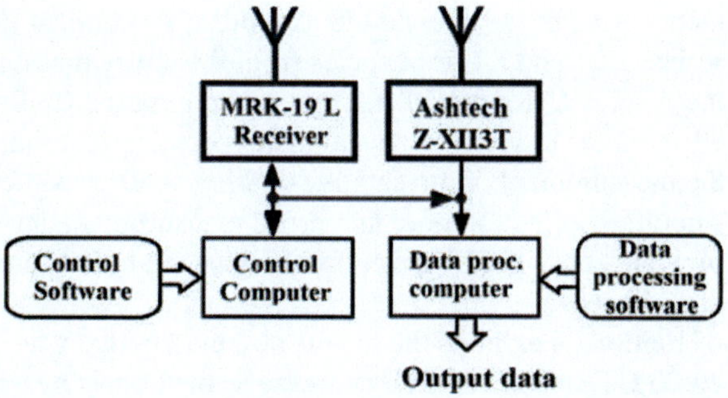

Fig. 7.4 Testing equipment stand to simultaneously observe TEC variations and the scintillation index

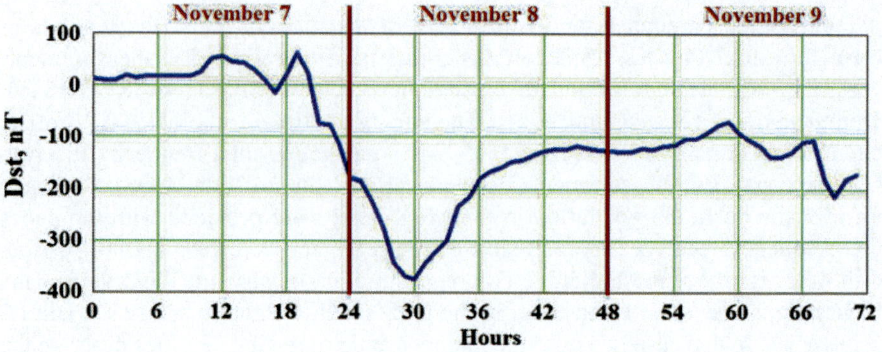

Fig. 7.5 Variations in the geomagnetic field index *Dst* during the 2004 Nov 9–10 geomagnetic disturbance. Reproduced based on the open assess data (https://wdc.kugi.kyoto-u.ac.jp/dst_final/200411/index.html)

$$\Delta S = \sqrt{\Delta X^2 + \Delta Y^2 + \Delta Z^2} \tag{7.8}$$

Separately, we detect and record the CNR measurements with a 4-s record period at the GPS L1 from the dataflow at the MRK-19L output. Those measurements are used to compute the *S*4-index. The computer also records the CNR measurements taken by the ASHTECH Z-XII3T with a 30-s record period at L1, L2 frequencies.

For observations, we chose 2004 Nov 9, when there was a complicated ionospheric geomagnetic disturbance caused by the 2004 Nov 3 (01:24–01:44 UT) and 2004 Nov 5 (11:26–11:41 UT) solar flares. Figure 7.5 presents the variations in the geomagnetic field index Dst within three days from 00:00 UT Nov 8 to 00:00 UT Nov 11, 2004.

During the 2004 Nov 9 first half, one observed moderate geomagnetic field disturbances after dramatic *Dst* variations at the previous main phase of the storm, from Nov 7 22:00 UT to Nov 8 12:00 UT. Within the Nov 9 18:00–21:00 UT interval, there was another powerful geomagnetic field disturbance caused by the Nov 5 solar flare.

The main expected manifestation of intensive ionospheric scintillations may be a decrease in the mean satellite signal in a receiving point, when averaging measurement series by all the visible satellites. To confirm this, we detected CNR hourly series at L1 and L2 frequencies from the entire measurement scope accumulated by the ASHTECH Z-XII3T per day, and computed CNRs averaged over each hour for all satellites in view. To reveal the expected effect more reliably, we addressed only the measurements from satellites with $\gamma > 60°$ elevation. Thus, one may eliminate a smoothing effect of more numerous measurements by satellites with low elevations, for which the signal fading effect may be related not only to ionospheric geomagnetic disturbances.

Figure 7.6 exhibits the results obtained in that way. Apparently, within 17:00 and 20:00 UT, one observed a decrease in the hourly average level of satellite signals at both operation frequencies. This phenomenon timely agrees with the geomagnetic disturbance main phase (Fig. 7.5).

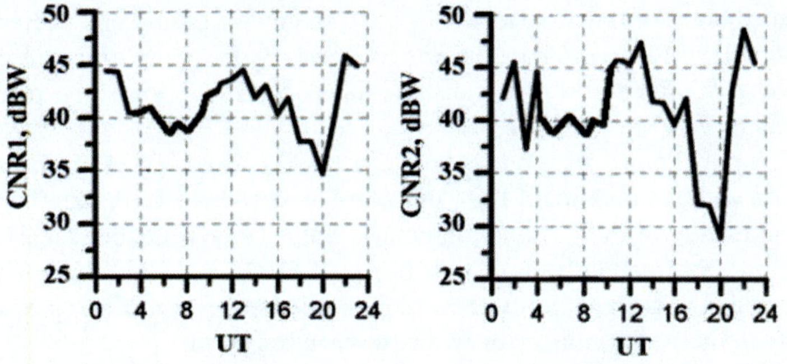

Fig. 7.6 Hourly average CNR values measured by the signals of the satellites with >60° elevation at the GPS L1 (left panel) and L2 (right panel) on Nov 9, 2004

Figure 7.7a provides series of the ISTP positioning errors obtained with the MRK-19L in the measurement mode by GPS signals. Within 16:00 and 21:00 UT (rectangle in Fig. 7.7a), one may observe a sharp increase in positioning errors.

For the observation point, this time corresponded to night side conditions, 23:00 through 04:30 LT, when one should have observed lower ionospheric values for the electron concentration and for TEC. Nevertheless, it is this period, when one observed periodic sharp increases in positioning errors. This period partially coincides in time

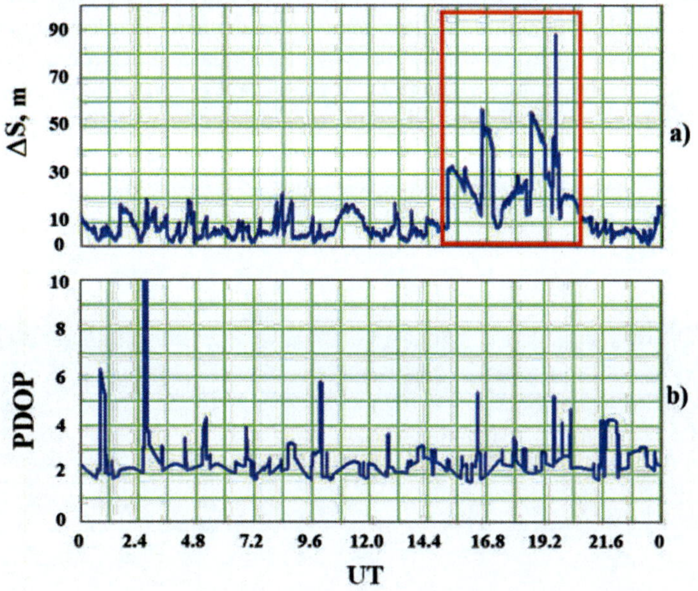

Fig. 7.7 Observations of spherical positioning error (panel **a**) and geometric factor (panel **b**) with the MRK-19L under the 2004 Nov 9 magnetic storm conditions (the period corresponding to the main phase of the storm is within the red rectangle)

with the magnetic storm main phase (Fig. 7.5), and with a period of a decrease in the mean CNR (Fig. 7.6). To separate the effects caused by systematic diurnal variations in the geometry of the observed constellation at the ISTP, Fig. 7.7b presents the PDOP dynamics.

Note that positioning error bursts between 16:30 and 20:00 UT, in this case, did not coincide with the maximal PDOP that were observed at 01:10, 02:40, and 10:00 UT (a local feature of the observation point). At the same time, between 16.30 and 20.00 UT, one observed only short-time bursts of PDOP > 4.5–5 values. Thus, TEC variations and GNSS signal ionospheric scintillations may contribute prevalently to an increase in the positioning error within a receiving point.

Figure 7.8 provides examples for simultaneous observations of small-scale TEC disturbances and the *S*4-index. The TEC values were reconstructed for PRN 18, 26, and 29 satellites LOSs. The *S*4-index was computed by the CNR measurements taken at the GPS L1 in the MRK-19L receiver by the signals from the above satellites. We averaged the measurements (when computing the scintillation index) over a 5-min interval, which corresponds to the mean period for small-scale ionospheric disturbances.

One can see that between ~18:30 and 21:00 UT, observed were intensive small-scale ionospheric disturbances leading to an emergence of GPS satellite signal intensity scintillations. These phenomena coincide with the magnetic storm main phase (Fig. 7.5), with a decrease in the mean CNR at the GPS L1, L2 (Fig. 7.6), and with a sharp increase in the positioning error (Fig. 7.7). Thus, one may ascertain a temporal coincidence between the decrease in the positioning quality and the disturbances of the GPS signal propagation medium during the magnetic storm main phase, that are sufficiently exactly reflected by the *S*4-index under the addressed conditions. Thereby, one may recommend this index for a real-time check of the measurement quality by signals from separate satellites in order to detect and eliminate, in real-time, the signals featuring a higher level of ionospheric scintillations from the positioning solution under magnetic storm conditions.

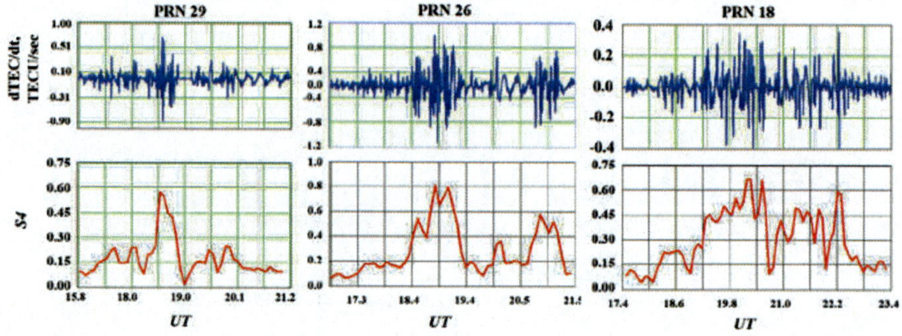

Fig. 7.8 Simultaneous observations of the TEC derivative obtained from measurements with the ASHTECH Z-XII3T (upper row of figures), and of the *S*4-index from the measurements with the MRK-19L (lower row of figures). The numbers of GPS satellites are indicated over the figures

7.3 Alerts in Case of Powerful L-band Solar Radio Bursts

As shown above, the $>10^3$ sfu solar radio bursts in the UHF may be dangerous for the radionavigation parameter measurement quality. Thereby, we should address the issue of a possibility to predict $>10^3$ sfu solar radio bursts and to alert (in real-time) GNSS users to similar heliophysical events.

In general, the known statistics of solar radio flare observations shows the following (Giersch et al. 2017):

(1) the occurrence frequency for radio flares substantially differs in different frequency bands;
(2) a solar activity cycle is not related to the occurrence frequency of powerful radio flares, and the emergence of powerful radio flares does not coincide with the solar activity maxima;
(3) the occurrence frequency of solar radio flares decreases with an increase in the flare power, and extreme radio flares are a rare phenomenon of non-periodic and hard-to-predict character.

The above conclusions impede forecasting dangerous and, especially, extreme radio flares based on historical data. Nevertheless, in (Huang et al. 2018), the authors propose a statistical empirical power-view dependence that associates an event occurrence frequency (radio flares within the ~1–2 GHz range) with its power. The dependence has the following view:

$$F(S) = 37102 \cdot S^{-1.791} \tag{7.7}$$

where S is a rate of flow for a possible flare radio emission. For example, at the expected flare power $S \geq 4000$ sfu, the occurrence frequency for such an event equals ~ 5 flares per year.

Using empirical dependences based on the statistics from long-term observations enables to forecast the emergence of extreme flares. However, the reliability of such forecasts has not been studied, and the exact time for the emergence of such a flare is impossible to forecast. Therefore, the worth of this approach to forecasting dangerous and extreme solar radio bursts with the goal of long-term or a real-time alert of GNSS/WAAS users is not satisfactory.

There is a great number of studies addressing mechanisms for generating different-class solar radio bursts within 300–3000 MHz (Dulk 1985; Kahler 1992). As a result of some investigations, revealed was a considerable correlation among the phenomena of radio bursts, coronal mass ejections, and solar radiation flares within the soft (0.1–0.8 nm) and hard (0.05–0.1 nm) X-ray. An especially strong correlation is known to be observed between microwave hard X-ray bursts. Microwave bursts are mainly concentrated within centimeter waves, but they often cross over the UHF as well (Bruzek et al. 1977).

Let us estimate a possibility to use observations of the Sun hard X-rays and CME observations to alert a GNSS user to powerful bursts of the Sun UHF emission. In

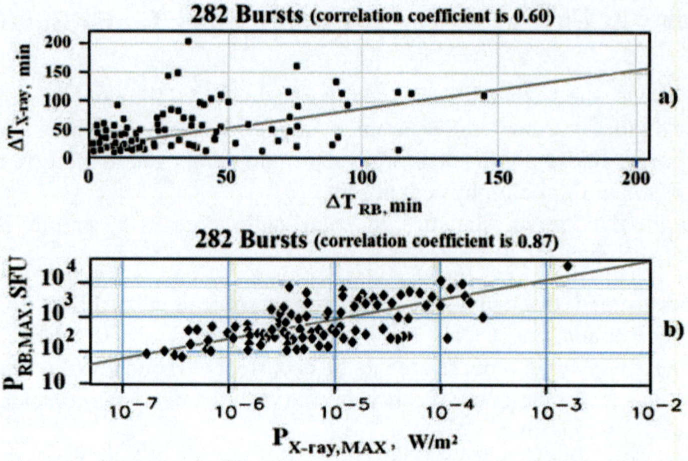

Fig. 7.9 Correlation between the emergence of solar radio bursts in the X-ray and in the UHF: duration of flares (**a**) and their power (**b**). Re-produced and modified figure based on the part of the figure published in (Michalek et al. 2009)

(Michalek et al. 2009), there are the results of long-term (1996–2004) observations that involved 419 different-power and different-duration flares within the X-ray and UHF ranges. The authors note the emergence of UHF emission bursts both before and after the X-ray bursts. The observations of the radio bursts emerged after the X-ray ones (282 flares of 419) corroborated a significant correlation between those events. Figure 7.9 illustrates the statistical interdependences between the solar burst power and duration in the UHF and the hard X-ray ranges.

As one can see, the correlation coefficient is 0.6 by flare duration and 0.87 by power. Herewith, the radio burst onset was revealed to be delayed (relative to the X-ray burst onset) by 3 min on average. However, the radio emission peaks 2 min earlier, than it does in the X-ray range. This allows us to believe that there are technical possibilities to alert GNSS users to dangerous or extreme solar radio burst in real-time.

Along with that, there are grounds to assume that the above correlative dependences, as well as flare comparative temporal characteristics, may be substantially violated with a growth in the flare power. Based on the analysis above (Sect. 3.4.1), one may note the following:

1. There is a great variance for the difference between an X-ray flare onset and a UHF radio burst onset (from about synchronous coincidence to a 40-min difference).
2. There is a mismatch between an X-ray flare power and the related radio burst. The least powerful 2003 Oct 28 UHF radio burst corresponds to the most powerful X-ray burst, while the inverse situation is observed for the 2005 Dec 6 and 13 events.
3. There is a mismatch between the hard X-ray burst durations and those of the UHF.

All this results from the fact that there are different mechanisms for flare generation in different ranges (see Sect. 3.2). Common is only the triggering mechanism: energy release in the volume of the Sun active region atmosphere.

Thus, the estimates for the Sun X-ray current power may not be accepted as a reliable indicator to alert GNSS users to a forthcoming powerful (>1000 sfu) solar radio burst. However, the results in (Michalek et al. 2009) may be regarded as a good start for investigations by engaging observations of most powerful radio- and hard X-ray solar flares from the measurement across several 11-yr solar activity cycles.

Additional possibilities to forecast dangerous solar radio bursts are related to CME observations. According to (Cane and Richardson 2003), radio bursts may manifest themselves both before and after a CME. Of the above 419 UHF radio bursts, detected were only 105 CME-related events: of the latter, 49 bursts occurred before CMEs, and 59 after. For the latter group of bursts, it is shown that, with a greatest probability, UHF radio bursts emerged, when the CMEs that generated them propagated as far as ~1.23 solar radii from the Sun center. The radio burst maximum is reached after a CME propagates to ~1.86 solar radii. This observation, in principle, gives a chance to forecast powerful UHF radio bursts, provided that a CME origin localization and its origin time are known at a sufficient accuracy. However, so far, CME observations cannot be addressed as a reliable precursor of solar radio emission powerful bursts within the GNSS operation frequency range.

In general, one may conclude that, for the time being, it is impossible to forecast (sufficiently accurately) powerful solar UHF radio bursts for the benefit of GNSS/WAAS users, based on the known up-to-date methods and facilities. The only recommended measure may be a continuous radio monitoring of the Sun in the UHF range and issuing real-time alerts for GNSS users, when >10^3 sfu radio bursts emerge (Demyanov and Yasyukevich 2014).

When arranging a 24-h radio spectrographic monitoring of the Sun, one should also record the radio-burst spectrum type, which provides additional possibilities to forecast these phenomena on the Earth. So, a type IV burst is always broadband and long-term; it accompanies an ejection of coronal mass volume from the Sun surface. However, this burst is preceded by type II bursts that are signatures for a shock generation. Type III bursts are indications of an expected high-energy particle fluxes. There exists the world net of radio spectrographs, there are also special solar radio observatories. However, these possibilities have been poorly used, so far, to build a system for monitoring and forecasting the operation quality of Global Positioning Systems and their applications.

7.4 CNR Power Improvement in Satellite Communication Links

A signal decrease at the tracking loop input during solar radio bursts was discussed in (Klobuchar et al. 1999; Chen et al. 2005). In (Demyanov and Yasyukevich

Table 7.1 Minimal signal power within the 20.46 MHz band at a navigation receiver antenna (Table 3-Va from (IS-GPS-200J 2018))

SV blocks	Channel	Signal	
		P(Y) (dBW)	C/A or L2C (dBW)
IIA/IIR	L1	−161.5	−158.5
	L2	−164.5	−164.5
IIR-M/IIF	L1	−161.5	−158.5
	L2	−161.5	−160.0
III	L1	−161.5	−158.5
	L2	−161.5	−158.5

2014; Zakharov et al. 2016), demonstrated signal phase losses-of-lock (LOL) during powerful solar radio bursts and under the geomagnetic disturbance conditions. Those LOLs occur, to a greater extent, at the auxiliary frequency L2 owing to two factors: (a) smaller transmitter power at that frequency band, and (b) CNR loss due to semi-codeless (codeless) processing of signals with the unknown code structure. Thus, increasing the transmitter power of GPS satellites may become an effective measure to enhance the GNSS functioning stability under space weather extreme conditions.

As per the GPS Interface Specification (IS-GPS-200J 2018), the power of the signal broadcast by new-generation GPS satellites (starting with an IIR-M block) was increased. The schedule for replacing the obsolete orbital blocks with new ones having an increased transmitting power is presented in the IGS archives at ftp://igs.org/pub/station/general/igs08.atx and http://igs.org/pub/station/general/igs08.atx.

Table 7.1 shows the minimal power of signals from different GPS blocks in a receiving point. Apparently, starting with IIR-M blocks, the minimal L2 signal power increased by 3–5 dBW. Potentially, such a power increase may enhance the noise immunity of radio-navigation parameter filtering in a GNSS receiver.

Let us analyze the real GNSS hardware noise immunity attained when processing signals by new-generation GPS satellites. The data from measuring radio-navigation parameters in the RINEX format contain parameter S (Pestana 2015). The latter, as per the RINEX 3.03 specification (ftp://igs.org/pub/data/format/rinex303.pdf), represents the carrier-to-noise ratio (CNR). We will further refer to this parameter as signal strength (a received signal power in the specified frequency band). RINEX 3.03 files contain the data on the signal strength at L1 for the open civil C/A code (S1C measurements), as well as at L2 for the encrypted P(Y) code at the semi-code signal processing based on the z-tracking technology (S2W measurements). Thus, we can compare the L2 signal strength when using the encrypted P(Y) code for GPS IIA and GPS IIF blocks. Also, we can estimate the of the signal strength when using a new civil open code L2C (S2X measurements) that new-generation satellites broadcast.

Figure 7.10 shows the dynamics of the CNR at L1 (S1) and at L2 (S2) of the GPS satellite PRN10 received at the IGS ZIM2 station on 2015 Apr 29 (black dots) and

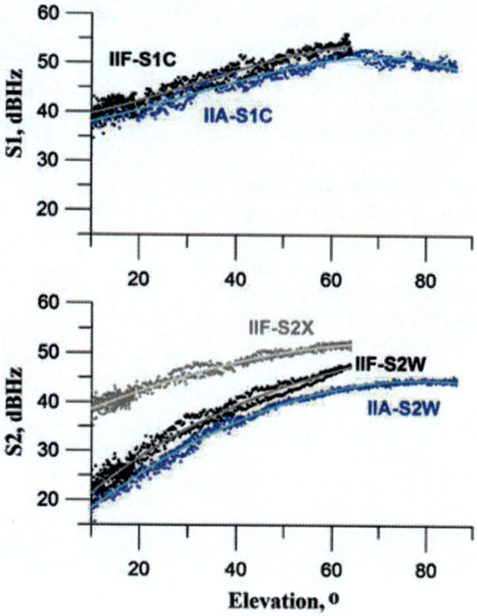

Fig. 7.10 Dependence between the L1 (upper panel) and L2 signal strengths (lower panel) and the PRN10 elevation on 2015 Apr 29 (dark blue dots) and on 2015 Dec 27 (black and gray dots) at the ZIM2. Dark blue dots present the IIA S1C (upper panel) and the S2W (lower panel); black dots present - the IIF S1C (upper panel) and the S2W (lower panel), gray dots - the IIF S2X (Yasyukevich et al. 2021)

on 2015 Dec 27 (dark blue dots). As per the official IGS information (http://www.igs.org/), at that time, the ZIM2 was equipped with a Trimble NETR9 receiver and a TRM59800.00 antenna.

On the addressed days of measurements, there were no replaces of the ZIM2 receiver hardware. At the GPS PRN10, the IIA block was replaced with the IIF between 2015 Aug 3 and 2015 Oct 31. Thus, the days selected for measurements were beyond the satellite hardware modernization period.

One can see that, as a result of the block replacement, there occurred an L1 strength increase recorded with S1C at ~1.5–2.5 dBHz. The L2 signal strength increase was even more essential: 3–4.5 dBHz for the P(Y) code (S2W). The open code L2C (S2X) signal strength increased most notably (up to 8.5–20 dBHz).

Let us analyze a possible effect of a similar signal strength augmentation on increasing the noise immunity when receiving signals from GPS satellites. For this purpose, we address the dynamics for the GPS signal strength during various radio bursts of the Sun for the same satellite. Below, there are the results for the 2011 Sep 24 (Sreeja et al. 2013) and 2017 Sep 6 radio bursts (Yasyukevich et al. 2018; Berdermann et al. 2018). Those bursts, by the classification from (Huang et al. 2018), refer to extreme and dangerous, respectively. For each burst, we studied the dynamics for the received signal strength at the fundamental (L1) and auxiliary (L2, L5) GPS frequencies at the ZIM2 station.

The analysis was performed for different-generation satellite blocks. We used the S1C and S2W measurements that can be obtained both for "obsolete" (IIA/IIR-A), and "new" (IIR-M/IIF) GPS blocks. Also, we used the S2X measurements (to estimate the benefits of using new open L2C codes) and the S5X (to analyze the

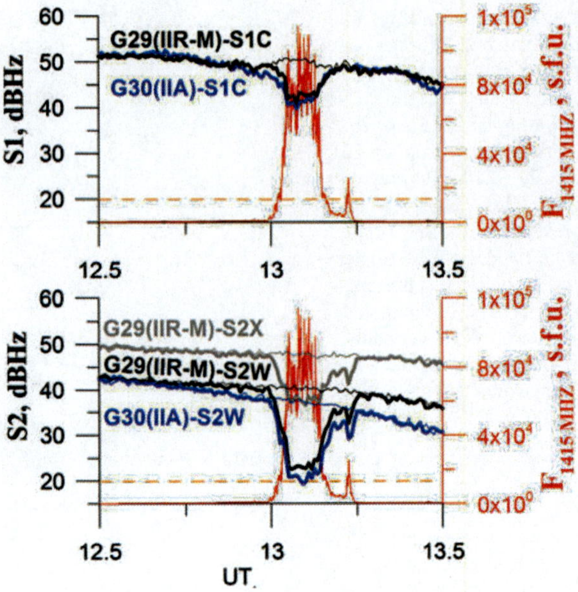

Fig. 7.11 1415 MHz solar radio emission power flow (red curve) during the 2011 Sep 24 radio burst, and the ZIM2 received signal strength at L1 (upper panel) and L2 (lower panel): dark blue curve depicts the S1C and S2W for PRN30 (IIA); black curve is for the S1C and S2W, gray curve traces the S2X for PRN29 (IIR-M). Thin lines mark the corresponding signal strengths on the previous day (2011 Sep 23) (Yasyukevich et al. 2021)

receiving signal strength at L5). During the first burst, the ZIM2 was equipped with a Trimble NETR5 receiver, but after 2015 Apr 27, it was replaced with a Trimble NETR9 version.

To analyze the radio emission flow, we used the Sagamore Hill spectrograph data, that is a part of the RSTN international network (ftp://ftp.ngdc.noaa.gov/STP/space-weather/solar-data/solar-features/solar-radio/rstn-1-second/sagamore-hill/). According to the Sagamore Hill data, the radio emission flow intensity at 1415 MHz started growing about 12:32 UT, and reached the peak of 114,144 sfu at 13:04:46 UT on 2011 Sep 24 (Fig. 7.11, red curve).

Figure 7.11 also shows the dynamics for the receiving signal strength: dark blue curves—for L1 (S1C) and L2 (S2W) from the IIA satellite; black curves—for L1 (S1C) and L2 (S2W) from the IIR-M, and gray curves—for L2C (S2X) from the IIR-M. According to the RINEX 3.03 specification, the strength threshold for a satisfactory quality of receiving and processing a signal is no less than 30 dBHz. For the L1 signal, the CNR threshold at the tracking phase loop input (when a LOL may occur) is about 28 dBHz (Hofmann-Wellenhof et al. 2008). Special post-processing procedures can reduce this value to about 20 dBHz (Psiaki 2001). Thereby, we chose 20 dBHz as a threshold for the minimal signal strength. The threshold level is marked by the orange dotted line in Fig. 7.11.

Apparently, the radio burst power flow maximum is accompanied by a decrease in the signal strength both at L1 and at L2. At L1, the signal strength from PRN29 and PRN30 practically coincides. At the same time, the PRN30 (IIA) L2 signal strength is substantially lower and drops below 20 dBHz, while, for the IIR-M block (PRN29), the minimal strength values appeared to be higher by ~5 dBHz.

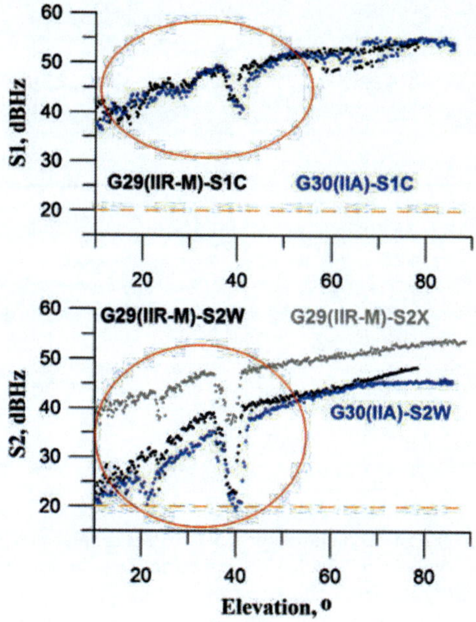

Fig. 7.12 Dynamics for the PRN29 and PRN30 signals at the ZIM2 at L1 (upper panel) and L2 (lower panel). Dark blue dots present the data for PRN29 (IIA), black and gray dots signify the data for PRN30 (IIR-M). Orange dotted line marks the 20 dBHz threshold (Yasyukevich et al. 2021)

Because a satellite elevation substantially affects the signal strength, it is more correct to compare the relations between the signal strength and the elevation. Figure 7.12 shows the signal strength dependence (when processing different signal components) on the satellite elevation for L1 (upper panel) and L2 (lower panel). Dark blue dots present the data for the S1C and S2W measurements from PRN30 (IIA block), black dots signify the S1C and S2W from PRN29 (IIR-M block), gray dots show the L2C (S2X) signal strength from PRN29.

Apparently, at the L1, the elevation dependences for two blocks practically coincide. During a burst, one observes a sharp synchronous decrease in the L1 (S1C) signal strength by about 7.5 dBHz. This decrease occurs for the signals from both PRN29 and PRN30 at close elevations. For the L2 signals, one can see a difference between the signal strength that is recorded from the IIR-M and IIA blocks. For the more obsolete IIA, the received signal strength is lower at all the elevations. During a burst, one observes a decrease in the L2 signal strength by more than 17 dBHz for both blocks. Due to primarily lower signal strength, the IIA minimal value appears below the 20 dBHz threshold, which potentially results in a greater LOL probability. At the same time, the decrease in the strength recorded when receiving a new open L2C (S2X) code is near 10 dBHz.

During the 2017 Sep 6 weaker radio burst, the maximal value for the radio emission flow from San Vito radio spectrograph (a part of the RSTN network, ftp://ftp.ngdc.noaa.gov/STP/space-weather/solar-data/solar-features/solar-radio/rstn-1-second/san-vito/) reached 19,604 sfu. Figure 7.13 provides the radio emission flow dynamics (red curve).

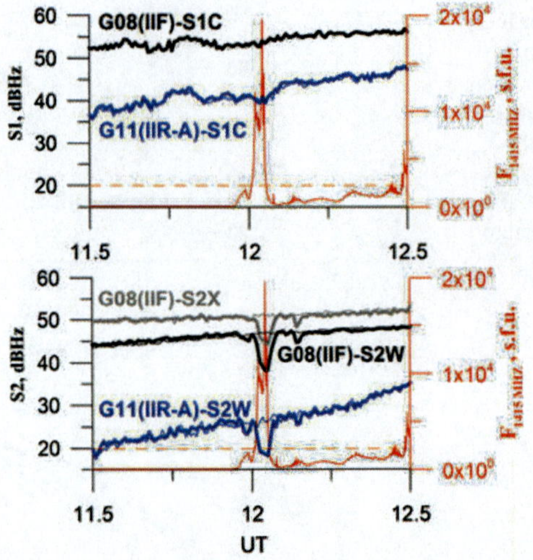

Fig. 7.13 Dynamics for the 1415 MHz solar radio emission from the SanVito spectrograph (red curve) during the 2017 Sep 6 radio burst, and for the strength of the GPS signals received by the ZIM2 at the L1 (upper panel) and L2 (lower panel) for two satellites: PRN11 (IIR-A block), the S1C and S2W measurements (blue curves), and PRN08 (IIF block), the S1C, S2W (black curves) and S2X measurements (gray curves). Thin lines show the corresponding signals' strengths on the previous day (Sep 5). Orange dotted line marks the 20 dBHz threshold (Yasyukevich et al. 2021)

Apparently, this burst did not affect the L1 signal strength, which, probably, is caused by the energy distribution features in the radio emission spectrum (Sato et al. 2019). At the same time, one can observe a notable signal strength drop at the L2 that reached ~8 dBHz. This result agrees with (Sato et al. 2019), where the authors showed a decrease in the L2, L5 signals down to 10 dBHz in the European region. The L2 signal strength decrease, when using the L2C (S2X) open code, appeared slightly lower and equaled ~7 dBHz.

Figure 7.14 shows the signal strengths depending on the elevation. Unfortunately, we did not find the data for different-type satellites at close elevations for this burst. However, one can see essential differences in the strength dynamics at L2. The solar flare effect is in a sharp decrease in the L2 signal.

Apparently, the PRN11 (IIR-A) signal has initially a lower mean value for the L2 signal strength (S2W, Fig. 7.14, lower panel). During the burst, this signal strength dropped below the 20 dBHz threshold. Note that, for PRN08 (IIF), the radio burst effect was recorded at a higher elevation, which does not make a direct comparison between the signal strengths from those satellites during the burst quite correct. The burst-related decrease in the total L2 signal strength (S2W) for both satellites is comparable by amplitude and equals about 8 dBHz, regardless of the elevation. The L2X signal from PRN08 is higher, particularly, at smaller elevations, and the burst-related effect for that signal is substantially weaker. Thus, the L2 signal strength increase, as well as applying a high-precision open code, may significantly improve

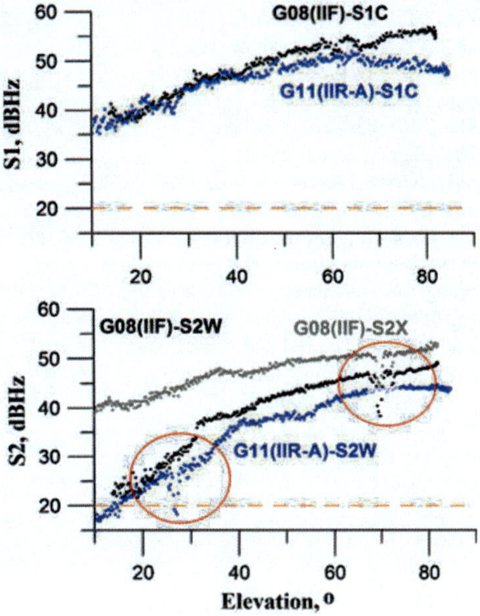

Fig. 7.14 Dynamics for the signal strength from satellites PRN08 and PRN11 at the ZIM2 at L1 (upper panel) and L2 (lower panel) during the 2017 Sep 6 radio burst. Dark blue points present the data for PRN11 (block IIR-A), black dots for PRN08 (unit IIF). Gray dots show the L2 signal strength (S2X, PRN08). Orange dotted line marks the 20 dBHz threshold (Yasyukevich et al. 2021)

the robustness of radio-navigation parameter measurements under powerful solar radio burst conditions.

The presented results show explicitly that, with a transmitter power increase during the replacement of GPS orbital blocks with up-to-date ones, the L2 signal strength in a receiver grows by 3–4.5 dBHz. This growth equals up to 20 dBHz, if a new open L2C (S2X) code is used. The solar radio burst effect leads to a noticeable drop in the strength of the signals recorded at L2 and L5, at both the semi-code processing (S2W), and when using an open code (S2X, S5X). Herewith, the L2 signal at the semi-code processing (z-tracking) featured the least total strength, and, accordingly, the lowest noise immunity against solar radio bursts.

The signal strength drop during solar radio bursts potentially leads to a greater probability of a signal (pseudo-delay) LOL (Afraimovich et al. 2009). To estimate the effect of the new-generation broadcast signal increase on the general GNSS noise immunity during solar radio bursts, we analyzed a change in the density of GPS signal phase LOLs for L1 and L2 within the addressed above bursts.

For computations, we used the GNSS global network data including about 5000 receivers (4659 in 2011 and 4939 in 2017). Figure 7.15 presents the map for the GNSS network station location and marks the sub-solar points for the 2011 Sep 24 (upper panel) and 2017 Sep 6 (lower panel) radio bursts.

To analyze the dynamics for the phase LOL density during bursts, we computed the mean LOL density separately for the updated and obsolete satellites. Used were the data from the stations that were in the sub-solar region with $\geq 60°$ Sun zenith angle at the burst instant. We accounted for only the measurements from the satellites,

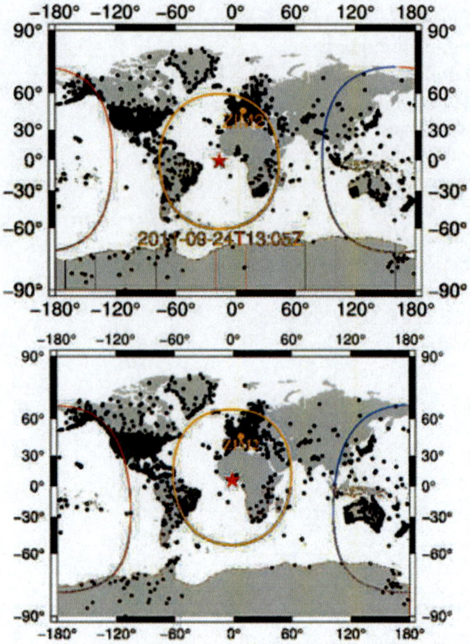

Fig. 7.15 Distribution of GNSS receivers to analyze the density of the signal phase LOLs during the 2011 Sep 24 (upper panel) and the 2017 Sep 6 (lower panel) radio bursts. The red asterisks on the panels mark the sub-solar point position at the burst maximum. The orange line restricts the region with <60° solar zenith angle. The red and dark blue lines show the positions of the dawn and dusk solar terminator, respectively (Yasyukevich et al. 2021)

whose elevation surpassed 10°. The measurements were recorded in 30-s increments. At each instant, we involved more than 3000 measurements in our statistics.

Figure 7.16 shows the dynamics for the mean phase LOL density at L1 (black color) and L2 (dark blue color) for the 2011 Sep 24 (left) and the 2017 Sep 6 (right) radio bursts. The top panels present the LOL dynamics for the updated satellites (IIR-M, IIF), the lower panels display the same for the obsolete ones (IIA, IIR-A, IIR-B). Gray curves show the corresponding distributions for the previous (quiet) days.

For all types of satellites, the phase LOL density at L2 is expected to be significantly higher, than that of L1 during the bursts. In both cases, recorded are sharp spikes (as compared with a quiet day) in the LOL dynamics. Those spikes coincide with the instants of the radio burst maxima (vertical dotted lines).

During the extreme 2011 Sep 24 burst, the phase LOL density increase at L2 reached 30% of the total measurements of obsolete satellites' signals (Fig. 7.16b). The updated satellites showed the maximal value for the L2 phase LOL density being ~10% (Fig. 7.16a). At L1, the maximal LOL during the 2011 Sep 24 burst was about 8% of all the measurements by the updated satellite signals and ~15% by the signals of obsolete ones.

During the 2017 Sep 6 burst, there was also an increase in the phase LOL density, however, to a lesser degree, than that in 2011. First of all, this is related to a smaller intensity of the burst. The maximal values for the phase LOL density at L2 during the burst reached about 1.6% by measurements of obsolete satellite signals and about

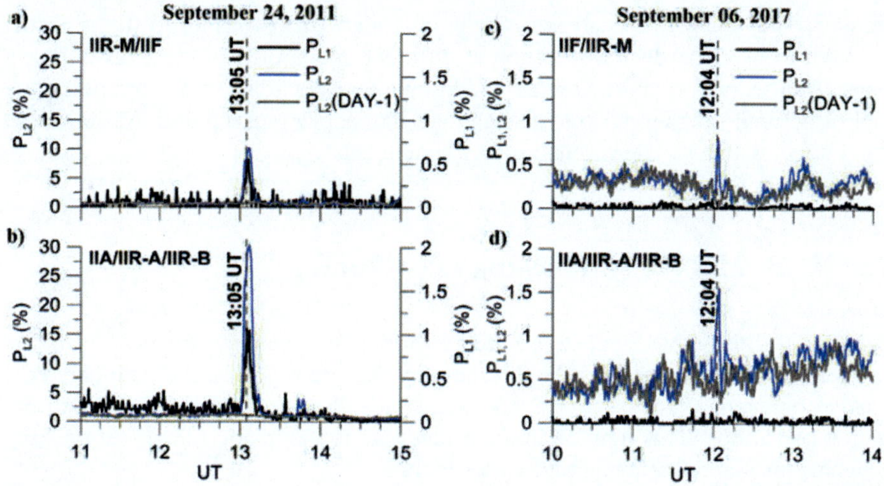

Fig. 7.16 Dynamics for the mean phase LOL density at L1 (black curve) and L2 (dark blue curve) during the 2011 Sep 24 (left) and the 2017 Sep 6 (right) solar radio bursts for the updated (panels **a**, **c**) and obsolete (panels **b**, **d**) satellites. Gray lines show the dynamics for the L2 phase LOL density on the previous quiet days. Vertical dotted lines mark the radio burst maxima (Yasyukevich et al. 2021)

0.8% for the updated satellites. There was no expressed effect on the L2 phase tracking for this burst.

Berdermann et al. (2018) investigated the phase LOL dynamics during the 2017 Sep 6 burst for the European region, and concluded that the burst did not cause changes in the phase LOL density at the GPS frequencies. The absence of effects on the global scale during this burst was also noted in (Yasyukevich et al. 2018). The reason for the difference in the results presented here from the previous studies is related to the selection of the stations for the analysis. The main advantage is introducing the limitations on the Sun zenith angle. During the bursts in Europe and America, a number of stations were on the Earth sunlit surface; however, for most of those stations, the Sun zenith angle was small (<30°), and the burst effect at those stations was hardly observed.

Our analysis shows that, under the addressed solar radio burst conditions, increasing the power of transmitters onboard GPS orbital blocks enabled to reduce the L2 phase LOL density by more than a factor of two. Using the open high-precision ranging code also yields an essential CNR gain at the tracking loop inputs. Using the above measures enables to increase the noise immunity for GNSS satellite signal receiving and processing under the strong and extreme solar radio burst conditions.

Another relevant conclusion is that the solar radio emission power distribution, as well as the prevalent polarization kind, is non-uniform within the GNSS signal radio spectrum. This provides additional possibilities to enhancing the GNSS noise immunity under the solar radio burst conditions. First, it is necessary to organize more detailed monitoring of the solar radio emission spectrum within the GNSS operation frequency range, which enables to define and alert (in real time) to the expected quality of GNSS hardware operation at different operational frequencies during a burst. And second, it is required that additional investigations into determining the GNSS optimal operational frequencies that may appear susceptible to solar radio burst effect with a smaller probability be conducted.

7.5 Code and Phase Tracking Loop Tuning

Using real-time alerts, such as scintillation indices, geomagnetic field variation indices, the alert to excessing the threshold of dangerous solar radio emission, etc., enable to adapt GNSS user receivers to dramatically changing radio wave propagation conditions. This may be implemented due to automatic tunings in tracking schemes for pseudo-delay and the carrier phase.

Figure 7.17 provides functional schemes for the phase lock loop (PLL) and for the code-delay lock loop (DLL). When one linearly approximates the discrimination characteristic effective part of a tracking loop, one determines the measurement accuracy for the carrier phase and for the pseudo-delay by using Expressions (2.35) and (2.36). From those expressions, it follows that the measurement accuracy depends on the accumulation time (T) in the correlator and on the noise band of the tracking loop for the DPP and PLL (ΔF_{DLL}, ΔF_{PLL}). Besides, code (phase) discriminator type defines the output voltage (U_D) and steepness of the discrimination characteristic (S_D).

On the schemes, the red circles highlight the spots for tuning those loop parameters. Correct adaptive tuning of those parameters to increase accuracy and robustness of the code-delay (carrier phase) tracking under intensive ionospheric scintillation conditions requires addressing the problem of their tuning tolerance limits.

For the pseudo-delay discriminator, one accepts the accumulation time minimum $T \geq 1$ ms, which corresponds to the duration of one C/A-code cycle (GLONASS 2005). A shorter accumulation time makes it impossible to estimate the code pseudo-delay. At the same time, to detect the navigation message from a satellite signal, one requires a more essential accumulation time that corresponds to the duration of one navigation message (20 ms).

When making the carrier phase discriminator, there emerge stricter limitations on the minimal accumulation time. Even at $T = 1$ ms, the phase change (due to the Doppler frequency drift) may attain 10 rad. Thus, a correct accumulation in a PLL discriminator requires a priori information on the Doppler frequency (ω_D). Here, we use the assumption that $\omega_D = \text{const}$ within the T. In the latter case, hence, one should define the requirements for the minimal time T. Those requirements will be related to the carrier phase approximation error between the accumulation start and end in the correlator, with allowance for mutual acceleration between a satellite and a receiver. The phase approximation error in the accepted measurement model is estimated as (GLONASS 2005):

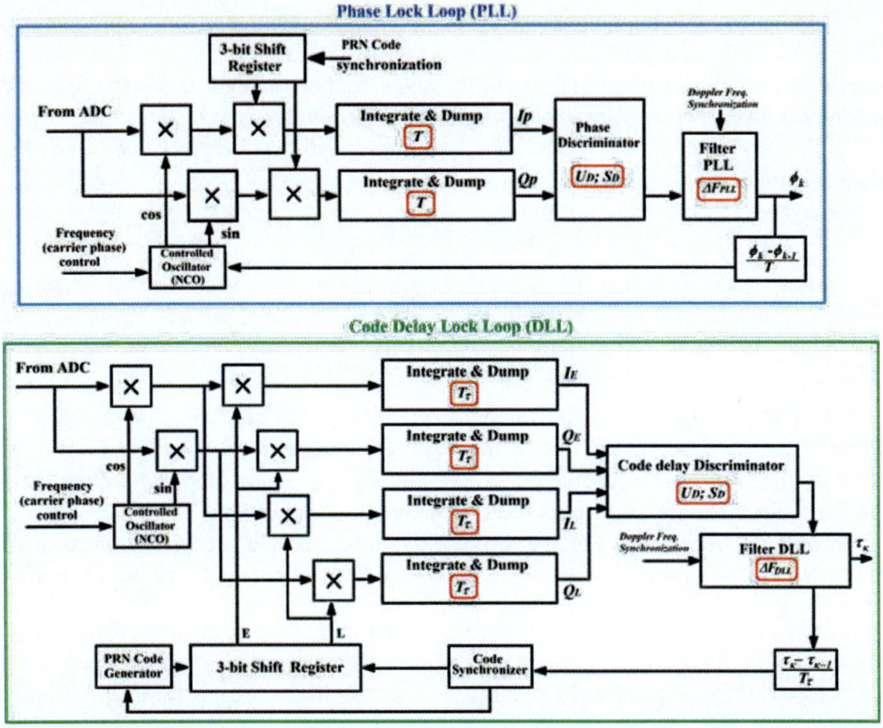

Fig. 7.17 The structure of the phase and code delay tracking loops. Re-produced, modified and translated from Russian based on the figure published in the book (GLONASS 2005)

$$\delta\varphi = \frac{\pi \cdot a \cdot T^2}{\lambda}, \tag{7.8}$$

where a is a satellite-receiver mutual acceleration; λ is a wavelength.

Based on (7.8), one may regard that, for example, at $a = 50\ \text{m/s}^2$ and $\lambda = 0.1903$ m (for the GPS L1), the corresponding phase approximation errors are ≈0.046° at T = 1 ms and ≈19° at T = 20 ms, respectively. Thus, when placing a receiver in a highly-mobile object, there are essential limitations on the minimal T.

When addressing the problem of the maximally admissible T, it is required that this time be less than the stationarity period of the estimated radio navigation parameter. Otherwise, the a priori model for that parameter change will not enable to determine correctly the optimal transfer ratio for a smoothing filter. Thus, not only the filtering accuracy will depend on determining the maximally admissible T, but also the tracking loop operation robustness. Let us analyze this problem in greater detail separately for the PLL and for the DLL schemes.

We address an optimal third-order filter for the PLL scheme, in which the a priori linear Markov's chain model for a phase change is used (GLONASS 2005):

$$\phi_k = \phi_{k-1} + T \cdot v_{k-1};\ v_k = v_{k-1} + T \cdot a_{k-1}\ a_k = a_{k-1} + \xi_{\phi,k-1} \tag{7.9}$$

where $\xi_{\varphi,k}$ is white Gaussian noise with a D_ξ variance; v_k and a_k are the first and second phase derivatives of the carrier phase, respectively.

At the filter output, the model (7.9) parameter estimates extrapolated onto the instant k are generated:

$$\begin{aligned} \hat{\phi}_k &= \tilde{\phi}_k + K_1 \frac{S_{D,\phi,REAL}}{S_{D,\phi,COMP}} \left(\lambda_k - \tilde{\phi}_k\right) & \tilde{\phi}_k &= \hat{\phi}_{k-1} + T \cdot \hat{v}_{k-1} \\ \hat{v}_k &= \tilde{v}_k + K_2 \frac{S_{D,\phi,REAL}}{S_{D,\phi,COMP}} \left(\lambda_k - \tilde{\phi}_k\right) & \tilde{v}_k &= \hat{v}_{k-1} + T \cdot \hat{a}_{k-1} \\ \hat{a}_k &= \hat{a}_{k-1} + K_3 \frac{S_{D,\phi,REAL}}{S_{D,\phi,COMP}} \left(\lambda_k - \tilde{\phi}_k\right) \end{aligned} \tag{7.10}$$

where $\lambda_k = \phi_k + \eta_{\phi,k}$ is the equivalent observation of the estimated parameter with white Gaussian observational noise $\eta_{\phi,k}$, and a variance $D_{\eta,\varphi}$; K_1, K_2, K_3 are filter gains by the model (7.9) parameters; $S_{D,\varphi,\,\mathrm{COMP}}$ and $S_{D,\,\varphi,\,\mathrm{REAL}}$ are the computed and the true values for the discrimination characteristic steepness.

Smoothing filter gains are determined by the current a posteriori variance of measurements (D_ξ) and observations ($D_{\eta,\varphi}$), that is proportional to the accumulation time T, so:

$$K_1 = 2 \cdot \left(\frac{D_\xi}{D_{\eta,\phi}}\right)^{\frac{1}{6}} \cdot T;\ K_2 = 2 \cdot \left(\frac{D_\xi}{D_{\eta,\phi}}\right)^{\frac{1}{3}} \cdot T;\ K_3 = \left(\frac{D_\xi}{D_{\eta,\phi}}\right)^{0.5} \cdot T \tag{7.11}$$

The computed discrimination characteristic steepness is determined proceeding from the expected CNR at the loop input, as well as from the accumulation time. In addition, it is determined with allowance for admissible errors in estimating the phase, code-delay, and frequency ε_ϕ, ε_τ, ε_ω, respectively.

$$S_{D,\varphi,COMP} = \frac{\partial U_D(\varepsilon_\varphi)}{\partial \varepsilon_\varphi}. \tag{7.12}$$

Here, the voltage at the phase discriminator output ($U_D(\varepsilon_\phi)$) is computed separately for the cases of a large and small CNR in the following way:

$$\begin{aligned} U_{D,S}(\varepsilon_\phi) &= 2 \cdot CNR^2 \cdot T^2 \cdot \rho^2(\varepsilon_\tau) \cdot \sin\left(2\varepsilon_\phi + \varepsilon_\omega \cdot T\right) \cdot \mathrm{sinc}\left(\frac{\varepsilon_\omega \cdot T}{2}\right) \\ U_{D,L}(\varepsilon_\phi) &= 2 \cdot CNR \cdot T \cdot \rho(\varepsilon_\tau) \cdot \sin\left(\varepsilon_\phi + \frac{\varepsilon_\omega \cdot T}{2}\right) \cdot \mathrm{sinc}\left(\frac{\varepsilon_\omega \cdot T}{2}\right) \end{aligned} \tag{7.13}$$

First formula in (7.13) is written considering that used is a phase discriminator optimal at a small CNR. Such a discriminator generates the output voltage as per the algorithm

$$u_{D,\varphi}(t_k) = -I_{P,k} \cdot Q_{P,k}, \tag{7.14}$$

where $I_{P,k}$, $Q_{P,k}$ are the quadrature components of the phase discriminator input signal (Fig. 7.17) at an instant k.

Second formula in (7.13) corresponds to using a phase discriminator optimal at a large CNR, whose output voltage is generated as per the algorithm

$$u_{D,\varphi}(t_k) = Q_{P,k} \cdot sign\left(I_{P,k}\right). \tag{7.15}$$

As one can see from (7.13), the phase discriminator output voltage depends on the loop input CNR. This is particularly essential under GNSS signal ionospheric scintillation conditions, when the signal dramatically decreases (by 25–30 dB) over a short observational interval. To reduce this dependence in the phase loop operation under the above conditions, one can use a phase discriminator, whose output voltage will not directly depend on the CNR. In such a discriminator, the output voltage is generated as per the algorithm

$$u_{D,\varphi}(t_k) = -\frac{Q_{P,k}}{I_{P,k}}. \tag{7.16}$$

For the latter discriminator, the output voltage equals

$$U_D(\varepsilon_\varphi) = tg\left(2\varepsilon_\varphi + \varepsilon_\omega \cdot T\right). \tag{7.17}$$

Considering the linear character of the phase change model (7.9), the maximal accumulation time T is determined relative to the phase stationarity time, at which the accepted linear approximation is met. Unfortunately, there are no statistically significant observations of the GNSS signal carrier-phase stationarity under ionospheric scintillation conditions. Therefore, we will estimate the maximal accumulation time, proceeding from general requirements for robustness of the phase tracking system described by Model (7.9).

In (GLONASS 2005), the carrier phase tracking meter operation stability is shown to be provided, when the following requirement is met

$$\frac{S_{D,\phi,REAL}}{S_{D,\phi,COMP}} \geq \frac{1}{4} \tag{7.18}$$

I.e., the computed value for the discrimination characteristic should meet the requirement $S_{D,\varphi,COMP} \leq 4S_{Д,\varphi,\min}$, where $S_{Д,\varphi,\min}$ is computed at the satellite signal minimal strength relation in a receiving point, at which a robust carrier phase lock is still provided.

Except a correct selection of a phase discriminator, defining an optimal noise band for the phase lock loop has a great significance for the phase measurement robustness. So, according to (GLONASS 2005), the noise band of the optimal third-order detecting filter may be determined as follows:

$$\Delta F_{PLL} = \frac{K_2 \cdot K_1^2 + K_2^2 - K_1 \cdot K_3}{4 \cdot (K_1 \cdot K_2 - K_3)}, \tag{7.19}$$

where the K_1, K_2, K_3 filter gains are determined in (7.11).

From (7.11) and (7.19), there comes a direct interdependence of the parameters determining the PLL robustness: accumulation time and noise bandwidth. Through the statistical test method, it was established that, to provide the PLL robust operation, the next requirement should be met (GLONASS 2005)

$$\Delta F_{PLL} \le \frac{0,7}{T}. \tag{7.20}$$

The recommended value ΔF_{PLL} in most practical applications is 18–20 Hz (Kaplan 1996). From (7.20), one may determine an admissible maximal T in the phase discriminator integrator, that appears to be $T_{\mathrm{MAX}} \le$ 30–35 ms. Practically, this value is selected so that, with allowance for a priori errors ε_ϕ, ε_τ, ε_ω and the CNR change limits at the loop input, requirement (7.18) is not violated.

Similarly, one can estimate the limits for the DLL tuning parameters. To filter this radio-navigation parameter, one uses second-order tracking filters, in which the following linear model of measurement is accepted (GLONASS 2005):

$$\tau_k = \tau_{k-1} + T_{\tau,COR} \cdot \upsilon_{\tau,k-1};\ \upsilon_{\tau,k} = \upsilon_{\tau,k-1} + \xi_{\tau,k-1} \tag{7.21}$$

where $\xi_{\tau,k}$ is white Gaussian noise with variance $D_{\xi,\tau}$.

At the pseudo-delay filter output, parameter model (7.21) estimates extrapolated for an instant k are generated:

$$\begin{gathered} \hat{\tau}_k = \tilde{\tau}_k + K_{\tau,1} \frac{S_{D,\tau,REAL}}{S_{D,\tau,COMP}} \cdot u_{,\tau,k} \qquad \tilde{\tau}_k = \hat{\tau}_{k-1} + T_\tau \cdot \hat{\upsilon}_{\tau,k-1} \\ \hat{\upsilon}_{\tau,k} = \tilde{\upsilon}_{\tau,k-1} + K_{\tau,2} \frac{S_{D,\tau,REAL}}{S_{D,\tau,COMP}} \cdot u_{,\tau,k}\ u_{D,\tau,k} = U(\tau_k, \tilde{\tau}_k) + \eta_{\tau,k} \end{gathered} \tag{7.22}$$

where $u_{D,\tau,k}$ is a voltage at a pseudo-delay discriminator output; T_τ is the accumulation time in the DLL correlator; $\eta_{\tau,k}$ is white Gaussian noise of code-delay observations with variance $D_{\eta,\tau}$; $K_{\tau,1}$, $K_{\tau,2}$ is a filter gain; $S_{D,\tau,COMP}$ and $S_{D,\tau,REAL}$ is the computed and real values of the discrimination characteristic steepness.

Filter gains, like in the previous case, are determined by a posteriori observation ($D_{\eta,\tau}$) and measurement ($D_{\xi\tau}$) noise variances (GLONASS 2005):

$$K_{\tau,1} = \left(4 \cdot \frac{D_{\xi,\tau}}{D_{\eta,\tau}}\right)^{0,25} \cdot T_\tau;\ K_{\tau,2} = \left(\frac{D_{\xi,\tau}}{D_{\eta,\tau}}\right)^{0.5} \cdot T_\tau \tag{7.23}$$

The computed discrimination characteristic steepness is also estimated proceeding from the expected CNR at a loop input, accumulation time, and with allowance for admissible errors in estimating the phase, pseudo-delay, and frequency. Herewith, like in the previous case, the relation between the computed and the real steepness of these characteristics depends on the discriminator operation algorithm at large and small CNRs. The corresponding expressions for the discriminator output voltage in the first and the second cases are as follows:

$$
\begin{aligned}
U_S(\varepsilon_\tau) &= 4 \cdot CNR^2 \cdot T_\tau^2 \cdot cos^2\left(\varepsilon_\phi + \frac{\varepsilon_\omega \cdot T_\tau}{2}\right) \cdot \text{sinc}^2\left(\frac{\varepsilon_\omega \cdot T_\tau}{2}\right) \cdot \rho(\varepsilon_\tau) \cdot \\
&\cdot \left(\rho\left(\varepsilon_\tau - \frac{T_d}{2}\right) - \rho\left(\varepsilon_\tau + \frac{T_d}{2}\right)\right) \\
U_L(\varepsilon_\tau) &= 2 \cdot CNR \cdot T_\tau \cdot cos\left(\varepsilon_\phi + \frac{\varepsilon_\omega \cdot T_\tau}{2}\right) \cdot \text{sinc}\left(\frac{\varepsilon_\omega \cdot T_\tau}{2}\right) \cdot \\
&\cdot \left(\rho\left(\varepsilon_\tau - \frac{T_d}{2}\right) - \rho\left(\varepsilon_\tau + \frac{T_d}{2}\right)\right)
\end{aligned}
\tag{7.24}
$$

where T_d is correlator spacing between early, prompt and late code replica in the element of a ranging code cheep τ_s.

The first of Expressions (7.24) corresponds to the case of using the following delay discriminator:

$$u_{D,\tau}(t_k) = I_{P,k} \cdot \left(I_{E,k} - I_{L,k}\right) \tag{7.25}$$

while the second expression corresponds to a discriminator

$$u_{D,\tau}(t_k) = sign(I_{P,k}) \cdot \left(I_{E,k} - I_{L,k}\right) \tag{7.26}$$

where $I_{P,k}$, $I_{E,k}$ and $I_{L,k}$ are samples of in phase (I)-components of the correlators corresponding to the prompt signal ($I_{P,k}$) and to its early ($I_{E,k}$) and late ($I_{L,k}$) replicas.

To eliminate the discrimination characteristic dependence on the variations in the loop input CNR, one can use a discriminator with the following operation algorithm:

$$u_{D,\tau}(t_k) = \frac{\sqrt{\left(I_{E,k}^2 + Q_{E,k}^2\right)} - \sqrt{\left(I_{L,k}^2 + Q_{L,k}^2\right)}}{\sqrt{\left(I_{E,k}^2 + Q_{E,k}^2\right)} + \sqrt{\left(I_{L,k}^2 + Q_{L,k}^2\right)}} \tag{7.27}$$

where Q_k, $Q_{E,k}$ and $Q_{L,k}$ are signal quadra phase (Q)-component samples that correspond to the prompt signal and its early and late replicas.

The second-order pseudo-delay filter noise bandwidth may be found from the following expression:

$$\Delta F_{DLL} = \frac{K_{\tau,1}^2 + K_{\tau,2}}{4 \cdot K_{\tau,1}}. \tag{7.28}$$

The optimal noise bandwidth and the accumulation time, at which the addressed discrete tracking system is in a steady state, is determined from the requirement

$$\Delta F_{DLL} \approx \frac{0.52}{T_\tau} < \frac{\sqrt{1 + \frac{S_{D,\tau,COMP}}{S_{D,\tau,REAL}}} - \sqrt{1 + \frac{2 \cdot S_{D,\tau,COMP}}{S_{D,\tau,REAL}}}}{T_\tau}, \tag{7.29}$$

In (Kaplan 1996; GLONASS 2005), the optimal pseudo-delay filter noise bandwidth should be within 0.8–8 Hz. Following from requirement (7.29), one can find the accumulation time maximal limit $T_{\tau,\text{MAX}}$ that will be $\geq$65 ms. However, as shown above, it is almost inexpedient to raise the $T_{\tau,\text{MAX}}$ above 20 ms.

From the obtained estimates, we test how the rearrangement in the accumulation time and in the noise bandwidth of radio-navigation parameters may affect the robustness of their measurements under ionospheric scintillation conditions.

The ionospheric scintillation effects can be set by the $S4$-index. We will analyze the radio-navigation measurement robustness variation, as ionospheric scintillations grow, by observing the evolution of the minimal CNR threshold, at which filtering is still provided at a satisfactory measurement accuracy.

A robust phase lock in a tracking loop is known to be provided as long as the phase filtration error RMS meets the condition in (Kaplan 1996)

$$\sigma_\varphi = \sqrt{\sigma_T^2 + \sigma_F^2 + \tfrac{\sigma_S}{3}} \leq \sigma_{\varphi,\max} = 15°. \tag{7.30}$$

Based on the known results from (Conker et al. 2003), and considering the dymamic and stress component σS = 0, we rewrite (7.30) with allowance for ionospheric scintillations presented by the $S4$-index as follows

$$\sigma_\varphi^2 = \sigma_F^2 + \sigma_T^2 = \left(160 \cdot \frac{\sigma_F(\tau) \cdot f_c}{\Delta F_{PLL}}\right)^2 + \left(\frac{360}{2\pi}\right)^2 \frac{\Delta F_{PLL} \cdot \left[1 + \frac{1}{2T \cdot 10^{0.1 \cdot CNR} \cdot (1 - 2S_4^2)}\right]}{10^{0.1 \cdot CNR} \cdot (1 - 2S_4^2)} \leq \sigma_{\varphi,\max}, \tag{7.31}$$

where σ_F and σ_T are carrier phase filtering error RMS (in deg), caused by the effect of thermal noise and short-time clock frequency instability ($\sigma_F(\tau)$).

Expression (7.31) is written for a third-order phase lock loop and without considering receiver vibrations and jerks ($\sigma_S = 0$). From (7.31, one can directly obtain an expression to determine the minimal threshold for a loop input signal (CN_{thr}) with allowance for the $S4$-index and PLL tunings. For this purpose, we equate Expression (7.31) to a maximally admissible error in the phase measurement ($\sigma_{\varphi,\text{MAX}} = 15°$), that does not exceed the discrimination characteristic linear segment limits:

$$\sigma_F^2 + \left(\frac{360}{2\pi}\right)^2 \cdot \frac{\Delta F_{PLL} \cdot \left[1 + \frac{1}{2T \cdot 10^{0.1 \cdot CN_{thr}} \cdot (1 - 2S_4^2)}\right]}{10^{0.1 \cdot CN_{thr}} \cdot (1 - 2S_4^2)} = \sigma_{\varphi,\text{MAX}}^2, \tag{7.32}$$

Whence, with allowance for (2.35), we can directly find the CNR_{thr}:

$$CNR_{thr} = 10\lg\left[\begin{array}{c}\dfrac{1}{4\cdot T\cdot(\sigma^2_{\varphi,\mathrm{MAX}}-\sigma^2_F)\cdot(1-2\cdot S4^2)^2}\cdot\\ \cdot\left[\sqrt{D}+2\left(\dfrac{360}{2\pi}\right)^2\cdot T\cdot\Delta F_{PLL}\cdot(1-2\cdot S4^2)\right]\end{array}\right]$$

$$D = 4\cdot\left(\frac{360}{2\pi}\right)^4\cdot T\cdot\Delta F_{PLL}\cdot(1-2\cdot S4^2)^2\cdot(T\cdot\Delta F_{PLL}-2(\sigma^2_{\varphi,\mathrm{MAX}}-\sigma^2_F)) \tag{7.33}$$

Based on (7.33), we made the plots for the $CNR_{thr}(S4)$ dependences at the PLL different tuning parameters (Fig. 7.18). As one can see, irrespective of the loop tuning parameters, the minimal CNR_{thr} grows as the scintillation intensity increases, which also leads to an increase in the probability of a phase LOL. At the $S_4 \le 0.5$ scintillations, this does not manifest itself so significantly: the of robust PLL threshold, CNR_{thr}, augments by no more than 2–3 dB. However, at the further growth in the scintillation intensity, the situation dramatically deteriorates. So, at $S_4 = 0.7$, the CN_{thr} increases by 10–12 dB. At the same time, the CNR necessary for a robust phase lock already exceeds the average input CNR. In this situation, one may expect (with a high probability) a signal phase LOL for the GNSS signals for the satellites that are observed at low elevations.

Due to loop tuning parameter changes, one can achieve an essential decrease in the CNR threshold and provide a more robust phase lock under the ionospheric scintillation conditions. As one can see from Fig. 7.18a, that the PLL noise bandwidth (ΔF_{PLL}) narrowing may provide the greatest positive effect here. When the noise bandwidth narrows tenfold, the CNR_{thr} threshold in a strong scintillation region ($S_4 > 0.5$) drops by 10 dB.

An increase in the integration time T at $S_4 > 0.5$ produces a smaller effect. If the CNR_{thr} increases tenfold, the CNR_{thr} drops only by 2–3 dB (Fig. 7.18b). And, finally, the clock frequency instability weakly affects the CNR_{thr} threshold, as long as $\sigma F(\tau) \le 10^{-10}$ (Fig. 7.18c, two lower curves). In contrast, at a low frequency stability ($\sigma F(\tau) \ge 10^{-10}$), this may become a main factor reducing the PLL robustness (Fig. 7.18c, upper curve).

Let us perform a similar analysis for a DLL. In the presence of ionospheric scintillations, the variance in the estimate of this parameter may be computed as follows (Conker et al. 2003)

$$\sigma_\tau^2 = \frac{\Delta F_{DLL}\cdot T_d\left[1+\frac{1}{T_\tau\cdot 10^{0.1\cdot CNR}\cdot(1-2S_4^2)}\right]}{2\cdot 10^{0.1\cdot CNR}\cdot(1-2S_4^2)} \le \sigma_{\tau,\mathrm{MAX}} = \frac{\tau_S}{6} \tag{7.34}$$

where τ_S is a ranging code pulse duration; T_d is a code-delay correlator spacing (in ranging code pulse duration units); $\sigma_{\tau,\mathrm{MAX}}$ is a maximally-admissible error that does not exceed the limits of the discrimination characteristic linear segment.

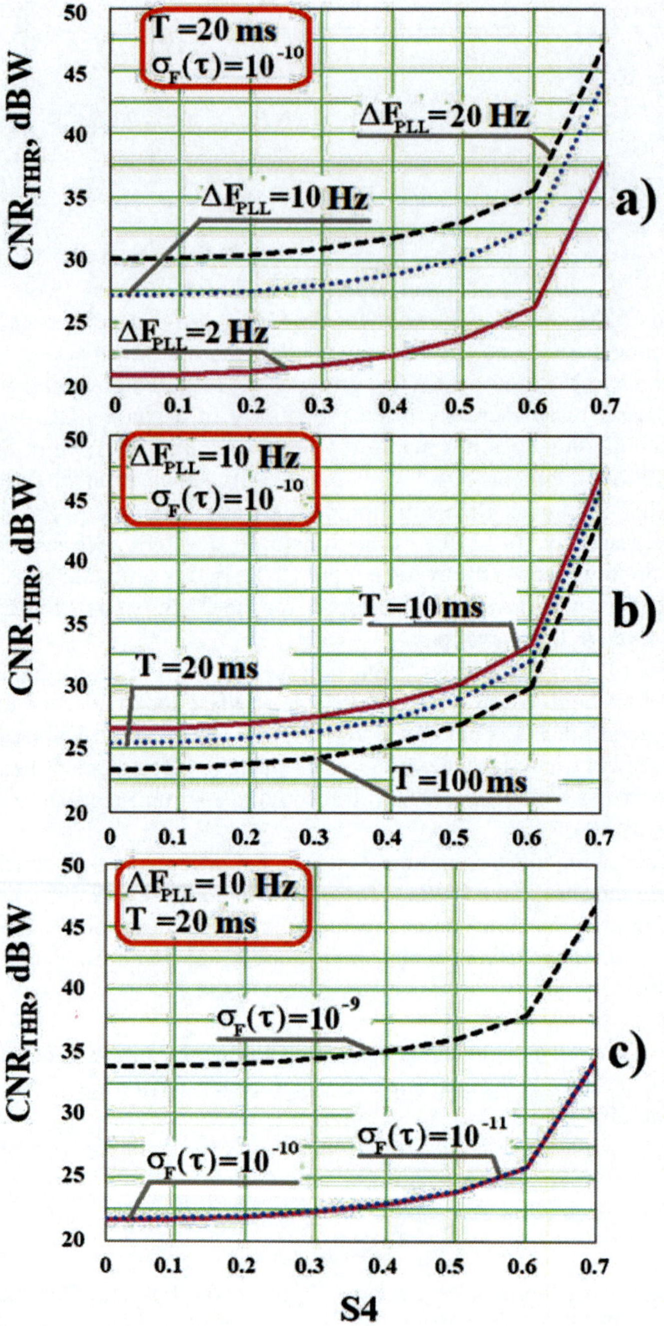

Fig. 7.18 Phase loop input threshold CNR dependence on the S4-index at different loop tunings

Let us equate this expression to the $\sigma_{\tau,\mathrm{MAX}}$ and, considering (2.35, Chap. 2.1.3), we obtain an expression to determine the DLL CNR threshold

$$CNR_{thr} = 10\lg\left[\frac{B - 8\cdot F1\cdot T_d^2\cdot(1-T_d)\cdot\Delta F_{DLL}\cdot(1-2\cdot S4^2)\cdot T_\tau}{2\cdot(1-2\cdot S4^2)^2\cdot T_\tau\cdot\left(\sigma_{\tau,\mathrm{MAX}}/Lc\right)}\right]$$

$$\begin{aligned} B = {} & (8\cdot F1\cdot T_d^2\cdot(1-T_d)\cdot\Delta F_{DLL}\cdot(1-2\cdot S4^2)\cdot T_\tau)^2 + \\ & + 4\cdot 16\cdot F1\cdot F2\cdot T_d^3\cdot\Delta F_{DLL}\cdot T_\tau\cdot(1-2\cdot S4^2)^2\cdot\left(\sigma_{\tau,\mathrm{MAX}}/Lc\right) \end{aligned} \quad (7.35)$$

where $F1, F2$ are parameters depending on a type of the correlator and of the discriminator of a tracking loop (typical values are $F1 = 0.5$, $F2 = 1$); Lc is a ranging code wavelength, m; T_τ is the accumulation time in the DLL correlator; ΔF_{DLL} is a DLL filter noise bandwidth.

Figure 7.19 presents dependences of the minimal CNR_{thr}, at which the condition $\sigma_{\Delta\tau} \le \sigma_{\Delta\tau,\max}$ is met, on the $S4$-index. Figure 7.19a shows a CNR_{thr} dependence on the noise bandwidth ΔF_{DLL} at $T_\tau = 20$ ms when using a serial correlator and discriminator determined by Expression (7.27. Figure 7.19 exhibits a CNR_{thr} dependence on the integration time T_τ, at the noise bandwidth $\Delta F_{DLL} = 8$ Hz and when using the same type of the correlator and the discriminator as in Fig. 7.19a. In all the cases, one assumed a tune-out of DLL correlator spacing is $T_d = 0.5\ \tau_S$.

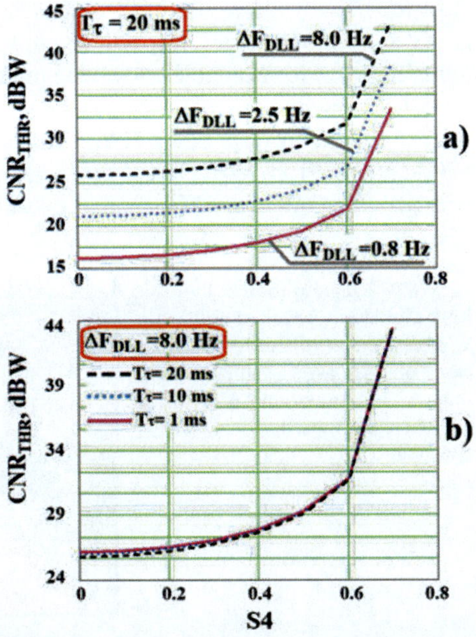

Fig. 7.19 Dependence of a CNR threshold at a DLL input on the $S4$-index, at different loop tunings

As one can see, the most effective way to achieve a DLL noise immunity increase under the ionospheric scintillation conditions is a rearrangement of the noise bandwidth ΔF_{DLL} (panel a). A change in the loop noise bandwidth within $\Delta F_{DLL} = 0.8$ … 8 Hz at $S_4 > 0.5$ provides a CNR_{thr} gain by 9–10 dB. The effect of ionospheric scintillations is known to lead to a short-time CNR troughs by 15–30 dB. Therefore, an adaptive rearrangement of the DLL noise bandwidth may become an effective way to adapt the navigation receiver operation to ionospheric scintillations.

Along with that, an integration time variation within admissible limits (7.29) does not lead to a notable increase in the DLL noise immunity with a forcing from $S_4 >$ 0.5 scintillations.

7.6 WAAS Improvement Under Space Weather Impact

A WAAS important function is the real-time check of RNP availability. The availability is regarded as a probabilistic function of two factors: geometric factor and total residual ranging error upon applying differential corrections (2.155). Under geomagnetic disturbance conditions, the ionospheric ranging error features the greatest spatial–temporal variability (Table 2.7). As shown in Sect. 5.2.2, a result of this is an increase in the probability of false alerts to deterioration of RNP availability. The WAAS RNPs are regarded available, as long as the current HPL and VAL persist within a domain that produces an «availability triangle» (Fig. 7.20).

In Fig. 7.20, the horizontal axis hosts the RMS values for the total residual ranging error (σ, (2.143)). Here, the vertical axis shows the HPL or VPL values in the horizontal and vertical planes, respectively. In Fig. 7.20, one can distinguish among three domains: «service available», «service unavailable» and «mismatch alert».

If the computed (based on the WAAS service information) current HPL (VPL) do not exceed the corresponding HAL (VAL), the RNPs are available. In case if HPL > HAL (or VPL > VAL), a user receives an alert to unavailability of positioning with a required accuracy. If it appears that a real value for the user's positioning error

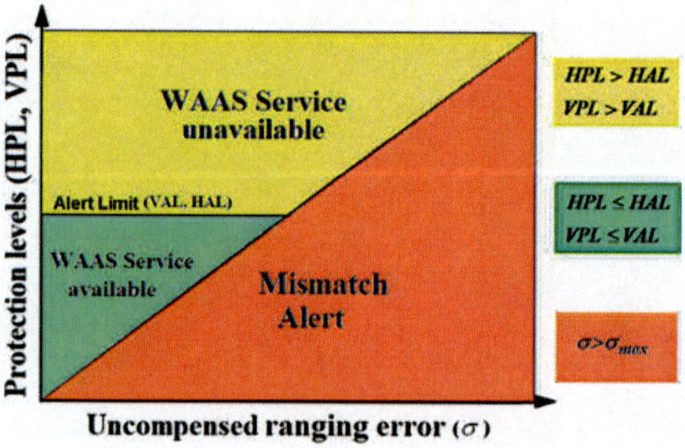

Fig. 7.20 Diagram for the RNP availability check of a WWAS user. Modified figure based on open-assess source (https://gssc.esa.int/navipedia/index.php/Integrity#cite_note-5)

exceeds the HAL (VAL), but, at the same time, the protective levels (computed based on the WAAS service information) have not exceeded the HAL/VAL threshold, then there occurs information dangerous mismatch. The latter situation is most dangerous and may be caused by that a non-offset ranging error exceeds a maximal value (σ_{max}), although the user is not alerted to this. This may occur, in particular, when a differential correction to an ionospheric ranging error, computed based on the WAAS service information, mismatches the error real status. Such situations were, indeed, observed under powerful geomagnetic disturbance conditions or under the forcing from strong solar radio bursts (See Sect. 5.1.3).

Thus, one of the most important trends to improve the WAAS operation is reducing the probability of false alerts and information dangerous mismatches concerning RNP availability.

A search for the solution of the above problem is the main goal in a number of studies (Sparks et al. 2022, 2011; Paredes et al. 2008; Walter et al. 2004; Blanch and Walter 2002). The principal development of those investigations advances toward an extensive way to update WAASs by increasing a number of observations through integrating several GNSSs (Walter et al. 2004). Besides, GNSS additional operational frequencies have been put in operation, the density of reference stations has been also increased (Indriyatmoko et al. 2008). A set of such measures enables to increase the measurement redundancy and to enhance the spatial resolution of the Grid Ionospheric Vertical Errors (GIVE). As a result, one should expect the spatial–temporal decorrelating factor of differential corrections to the ionospheric ranging error to better correspond to the real spatial–temporal dynamics of the ionosphere.

As a supplement to the above measures, using more sophisticated methods to interpolate the spatial distribution of ionospheric ranging errors within a WAAS operation area may become an effective way to solve the problem (Blanch 2003; Sparks et al. 2011).

However, all the listed above solutions maintain the same principal disadvantage: using interpolated (i.e. smoothed) and, consequently, not always sufficiently reliable estimates for ionospheric errors at the GIVE nodes. Besides, the decisive rule of RNP availability/unavailability excludes one or several GIVE cells (i.e. a sufficiently large domain) from the WAAS operation area. This inflexible decisive rule essentially roughens the possibility of differential correction without good cause. As shown above, the presence of local ionospheric disturbances is not always accompanied by a significant deterioration of positioning accuracy, particularly, within a vast space.

Developing GNSS networks enables to substantially increase a number and density of radio-navigation parameter measurements within an extended territory. Let us address a possibility to check RNP availability within an extended operation area, based on using direct measurements of ranging errors and errors in determining coordinates on a dense regional network of reference stations (CORS, for example).

We understand by RNP *availability* to a user a full probability of GNSS navigation hardware health and maintaining the set requirements for *accuracy* and *continuity* of determining the object state vector (2.30) during the problem solution.

A direct measure to estimate *accuracy* is the current absolute error in determining user's spatial coordinates (2.162). Also, we assume that the distribution of this error values obeys the normal distribution law (Khivrich et al. 1984).

To estimate *continuity* of the GNSS user service, we introduce the notion of *coordinate determination failure*. We understand by the failure such an event that meets the following condition:

$$(|\Delta X_i| > \Pi_{\Delta X}) \vee (|\Delta Y_i| > \Pi_{\Delta Y}) \vee (|\Delta Z_i| > \Pi_{\Delta Z}) \tag{7.36}$$

where $\Pi_{\Delta X}$, $\Pi_{\Delta Y}$, $\Pi_{\Delta Z}$ are corresponding thresholds for errors in determining the right rectangular geocentric coordinates, that are specified by the user's RNPs.

Besides, we introduce the notion of *potential coordinate determination failure*. We understand by this event an *expected* deterioration of positioning accuracy due to a dramatic DOP change when loose tracking one or several satellites. As a potential slip indicator, we will consider the current geometric factor computed in a user's receiver.

Considering the specified criteria, we present a quantitative estimate for RNP availability like a full probability of determining a GNSS user's positioning throughout its certain time interval ΔT as follows (Khivrich et al. 1984)

$$W = P(B/A, C) \cdot (1 - W_{\Delta T}) \tag{7.37}$$

where A is an event of navigation system health operation; C is an event of the navigation receiver health status; B is an event of keeping RNP well; (1-$W_{\Delta \mathrm{T}}$) is a probability of positioning failure (due to any reasons) absence when determining coordinates during the ΔT interval.

Keeping the receiver and the GNSS healthy, the first factor in (7.37) represents a probability of finding an object (user) within a safety parallelepiped (or another figure). The probability, in the general case, may be specified as follows

$$P(B/A, C) = P_{TX} P_{TY} P_{TZ}. \tag{7.38}$$

where P_{TX}, P_{TY}, P_{TZ} are probabilities of finding the error in coordinate determination within the set limits by the vertical, longitudinal, and lateral axes.

Describing the probability functions P_{TX}, P_{TY}, P_{TZ} depends on the second-order surface kind that restricts the safety domain. For example, if this domain is described as a parallelepiped, the probabilities P_{TX}, P_{TY}, P_{TZ} can be estimated as follows (Khivrich et al. 1984)

$$P_{TX}\{\alpha_X < \Delta X_i < \beta_X\} = 0.5\left[\Phi\left(\frac{\beta_X - M(\Delta X)}{\sigma_X}\right) - \Phi\left(\frac{\alpha_X - M(\Delta X)}{\sigma_X}\right)\right]$$
$$P_{TY}\{\alpha_Y < \Delta Y_i < \beta_Y\} = 0.5\left[\Phi\left(\frac{\beta_Y - M(\Delta Y)}{\sigma_Y}\right) - \Phi\left(\frac{\alpha_Y - M(\Delta Y)}{\sigma_Y}\right)\right]$$
$$P_{TZ}\{\alpha_Z < \Delta Z_i < \beta_Z\} = 0.5\left[\Phi\left(\frac{\beta_Z - M(\Delta Z)}{\sigma_Z}\right) - \Phi\left(\frac{\alpha_Z - M(\Delta Z)}{\sigma_Z}\right)\right], \tag{7.39}$$

where Φ (…) is the Laplace function (probability integral); $M\ (\Delta X)$, $M\ (\Delta Y)$, $M\ (\Delta Z)$, $\sigma_{\Delta X}$, $\sigma_{\Delta Y}$, $\sigma_{\Delta Z}$ are a standard mean and RMS for the positioning errors by the X-, Y-, and Z-axes, respectively; α and β are upper and lower confidence bounds for the positioning errors by the vertical, longitudinal, and lateral axes.

To estimate the second factor of Expression (7.37), we will regard that, at a sufficiently great number of coordinate measurements, and at a relatively small probability of a separate event (positioning failure), the distribution of the probability of the failure number within ΔT obeys the Poisson law, therefore:

$$W_{\Delta T} = P(k) = \frac{a^k}{k!} \cdot \exp(-a) \tag{7.40}$$

where $k = 0,1,2 \ldots n$ is a possible number of failures when determining coordinates over elementary intervals Δt_i being a part of the observational period ΔT; a is the Poisson distribution parameter.

To estimate the probability $W_{\Delta \mathrm{T}}$, one should bear in mind that the slip distribution within the observational interval ΔT may appear non-stationary in time. Thereby, determining the unknown parameter a is expedient to do by smoothing the histogram for slip occurrence frequency k (observed within ΔT) by the Poisson law through the least square method:

$$\sum_{\mathrm{k}=0}^{n}\left(P_{\mathrm{k}} - \tfrac{a^k}{k!}e^{-a}\right)^2 \to \min \tag{7.41}$$

where P_k is an experimental (opriori observed) value for the probability of emergence of k positioning failures; n is the greatest possible number of the failures in the observational series.

To account for the emergence of "potential slips," we introduce an additional factor like the following switching function into Expression (7.37):

$$G_{DOP} = \begin{cases} 1, & PDOP \le \Pi_{DOP} \\ 0, & PDOP > \Pi_{DOP} \end{cases}, \tag{7.42}$$

where Π_{DOP} is a geometric factor maximal threshold that a user sets individually as per Criterion (7.36).

Thus, the final expression to quantitatively estimate *availability* of positioning, considering the requirements (set for a particular user) for positioning *accuracy* and *continuity*, may be written as follows:

$$W = P(B/A, C) \cdot (1 - W_{\Delta T}) \cdot G_{DOP} \tag{7.43}$$

A practical implementation of the proposed estimate for RNP availability is based on using the regional network of reference stations that has a sufficient density and time to update the service-correcting information. At each reference station, its status vector $\mathbf{\Pi} = |x, y, z, T|$ is periodically computed. By comparing the components of vector $\mathbf{\Pi}$ with the reference station known coordinates, the vector of the current errors in positioning the station $\mathbf{\Delta\Pi} = |\Delta x, \Delta y, \Delta z|$ is generated. Besides, at each reference station, one estimates the current ranging errors (ΔR_{SV}) by visible navigation satellites.

The set of observations obtained thereby is transmitted to a regional analytical processing center, where the following operations are performed:

- screening and remove anomalies within the observation set, detecting failures in determining coordinates, and generating a histogram for positioning *k*- failures occurrence frequency over the ΔT (P_k in Formula 7.41);
- computing the parameter *a* to distribute the current probability of positioning *k*-failures (7.41);
- estimating statistics for the general set of errors in determining spatial coordinates of a network reference stations ($M\ (\Delta X)$, $M\ (\Delta Y)$, $M\ (\Delta Z)$, $\sigma_{\Delta X}$, $\sigma_{\Delta Y}$, $\sigma_{\Delta Z}$) over the ΔT;
- estimating RMS for the series of slant ranging errors ($\sigma_{\Delta\, RSV}$) by all visible satellites at all the network reference stations;
- determining thresholds for the horizontal (HDOP) and vertical (VDOP) (Π_{VDOP} and Π_{HDOP}, respectively) by RNP groups of system users.

Determining thresholds Π_{VDOP} and Π_{HDOP} by categories of requirements for positioning accuracy can be done as follows. From the network measurement results, the analytical center (master station) generates a set of measurements $S = \left|\sigma_{\Delta Rsv,1}\sigma_{\Delta Rsv,2}\ldots\sigma_{\Delta Rsv,n}\right|$ of *n* reference stations during the ΔT time interval. Further, one estimates a ranging error RMS ($\sigma_{\Delta R}$) for the entire general set of measurements *S*. After this processing, the current estimates for the thresholds of the HDOP and VDOP are generated as:

$$\begin{aligned} \Pi_{HDOP} &\leq \frac{\left[\sqrt{\sigma_X^2 + \sigma_Y^2}\right]_{\max}}{\sigma_{\Delta R}}, \\ \Pi_{VDOP} &\leq \frac{[\sigma_Z]_{\max}}{\sigma_{\Delta R}}, \end{aligned} \tag{7.44}$$

where the max index marks the corresponding maximally admissible RMS of the errors in determining the coordinates within a plane and by height, respectively.

Maximally admissible RMS for the coordinate determination errors are assigned (in plane and by height) in correspondence with RNPs when solving a particular task. Therefore, the final expression for inclusion Function (7.42) at a current instant k should have the following view:

$$G_{DOP,k} = \begin{cases} 1, & (HDOP_k \leq \Pi_{HDOP,k}) \wedge (VDOP_k \leq \Pi_{VDOP,k}) \\ 0, & (HDOP_k > \Pi_{HDOP,k}) \vee (VDOP_k > \Pi_{VDOP,k}) \end{cases}. \tag{7.45}$$

Estimating the positioning availability in a user receiver is done based on the service-correcting information transmitted from an analytical processing center (master station) in the following order:

- transpating current parameters $M(\Delta X)$, $M(\Delta Y)$, $M(\Delta Z)$, $\sigma_{\Delta X}$, $\sigma_{\Delta Y}$, $\sigma_{\Delta Z}$ to a user to estimate the probability of finding an object within the set safety parallelepiped or another figure (7.38–7.39);
- translating the parameter a to estimate the probability of emergence of k-slips over the ΔT interval from Expression (7.40);
- translating the parameter $\sigma_{\Delta R}$ to determine the HDOP/VDOP thresholds Π_{VDOP} and Π_{HDOP}, based on Expressions (7.44).

In a user receiver, depending on RNPs, the following should be assigned:

- requirements for an admissible error in coordinate determination (α and β, $\left[\sqrt{\sigma_X^2 + \sigma_Y^2}\right]_{\max}$ and $[\sigma_Z]_{\max}$);
- admissible number of positioning slips (k) over the ΔT interval.

Upon determining the requirements for positioning accuracy and continuity, a user may quantitatively estimate the current positioning availability, by using general Formula (7.43).

The reliability of RNP availability estimate directly depends on how much the spatial–temporal decorrelating factor of coordinate determination error corresponds to the real distribution of that error. In (Demyanov et al. 2007), the error $\mathbf{\Delta\Pi} = |\Delta x, \Delta y, \Delta z|$ decorrelation spatial radius (at a 0.707 correlation) ranges within wide limits and equals 40–300 km for quiet geomagnetic conditions and 38–200 km for a powerful magnetic storm. Herewith, the recommended time to update the service-correcting information should not exceed 30 s.

Summarizing, we now switch to addressing the proposals to build up-to-date GNSS applications. Possibilities of such systems should be enhanced by using additional tools accounting for the effect of space weather irregular factors. Figure 7.21 presents a block diagram for a complex technical system of differential correction, prediction, and monitoring of the positioning quality, with allowance for forcing from geomagnetic disturbances and solar radio bursts.

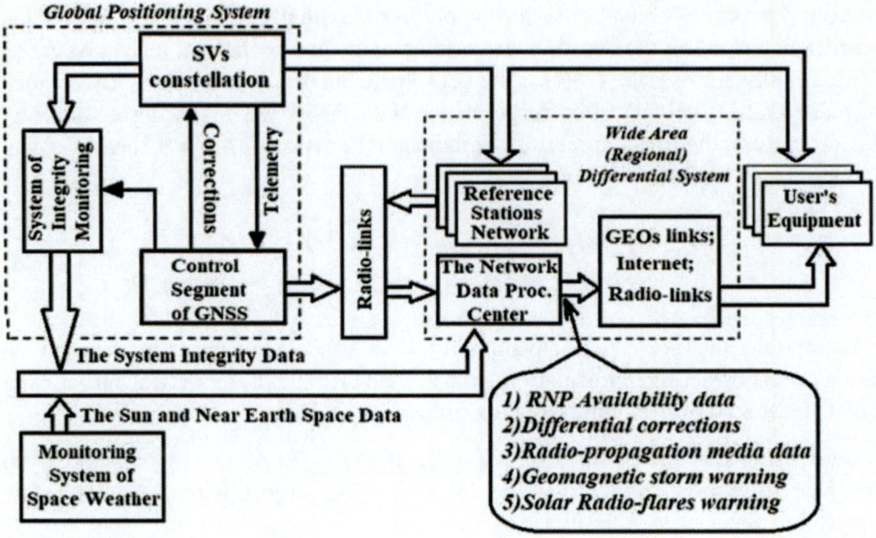

Fig. 7.21 Complex system of differential correction, prediction, and monitoring of positioning quality

The basis of the complex system is a network of reference stations. WAAS reference stations network supplemented by local measuring CORS networks or by segments of international/national measuring networks (Sect. 4.2.1) is able to be such a network. One may also address a combination of several various networks based on the common analytical centers of information processing in addition to the WAAS main information processing centers.

Another relevant component of the system is a suite of facilities to monitor the status of the Sun and of the near-Earth space (Space Weather Suite). Figure 7.22 presents the composition and the structure of such a suite (Williamson et al. 2000).

This system includes a suite of measuring facilities placed at several levels relative to the Earth surface.

A ground-based sensor level involves ground-based coronagraphs, solar optical and radio observatories, a network of magneto-variation stations (INTERMAGNET), etc.

Coronagraphs obtain solar corona images, enable to observe coronal mass ejections into the interplanetary space, i.e. to record an event that may potentially lead to a geomagnetic disturbance. Radio telescopes produce information on the solar radio emission power level, as well as allow one to determine, what type radio bursts relate to (Type I through Type V) (Dulk 1985).

Optical telescopes are used to obtain Sun images in the visible and in the H_α spectra, as well as to study localizations and structures of sunspots on the Sun surface. The sunspot shape and size report evidence the evolution stage of an active region (AR), and, hence, an expected solar flare. Besides, CME energetic particles from the western ARs were established to reach the Earth with a much higher probability than

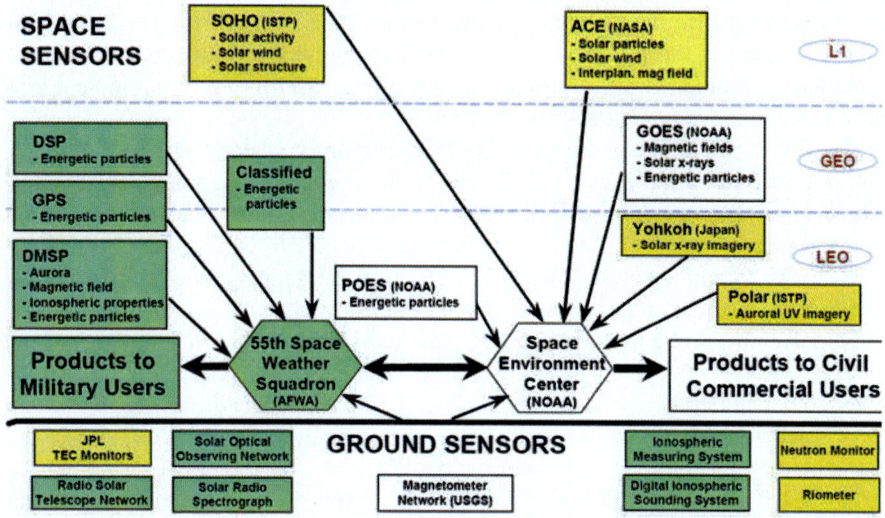

Fig. 7.22 Structure of a system to monitor the status of the Sun and of the Near-Earth Space (space weather suite) (Williamson et al. 2000)

those of the eastern ARs. Thus, observing the Sun surface in the optical and in the H_α ranges provides additional possibilities for a long-term forecast of geomagnetic disturbances (24–72 h prior to the disturbance onset on the Earth).

At magneto-variation stations composing the USGS international network, one measures variations in the Earth magnetic field strength. The geomagnetic field data standard for the USGS represent time series for three components of the geomagnetic field (XYZ or HDZ), and the modulus of the full magnetic-field vector *F*. The X-, Y-, and Z-axes produce an orthogonal basis; at that, the Z-axis is directed to the Earth center, the X-axis is aimed at the northern magnetic pole, and the Y-axis shows to the magnetic east. The geomagnetic field *H*-component is produced as a superposition of the X- and Y-components of the Earth magnetic field.

A combination of the above measurements is used to determine the global indices for magnetic activity:

- *Kp* and *Ap,* as well as local characteristics for the geomagnetic field intensity;
- *Dst*-index in the equatorial ring current region;
- 3-h *a*- and *K*-indices for local geomagnetic activity;
- Strength of the geomagnetic field longitudinal (northern) component, *Ht*, measured at separate magneto-variation stations.

When selecting the magnetic activity indices for fore- and nowcasting of GNSS operation quality under magnetic storm conditions, one should consider the following. The *Kp*-index is estimated as a mean value for local *K*-indices of the magnetic activity obtained from magneto-variation stations located preferentially in the sub-auroral region. Thereof, this index does not completely account for the geomagnetic field global disturbance structure (particularly, at lower latitudes), because it forms primarily by polar geomagnetic disturbances. Thus, the *Kp*-index (or its modification, *Ap*-index) may appear useful only as an indicator to mark geomagnetically disturbed/undisturbed days.

The *Dst*-index reflects geomagnetic field variations in greater detail due to a better temporal resolution. However, this index, on the contrary, best describes geomagnetic field variations at the equatorial and subequatorial latitudes, because the index is obtained based on the measurements in the region of equatorial ring currents. Therefore, the *Dst* variations are not well suited to estimate geomagnetic field variations at high and auroral latitudes when a magnetic storm transits.

To solve the problem to nowcast GNSS hardware functioning under various geomagnetic conditions, it is expedient to use direct measurements of the *Ht* component at the magneto-variation stations located in the investigated Earth region. Such measurements most adequately reflect the character of the local processes in the ionosphere during geomagnetic disturbances.

Low Earth Orbiting (LEO) satellites provide observations of the medium around the polar caps and the auroral region. Polar Orbiting Operational Environmental Satellites (NOAA, POES) with a polar (98° inclination) orbit are placed at 850 km high and measure the full and mean energy of solar protons and electrons in that region (Space Enviromental Monitor, or SEM). Herewith, recorded are flux power values during each single flyby of a satellite over the polar region (the measurement session takes about 25 min). By accumulating such measurements, one can estimate the integral energy detected in the Northern hemisphere as a result of bombing the upper atmosphere by particle fluxes. Besides, single flyby of satellites enable to estimate the current position of the auroral oval boundary. For the intervals with a maximal level of a geomagnetic disturbance, the temporal resolution for the auroral oval boundary position measurements is ~2 min. The area inside the auroral oval features a higher intensity of ionospheric scintillations. Therefore, the estimate for the position of its southern boundary on the time scale close to the real enables to localize the area of higher hazards for a WAAS user, within which one may expect a higher probability of real navigation characteristics deterioration (Section 5.3.3).

Geostationary Earth Orbit (GEO) satellites, in particular, the GOES satellite, measure the power of the Sun X-rays and the flux density of protons, electrons, and other solar energetic particles in a broad band of energies at the altitude of 40,000 km. These measurements enable to define nowcasts of the geomagnetic disturbance level and dangerous solar radio bursts.

Space sensors. This group of facilities comprises the Solar Heliospheric Observatory (SOHO) and the Advanced Composition Explorer (ACE) satellites. The SOHO serves to determine the time, when a plasma ejection leaves the Sun, and to measure the integral ultraviolet (EUV) solar flux within the 0.1–50 nm range. This parameter is one of the most significant to forecast the estimates for the electron concentration profile and for the TEC in the ionosphere.

The ACE is located at the Lagrange point (L1) about 1.5 min km from the Earth toward the Sun. The Real-Time Solar Wind (RTSW) system installed at the ACE monitors continuously the solar wind and enables to nowcast strong geomagnetic activity no more than 1 h prior to its onset on the Earth. The RTSW processes measurements from the ACE subsystems (MAG—IMF vectors, SWEPAM—solar wind ions, EPAM—energetic ions, electrons, and high-energetic particles). Those data are used in the model developed to nowcast geomagnetic *Kp*-indices two hours before they are observed on the Earth.

Thus, using the Space Weather Suite enables to measure the following set of parameters reflecting the Sun status and the Near-Earth Space medium conditions:

- indices of geomagnetic activity and parameters for the geomagnetic field variations strength (*Kp*, *Dst*, *Ht*, etc.);
- integral density for fluxes of solar different-energy protons and electrons within the auroral oval;
- integral density for the flux of solar radio emission at the GNSS operational frequencies;
- integral flux of the *EUV* solar radiation within 0.1–50 nm and the solar X-ray data within 0.05–0.8 nm.

A set of the above parameters may be used for fore- (1 h through several days) and nowcasts (<1 h), as well as for a real-time alert of GNSS users to dangerous geomagnetic and solar events.

Thus, a center (or centers) for information processing, predicting, and monitoring the positioning quality implements the following functions:

- forecast of a geomagnetic disturbance and generation of alerts for users by their RNP categories;
- nowcast of the differential system integrity and RNP availability, based on the analysis of geomagnetic activity indices;
- computation of the "vertical" TEC, spatial–temporal TEC gradients a 1–60 min period within a WAAS operation area, and formation of the current GIVE;
- formation of real-time alerts for users to expected positioning deterioration based on analyzing the current measurements of the geomagnetic field variations strength, and of the solar radio emission flux within the L1, L2, and L5 frequency bands;
- acquisition (from international analytical data processing centers) and accumulation of statistically verified statistics for the errors in coordinate determination, and for the density of positioning failures under various helio-geophysical conditions, as well as estimates for maximally-observed positioning errors under the specified helio-geophysical conditions (a posteriori monitoring);
- RNP availability control and RNP availability failure alerts for users, based on the current statistics of measurements and on the results of long-term a posteriori monitoring.

A set of all the corrections and alerts enters a GNSS user receiver or an Integrated Navigation Complex at a traffic control system and serves to increase positioning accuracy and continuity under unfavorable helio-geophysical conditions.

Besides the proposed measures, artificial improvement of the constellation geometry may become an important way to recover (in real time) disruptions in the GNSS radio-navigation field due to random or systematic DOP deterioration. One well-known way to artificially improve DOP is using pseudolutes for differential GPS (Parkinson and Fitzgibbon 1987).

Another solution to improve geometry of the navigation solutions is to choose navigation function form. The spheres centered at the GNSS satellite locations (2.32) are to be "natural" navigation functions relating radio-navigation parameter measurements to the object state vector components. However, in (Maryuknenko et al. 2014), for a case of a positional navigation on a plane, offered was using second-order non-spherical surfaces as navigation functions. In this case, the listing of possible position lines comprises *m* curves that transit through a point, where a navigation object is located. The parameters for those curves may be determined based on the measured navigation parameters (ranges, baselines, and mutual angles between a navigation object and radio-navigation signal sources). Thus, optimizing the geometry coefficient (when using an arbitrary set of *m* navigation functions like cuts of spherical and non-spherical figures) reduces to such a combination of minimally necessary set of position lines, at which the intersection angle of the tangents to the latter converges to $\pi/2$ (Maryuknenko et al. 2014). The stated may be also spread to non-spherical position surfaces that may be used to solve the problem of minimizing DOPs at their sudden deterioration under Space Weather impact.

Summary for this Chapter

Considering the GNSS operation under unfavorable helio-geophysical conditions, we recommend a necessary complex of measures aimed at providing GNSS RNPs. The measures involve:

1. Fore- and nowcasts of the positioning quality under geomagnetic disturbance conditions;
2. Real-time alert for users to a geomagnetic disturbance in the GNSS satellite signal propagation medium;
3. Real-time alert to powerful solar radio bursts at the GNSS frequencies, with allowance for the radiation energy distribution spectrum relative to the GNSS operational frequency plan;
4. Use of a complex technical system for differential correction, prediction, and monitoring of the GNSS positioning quality under unfavorable helio-geophysical conditions;
5. Adaptation of PLLs and DLLs to dramatic changes in signal propagation conditions;
6. Measures for DOP artificial improvement.

To perform an alert, one should develop forecast methods and select optimal indices of geomagnetic activity. The issue of selecting the index (a set of indices) optimal for solving the problem by the false-alert minimum criterion has remained open and requires further investigations. Using the *Dst*- and *Kp*-indices provides a sufficiently high probability of a false alert. Therefore, there is no possibility to apply those indices to a real-time alert to the current level of RNP availability. However, the above indices may be recommended for forecasts ($\geq$1 h) of the expected positioning availability deterioration.

The up-to-date state of investigations into mechanisms for solar radio burst formation does not allow one to precisely forecast dangerous solar radio bursts in the UHF for GNSS users. The only recommended measure may be continuous radio monitoring of the Sun in the UHF with issuing a real-time alert for GNSS users at the emergence of radio bursts stronger than $>10^3$ sfu. Additionally, we can recommend to comparatively monitor the energy distribution of a radio burst within the radiation spectrum and the GNSS operational frequency plan in order to define a real-time alert to availability of measurements within separate GNSS frequency channels.

Using the $S4$-index is proposed for a real-time check of radio-navigation parameter measurement integrity by signals from separate satellites under geomagnetic disturbance conditions. At the same time, one should define the measurement accumulation time to compute the current $S4$-index values. The accumulation time should be optimal to detect ionospheric scintillations that, most likely, may lead to the emergence of abnormal errors in ranging measurements. Open has remained the issue of searching for an alternative index (a set of indices) enabling, with a high probability, to detect an anomalous status of the wave propagation channel and to exclude abnormal ranging measurements within the modified RAIM algorithm.

Using real-time alerts to the status of the wave propagation channel may enable to adapt the hardware to dramatically changing conditions by tuning automatically the DLL and PLL settings. An optimal selection of such tuning parameters (mainly, the smoothing filter noise band) enables to decrease the effect of strong scintillations on the DLL and PLL quality. An efficient measure is also an increase in the power potential of the satellite-receiver links due to augmenting a GNSS satellite transmitter power. One can achieve the same effect by reducing the RF front-end inner thermal noise and improving the spatial-selective directional properties of the receiver antenna.

Nowadays, there are all the necessary elements to deploy a complex system of differential correction, prediction, and monitoring of the GNSS positioning quality under unfavorable helio-geophysical conditions. Ground- and space-based facilities to monitor the status of the Sun and of the near-Earth space have been developing. International and regional networks of constantly operating stationary points to measure radio-navigation parameters have been deployed. There exist analytical centers to process data from measurements of the above networks that determine time–frequency and ephemeris errors, as well as ionospheric and tropospheric ranging errors.

These elements enable to generate the structure of such a complex system that will provide the GNSS RNP under unfavorable helio-geophysical conditions. The system may be regarded as a basis for a perspective service of global monitoring and forecasting of the integrated uniform GNSS current operation quality.

References

Afraimovich EL, Perevalova NP (2006) GPS-monitoring of the earth upper atmosphere. SB RAS, Institute of Solar-Terrestrial Physics. Irkutsk: SI RC RS ESRC SB RAMS, Russian, p 479

Afraimovich EL, Demyanov VV, Smolkov GY (2009) The total failures of GPS functioning caused by the powerful solar radio burst on December 13, 2006. Earth Plan Space 61:637–641. https://doi.org/10.1186/BF03352940

Akasofu SI, Chapman S (1972) Solar-terrestrial physics. International Series of Monographs on Physics, Clarendon Press, Oxford

Bazarzhapov AD, Matveev MI, Mishin VM (1979) Geomagnetic variations and storms. Novosibirsk: Nauka (Science) Publishers, p 287, in Russian

Berdermann J, Kriegel M, Banys D, Heymann F, Hoque MN, Wilken V, Borries C, Heßelbarth A, Jakowski N (2018) Ionospheric response to the X9.3 Flare on 6 September 2017 and its implication for navigation services over Europe. Space Weather 16:1604–1615. https://doi.org/10.1029/2018SW001933

Berghmans D, van der Linden RAM, Vanlommel P, Warnant R, Zhukov A et al (2005) Solar activity: nowcasting and forecasting at the SIDC. Annal Geophys Europ Geosci Union 23(9):3115–3128

Bhattacharrya A (1999) Deterministic retrieval of ionospheric phase screen from amplitude scintillations. Radio Sci 34:229–240

Blanch J, Walter T (2002) Application of spatial statistics to ionosphere estimation for WAAS. In: Proceedings of the 2002 national technical meeting of the institute of navigation, San Diego, CA, pp 719–724. https://www.ion.org/publications/abstract.cfm?articleID=266

Blanch J (2003) Using kriging to bound satellite ranging errors due to the ionosphere: a dissertation of Ph.H.D: December 2003. Stanford University, p 134

Bruzek A, Durrant CJ, Fraungofer I, Reibel FD (1977) Illustrated glossary for solar and terrestrial physics. Publishing Company. Dordrecht-Holland/Boston-USA

Cane HV, Richardson IG (2003) Interplanetary coronal mass ejections in the near-earth solar wind during 1996–2002. J Geophys Res 108(A4):SSH 6–1

Chen Z, Gao Y, Liu Z (2005) Evaluation of solar radio bursts' effect on GPS receiver signal tracking within international GPS service network. Radio Sci 40. https://doi.org/10.1029/2004RS003066

Conker RS, El-Arini MB, Hegarty CJ, Hsiao T (2003) Modeling the effects of ionospheric scintillation on GPS/satellite-based augmentation system availability. Radio Sci 38(1):1001. https://doi.org/10.1029/2000RS002604

Datta-Barua S, Walter T, Blanch J, Enge P (2005) Can WAAS availability be inferred from geomagnetic data? An analysis. In: Proceedings of the beacon satellite symposium BSS 2004, Trieste, October 18–22. Italy, 2004. https://web.stanford.edu/group/scpnt/gpslab/pubs/papers/DattaBarua_IES_2005.pdf

Demyanov VV, Gamayunov IF, Afraimovich EL, Alyoshechkin AM (2007) The space-time correlation of absolute errors in determination of coordinates of SRNS users in various geophysical conditions. Radioelectron Commun Syst 50(5). https://doi.org/10.3103/S0735272707050044

Demyanov VV, Yasyukevich YV (2014) Mechanisms for effect of irregular geophysical factors on the satellite radio-navigation systems operation: monograph. ISU Publishers, Irkutsk, p 349 (Solar-terrestrial physics), in Russian. ISBN 978-5-9624-1098-2

Dow JM, Neilan RE, Rizos C (2009) The international GNSS service in a changing landscape of global navigation satellite systems. J Geodesy 83:191–198. https://doi.org/10.1007/s0019000803003

Dulk GA (1985) Radio emission from the sun and stars. Ann Rev Astron Astrophys 23:169–224

Giersch OD, Kennewell J, Lynch M (2017) Solar radio burst statistics and implications for space weather effects. Space Weather 15:1511–1522. https://doi.org/10.1002/2017SW001658

Hofmann-Wellenhof B, Lichtenegger H, Wasle E (2008) GNSS—global navigation satellite systems: GPS, GLONASS, Galileo. Springer, Vienna. https://doi.org/10.1007/978-3-211-73017-1

Huang W, Aa E, Shen H, Liu S (2018) Statistical study of GNSS L-band solar radio bursts. GPS Sol 22:114. https://doi.org/10.1007/s10291-018-0780-4

Indriyatmoko A, Kang T, Lee YJ, In JG, Cho YB, Kim J (2008) Artificial neural networks for predicting DGPS carrier phase and pseudorange correction. GPS Sol 12(4):237–247. https://doi.org/10.1007/s10291-008-0088-x

IS-GPS-200J (2018) Global positioning systems directorate. Syst Eng Integr Interf Specifi IS-GPS-200. Available online: https://www.gps.gov/technical/icwg/IS-GPS-200J.pdf

Kahler SW (1992) Solar flares and coronal mass ejections. Ann Rev Astron Astrophys 30:113

Kaplan ED (1996) Understanding GPS: principles and applications. Artech House Publisher, Boston, London, p 556

Khivrich IG, Mironov NF, Belkin AM (1984) Aero-navigation. Moscow: Transport., p 328, in Russian

Klobuchar JA, Kunches JM, Van Dierendonck AJ (1999) Eye on the ionosphere: potential solar radio burst effects on GPS signal to noise. GPS Sol 3(2):69–71. https://doi.org/10.1007/PL00012794

Lekshmi VD, Balan N, Tulasi SR, Liu J-Y (2011) Statistics of geomagnetic storms and ionospheric storms at low and mid latitudes in two solar cycles. J Geophys Res Space Phys 116:A11328. https://doi.org/10.1029/2011JA017042

Maryuknenko VS, Mukhopad YF, Demyanov VV, Mironov BM (2014) Information support of transportation on a basis of integrated navigation systems. Novosibirsk Sci 256. ISBN 978-5-02-019157-0. In Russian

Michalek K, Puchowska P, Rams A (2009) Statistical analysis of decimetric radio bursts, flares and coronal mass ejections. G Sol Phys 257:113–124

MRK19L receiver (1999) Operation manual. UE 2.517.011 RE. Krasnoyarsk, in Russian

Parkinson BW, Fitzgibbon KT (1987) Optimal locations of pseudolutes for differential GPS. Navigation (USA) 33(4):259–283

Paredes E, Pandya N, Sparks L, Komajathy F (2008) Reconstructing the WAAS underestimated ionospheric gradient threat model for the WAAS expansion into Mexico. In: Proceedings of the 21st international technical meeting of the satellite division of the institute of navigation (ION GNSS 2008), Savannah, GA, pp 1938–1947. https://www.ion.org/publications/abstract.cfm?articleID=8107

Perov AI, Kharisov VN (2005) GLONASS: basics and principles of functioning. Radiotechnica, Moscow, p 688, in Russian. ISBN 5-93108-076-7

Pestana A (2015) Technical report: reading RINEX 2.11 observation data files. https://doi.org/10.13140/RG.2.1.4888.4087

Psiaki ML (2001) Block acquisition of weak GPS signals in a software receiver. ION GPS, Salt Lake City, UT, pp 2838–2850

Sato H, Jakowski N, Berdermann J, Jiricka K, Heßelbarth A, Banyś D, Wilken V (2019) Solar radio burst events on 6 September 2017 and its impact on GNSS signal frequencies. Space weather 17:816–826. https://doi.org/10.1029/2019SW002198

Sparks L, Altshuler E, Pandya N, Blanch J, Walter T (2022) WAAS and the ionosphere—a historical perspective: monitoring storms. Navigation 69(1). https://doi.org/10.33012/navi.503

Sparks L, Blanch J, Pandya N (2011) Estimating ionospheric delay using kriging: 1. Metodol Radio Sci 46(1). https://doi.org/10.1029/2011RS004667

Sreeja V, Aquino M, Long K (2013) Impact of the 24 September 2011 solar radio burst on the performance of GNSS receivers. Space Weather 11:306–312. https://doi.org/10.1002/swe.20057

Walter T, Hansen A, Blanch J, Enge P, Mannucci T, Pi X, Sparks L, Iijima B, El-Arini B, Lejeune R, Hagen M, Altshuler E, Fries R, Chu A (2001) Robust detection of ionospheric irregularities. Navigation 48(2):89–100. https://onlinelibrary.wiley.com/toc/21614296/2001/48/2. https://doi.org/10.1002/j.2161-4296.2001.tb00231.x

Walter T, Enge P, Reddan P (2004) Modernizing WAAS [electronic resource]. Stanford, 2004. Mode of access: http://waas.stanford.edu/~wwu/papers/gps/PDF/WalterIONGNSS04.pdf

Williamson SP, Evans DL, Neyland MA, Waldron RE Devine JF, Whatley D, Weiler EJ, Leinen MS, Babcock MR (2000) National space weather program implementation plan, 2nd edn. https://www.nasa.gov/sites/default/files/atoms/files/o45558847.pdf

Yasyukevich Y, Astafyeva EEI, Padokhin A, et al (2018) The 6 September 2017 X-class solar flares and their impacts on the ionosphere, GNSS, and HF radio wave propagation. Space Weather 16:1013–1027. https://doi.org/10.1029/2018SW001932

Yasyukevich YV, Yasyukevich AS, Astafyeva EI (2021) How modernized and strengthened GPS signals enhance the system performance during solar radio bursts. GPS Sol 25:46. https://doi.org/10.1007/s10291-021-01091-5

Zakharov VI, Yasyukevich YV, Titova MA (2016) Effect of magnetic storms and substorms on GPS slips at high latitudes. Cosmic Res 54(1):20–30. https://doi.org/10.1134/S0010952516010147